AF180495

Solid State Lighting Technology and Application Series

Series Editor
G. Q. Zhang

For further volumes:
http://www.springer.com/series/8864

Clemens J. M. Lasance • András Poppe
Editors

Thermal Management for LED Applications

Springer

Editors
Clemens J. M. Lasance
Philips Research Emeritus
Eindhoven, The Netherlands

SomelikeitCool
Eindhoven, The Netherlands

András Poppe
MicReD unit, Mentor Graphics
Mechanical Analysis Division
Budapest, Hungary

Budapest University of Technology
and Economics
Department of Electron Devices
Budapest, Hungary

ISSN 2196-4203
ISBN 978-1-4939-4133-9 ISBN 978-1-4614-5091-7 (eBook)
DOI 10.1007/978-1-4614-5091-7
Springer New York Heidelberg Dordrecht London

Printed on acid-free paper

Springer is part of Springer Science+Business Media (www.springer.com)

Preface

Light emitting diodes (LEDs) are a solid-state lighting source increasingly being used in display backlighting, communications, medical services, signage, and general illumination. LEDs offer design flexibility, have small exterior outline dimensions, higher energy efficiency, have longer life, provide higher performance, a wider range of controllable color temperatures, are eco-friendly products, and don't suffer from low-temperature startup problems, as compared to conventional lighting sources. Unfortunately, they produce heat, and contrary to conventional lighting sources this does cause serious problems (please keep in mind that if the bulb would have been invented today it would never have been legalized because of the excessive touch temperatures).

This book is about thermal management of LEDs and especially LED applications. The main question to be addressed is why we need thermal management in the first place. As Christian Belady put it eloquently in 2001:

> The ultimate goal of system thermal design is not the prediction of component temperatures, but rather the reduction of thermally associated risk to the product.

Hence, if your boss asks you: "Take care that the junction temperature does not exceed 125 °C", you may answer: "Why? Do we sell temperatures?" and then you may educate your boss: "The real objectives to realize are: we want to keep the Lifetime beyond x years, we want to keep the Color Point within margin y, and we want to raise the Efficiency to z %. And yes, these objectives that determine the quality of our LED-based products are linked to the junction temperature, but never as a goal in itself."

The book is divided in four parts: Part A: Basic Physics, Part B: Testing and Standardization, Part C: Advances in Cooling Technologies, and Part D: Applications.

Part A: Basic Physics

The evident link between temperature and various quality-related issues is the rationale behind Chap. 1, an Introduction to LED Thermal Management and Reliability, presenting a general overview of LED basics, LED manufacturing and LED failure mechanisms. Obviously, the first step in reaching the aforementioned goal of LED thermal design is to understand the basics of LED physics: what are the reasons that the temperature is rising anyway? This subject is treated in Chap. 2 on the basics of solid-state physics of LEDs. To be able to perform back-of-the-envelope calculations to get a rough idea about the feasibility of your design from a thermal point of view is an important first step: this topic is treated in Chap. 3.

Part B: Testing and Standardization

This part starts with Chap. 4 dealing with basic thermal characterization and testing. How to test LEDs from a lighting point of view is the subject of Chap. 5. Chapter 6 discusses the increasing need for a more sophisticated thermal characterization of LEDs and LED-based products. Unfortunately, the progress in thermal characterization has not kept pace with the exponential growth of the LED-business. Due to the lack of worldwide-accepted standards a manufacturer can publish whatever thermal information she/he wants. Hence, it becomes a problem for the experienced user because the thermal data that are published are often rather useless in practice when accuracy is at stake.

Part C: Advances in Cooling Technologies

Air cooling is by far the most frequent cooling method and will be for a long time to come. Natural and forced convection cooling, including synthetic jet cooling, is discussed in Chap. 7. Another important thermal management control option is the thermal interface material (TIM). For high-performance LED-based products it may even be the limiting factor in the thermal resistance chain. Current and future thermal interface materials are dealt with in Chap. 8. An even more important thermal control option is area enlargement, either by heat sinks attached to the LED substrate or by first transferring the heat through heat pipes to a location where area enlargement is easier to handle. Chapter 9 deals with the fundamentals a designer should master to enable an optimal choice out of the thousands of heat sinks available on the market.

Part D: Applications

Chapter 10 is related to Chap. 9 but focuses on applications in practice. Problems inherent to LED manufacturing from a thermal point of view are the subject of Chap. 11. Thermal management of sophisticated LED solutions is treated in Chap. 12, while another important application field, namely LED-driven display technologies, is discussed in Chaps. 13 and 14, the first providing a historical overview, and the second the state-of-the-art. Many LED applications are designed for use in harsh environments, for example automotive, aircraft, outdoor lighting, and signage everywhere on earth. Chapter 15 provides insight in the challenges these applications are facing. The book closes with Chap. 16 showing an overview of future directions in LED applications.

In summary, the editors are convinced that this book covers (almost) all aspects of thermal management that are relevant to the design of LEDs and LED–based systems.

In closing this preface, the editors would like to express their sincere thanks to all authors who made this book possible.

Philips Research Emeritus, The Netherlands — Clemens J. M. Lasance
Budapest University of Technology and Economics and Mentor Graphics, Hungary — András Poppe

Contents

Part IV Applications

Contributors

Cathy Biber, PhD Biber Thermal Design Ltd., Portland, OR, USA
e-mail: biberthermal@gmail.com

Moon-Hwan Chang, PhD Center for Advanced Life Cycle Engineering (CALCE), University of Maryland, Building 89, Room 1103, College Park, MD 20742, USA
e-mail: mhchang@calce.umd.edu

Péter Csuti Department of Electrical Engineering and Information Systems, University of Pannonia, Egyetem u. 10, 8200 Veszprém, Hungary
e-mail: csutip@vision.uni-pannon.hu; csuti.peter@virt.uni-pannon.hu

Diganta Das, PhD Center for Advanced Life Cycle Engineering (CALCE), University of Maryland, Building 89, Room 1103, College Park, MD 20742, USA
e-mail: digudas@calce.umd.edu

Norbert P. Engelberts Optimal Thermal Solutions BV, Bussum, The Netherlands
e-mail: nengelberts@ots-eu.com

Gábor Farkas, PhD Mentor Graphics MAD MicReD unit, Infopark D, Gábor Dénes utca 2, Budapest 1117, Hungary
e-mail: gabor_farkas@mentor.com

Robert F. Karlicek, Jr., PhD Smart Lighting Engineering Research Center, Rensselaer Polytechnic Institute, CII7015, 110 8th Street, Troy, NY 12180, USA
e-mail: karlir@rpi.edu

Clemens J. M. Lasance, Ir Philips Research Laboratories Emeritus, Consultant at SomelikeitCool, Nuenen, The Netherlands
e-mail: lasance@onsnet.nu

Jinmin Li Institute of Semiconductors, Chinese Academy of Science, A35, QingHua East Road, Haidian District, 912, 100083 Beijing, P. R. China
e-mail: jmli@semi.ac.cn

Zhe Liu Institute of Semiconductors, Chinese Academy of Science, A35, QingHua East Road, Haidian District, 912, 100083 Beijing, P. R. China

G. A. Luiten, Ir Philips Research, Eindhoven, Netherlands
e-mail: wendy.luiten@philips.com

Raghav Mahalingam, PhD Nuventix, 4635 Boston Lane, Austin, TX 78735 USA
e-mail: raghav@nuventix.com

Michael Pecht, PhD Center for Advanced Life Cycle Engineering (CALCE), University of Maryland, Building 89, Room 1103, College Park, MD 20742, USA
e-mail: pecht@calce.umd.edu

Jim Petroski Cedar Rapids, IA 52402, USA
e-mail: jpetroski@rambus.com

András Poppe, PhD Mentor Graphics Mechanical Analysis Division, MicReD unit, Budapest XI, Infopark D, Gábor Dénes utca 2, 1117, Budapest, Hungary
e-mail: andras_poppe@mentor.com

Department of Electron Devices, Budapest University of Technology and Economics, Budapest, Hungary
e-mail: poppe@eet.bme.hu

David L. Saums DS&A LLC, 100 High Street Amesbury MA 01913, USA
e-mail: dsaums@dsa-thermal.com

János Schanda, Professor Emeritus Department of Electrical Engineering and Information Systems, University of Pannonia, Egyetem u. 10, 8200 Veszprém, Hungary
e-mail: schanda@vision.vein.hu; schanda.janos@virt.uni-pannon.hu

Ferenc Szabó, PhD Department of Electrical Engineering and Information Systems, University of Pannonia, Egyetem u. 10, 8200 Veszprém, Hungary
e-mail: szabof@vision.uni-pannon.hu; szabo.ferenc@virt.uni-pannon.hu

Theo Treurniet, PhD LED Platform Development, Philips Lighting, 80020, 5600 JM, Eindhoven, The Netherlands
e-mail: theo.treurniet@philips.com

Junxi Wang Institute of Semiconductors, Chinese Academy of Science, A35, QingHua East Road, Haidian District, 912, 100083 Beijing, P. R. China

Ross Wilcoxon, PhD Cedar Rapids, IA 52402, USA
e-mail: ross.wilcoxon@gmail.com

Kazuaka Yazawa, PhD Birck Nanotechnology Center, Purdue University, 1205W. State Street, West Lafayette, IN 47907-2057, USA
e-mail: kyazawa@purdue.edu, kaz@soe.ucsc.edu

University of Santa Cruz, Santa Cruz County, CA, USA

Richard Young, PhD Instrument Systems GmbH, Munich, Germany
e-mail: young@instrumentsystems.com

Part I
Basic Physics

Chapter 1
Introduction to LED Thermal Management and Reliability

Michael Pecht, Diganta Das and Moon-Hwan Chang

Abstract *Some like it hot, others do not.* And those others for sure include the designers of products that contain light-emitting diodes (LEDs). This book is about thermal management of LEDs and especially LED applications. The main question to be addressed is: Why do we need thermal management? As Belady put it eloquently in 2001 [Belady and Minichiello, Electronics Cooling Magazine, May issue, 2003]:

> The ultimate goal of system thermal design is not the prediction of component temperatures, but rather the reduction of thermally associated risk to the product.

Hence, the objectives of a designer are not in the first place to calculate or measure temperatures, but to keep the lifetime beyond x years, to keep the color point within margin y, and to raise the efficiency to z %. And indeed, these objectives, determining the quality of LED-based products, are linked to the junction temperature. This is the main reason why a book on LED thermal management starts with an introductory chapter on LED reliability issues.

Parts of this chapter have been sourced from a chapter in a book on Solid State Lighting Reliability [Pecht and Chang, Solid state lighting reliability: components to systems, Springer, New York, pp. 43–110, 2013].

1.1 Introduction to Light-Emitting Diodes

Light-emitting diodes (LEDs) are solid-state lighting sources increasingly being used in display backlighting, communications, medical services, signage, and general illumination. Due to their versatility, LED application areas include liquid-crystal display (LCD) backlights, displays, transportation equipment lighting, and general lighting. LEDs are used as a light source for LCD backlights, including in mobile

M.-H. Chang (✉) · M. Pecht · D. Das
Center for Advanced Life Cycle Engineering (CALCE), University of Maryland, Building 89, Room 1103, College Park, MD 20742, USA
e-mail: mhchang@calce.umd.edu

M. Pecht
e-mail: pecht@calce.umd.edu

D. Das
e-mail: digudas@calce.umd.edu

C. J. M. Lasance, A. Poppe (eds.), *Thermal Management for LED Applications*,
Solid State Lighting Technology and Application Series 2,
DOI 10.1007/978-1-4614-5091-7_1,

phones, cameras, portable media players (PMPs), notebooks, monitors, and televisions (TVs). Display areas include LED electric scoreboards, outdoor billboards, and signage lighting, such as LED strips and lighting bars. Examples of transportation equipment lighting areas are vehicle/train lighting (e.g., meter backlights, tail and brake lights) [3] and ship/airplane lighting (e.g., flight error lighting and searchlights). General lighting applications are divided into indoor lighting (e.g., LED light bulbs, desk lighting, and surface lighting) [4, 5], outdoor lighting (e.g., decorative lighting, street/bridge lighting, and stadium lighting), and special lighting (e.g., elevator lighting and appliance lighting) [6, 7]. The use of LEDs in general lighting has increased, starting with street lighting in public areas and moving on to commercial/business and consumer-level lighting.

LEDs offer design flexibility, ranging from zero-dimensional lighting (dot-scale lighting) to three-dimensional lighting (color dimming using a combination of colors), with one-dimensional lighting (line-scale lighting) and two-dimensional lighting (local dimming, i.e., area-scale lighting) in between. LEDs have small exterior outline dimensions and offer high energy efficiency that results in lower power consumption with low voltage (<4 V) and low current operation (<700 mA). They have longer life—up to 50,000 h and provide higher performance, such as ultra-high speed response time (microsecond-level on-off switching), a wider range of controllable color temperatures, a wider operating temperature range, and no low-temperature startup problems. In addition, LEDs have better impact resistance. LEDs are also eco-friendly products, with low ultra-violet (UV) radiation (higher safety) and no mercury. Interested readers may consult references [8–13] for more details.

LEDs range from a narrow spectral band emitting a single-colored light, such as red, yellow, green, or blue, to white, to a distribution of luminous intensity and various types and shapes, depending on the color mixing and package design. White light is a mixture of all visible wavelengths, as shown in Fig. 1.1. Every LED color is represented by unique $x-y$ coordinates, as shown in Fig. 1.2. Red is on the far right, green is on the top left, and blue is on the bottom left. The International Commission on Illumination (CIE) chromaticity coordinates of x, y, and z are a ratio of the red, green, and blue stimulation of light compared to the total amount of the red, green, and blue stimulation. By definition, the sum of the normalized tri-stimulus values $(x+y+z)$ is equal to 1. The white area of the chromaticity diagram can be expanded, and boundaries can be added to create each color range. The color temperatures and the Planckian locus (black-body curve) show how they relate to the chromaticity coordinates [14].[1] As the temperature of the black body increases, the chromaticity location moves from the red wavelength range toward the center of the diagram in Fig. 1.2.

[1] The color temperature of a white light is the temperature of an ideal Planckian black-body radiator that radiates a light of comparable hue to that light source. Thus, the color temperature of a white light of thermal radiation from an ideal black-body radiator is defined as equal to its surface temperature in kelvins. When the black-body radiator is heated to high temperatures, the heated black body emits the color, going from red, to orange, to yellow, to white, and finally to bluish white. The Planckian locus starts out in the red, then moves through the orange and yellow, and finally to the white region. The color temperature of a light source is regarded as the temperature of a Planckian black-body radiator that has the same chromaticity coordinates.

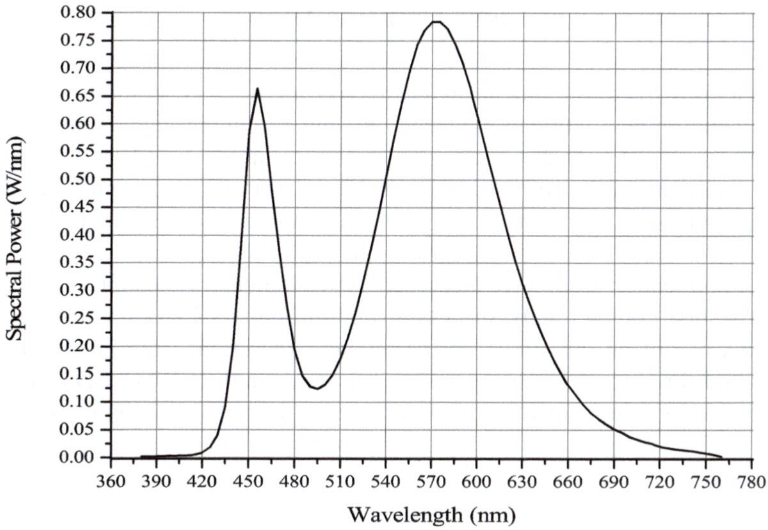

Fig. 1.1 Spectral power distribution—white light-emitting diode (LED)

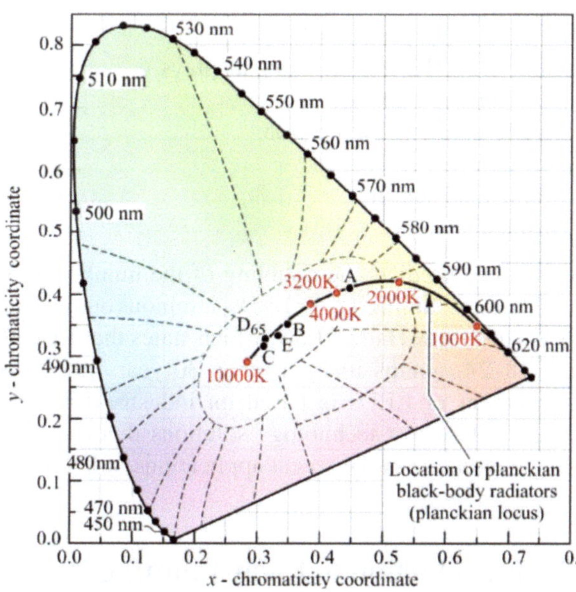

Fig. 1.2 International Commission on Illumination (CIE) 1931 chromaticity diagram. (Source: [15] © Cambridge University Press, reprinted with permission)

Color change should be considered in LED applications because LED degradation not only results in reduced light output, but also in color changes. LED modules are composed of many LEDs. This means that if some number of LEDs experience color changes, it will be recognized by users. Even if all LEDs degrade at the same level, LED modules need to maintain their initial color, especially for indoor lighting and backlight applications.

1.2 Short History of Light-Emitting Diodes

The history of LED development can be divided into three generations with distinct advancements in new fabrication technology and equipment, new phosphor materials, and improved heat dissipation packaging technologies. LEDs have become brighter and color variance more flexible. Also, light efficiency and light efficacy have been improved. The first commercialized LED was produced in the late 1960s. This first generation of LEDs lasted from the 1960s until the 1980s. In this period, the major application areas were machinery status indicators and alpha-numeric displays. The first commercially successful high-brightness LED was developed by Fairchild Co. Ltd. in the 1980s. In the second generation, from the 1990s to the present, high-brightness LEDs became very popular in the world LED market. The main application areas for the second generation included motion displays, LED flashes, LED back light units (BLU), mobile phones, automotive LED lighting, and architecture.

The third generation is now arriving in the market. These LEDs have been developed for substantial savings in electrical energy consumption and a reduction in environmental pollution. Future LED application areas are expected to include general lighting, lighting communication [16], medical/environmental purposes, and critical applications in system controls. Some examples are:

- Portable LED projector,
- Large-size LED backlighting displays,
- LED general lighting,
- Visible light communication,
- Purifiers,
- Biomedical sensors,
- Artificial sun.

Moore's law predicts the doubling of the number of Si transistors in a chip every 18–24 months. Similarly, for LEDs, luminous output (luminous flux, measured in lm) appears to follow Haitz's Law, which states that LED flux per package has doubled every 18–24 months for more than 30 years [9]. This trend in the technological advancement of LEDs is based on industry-driven R&D efforts targeting high-efficiency, low-cost technology solutions that can successfully provide an energy saving alternative to the recent applications of LEDs.

1.3 Introduction to Light-Emitting Diode Manufacturing

The LED supply chain starts from an LED chip and progresses to an LED package, an LED module, and then to a system. LED production starts with a bare wafer, such as sapphire, GaN, SiC, Si, or GaAs. Many thin epilayers are then grown on the bare wafer. Different colors of LEDs can be made by using different types of epiwafers. The types of epiwafer are InGaN/AlGaN for producing blue, green, and UV-range

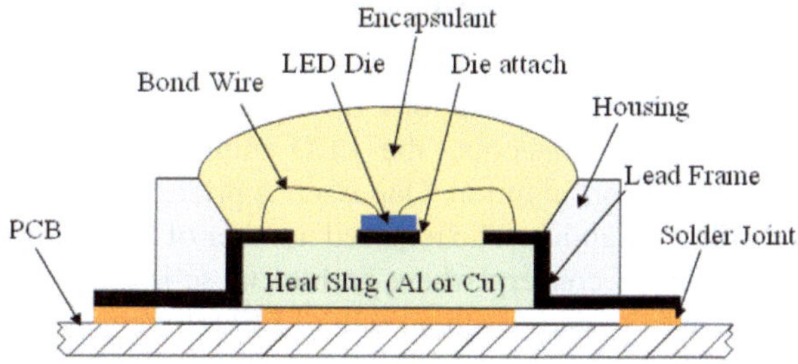

Fig. 1.3 LED package assembled with printed circuit board (PCB)

light; InAlGaP for producing red and yellow light; and AlGaAs for producing red or infrared-range light. The LED chip fabrication process involves attaching electric contact pads to an epiwafer and cutting the epiwafer into LED dies that are then packaged.

LEDs are classified into two types: white LEDs and red-green-blue (RGB) LEDs. White LED packages can use red/green/blue/orange/yellow phosphors with blue LED chips to produce white light. The phosphors comprise activators mixed with impurities at a proper position on the host lattice. The activators determine the energy level related to the light emission process, thereby determining the color of the light emitted. The color is determined by an energy gap between the ground and excitation states of the activators in a crystal structure. RGB LED packages represent red, green, and blue LED packages, as well as LED packages with multi-dies in a single package that produce white light using a combination of red, green, and blue LED dies.

A cross-sectional side view of a white LED is shown in Fig. 1.3. An LED package mounted on a printed circuit board is composed of housing, encapsulant, die, bond wire, die attach, lead frame, metal heat slug, and solder joint. The housing is a body for supporting and protecting the entire structure of an LED device. The housing is usually formed from materials such as polyphthalamide (PPA) or liquid crystal polymer (LCP). The encapsulant positioned over the housing is a resin material for the LED package in the shape of a dome. The typical material types for the resin are epoxy or silicon. The die is a compound semiconductor. The lead frame is used to apply external power to the LED die. The die attach is used to mechanically and thermally connect the chip onto the lead frame. Typical types of die attaches are Ag paste and epoxy paste. Phosphors dispersed in the encapsulant are used to emit the white light excited by absorbing a portion of the light from the LED dies.

LED types are placed in the following major categories depending on LED electrical power: low power LEDs are under 1 W of power (currents typically near 20 mA); medium power LEDs (high-brightness LEDs) between 1 and 3 W of power (currents typically in the 30 mA/75 mA/150 mA range); and high power LEDs (ultra–high brightness LEDs) have more than 3 W of electrical power (currents typically in the 350 mA/750 mA/1000 mA range). The LEDs vary because of the LED current–voltage curves, which differ between materials.

1.4 Introduction to Light-Emitting Diode Reliability

Despite exciting innovations driven by technological advances and ecological/energy-saving concerns, the LED industry still faces challenges in attracting widespread consumption. One issue is price, and another is the lack of information regarding reliability. The required number of LEDs for general lighting applications is a matter of concern where both of these issues converge. It may take anywhere from tens to sometimes thousands of LEDs to replace one conventional lamp because the emission of a single LED covers a limited area. If one single LED fails, then the final product is sometimes treated as a failure. For example, the failure of LEDs in an LCD display is very critical, even when only a single LED package experiences changes in optical properties [17]. Failures of an LED or LEDs appear as dark spots, dark areas, or rainbow areas.

The LED die is a semiconductor and the nature of manufacturing of LED packages is similar to that of microelectronics. However, there are unique functional requirements, materials, and interfaces in LEDs resulting in different failure modes and mechanisms. The major causes of failures can be divided into die-related, interconnect-related, and package-related. The die-related failures include severe light output degradation and burned/broken metallization on the die. The interconnect failures of LED packages are electrical overstress-induced bond wire fracture/wire ball bond fatigue, electrical contact metallurgical interdiffusion, and electrostatic discharge, which leads to catastrophic failures of LEDs. Package-related failure mechanisms include carbonization of the encapsulant, encapsulant yellowing, delamination, lens cracking, phosphor thermal quenching, and solder joint fatigue that results in optical degradation, color change, electrical open/short, and severe discoloration of the encapsulant.

LED manufacturers usually perform tests in the product development cycle during the design and development phases. Typical qualification tests of LEDs are categorized into operating life tests and environmental tests by using industrial standards such as Joint Electron Devices Engineering Council (JEDEC) or Japan Electronics and Information Technology Industries Association (JEITA) [18, 19]. Readers interested in life testing are encouraged to consult [20].

A lifetime estimate is generally made using the Arrhenius model. Activation energy is sensitive to the test load condition, types of materials, and mechanical design of LED packages. This estimate is life with uncertainties such as exponential extrapolation of lifetime, assumption of activation energy, possible failure mechanism shift between test and usage conditions, and discount of all other failure causes besides temperature.

LED core technology in terms of structural and reliability analysis is shown in Fig. 1.4. To develop a final LED product, manufacturers are required to consider each level in Fig. 1.4 (composed of LED die, LED package, LED module, and system), because market share power is based on optimal thermal dissipation, high external quantum efficiency, high electrical power conversion efficiency, enhanced performance, low cost, advanced opto-mechanical design (minimizing rainbow or glare effects), and high reliability.

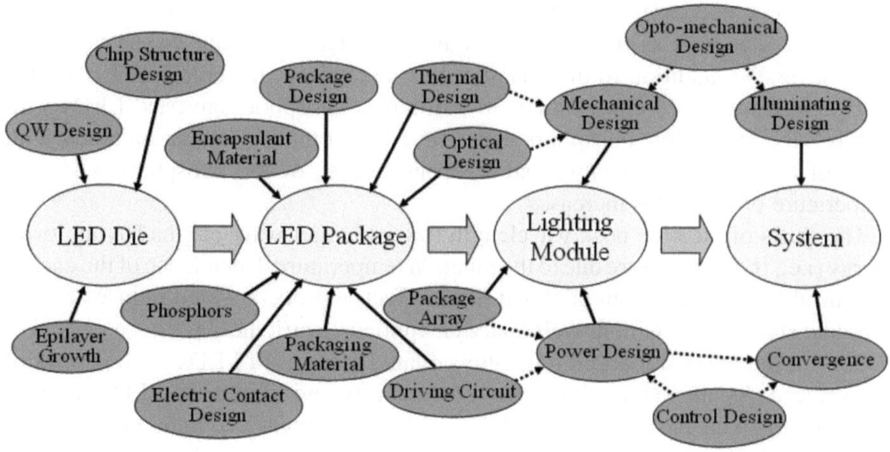

Fig. 1.4 Light-emitting diode (LED) core technology

LED package reliability is important for improving LED lighting system reliability, since all other parts, including mechanical parts, power, and electric circuits, can be repaired or replaced by scheduled maintenance before the system experiences failure. Once an LED package or LED module encounters failures, this means that the LED system needs to undergo unscheduled maintenance, which causes end-users high cost for replacement. Many LED failure modes and mechanisms are related to thermal, electrical, and humidity stress.

1.5 The Rationale of Light-Emitting Diode Thermal Management

End-product manufacturers that use LEDs expect the LED industry to guarantee the lifetime of LEDs in their usage conditions. Such lifetime information would allow LED designers to deliver the best combination of purchase price, lighting performance, and cost of ownership for the life of the products. One barrier to the acceptance of LEDs in traditional applications is the relatively sparse information available on their reliability [20]. When a higher drive current is applied to LEDs, there is increased light output but that typically comes with increased heat generation. The light output can change as a result of the operating conditions, temperature in particular [20–27]which is impacted by heat generation and depends on the methods of dispersion of the heat. For example, light output decreases with a temperature rise in the LEDs, since the quantum efficiency decreases at higher temperature that contributes to more nonradiative recombination events in LEDs [20, 28]. Temperature increase results in forward voltage drop due to the decrease of the bandgap energy of the active region of LEDs and also results in the decrease in series resistance.

The resistance decrease is due to higher acceptor activation occurring at elevated temperatures as well as the resulting higher conductivity of the *p*-type layer and active layers. In addition to the quantum efficiency drop, the colors of LEDs also change with increased temperature. In particular, phosphor-converted LEDs with blue InGaN and yellow phosphors experience light output degradation along with shifts of blue peak wavelength and the peak energy of the phosphors when the temperature of the LEDs increases.

The shifts of the blue peak wavelength toward longer wavelengths having lower energy (i.e., redshifting) are due to the junction temperature dependence of the energy gap shrinkage and quantum confined Stark effect, a process which reduces energy of bound states in a quantum well under an applied electric field [29].

To sum up, many important reliability- related features of LEDs are functions of temperature. As an example, the long-term stability and lifetime of LEDs are typically judged on the basis of measured light output. The measured light output mostly depends on the junction temperature. Hence, the correctness of light output measurements is dependent on the temperature stability of the light output measurement setup and by the accuracy of the temperature measurement (details of laboratory measurement of light output are provided in Chap. 5). In the daily field practice there are associated uncertainties with prediction of the junction temperature because there are only indirect ways of measuring and converting temperatures from reference points to the junction temperature. (Chap. 4 provides details of thermal testing along with a laboratory method of indirect measurement of LEDs' junction temperature. The detailed description of a standardized laboratory thermal testing procedure is provided by the JEDEC JESD51–51 document [30]). Long-term stability analyses of LEDs need to demonstrate that the thermal conditions of the LEDs have not changed during the entire aging/testing process in order to enable correct correlation between light output characteristics and R_{thJA} (LED's junction-to-ambient thermal resistance). Little information has been published about how the light output measurements in reliability studies are performed, but it is suspected that the current R_{thJA} of the LEDs during aging test measurements is often uncontrolled and changes over time. As a consequence, some of the reported light output variations could be attributed to R_{thJA} variations of the test setup. The in-situ measurements briefly described in Sect. 5.4.2 of Chap. 5 were aimed to assure this thermal stability requirement and the thermal transients measurements accompanying the light-output measurements described there, provided means of following the changes of the overall R_{thJA} of the LEDs being tested. (Further details on this experiment are provided in [31]). Another way to prevent this uncertainty is to eliminate the potential changes in R_{thJA} by ensuring that all light output characteristics are presented as a function of the real junction temperature [32]. The only way that the reliability data provided by different vendors can be assured is by standardizing all relevant measurements and definitions.

Besides the standardization of reliability-related tests, an important source of information for a designer is the published data in the data sheets, especially thermal data such as junction-to-ambient and junction-to-case thermal resistances. The designer needs these data to ensure that the maximum allowable temperatures prescribed by the vendors are not exceeded. It is necessary for these data to be standardized, because lower thermal resistance is a major selection criterion. Lasance and Poppe [33–36] and Poppe et al. [32] discuss the need for more sophisticated thermal characterization and standardization of LEDs and LED-based products. The reason is that progress in these fields has not kept pace with the exponential growth in applications. This situation became a serious problem for leading manufacturers who are focusing on a sustainable business for the future and are willing to publish reliable thermal data. Unfortunately, due to the lack of globally accepted standards, manufacturers could publish whatever they wanted. The lack of standards also meant a problem for the experienced user because the thermal data that are published are often of limited use in practice when accuracy is at stake, and accuracy is needed for estimation of expected performance and lifetime. Remarkably, the situation was not much different from the one that the integrated circuit (IC)-world was facing almost 20 years ago [32, 37–41]. Around 1990 it became clear that thermal characterization of IC packages was problematic. Manufacturers all over the world were using different standards. Even within a single manufacturer, intolerable differences showed up. To solve the thermal characterization problems, manufacturers must publish thermal data in such a way that the end-user can use this data. End-users are responsible for the specifications of the thermal environment to which the LEDs are exposed. Provided that the manufacturers want to cooperate, it would be easy to apply the standard protocols used by IC business.

In addition to standardization itself and suggestions for improved test setups, Poppe and Lasance discussed [33–36] the role of thermal characterization, the definition of thermal resistance, the different goals of manufacturers and system designers, the similarities and differences between LED and IC thermal characterization, the drawbacks of the current thermal data in data sheets, and an overview of the questions that an LED thermal standardization body should address. The first steps in this regard have been recently made by JEDEC through the publication of the JESD51–5x series of standards [30, 42, 43]. Further information about standardization of LED thermal characterization is provided in Chap. 6.

1.6 Conclusions

The conventional way to predict the lifetime of LEDs employs the Arrhenius model to extrapolate test results at high temperature to expected operating temperatures. A major problem is that the Arrhenius model is not adequate to represent the failures of LEDs. Light output degradation is the major failure mode of LEDs, and it results from hygromechanical and electrical stresses, in addition to thermal stresses. A more realistic method of LED lifetime estimation needs to reflect total consideration of temperature, the level of forward current, relative humidity, mechanical stress, and materials.

The overall reliability of LED packages is related to interconnect failures, semiconductor failures, and package failures. Interconnect failures are responsible for broken bond wire/lifted ball, electrical metallurgical interdiffusion, and electrostatic discharge. LED semiconductor failures are manifested as die cracking, defect and dislocation generation and movement, dopant diffusion, and electromigration. Package failures involve mechanical interaction with LED chips, die adhesives, heat slugs, lead frames, and encapsulants. The failure mechanisms responsible for package failures include carbonization of the encapsulant, delamination, encapsulant yellowing, phosphor thermal quenching, and lens degradation.

Cooperation between thermal, electrical, and optical standards bodies and professional societies is required to arrive at globally accepted thermal standards to measure junction and reference temperatures to ensure a fair comparison of published performance and reliability data. Since the end user needs total reliability of the final products, reliability research of LED packages has to be expanded to the reliability study of the complete LED-based system, including the luminaires and electronics.

References

1. Belady C, Minichiello A (2003) Effective thermal design for electronic systems. Electronics Cooling Magazine 9(2):16–21. http://www.electronics-cooling.com/2003/05/effective-thermal-design-for-electronic-systems/
2. Pecht MG, Chang MH (2013) Failure mechanisms and reliability issues in LEDs. In: van Driel WD, Fan XJ (eds) Solid state lighting reliability: components to systems. Springer, New York, pp 43–110
3. Huang MS, Hung CC, Fang YC, Lai WC, Chen YL (2009) Optical design and optimization of light emitting diode automotive head light with digital micromirror device light emitting diode. Optik-Int J Light Electron Opt 121(10):1–9
4. Lee SWR, Lau CH, Chan SP, Ma KY, Ng MH, Ng YW, Lee KH, Lo JCC (2006) Development and prototyping of a HB-LED array module for indoor solid state lighting. High density microsystem design and packaging and component failure analysis, HDP'06, pp 141–145
5. Peon R, Doluweera G, Platonova I, Irvine-Halliday D, Irvine-Halliday G (2005) Solid state lighting for the developing world—the only solution. Optics & Photonics 2005, Proceedings of SPIE, vol 5941, pp 109–123, San Diego
6. Pinto RA, Cosetin MR, da Silva MF, Denardin GW, Fraytag J, Campos A, do Prado RN (2009) Compact emergency lamp using power LEDs. Industrial Electronics, 2009, IECON'09. 35th Annual Conference of IEEE pp 3494–3499
7. Shibata S-I, Oyabu T, Kimura H (2009) Bioelectric potential characteristic of pothos under light emitting diode. ICCAS-SICE pp 4663–4668
8. Krames MR, Shchekin OB, Mueller-Mach R, Mueller GO, Zhou L, Harbers G, Craford MG (2007) Status and future of high-power light-emitting diodes for solid-state lighting. J Disp Technol 3(2):160–175
9. Steigerwald DA, Bhat JC, Collins D, Fletcher RM, Holcomb MO, Ludowise MJ, Martin PS, Rudaz SL (2002) Illumination with solid state lighting technology. IEEE J Sel Top Quantum Electron 2:310–320
10. Steranka FM, Bhat J, Collins D, Cook L, Craford MG, Fletcher R, Gardner N, Grillot P, Goetz W, Keuper M, Khare R, Kim A, Krames M, Harbers G, Ludowise M, Martin PS, Misra M, Mueller G, Mueller-Mach R, Rudaz S, Shen YC, Steigerwald D, Stockman S, Subramanya S, Trottier T, Wierer JJ (2002) High power LEDs—technology status and market applications. Phys Status Solidi (a) 194(2):380–388
11. Schubert EF, Kim JK, Luo H, Xi J-Q (2006) Solid-state lighting-a benevolent technology. Rep Prog Phys 69(2):3069–3099

12. Aoyama Y, Yachi T (2008) An LED module array system designed for streetlight use. Energy 2030 Conference, 2008. Energy 2008. IEEE, pp 1–5
13. Vittori R, Scaburri A (2009) New solid state technologies and light emission diodes as a mean of control and lighting source applicable to explosion proof equipment, with the scope to reduce maintenance, to limit the risk of bad maintenance and to expand the plants' life. PCIC Europe, 2009. Conference Record, pp 193–198
14. King M (2010) Characteristics of high brightness LEDs. Electronic Design Online Conference Series, session 5, 22 June, pp 1–16
15. Schubert EF (2006) Light-emitting diodes. 2nd edn. Chap. 18. Cambridge University Press, New York, pp 308–309
16. Wipiejewski T, Moriarty T, Hung V, Doyle P, Duggan G, Barrow D, McGarvey B, O'Gorman M, Calvert T, Maute M, Gerhardt V, Lambkin JD (2008) Gigabits in the home with plugless plastic optical fiber (POF) interconnects. Electronics System-Integration Technology Conference (ESTC), 2008. 2nd, pp 1263–1266
17. Chang YN, Hung CC, Tung SC (2009) Auto mixed light for RGB LED backlight module, Industrial Electronics, 2009. ISIE 2009. IEEE International Symposium, pp 864–869
18. Cree (2009) Cree Xlamp XR Family LED Reliability, CLD-AP06 Rev. 7, Cree, Inc., pp 1–6. http://scn.cree.com/products/pdf/XLamp_Reliability.pdf
19. Nichia (2012) Specifications for White LED Model: NCSW119T-H3, Nichia STS-DA1-2296, Nichia Corporation, pp 1–16. http://www.nichia.co.jp/specification/en/product/led/NCSW119-H3-E.pdf
20. Chang M-H, Das D, Varde PV, Pecht M (2012) Light emitting diodes reliability review. Microelectron Reliab 52(5):762–782
21. Bar-Cohen A, Kraus AD (1998) Advances in thermal modeling of electronic components and systems, vol 4. ASME Press
22. Gao S, Hong J, Shin S, Lee Y, Choi S, Yi S (2008) Design optimization on the heat transfer and mechanical reliability of high brightness light emitting diodes (HBLED) package. Electronic Components and Technology Conference (ECTC), 2008. 58th, pp 798–803
23. Jayasinghe L, Gu Y, Narendran N (2006) Characterization of thermal resistance coefficient of high-power LEDs. 6th International Conference on Solid State Lighting, Proceedings of SPIE, pp 1–10
24. Gu Y, Narendran N, (2004) A non-contact method for determining junction temperature of phosphor-converted white LEDs. Third International Conference on Solid State Lighting, Proceedings of SPIE 5187, pp 107–114
25. Sanawiratne J, Zhao W, Detchprohm T, Chatterjee A, Li Y, Zhu M, Xia Y, Plawsky JL (2008) Junction temperature analysis in green light emitting diode dies on sapphire and GaN Substrates. Phys Status Solidi (c) 5(6):2247–2249
26. Chhajed S, Xi Y, Li Y-L, Gessmann Th., and Schubert EF (2005) Influence of junction temperature on chromaticity and color-rendering properties of trichromatic white-light sources based on light-emitting diodes. J Appl Phys 97:054506–054508
27. Chen ZZ, Liu P, Qi SL, Lin L, Pan HP, QinZ. X, Yu TJ, He ZK, Zhang GY (2007) Junction temperature and reliability of high-power flip-chip light emitting diodes. Mat Sci Semicon Proc 10:206–210
28. Liu J, Tam WS, Wong H, Filip V (2009) Temperature-dependent light-emitting characteristics of InGaN/GaN Diodes. Microelectron Reliab 49:38–41
29. Peng L-H, Chuang C-W, Lou L-H (1999) Piezoelectric effects in the optical properties of strained InGaN quantum wells. Appl Phys Lett 74(6):795–797
30. JEDEC JESD51–51 standard Implementation of the electrical test method for the measurement of the real thermal resistance and impedance of light-emitting diodes with exposed cooling surface. http://www.jedec.org/sites/default/files/docs/JESD51-51_1.pdf
31. Poppe A, Molnár G, Csuti P, Szabó F, Schanda J. Aging of LEDs: a comprehensive study based on the LM80 standard and thermal transient measurements, CIE 27th Session-Proceedings, CIE 197:2011: (Vol 1, Part 1–2). Sun City, South-Africa, 10–15 July 2011. pp 467–477. Paper OP57. (ISBN: 978 3 901906 99 2)

32. Poppe A, Molnár G, Temesvölgyi T Temperature dependent thermal resistance in power LED assemblies and a way to cope with it. In: Proceedings of the 26th IEEE Semiconductor Thermal Measurement and Manage-ment Symposium (SEMI-THERM'10), 21-25 February 2010, Santa Clara, USA, pp 283-288. (ISBN: 978-1-4244-6458-1)
33. Lasance CJM, Poppe A (2009) Challenges in LED thermal characterization. 10th International Conference on thermal, mechanical and multi-physics simulation and experiments in microelectronics and microsystems, EuroSimE 2009 pp 1–11, Delft
34. Lasance C, Poppe A (2009) On the standardization of thermal characterization of LEDs. Pro1647 ceedings of the 25th IEEE semiconductor thermal measurement and management symposium 1648 (SEMI-THERM'09), 15–19 March 2009, San Jose, USA, pp 151–158
35. Poppe A, Lasance CJM (2008) On the Standardisation of thermal characterisation of LEDs Part II: Problem Definition and Potential Solutions, THERMINIC 2008, Rome, Italy, pp 213–219
36. Poppe A, Lasance CJM (2009) Hot topic for LEDs: standardization issues of thermal characterization. Light and lighting, Conference with special emphasis on LEDs and Solid State Lighting, Budapest, Hungary, CIE, pp 43–46. May
37. Lasance CJM (2003) Thermally driven reliability issues in microelectronic systems: status-quo and challenges. Microelectron Reliab 43:1969–1974
38. Joshi Y, Azar K, Blackburn D, Lasance CJM, Mahajan R, Rantala J (2003) How well can we assess thermally driven reliability issues in electronic systems today? Summary of panel held at the THERMINIC 2002. Microelectron J 34(12):1195–1201
39. Lasance CJM (2008) Ten years of boundary-condition-independent compact thermal modeling of electronic parts: a review. Heat Transf Eng 29:149–168
40. Lasance CJM (2002) The Conceivable Accuracy of Experimental and Numerical Thermal Analyzes of Electronic Systems'. IEEE CPT 25:366–382
41. Lasance CJM (2001) The European project PROFIT: prediction of temperature gradients influencing the quality of electronic products. In: Proceedings of the 17th IEEE Semiconductor Thermal Measurement and Management Symposium (SEMI-THERM'01), San Jose, USA, 20–22 March 2001, pp 120–125
42. JEDEC JESD51–50 standard Overview of methodologies for the thermal measurement of single-and multi-chip, single-and multi-PN-junction light-emitting diodes (LEDs). http://www.jedec.org/sites/default/files/docs/JESD51-50_1.pdf
43. JEDEC JESD51-52 standard Guidelines for combining CIE 127-2007 total flux measurements with thermal measurements of LEDs with exposed cooling surface. http://www.jedec.org/sites/default/files/docs/JESD51-52_1.pdf

Chapter 2
Solid-State Physics Fundamentals of LED Thermal Behavior

Jinmin Li, Junxi Wang, Zhe Liu and András Poppe

Abstract This chapter provides the basics on the physics of light-emitting diode (LED) operation: band structures, carrier transport, different recombination mechanisms, etc. and presents the Shockley model of ideal semiconductor diodes. Device construction techniques, LED packaging styles—all affecting LED efficiency/efficacy are also discussed.

2.1 Introduction

In 1907 electrical engineer Henry J. Round touched a crystal of silicon carbide found near Niagara Falls with two wires connected to a battery. As he subsequently reported: "On applying a potential of ten volts between two points on a crystal of carborundum, the crystal gave out a yellowish light." Though he may not have realized it at that time, Round had operated the first crude solid-state lamp, the light emitting diode.

The discovery of light emission in silicon carbide was important for several reasons, but the most significant reason concerned the mechanism that caused the carborundum crystal to emit light. Most electric light sources such as incandescence of a hot electrode, luminescence of a glowing plasma, or fluorescence of a phosphor coating, are capable of producing exceedingly high luminous outputs, but each of the conventional light sources is unsuited to a broad range of potential applications because of their slow response time, inherent fragility, and short lifetime.

The light-emitting diode (LED) conveniently eliminates the limitations of conventional lamps with its solid-state reliability, speed, and compact size. Since semiconductor light sources (without any further structures like phosphor layers) emit a narrow-wavelength spectrum, various colored sources can be fabricated without the need for filters. Understanding their physical operation gives the motivation

J. M. Li (✉) · J. X. Wang · Z. Liu
Institute of Semiconductors, Chinese Academy of Science, A35, QingHua East Road, Haidian District, 912, 100083 Beijing, P. R. China
e-mail: jmli@semi.ac.cn

A. Poppe
Budapest University of Technology and Economics, Department of Electron Devices, Magyar tudósok körútja 2, Bldg. Q, 1117 Budapest, Hungary
e-mail: poppe@eet.bme.hu

C. J. M. Lasance, A. Poppe (eds.), *Thermal Management for LED Applications*,
Solid State Lighting Technology and Application Series 2,
DOI 10.1007/978-1-4614-5091-7_2,

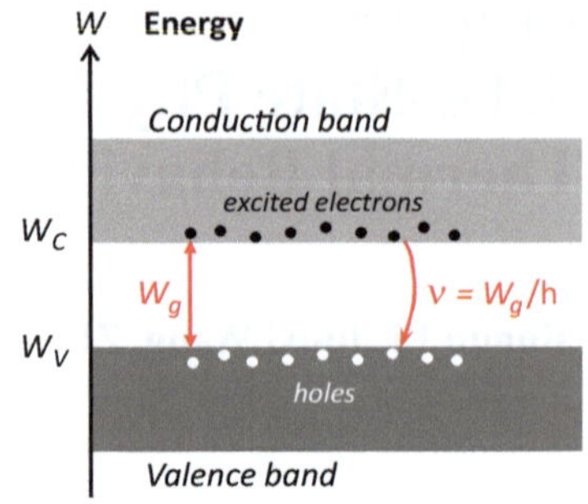

Fig. 2.1 Light emission occurs when an excited electron falls back from the conduction band with a direct energy state transition into the valence band and recombines with a hole. The frequency of the emitted light is determined by the W_g bandgap energy

and the necessary background knowledge to design proper cooling solutions for LEDs. The purpose of this chapter is to provide this background for the reader of this book.

This chapter is built up as follows. In Sect. 2.2, the basic concepts of solid-state physics that cause certain materials to emit light are described. Section 2.3 introduces the basic elements of bandgap engineering. Section 2.4 provides details on physics of LED operation including carrier transport, recombination mechanisms, and the I-V characteristics. Section 2.5 is dedicated to the description of temperature dependence of some of the fundamental processes inside an LED. Sections 2.6 and 2.7 describe chip level and package-level construction issues of LEDs and the chapter concludes with a summary.

For a more detailed study on the semiconductor physics, the reader is advised to refer to the classical handbooks of the topic [10, 12]. Detailed description of operation of LEDs can be also found in Schubert's widely known book [13].

2.2 Basic Concept of Monochromatic LEDs: A Qualitative Overview of the Operation

A simple way to efficiently generate light in many semiconductors is to fabricate the material as a pn-junction diode. In this manner, stimulated electrons can be efficiently and directly used to generate light, and a great deal of control can be exercised over the fabrication of the light-emitting region and the efficiency of the device.

Generally speaking, light generation in a semiconductor occurs when an excited electron resumes equilibrium through a direct *conduction band* to *valence band* transition by recombining with a hole in the valence band (see Fig. 2.1). The frequency and the wavelength of the emitted photon is determined by the $W_g = W_c - W_v$ bandgap energy as follows:

$$\nu = W_g/h \tag{2.1a}$$

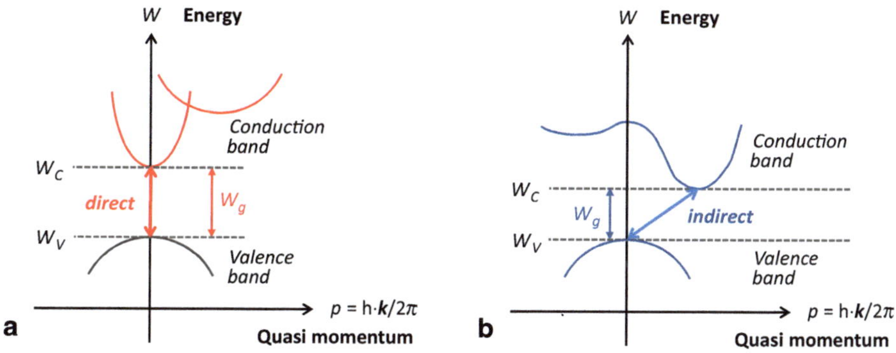

Fig. 2.2 Band structure of direct and indirect bandgap materials

$$\lambda = c/\nu = c \cdot h/W_g \tag{2.1b}$$

where h is Planck's constant and c denotes the speed of light.

As seen from Eqs. (2.1a) and (2.1b), the most important parameter to determine the properties of the emitted light is the energy band structure of the semiconductor material used. Semiconductor materials are classified as *direct bandgap* or *indirect bandgap* materials. Light emission takes place in pn-junctions formed in direct bandgap materials. Direct bandgap materials are characterized by an energy band structure such that the minimum energy level of the conduction band (where unlocalized electrons are free to move in the real physical space formed by the semiconductor lattice) and the maximum of the valence band (where vacancies of excited electrons called holes are left, acting as positively charged quasi-particles) are located at the same $\boldsymbol{k}$-wavelength number (quasi momentum), as shown in Fig. 2.2a. In the Brillouin zone, electrons' and holes' energy state is characterized by $\boldsymbol{k}$-wavelength number. In direct bandgap semiconductors, $\boldsymbol{k}$-wavelength numbers of electrons and holes are the same. They have the same momentum. But in indirect bandgap semiconductors, $\boldsymbol{k}$-wavelength numbers of electrons and holes are no longer the same, their momenta are different. Without the involvement of phonons, the electrons and holes cannot recombine in indirect bandgap semiconductors.

The junction diode provides a simple and efficient technique for placing large number of electrons in an excited state. In order for an electrical current to cross the potential barrier formed by the pn-junction, electrons must be raised from the valence to the conduction band. After the junction has been crossed, the electrons recombine with holes on the p side and this results in photon emission. Such a process is called *radiative recombination.* In contrast to this, in the so-called *nonradiative recombination* processes, as shown in Fig. 2.2b instead of photon emission, the energy of the recombining electron is transferred to the vibration energy of the semiconductor lattice by means of *phonons*, resulting in heat generation. Nonradiative recombination is facilitated by so-called *deep levels* or *traps* (energy levels close

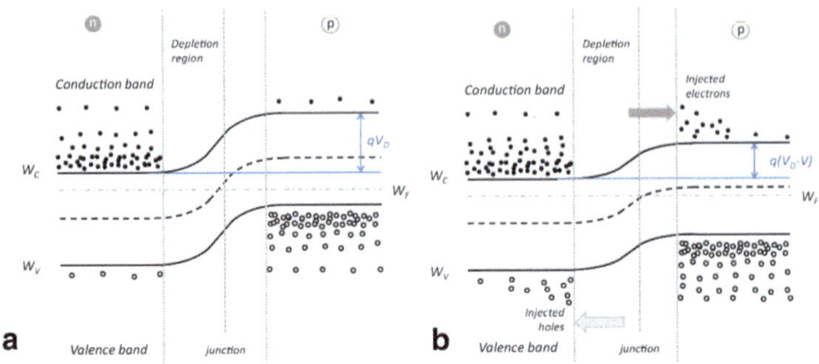

Fig. 2.3 Simplified energy-level diagrams (also known as simplified band diagrams) of a pn-junction, **a** No current flow, **b** Current flow

to the middle of the bandgap) which are called *luminescence killers*. Details of the different recombination processes will be discussed in Sects. 2.4.1–2.4.3.

A major goal of LED device construction is to assure the highest possible ratio of the electrical current associated with radiative recombination with respect to the total current of the LED device. The second goal is to assure that most of the photons generated at a pn-junction of an LED are coupled out to the outer space. These two aspects of LED device design will be discussed in detail in the last section of this chapter.

In this section, we start our discussion with the so-called *homojunctions*, i.e., pn-junctions consisting of a single material. Later in this chapter, *heterojunctions* and *quantum-well* structures used in today's modern power LEDs will be discussed in detail.

We further restrict our discussion now to the theory of *abrupt pn-junctions* with a homogeneous N_D donor concentration on the n-side and with a homogeneous N_A acceptor concentration on the p-side. Assuming fully ionized dopants, the concentration of the free electrons on the *n*-side is $n = N_D$, the concentration of the free holes on the p-side is $p = N_A$. On the n-doped side of the junction the electrons in the conduction band are the majority carriers while on the p-doped side the holes in the valence band are the majority carriers and both on the n- and p-sides of the junction there are some minority carriers also present.

In the close neighborhood of the unbiased pn-junction, the electrons of the n-side diffuse over to the p-side where they recombine with the holes. Similarly, holes diffuse to the n-side. As a result, the vicinity of the junction gets depleted from free carriers and a so-called *depletion region* is formed as indicated in Fig. 2.3. As there are no free carriers left in the depletion region, the only charge remaining there is from the ionized dopants (donors, acceptors). These dopants form a so-called *space charge region* which produces a potential that is called *diffusion potential* and which is denoted by V_D. Its value is determined by the N_A and N_D dopant concentrations and the n_i intrinsic carrier concentration of the semiconductor material:

$$V_D = \frac{kT}{q} \ln\left(\frac{N_A \cdot N_D}{n_i^2}\right) \tag{2.2}$$

where k is Boltzmann's constant, T is the absolute temperature of the semiconductor, and q is the elementary charge (charge of an electron).

Figure 2.3 shows graphs known as an *energy-level diagrams* or *band diagrams*. The valence and conduction bands on each side of the pn-junction are connected by sloping lines representing the continuous change of the energy levels as the effective doping level changes across the junction. In case of an unbiased pn-junction the diffusion potential forms an energy barrier of qV_D which blocks the flow of carriers across the junction, therefore there is no significant current flow through the junction (Fig. 2.3a). If an external V voltage is applied which results in reduction of the potential barrier electrons are injected into the p-side and holes are injected into the n-side and current flow occurs (Fig. 2.3b). This is a positive voltage between the p-side (also known as *anode*) and the n-side (referred to as *cathode)* of the pn-junction. While the injected carriers—now as minority carriers—are moving forward by diffusion, they are recombining with the majority carriers. If, for example, the recombination process is a radiative recombination (such as a direct bandgap transition), photon emission takes place.

The wavelength of the emitted radiation in such an arrangement is related to the bandgap of the semiconductor material making up the diode. *Bandgap engineering* means designing the material composition and the doping levels of an LED—which will determine the properties of the emission spectra (such as the peak wavelength) and device efficiency. Nowadays there are two widely used material systems: AlGaInP (aluminum-gallium-indium-phosphide) for producing red to yellow LEDs and InGaN/GaN (indium-gallium-nitride/gallium-nitride) for building green to blue LEDs [5–7], see Fig. 2.4.

White LEDs are typically so-called phosphor converted devices in which an LED chip with short wavelength (blue or UV) primary emission is covered by a phosphor layer which absorbs some of the short wavelength light and converts it to longer wavelength light in a wider spectral range. The fundamental physical operation of such LEDs—except the wavelength conversion process taking place in the phosphor—is the same as the operation of monochromatic LEDs, therefore it will not be discussed here separately.

2.3 Device Constructions Influencing the Energy Band Structure

The aim of this section is to provide an overview of bandgap engineering concepts resulting higher light generation efficiency.

2.3.1 Homojunction Devices

Correctly designed p and n materials are crucial to LEDs. In *homojunction* LEDs, the p and n materials are composed of the same energy gap semiconductor, as shown

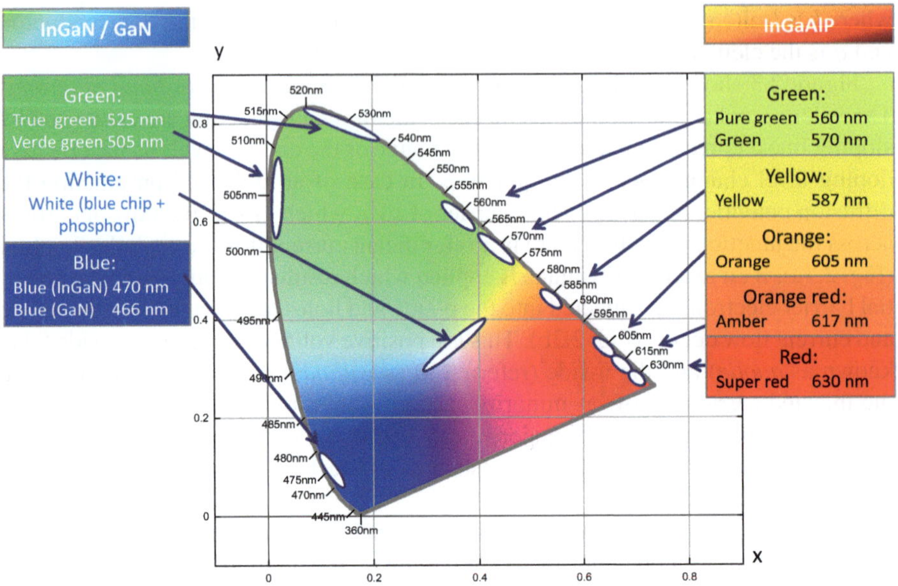

Fig. 2.4 Different LED colors realized by different material systems

in Fig. 2.3a. The *p* and *n* materials have the same bandgap structure. Due to the different doping type, the *p* and *n* material have different *Fermi levels*. When they are joined together, uniform *Fermi levels* are established which cause energy band bending (shown as W_F in Fig. 2.3). When an electron and hole go through the *space charge region* and recombination, a photon is generated. As the diffusion distance for electrons is larger than that of the holes, the injection ratio between electrons and holes is more than one in case the dopant concentrations in the *p* and *n* material are kept the same, which may lead to a serious thermal problem during operation. Furthermore, self-absorption from the active region lowers the output efficiency in homojunction LEDs.

2.3.2 Heterojunction Devices

A *heterojunction* is defined as a semiconductor structure in which the chemical composition changes with position. Heterojunctions are able to improve the performance of semiconductor devices because they permit the device designer to modify locally the energy-band structure of the semiconductor and so control the motion of the charge carriers. In the simplest case in a heterojunction two materials of different bandgaps are used.

Bandgap engineering [1] allows the LED designer to achieve higher efficiencies in converting electric current to light than conventional homostructures. Two

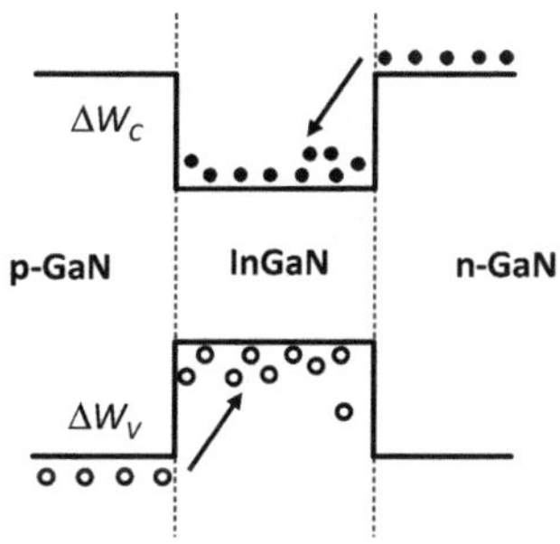

Fig. 2.5 Simplified energy band diagram for a heterojunction. LEDs in which the conduction band discontinuity ΔW_C and valence band discontinuity ΔW_V provide a barrier to confine carriers into the active region

major benefits of heterostructure LEDs are the increased minority carrier injection efficiency and the ability to use wide bandgap material. The wide bandgap layers are transparent to the photon reabsorption. The radiative efficiency in heterostructure LEDs is higher than that of the homostructure LEDs because of the increased current injection efficiency at the heterojunction. Since the radiative efficiency of the p-side is usually much higher than that of the n-side, higher efficiency devices are made by eliminating the minority carrier injection into the n-type material. By creating a heterostructure at the p/n interface, a valence band discontinuity is formed which provides a hole potential barrier. Figure 2.5 illustrates the energy band diagram for heterojunction flat-band conditions where the applied bias just cancels the built-in field. The conduction band discontinuity ΔW_C and the valence band discontinuity ΔW_V, produce a barrier to electron and hole flow [2], respectively.

Due to the barrier height limitation of the bandgap discontinuity, more carriers come across the pn-junction as leakage current. On the other side, nonradiative recombination through deep energy levels at the interface will reduce the radiative efficiency and causes a potential heating at the same time.

2.3.3 *Quantum Well Structure*

If one makes a heterostructure with sufficiently thin layers, quantum interference effects begin to appear prominently in the motion of the electrons. The simplest structure in which these may be observed is a *quantum well*, which simply consists of a thin layer of a narrower bandgap semiconductor between thicker layers of a wide bandgap material [3]. The band profile then shows a "rectangular well," as illustrated in Fig. 2.6. The electron wave functions in such a well consist of a series of standing waves, which might be found in a resonant cavity in acoustic, optical, or microwave technologies. The energy separation between these stationary states is enhanced by the small effective mass of electrons in the conduction bands of direct-gap semiconductors. With advanced epitaxial techniques, the potential profile of the quantum well does not need to be rectangular. Because the band edge energy is usually linear in the composition, W_V and W_C will follow the functional form of the composition. The quantum states in two parabolic wells [4] are illustrated in Fig. 2.7.

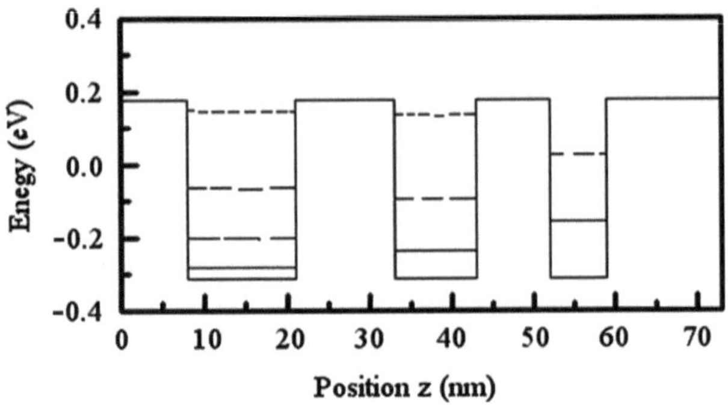

Fig. 2.6 Energy band profile of a structure containing three quantum wells, showing the confined state energy levels in each well. The gaps in the lines indicating the confined state energies show the locations of nodes of the corresponding wave functions

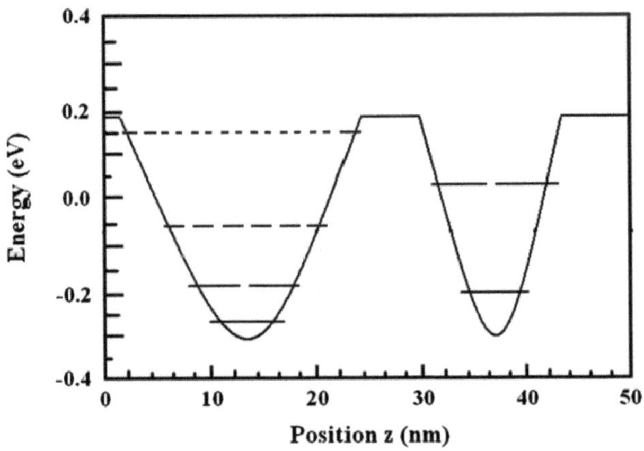

Fig. 2.7 Energy-band profile of a structure containing two parabolic quantum wells. The composition is similar to that of Fig. 2.3, and the overall widths of the wells are 20 and 8 nm

Quantum well heterostructures are key components of many optoelectronic devices, because they can increase the strength of electrooptical interactions by confining the carriers to small regions.

As for quantum well LEDs, they are a special class of heterojunction LEDs in which the thickness of the active region is less than the de Broglie wavelength of the electron in the semiconducting material. Basically, the electrons and holes are confined so tightly in space that their allowed states have discrete energy levels due to the localization. Recombination between such discrete energy levels results in very narrow emission spectra—characteristic to laser diodes.

2.4 Fundamental Physical Processes Determining the LED Behavior

This section provides a brief summary of the basic processes taking place in semiconductor pn-junctions.

2.4.1 *Carrier Transport*

When we talk about the carriers in the semiconductor, it means that electrons and holes are not treated separately, but are considered as macroscopic charge carrier populations or carrier concentrations. In semiconductor devices, carriers move due to different physical processes. The most important ones are: diffusion, drift caused by electric field, heat-emission, and tunneling. These processes—with a focus on recombination—are also discussed [10, 11].

2.4.1.1 Diffusion Current

Any kind of microscopic particles, be it molecules, atoms, or electrons, could generate diffusion movement, which is merely caused by the inhomogeneity of the density of the distribution of the particles. They will diffuse from regions of high concentration to regions of low concentration. The current in this case is called *diffusion current*. Carrier diffusion could happen, whether an external field is applied or not. The basic driving force of the carrier diffusion originates from the thermal energy of the majority carriers, which leads to the constant random motion of carriers. For a piece of uniformly doped semiconductor material, carriers homogeneously distribute throughout the physical space and the random motion of carriers in all directions results in zero net macroscopic current. When the carrier distribution is not uniform in the material, which is often caused by external electrical injection in LEDs, then a carrier concentration gradient exists between the region of high carrier density and the region of low carrier density. Then the random motion shows a pronounced tendency in a particular direction resulting in a net carrier flow in that direction. Thus, this carrier flow is called *diffusion current,* the quantity of which is closely related to the gradient of the carrier concentration.

Consider a piece of homogeneously p-doped semiconductor material with a metalized surface at its left side through which we inject electrons with a current density of J_n into the material, as shown in Fig. 2.8. This situation corresponds, e.g., to the situation, when electrons from the cathode side of a forward biased pn-junction reach the p-doped, anode side of the diode and are injected into the p-type semiconductor material (as already discussed with the help of Fig. 2.3). A concentration gradient of the electrons diffusing into the p-type material develops as the diffused electrons (as minority carriers) recombine with the majority carrier holes. The concentration of the injected electrons shows an exponential decay. The average L_n penetration depth of the electrons is called *diffusion length*—see Fig. 2.8.

It is easy to understand that carriers will diffuse from the region where their concentration is highest to the region where it is lowest. The diffusion coefficients D_n and D_p are used to represent the diffusion ability for electrons and holes, respectively. Higher diffusion coefficient and longer average *carrier lifetime* result in longer diffusion length. For the electrons:

$$L_n = \sqrt{\tau_n D_n} \tag{2.3}$$

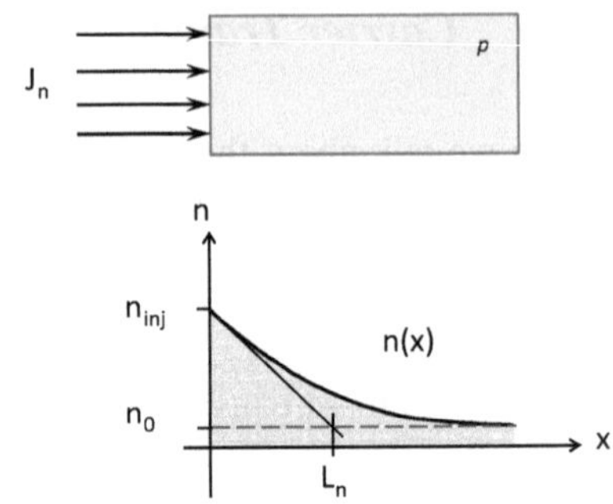

Fig. 2.8 Illustration of carrier diffusion

where τ_n denotes the average lifetime of the minority electrons in the p-type material. This lifetime depends on the net generation-recombination rate which will be discussed later. (Similar relationship between lifetime, diffusion coefficient, and diffusion length holds for the holes.)

As charge carriers diffuse from the high concentration regions to the low concentration regions along the concentration gradient, they form an electric current referred to as *diffusion current*. The diffusion current density of the electrons (for the sake of simplicity assuming only one-dimensional flow as in case in Fig. 2.8) could be expressed as:

$$J_n = qD_n \frac{dn(x)}{dx} \tag{2.4}$$

Similarly, we could obtain the diffusion current density of the holes:

$$J_p = -qD_p \frac{dp(x)}{dx} \tag{2.5}$$

In the expressions above, n and p are the densities (concentrations) of electrons and holes, respectively, q is the unit charge.

2.4.1.2 Drift Current

When an external voltage is applied on the semiconductor, the corresponding electric field would also affect the random movement of electrons: in average they will drift along the opposite direction of the electric field. The so-called *mobility* is used to represent the ability of the charge carriers to be drifted by the electric field, denoted by μ_n and μ_p for the electrons and holes, respectively (representing the average drift velocity of electrons and holes per unit electric field). The macroscopic electric current associated with this is called *drift current*. The drift current density for electrons is:

$$J_n = qn\mu_n E \tag{2.6}$$

and for holes is:

$$J_p = qp\mu_p E \tag{2.7}$$

where E denotes the absolute value of the electric field vector.

The mobility and diffusion coefficient in a semiconductor are related to each other. This relation is the so-called *Einstein relationship* [15]:

$$\frac{D_n}{\mu_n} = \frac{kT}{q}, \quad \frac{D_p}{\mu_p} = \frac{kT}{q} \tag{2.8}$$

where T is the absolute temperature of the semiconductor lattice, and k is Boltzmann's constant.

Based on the equations above, we could obtain the total current density in the semiconductor:

$$J = J_n + J_p = q\mu_n \left(nE + \frac{kT}{q}\frac{dn}{dx} \right) + q\mu_p \left(pE - \frac{kT}{q}\frac{dp}{dx} \right) \tag{2.9}$$

This is the basic current density equation for a semiconductor in which charge carriers flow as a result of drift and diffusion. This expression is generic to semiconductor diodes and applies to the carrier transport in LEDs as well.

2.4.2 *The I-V Characteristics of LEDs*

The equation describing the I-V characteristic of a semiconductor diode was first developed by Shockley. This so called Shockley equation for a pn-junction with a cross-sectional area of A looks as follows:

$$I = Aqn_i^2 \left(\frac{D_n}{L_n \cdot N_A} + \frac{D_p}{L_p \cdot N_D} \right) \cdot [\exp(V/V_T) - 1] \tag{2.10}$$

where $V_T = kT/q$ is called the *thermal voltage*—roughly 26 mV around room temperature (300 K). This equation can be derived from Eq. (2.9) and the energy level dependence of the carrier concentrations. (Readers interested in details of this derivation are advised to refer to any fundamental textbook such as [12] or [13]). Introducing the notation

$$I_0 = Aqn_i^2 \left(\frac{D_n}{L_n \cdot N_A} + \frac{D_p}{L_p \cdot N_D} \right) \tag{2.11}$$

formula Eq. (2.10) describing the *ideal diode characteristic* can be rewritten in the widely known form of

$$I = I_0 \cdot [\exp(V/V_T) - 1] \tag{2.12}$$

where I_0 is called the saturation current of the ideal diode characteristic. This equation is called the *Shockley model* of ideal pn-junctions.

As the useful LED operation (light emission) takes place under forward bias conditions, the widely used index F is applied in Eq. (2.12) to denote *forward current* and *forward voltage*:

$$I_F(V_F) = I_0 \cdot [\exp(V_F/V_T) - 1] \tag{2.13}$$

Deviation from the ideality is expressed by a so-called *emission factor* or *ideality factor* denoted by m:

$$I_F(V_F) = I_0 [\exp(V_F/mV_T) - 1] \tag{2.14}$$

As discussed in the introduction, significant forward current flows across a pn-junction if its anode-to-cathode voltage reaches a threshold value close to the diffusion voltage: $V_{th} \approx V_D$ This voltage value is published on LED datasheets as "forward voltage." This voltage value is closely related to the bandgap energy of the semiconductor material. The diffusion voltage should satisfy that $qV_D - W_g + (W_F - W_V) + (W_C - W_F) = 0$. In highly doped semiconductors, the separation between the band edges and the Fermi level is small compared with the bandgap energy, i.e., $(W_C - W_F) \ll W_g$ on the n-type side and $(W_F - W_V) \ll W_g$ on the p-type side. Thus, the diffusion voltage can be approximated by the bandgap energy divided by the elementary charge $V_{th} \approx V_D \approx W_g/q$. Thus, LEDs with short wavelength emission (blue LEDs/high bandgap energy) have higher forward voltage (close to 4 V), LEDs with long wavelength emission (red LEDs/low bandgap energy) have lower forward voltage (close to 2 V).

At high forward current levels, the internal series resistance of the LED causes the I-V characteristic to deviate from the Shockley model provided by Eq. (2.14), thus should not be neglected. The overall forward voltage measured between the anode and cathode contacts of an LED is the sum of the voltage drop on the internal electrical series resistance and the internal junction voltage denoted by V_{Fpn} in Fig. 2.9. Thus, using the notations of the figure:

$$V_{Fpn} = V_F - V_R = V_F - I_F \cdot R_S \tag{2.15}$$

With this, approximating [exp(V_{Fpn}/mV_T) - 1] with exp(V_{Fpn}/mV_T) Eq. (2.14) reads as follows:

$$I_F = I_0 \cdot \exp\left[\frac{V_F - I_F R_S}{mV_T}\right] \tag{2.16}$$

The effect of the internal electrical series resistance has to be considered in modeling of high power LEDs, especially when their high current operation (in the 350 mA, 700 mA or 1,500 mA range) is to be described. See more on LED modeling in Sect. 6.5.2.2. of Chap. 6. For further reading on deviations from the ideal diode characteristic, please refer to, e.g., Schubert's text book on LEDs [13].

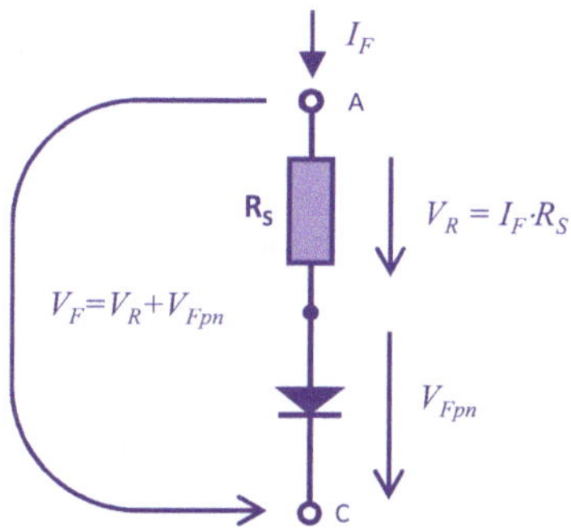

Fig. 2.9 Circuit diagram of a nonideal diode with the internal pn-junction and the internal series electrical resistance

2.4.3 *Temperature Dependence of the I-V Characteristics of LEDs*

As in Eq. (2.10) many parameters are temperature dependent, the diode characteristic is strongly temperature dependent. One of the most important factors determining the temperature dependence is n_i^2 which is the square of the intrinsic carrier concentration.

Figure 2.10 illustrates the temperature dependence of the forward voltage of pn-junctions when a fixed forward current is forced across the junction. (This electrical configuration is the most common one in case of DC-driven LEDs.) In the figure, this current is denoted by I_M as in thermal testing (see Chaps. 4 and 6); this current is called *measurement current.*

With this configuration the question of interest is the temperature dependence of the V_F forward voltage. The temperature sensitivity of the forward voltage of pn-junctions can be derived from the ideal diode characteristic (2.14)—provided that the operation is in the range of small forward currents when the effect of the internal series electrical resistance is still negligible. Assuming that $\exp(V_F/mV_T) \gg 1$ and rearranging it for forward voltage, one obtains

$$V_F = mV_T \ln(I_F/I_0) \tag{2.17}$$

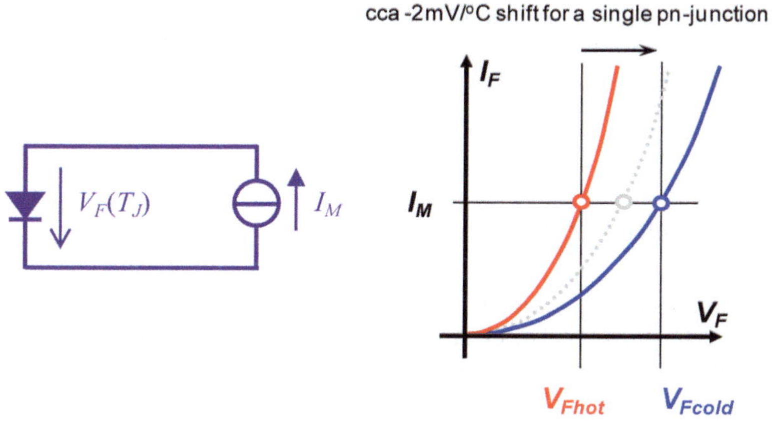

Fig. 2.10 Illustration of the temperature dependence of the forward voltage at constant forward current

The final result of the calculation of the $S_{VF} = dV_F/dT_J$ differential is as follows:

$$\frac{\mathrm{d}V_F}{\mathrm{d}T_J} = \frac{V_F}{T_J} - \frac{V_{G0} + m \cdot l \cdot V_T}{T_J} \tag{2.18}$$

where T_J is the absolute temperature of the semiconductor (referred to as *junction temperature* in all other chapters of this book), l is the power factor in the temperature dependence of n_i^2 in the formula of the I_0 saturation current (typical value is 3), and V_{GO} is the nominal value of the bandgap voltage of the semiconductor material: $V_{GO} = q \cdot W_g$. (The bandgap energy of materials used in LEDs is around ~ 1.4–4 eV, the corresponding bandgap voltage is ~ 1.4–4 V, see, e.g., [13]). Assuming an ideal diode ($m = 1$), this results in

$$S_{VF} = \frac{\mathrm{d}V_F}{\mathrm{d}T_J} = \frac{V_F - 3V_T - V_{G0}}{T_J} \tag{2.19}$$

For practical cases at usual temperatures, this differential has a fairly constant value. For a single pn-junction, the value is equal roughly to −1 to −3 mV/°C and is called the *temperature sensitivity of the forward voltage*. (S_{VF} is the reciprocal of the so-called *K-factor* defined in semiconductor device thermal testing standards).

2.5 Recombination Processes

The recombination processes are the essential basics of the useful LED operation: light emission. In the following, we describe these processes in detail.

As discussed in Sect. 2.1, after electrons and holes are injected into the active region of a diode, they can recombine either *radiatively* or *nonradiatively*. The radiative recombination will emit photons, which is preferred for light-emitting diodes. The nonradiative recombination will not emit photons, and the energy of the excited electrons will be relaxed to the semiconductor lattice resulting in self-heating. Self-heating through elevated temperatures accelerates all failure mechanisms of an LED, leading to reduced lifetime. Furthermore, if temperature increases, the nonradiative recombination loss will increase as well, resulting in the reduction of energy conversion efficiency of an LED. Detailed introduction of recombination types is summarized in Fig. 2.11 and their thermal behavior will be discussed in the subsequent subsections.

2.5.1 *Radiative Recombination*

If an electron recombines with a hole and a photon is emitted, the process is named *radiative recombination*. The electron-hole radiative recombination rate is proportional to the n electron and p hole concentrations, and is generally given by the

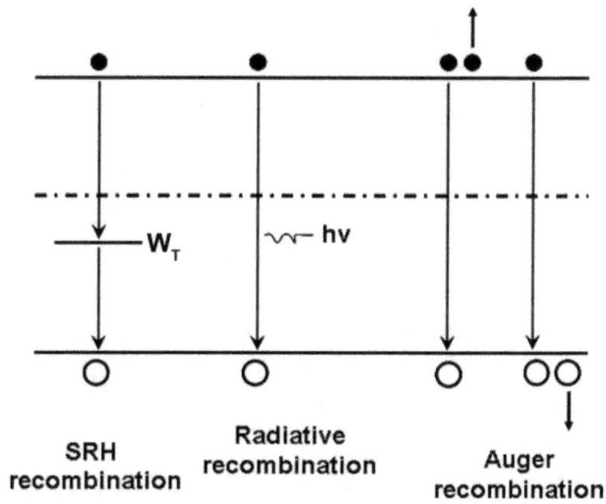

Fig. 2.11 Illustration of carrier recombination mechanisms in semiconductors

so-called *bimolecular rate equation:*

$$R_R = -\frac{dn}{dt} = -\frac{dp}{dt} = Bnp \tag{2.20}$$

where B is called the *bimolecular recombination coefficient.*

At the high excitation level, considering the injected carrier concentration is much larger than the equilibrium carrier density, the total electron/hole concentration is approximately equal to the excess carrier concentration.

$$n = n_0 + \Delta n \approx \Delta n \tag{2.21}$$

$$p = p_0 + \Delta p \approx \Delta p \tag{2.22}$$

where n_0 and p_0 is the equilibrium carrier concentration (without carrier injection).

If $\Delta n = \Delta p$, then it gives

$$R_R = Bn^2 \tag{2.23}$$

If the active region of an LED is composed of multiple quantum wells (MQWs), we can write

$$R = \sum_i B \frac{n^{2D}}{L_{QW}} \frac{p^{2D}}{L_{QW}}, \quad (\mathrm{i} = 1, 2, 3 \ldots \mathrm{m}) \tag{2.24}$$

where i denotes the quantum well (QW) labels from the p-side, and m is the total number of QWs, n^{2D} and p^{2D} denote the two dimensional electron and hole concentrations in the QWs, L_{QW} is the QW length.

The bimolecular recombination coefficient can be calculated using the van Roosbroeck-Shockley model [13].

$$B = \frac{R_0}{n_i^2} \tag{2.25}$$

where the R_0 radiative recombination rate at equilibrium is given by the van Roosbroeck-Shockley equation,

$$R_0 = \int_0^\infty R_0(\upsilon)d\upsilon = \int_0^\infty \frac{N(\upsilon)}{\tau(\upsilon)}d\upsilon = \int_0^\infty \frac{8\pi \upsilon^2 \bar{n}^2}{c^2} \frac{\alpha(\upsilon)}{e^{h\upsilon/kT} - 1}d\upsilon \tag{2.26}$$

where ν denotes the frequency of the photon involved, h is Planck's constant, c is the speed of light, $N(\nu)$ is the number of photons with frequency ν, $\tau(\nu)$ is the average photon lifetime, $\bar{n}$ is the refractive index, and $\alpha(\nu)$ is the absorption coefficient of photons. From Eq. (2.23) and Eq. (2.24), the coefficient B is estimated to be about $4.7 \times 10\text{–}10\,\text{cm}^3/\text{s}$ for GaN using its typical parameters. For further details on this, the reader is advised to refer to, e.g., Chap. 3 of Schubert's book *Light-Emitting Diodes* [13].

A more accurate model is based on quantum mechanics. The transition rate (i.e., transition probability per unit time) from a quantum mechanical state j to a state m is given by Fermi's Golden Rule using perturbation theory,

$$W_{j\to m} = \frac{d}{dt}\left|\alpha'_m(t)\right|^2 = \frac{2\pi}{\hbar}\left|H'_{mj}\right|^2 \rho(E = E_j - \hbar\omega_0) \tag{2.27}$$

where H'_{mj} is the transition matrix element that connects the initial j-th state with the final state m via the perturbation Hamiltonian H', and $\rho(E = E_j + \hbar\omega_0)$ is the density of final states. For the one-dimensional case, H'_{mj} is

$$H'_{mj} = \left\langle \psi_m^0 \right| H' \left| \psi_j^0 \right\rangle \tag{2.28}$$

Several characteristics should be noted for radiative recombination:

(i) A necessary condition for recombination is the spatial overlap between electron and hole wave function. Assuming H' is independent from the spatial position, the transition rate is proportional to the overlap integral Γ,

$$\Gamma = \Gamma_0 \left| \int_{-\infty}^{\infty} \phi_e(z)\phi_h(z)dz \right|^2 \tag{2.29}$$

(ii) The electron momentum should be conserved, since the photon momentum is negligibly small. The conservation of momentum condition is known as the ***k**-selection rule*.

As temperature increases, what will radiative recombination do? With increasing temperature, the carrier distribution in **k**-space will be more spread. Because of momentum conservation, the recombination rate of an electron/hole is proportional to the number of holes/electrons with equal momentum, thus the recombination probability (B) will decrease. The coefficient B has been calculated using several approaches, which result in $B \propto T^{-3/2}$ [14, 15]. Hence radiative recombination rate will decrease as $R_R \propto T^{-3/2}$.

2.5.2 Defect-Mediated SRH Recombination

There are several nonradiative recombination mechanisms. One typical mechanism is defect-mediated SRH (Shockley-Read-Hall) recombination.

The recombination of free carriers via deep levels was first analyzed by Shockley, Read, and Hall. The electron in transition between bands passes through the impurity level within the bandgap. Phonons could participate in this process, so the momentum of the electron and the hole is not to be the same. The nonradiative recombination rate through a deep level with trap energy W_T and trap concentration N_T is given by [16, 17, 19, 20]:

$$R_{SRH} = \frac{p_0 \Delta n + n_0 \Delta p + \Delta n \Delta p}{(N_T v_p \sigma_p)^{-1}(n_0 + n_1 + \Delta n) + (N_T v_n \sigma_n)^{-1}(p_0 + p_1 + \Delta p)} \tag{2.30}$$

where v_n and v_p are the electron and hole thermal velocities, σ_n and σ_p are the capture cross section of the traps, n_1 and p_1 are the electron and hole concentrations if the W_F Fermi energy is located at the trap level.

$$n_1 = n_i \exp\left(\frac{W_T - W_{Fi}}{kT}\right), \quad p_1 = n_i \exp\left(\frac{W_{Fi} - W_T}{kT}\right) \tag{2.31}$$

where, W_{Fi} is the Fermi level in the intrinsic semiconductor.

For n-type semiconductor, it should satisfy $n_0 \gg p_0$ and $n_0 \gg n_1$. At low excitation level, $\Delta n \ll n_0$, so

$$R_{SRH} = \frac{\Delta p}{(N_T v_p \sigma_p)^{-1}} \tag{2.32}$$

$$\tau_{SRH} = \frac{\Delta p}{R_{SRH}} = (N_T v_p \sigma_p)^{-1} \tag{2.33}$$

It indicates that the SRH recombination rate at low excitation level is dominated by the capture rate of minority carriers. At high excitation level, $\Delta n \gg n_0$, so

$$R_{SRH} = \frac{1}{(N_T v_p \sigma_p)^{-1}/\Delta p + (N_T v_n \sigma_n)^{-1}/\Delta n} \tag{2.34}$$

$$\tau_{SRH} = \frac{\Delta p}{R_{SRH}} = (N_T v_p \sigma_p)^{-1} + (N_T v_n \sigma_n)^{-1} = \tau_{SRH,p} + \tau_{SRH,n} \tag{2.35}$$

It suggests that the SRH recombination rate at high excitation level is limited by the smallest one of the electron or hole capture rate. Obviously, the SRH recombination rate is proportional to Δp. As temperature increases, the carriers' thermal velocities will increase, so there are more probabilities for carriers to be captured by traps. Therefore, defect-mediated recombination will increase. In addition, more phonons can participate in this process, which will also lead to an increase of the defect-assisted recombination.

2.5.3 *Auger Recombination*

Another typical mechanism is the *Auger recombination*. In this process, the energy released by electron-hole recombination is delivered to another electron or hole. The momentum conservation of the electron and the hole is also not necessary since phonons could participate in this process. The Auger recombination rate is given by:

$$R_{Auger} = C_p n p^2 + C_n n^2 p \tag{2.36}$$

At high excitation levels and $\Delta n = \Delta p$, the equation reduces to:

$$R_{Auger} = (C_n + C_p) n^3 = C n^3 \tag{2.37}$$

The Auger coefficient often increases with increasing temperature. As mentioned in reference [18], this coefficient is proportional to $(kBT/W_g)^{3/2}\exp(-W_g/kBT)$. Generally, the Auger recombination loss increases with temperature. Nevertheless, because Auger recombination is related to the energy band, the dependence of Auger recombination on the temperature is sometimes complicated.

2.5.4 *Radiative Recombination Efficiency Given by the ABC Model*

If $\Delta n = \Delta p$, the SRH, radiative, and Auger recombination can be simplified to be $A{\cdot}N$, $B{\cdot}N^2$, and $C{\cdot}N^3$, respectively. So, the carrier recombination loss rate is given by a classic equation,

$$R = AN + BN^2 + CN^3 \tag{2.38}$$

where A, B, C denotes the coefficient for SRH, radiative, and Auger recombination, respectively, N stays for carrier concentration. Hence the *radiant efficiency* can be given by

$$\eta_e = \frac{BN^2}{AN + BN^2 + CN^3} \tag{2.39}$$

The relation between carrier density N and the injection current density J is deduced as follows.

The carrier generation rate is

$$G = J/qd \tag{2.40}$$

where d is the effective thickness of the active region.

At steady state, $G = R$ should be satisfied, so

$$J = qd(AN + BN^2 + CN^3) \tag{2.41}$$

The carrier lifetime is

$$\tau = \frac{N}{R} = \frac{N}{AN + BN^2 + CN^3} \tag{2.42}$$

so

$$N \approx \Delta n = \frac{J}{qd} \cdot \tau \tag{2.43}$$

Generally, as temperature increases, the coefficient B will decrease, while A and C will increase. Therefore, the radiative recombination coefficient will decrease as temperature increases.

It should be pointed out that in some cases, $\Delta n \neq \Delta p$. For example, in MQW active region of GaN LEDs under electrical injection, the electrons and holes are injected separately from the n- and p-sides. For QWs from p-side to n-side, the hole density gradually decreases while the electron density gradually increases. In this condition, the local density difference of electrons and holes should be considered. The recombination rate cannot be simplified to be AN, BN^2, and CN^3, instead, it should be calculated as An or Ap (An when $n < p$, Ap for $p < n$), Bnp, $C(n^2p + np^2)$.

It should be noted that, when the carrier density is high enough, the ABC model might demand some amendments. It has been brought out that the density-dependent increase of the spontaneous emission changes from quadratic to linear, while the increase of the Auger recombination is reduced from cubic to approximately quadratic or even less, by calculation using a microscopic model [19].

2.6 Current Spread

Recently, GaN-based light-emitting diodes have attracted much attention due to their important applications in solid-state lighting. Typical GaN-based LEDs grown on a sapphire substrate have an etched mesa shape and a side-by-side contact configuration due to the insulating substrate. These diodes grown on an insulating sapphire substrate employ a lateral as well as a vertical current flow. In particular, the lateral current flow can lead to nonuniform current spreading, which locally produces light emission and heat generation. It was reported [20] that the nonuniform current spreading, the so-called *current crowding* effect, was strongly related to chip reliability in addition to local light emission. To be equidistant, the design of p-pad and n-pad electrodes frequently used heuristics to reduce the current crowding effect. The location and size of ohmic contacts are relevant to light extraction, because a portion of the injected carriers will still be confined underneath the thick metal bonding pad, even with the transparent current spreading indium tin oxide (ITO) layer. This section discusses the current flow patterns of different structures aimed at high extraction efficiency.

2.6.1 Common Contact Geometry and Current Crowding Problem

Current crowding easily occurs in mesa-structures on insulating sapphire substrates. In these LEDs, the p-type contact is usually located on the top of the mesa, and the n-type contact is located on an n-type buffer layer at the bottom of the mesa. As a result, the current tends to crowd at the edge of the mesa contact adjoining the n-type contact. Additionally, the geometric pattern of the electrodes also influences the current spreading due to potential difference in the distribution of the intrinsic diode causing the pn-junction current.

A lateral p-side-up mesa LED grown on sapphire is shown in Fig. 2.11a. It is intuitively clear that the pn-junction current crowds near the edge of the mesa as indicated in the figure. An equivalent circuit model is shown in Fig. 2.11b and includes the p-type contact resistance and the resistance of the n-type and p-type cladding layers. Assuming that the p-type metal contact has the same potential and that the voltage drop across the vertical series resistance (R_v) is much larger than kT/e, i.e., $I \cdot R_v \gg kT/q$, where I is the current and J the current density, 1D current distribution reported by Guo and Schubert is given by [20–23]:

$$J_G\left(x\right) = J_0 exp\left(-\frac{x}{L_{s,G}}\right) \tag{2.44}$$

where J_0 is the injected current density at the mesa edge and $L_{s,G}$ the current spreading length, which is written as:

$$L_{s,G} = \sqrt{(\rho_c + \rho_p)t_n/\rho_n} \tag{2.45}$$

where ρ_c is the specific p-type contact resistance, and ρ_p, ρ_n and t_p, t_n are the electrical resistivity and the thickness of p-type and n-type layers, respectively. An integral of Eq. (2.44) with respect to the mesa length (L) and the device width (W) gives the injection current (I):

$$I = \int_0^W \int_0^W J(x)dxdy = J_0 L_{s,G} W \left[1 - \exp\left(-\frac{L}{L_{s,G}}\right)\right] \tag{2.46}$$

Division of Eq. (2.46) by total device area (LW) thus yields average current density (J_G):

$$J_G(L) = J_0\left(\frac{L}{L_{s,G}}\right)^{-1}\left[1 - exp\left(-\frac{L}{L_{s,G}}\right)\right] \tag{2.47}$$

For low diode current densities or small vertical series resistances, i.e., $I \cdot R_v \approx kT/q$, the current distribution also can be expressed as [21, 22]:

$$J_T(x) = 2J_0\left(\frac{x}{L_{s,G}} + \sqrt{2}\right)^{-2} \tag{2.48}$$

The current spreading length ($L_{s,T}$) reported by Thompson is given by:

$$L_{s,T} = \left(\frac{m \cdot kTt_n}{q\rho_n J_0}\right)^{1/2} \tag{2.49}$$

where m is the diode ideality factor and has a typical values of $1.05 < m < 1.35$ [13]. The integral of (2.48) divided by the total device area also gives the average current density (J_T):

$$J_T(L) = 2J_0\left(\frac{L}{L_{s,G}}\right)^{-1}\left[\frac{1}{\sqrt{2}} - \left(\frac{L}{L_{s,G}} + \sqrt{2}\right)^{-1}\right] \quad (2.50)$$

The contact geometry of electrode is critical to the achievement of efficient current injection. For the determination of actual lateral mesa lengths (L), it is important to understand the exact value of L_s. According to Eq. (2.45), L_s is predictable when device parameters are known. In the case of low diode current densities or small vertical series resistance, which is a typical case for recently developed LEDs, the prediction of L_s by using Eq. (2.49) is not easy because of the difficulty in obtaining the parameter J_0. This indicates that the precise measurement of L_s is a key issue for realization of high-efficiency designs.

2.6.2 *Influence of Current Crowding Effect*

Usually, the current is mainly crowded near the mesa edge of the n-GaN layer, which results in a significantly inhomogeneous carrier distribution in the plane of the QWs. The nonuniform carrier distribution causes large carrier loss at high current and hence increases the drop of the *internal quantum efficiency* (IQE). Moreover, the current spreading length decreases as injection current increases, which intensifies the efficiency drop problem at high current. Self-heating of an LED leads to remarkable degradation of its external efficiency under high-current operation, i.e., the self-heating is one of the mechanisms responsible for the commonly observed efficiency decreasing with current.

An experimental result of the current-crowding effect found in an InGaN/GaN LED grown on sapphire substrate is shown in Fig. 2.12 [20]. A micrograph of the optical emission from an InGaN LED is shown in Fig. 2.13a. The picture is taken from the sapphire side of the LED and shows the intensity of blue light emission. The micrograph clearly reveals that the emission intensity decreases with increasing distance from the mesa edge. Figure 2.13b shows the experimental intensity as a function of the distance from the mesa edge. A theoretical fit to the experimental data using the exponential decrease in current density derived above is also shown in the figure. Experimental and theoretical data exhibit very good agreement if a current-spreading length of 550 μm is used in the calculation. With the expected future improvement of the contact and p-type doping in GaN devices, the requirement of larger device and contact sizes, current crowding effects will increase. Novel contact geometries are induced to alleviate the problem. For devices with dimensions much smaller than L_s, the current-crowding effect becomes irrelevant.

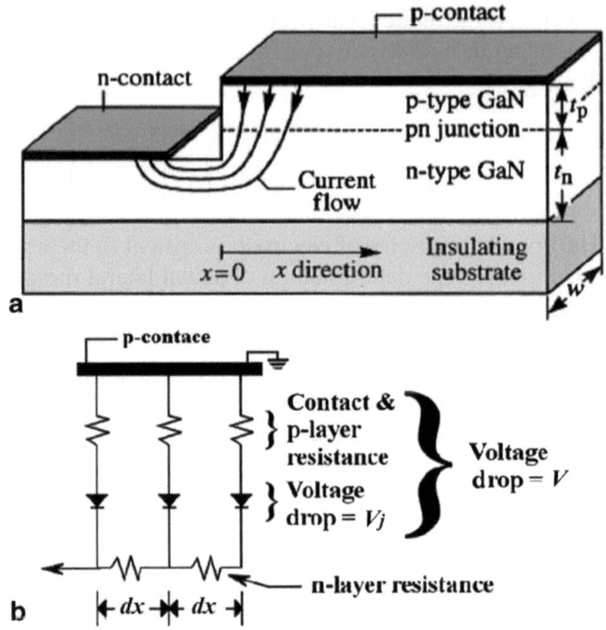

Fig. 2.12 **a** Current crowding in a mesa-structure GaN-based LED grown on sapphire substrate. **b** Equivalent circuit consisting of n-type and p-type layer resistances, p-type contact

2.6.3 Some Approaches to Spread Current

Current spreading in a lateral-injection LED can be improved by the modification of the electrode pattern. Guo et al. improved the design by the incorporation of interdigitated mesh patterns for more uniform current spreading [20]. The current crowding effect can also be reduced by the insertion of a *current blocking layer* (CBL) under the p-electrodes [23]. This can be attributed to the use of an insulated CBL structure, which leads to better current spreading outside the region beneath the opaque n-pad and less photon absorption at the n-electrode. In addition, an ITO layer is used as transparent current spreading layer.

2.7 Influence of LED Thermal Behavior on LED Performance

2.7.1 Self-Heating

Energy conservation demand for high-performance solid-state lighting has led to a sharp escalation in LED industries. Increasing the power density and the current level in LED chips has been used for minimizing cost. However, there are side effects in the LED due to self-heating. The main heat source in the LED package is the heat generation at the junction due to nonradiative recombination processes and Joule

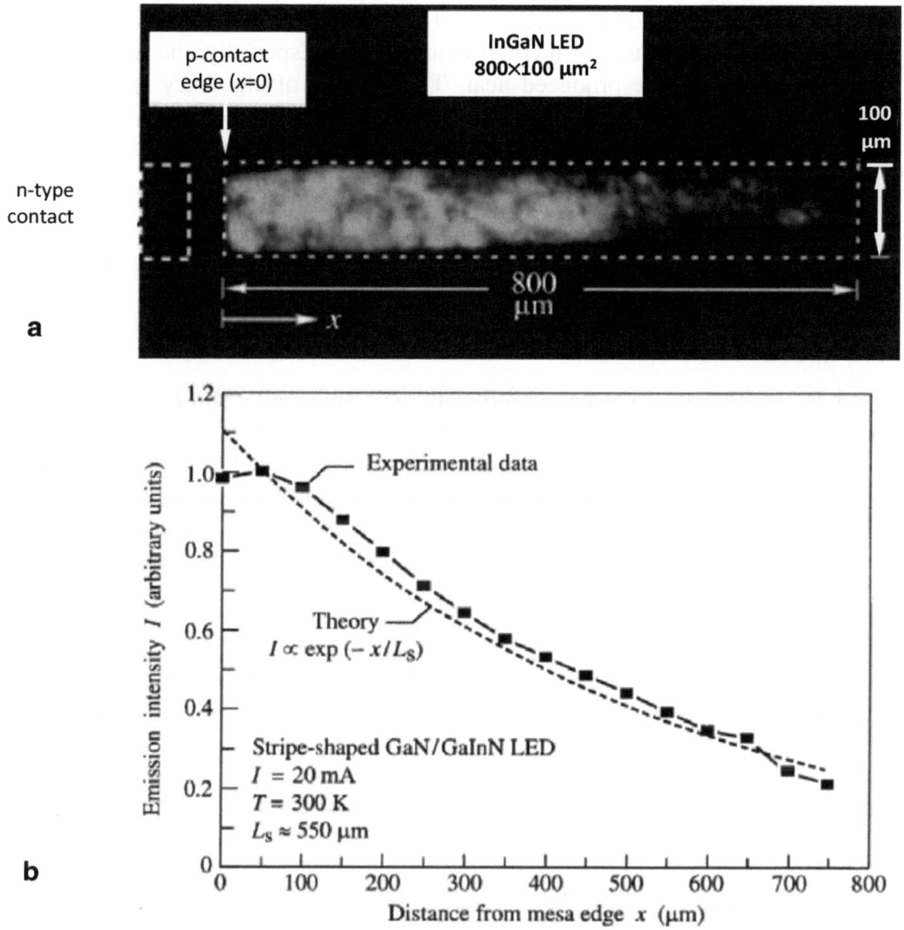

Fig. 2.13 **a** Micrograph of optical emission from mesa-structure InGaN/GaN LED grown on an insulating sapphire substrate. The LED has a stripe-shaped 800 × 100 µm p-type contact. **b** Theoretical and experimental emission intensity versus the distance from the mesa edge [20]

heating at the series electrical resistance of the diode and possible Joule heating at the interconnects. At the same time, the light absorption in the material and in the interface is another heat source. The induced temperature increase will affect the operation of the active layer and conductive layer in the LEDs. If the temperature increases to a high level, those layers may either fail temporarily or even permanently degrade. Therefore, it is crucial to minimize the temperature increase with proper design of LEDs both on chip and package level, not to mention system-level thermal management solutions.

In contrast to conventional semiconductor devices used in electronics/microelectronics, LEDs release a large part of energy in the form of light. The rest part of energy would be in the form of heat. Hence, the efficiency of an LED is one of the critical factors in thermal management. Especially, the wall-plug efficiency is related to the produced heat. The concept of efficiency is introduced first.

The active region of an ideal LED emits one photon for every electron injected. The *internal quantum efficiency* (IQE) is defined as:

$$\eta_{int} = \frac{no.\ of\ photons\ emitted\ from\ active\ region\ per\ second}{no.\ of\ electrons\ injected\ into\ led\ per\ second} = \frac{P_{int}/\,(h\nu)}{I/q} \tag{2.51}$$

where P_{int} is the optical power emitted from the active region and I is the injection current.

Photons generated in the active region could not escape into free space completely. The *extraction efficiency* is below unity due to several possible loss mechanisms, such as reabsorption in the substrate, absorption by the metal electrode, and absorption by the bulk material because of the total internal reflection. The extraction efficiency is defined as:

$$\eta_{extraction} = \frac{no.\ of\ photons\ emitted\ into\ free\ space\ per\ second}{no.\ of\ photons\ emitted\ from\ active\ region\ per\ second} = \frac{P_{opt}/(h\upsilon)}{P_{int}(h\upsilon)} \tag{2.52}$$

where P_{opt} is the optical power emitted into free space from an LED.

The *external quantum efficiency* is defined as:

$$\eta = \frac{no.\ of\ photons\ emitted\ into\ free\ spaceper\ second}{no.\ of\ electrons\ injected\ into\ LED\ per\ second} = \frac{P/\,(h\upsilon)}{I/q} = \eta_{int}\eta_{extraction} \tag{2.53}$$

The external quantum efficiency gives the ratio of the number of useful photons to the number of injected charge particles.

Finally, the *radiant efficiency* (also known as *power efficiency, energy conversion efficiency* or *wall-plug efficiency*—WPE in short) is defined as:

$$\eta_e = \frac{P_{opt}}{I_F \cdot V_F} \tag{2.54}$$

where $I_F \cdot V_F = P_{el}$ is the electrical power supplied to the LED. Therefore, the heating power in an LED is:

$$P_H = (1 - \eta_e) \cdot I_F \cdot V_F = I_F \cdot V_F - P_{opt} \tag{2.55}$$

The heating corresponds to the internal quantum efficiency and the extraction efficiency. The heat generated leaves the LED's active (epitaxial) layer mainly by conduction through the substrate.

2.7.2 *Effect of Self-Heating on Material Quality*

An LED package comprises various conducting and insulating materials that have different *coefficients of thermal expansion* (CTE). When the chip is heating up, each material deforms at a different rate. This nonuniform CTE induces mechanical stresses within the package. In an LED package, this nonuniformity not only exists in the package but also in the epitaxial system based on the heteroepitaxially grown gallium nitride on the sapphire substrate. To reduce this stress or to eliminate thermal failure, proper material selection and the minimization of temperature differences by proper thermal management is required.

The self-heating could induce degradation of LED material and this phenomenon will increase the nonradiative recombination and therefore reduce the internal quantum efficiency. For example, the highly strained GaN/InGaN interfaces in LEDs are responsible for a faster increase in the defect density due to hot-electron injection. These defects enhance the trap-assisted tunneling in the MQWs and results in the development of hot spots among devices after high current stressing of the MQWs. This in turn leads to an increase in the defect generation rate, which results in a thermal run-away. This phenomenon would ultimately result in the failure of the device. The reduction of dislocation densities in LEDs will reduce this thermally assisted mechanism and will increase the device reliability [24].

The self-heating may also lead to degradation of transparent package materials and this phenomenon will increase the absorption in the package. Hence, the power efficiency of the device is reduced. This can be observed, for example, as the optical degradation of phosphor-converted white GaN-based LEDs exposed to electrothermal stress. The optical output degrades rapidly with operating time, accompanied by the change of the spectral power distribution of the emitted light. After some stress time, the luminous flux is reduced by a factor and the color temperature is shifted. Optical degradation is obviously originating from the temperature-driven darkening of packaging materials and the injected current-driven generation of nonradiative recombination defects [25].

Another example is that the LED modules encapsulated with a hemispherical-shaped plastic lens exhibited a better lifetime due to their better thermal dissipation than those with cylindrical or elliptical plastic lenses. Results also showed that the optical power of the LED modules increased after removing the plastic lens, because the degradation of the material of the plastic lens decreased the amount of the emitted light. The key module package-related failure modes under thermal-aging were identified as the degradation of the plastic lens and lens material [26].

Silicone materials with a relatively high refractive index have been introduced for the encapsulation of high-power LEDs and LEDs with relatively short

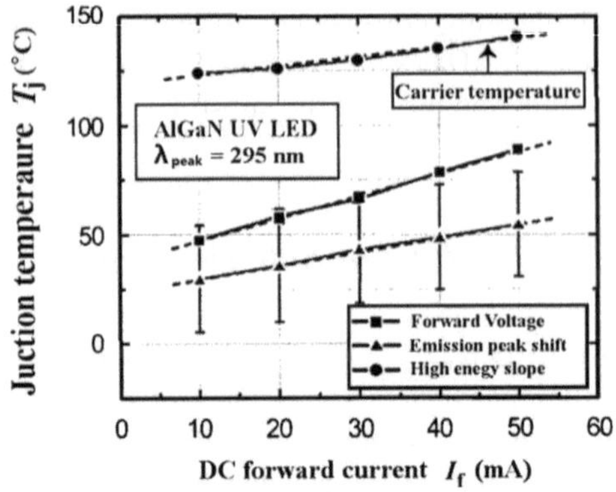

Fig. 2.14 Junction temperature inferred from emission peak energy as a function of DC forward current [28]

wavelengths. However, most of the existing silicone encapsulants still suffer from thermal and radiation-induced degradations and thus leads to reliability issues and a short lifetime [27].

2.7.3 *Effect of Self-heating on Efficiency/Efficacy*

Solid-state lighting applications normally require LEDs operating under high current injection. However, in this case, considerable heat fluxes can be generated within the device package, which will lead to strong self-heating effects and the temperature of active region will be much higher than the case temperature. It has been shown that the thermal stress induced by self-heating in high power LEDs can cause a decline in LED intensity and is generally considered to be one important factor of long-term reliability [13]. In this section, we will discuss the effect of self-heating on light emitting efficiency and reliability.

At low current levels, heat generation is dependent on Joule heating: the dominant heat source at low current levels is the active region. At high current levels, the contribution of parasitics becomes increasingly important (see, e.g., the series electrical resistance mentioned in Sect. 2.3.2). Figure 2.14 shows the dependence of junction temperature on current level. It is often described by the equation below:

$$T_J = T_C + R_{thJC} \cdot (I_F \cdot V_F - P_{opt}) \tag{2.56}$$

where T_C is the case temperature, R_{thJC} is the *junction-to-case thermal resistance*.

Radiative efficiency of recombination processes in LEDs is controlled by two processes: the radiative and the nonradiative process, as stated above. The radiative recombination rate can be expressed as $R = Bnp$. With self-heating effect, the induced thermal energy will dissociate carriers from radiative state and decrease the number of available carrier concentration and the subsequent light output.

In addition, with temperature increasing, the number of carriers in the $\boldsymbol{k}$-space interval decreases. Radiative recombination requires momentum conservation according to the k-selection rule. The recombination probability of an electron is proportional to the number of holes available at the same momentum, therefore, the recombination probability and the light efficiency decreases with increasing temperature.

Nonradiative recombination due to the generation of defects and Auger recombination is also temperature related. With temperature increasing, the generation of nonradiative defects and Auger effects will decrease the internal quantum efficiency of the devices and influence the final light output.

Under steady-state conditions, the rate equation of an LED can be expressed as:

$$\frac{dn}{dt} = \frac{J}{qd} - AN_T n - Bnp = 0 \tag{2.57}$$

where, J is the current density, q is the electron charge, d is the thickness of the active layer, A is the nonradiative recombination coefficient, B is the radiative recombination coefficient, N_T is the density of the defects responsible for nonradiative recombination, and n and p are the concentrations of electrons and holes in the active layer, respectively.

If radiative recombination dominates over nonradiative processes ($AN_T n \gg Bnp$), Eq. (2.57) can be simplified as:

$$L = Bnp \approx \frac{J}{qd} \tag{2.58}$$

In this case, light output is proportional to the injected current, resulting in a slope of one for L-I curve. On the other hand, if the nonradiative recombination process dominates over radiative recombination ($AN_T n \gg Bnp$), and therefore:

$$AN_T n \approx \frac{J}{qd} \tag{2.59}$$

which shows that n is proportional to the injected current and light output is proportional to the square of the injected current, so the slope of the L-I curve is about two.

Comparing the LED before and after heating stress, as shown in Fig. 2.15, the slope of the log-log L-I curve moves toward two and indicates that the concentration of nonradiative centers increases after the heating stress [29].

2.7.4 *Effect of Self-Heating on Reliability*

LEDs typically exhibit very long operational life. Their light output degrades slowly over time and generally do not suffer catastrophic failure. For the LEDs, the MTTF (mean time to failure) is defined as the point at which the light output decreases to a given factor, usually taken as 70 % of the initial intensity [31].

In most of LED reliability works, the main enhancement factor for degradation is excessive heat at the pn-junction as a function of the ambient temperature and

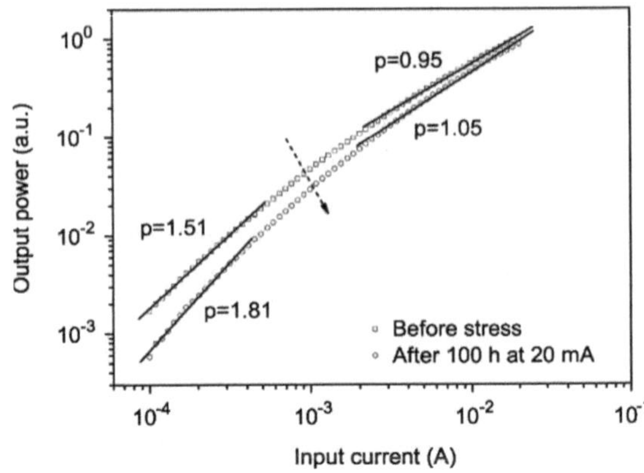

Fig. 2.15 L-I curve measured before and after stress [29]

self-heating generated by the electrical current. The relation is normally expressed by the Arrhenius equation:

$$MTTF_{70\%} = A_0 \exp\left(-\frac{\Delta W}{kT}\right) \tag{2.60}$$

where k is the Boltzmann constant, T is the temperature, and ΔW is the activation energy.

The degradation of LEDs is a multifaced problem including the generation of nonradiative defect, degradation of ohmic contact, coloration of clear epoxy, etc.

Figure 2.16 shows a comparison of I-V characteristics of an LED before and after degradation. At low bias regions, the current increases after degradation. This indicates the generation of defects, which cause additional channels in the active layers. At high bias voltage, the aged device shows an increase of series resistances, which is mainly due to the increase of the parasitic resistive components of the semiconductor and contact layers and accordingly decrease the ability of the current and emission spreading at the device surface, and the subsequent optical power [29].

During degradation, LEDs may also undergo gradual shifts in emitted spectra over time and these changes can also affect the maintenance of their luminous flux due to the changes in spectral power distribution. A typical variation of spectra during stress is shown in Fig. 2.17. The phosphor-related yellow emission decreased with respect to the main blue peak. The yellow component is more affected by thermal treatment than the blue component.

A cross-section of the aged device shows that it has a uniform browning of packaging material and degradation took place mainly beyond the chip region (shown in Fig. 2.18).

Considering the important thermal influence on LED light efficiency and reliability, many types of LED have been designed to achieve lower thermal resistance as shown in Fig. 2.19. Packages using heat sink slugs can effectively transfer heat from the chip directly to a printed circuit board with thermal resistances of 0.4 K/W.

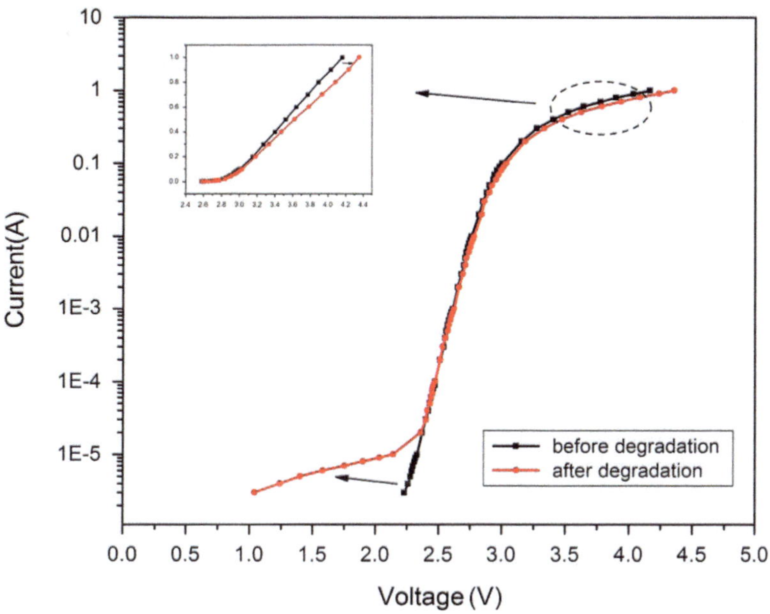

Fig. 2.16 Comparison of I-V characteristics before and after degradation

Fig. 2.17 Electroluminescence spectra during temperature stress at 200 °C [29]

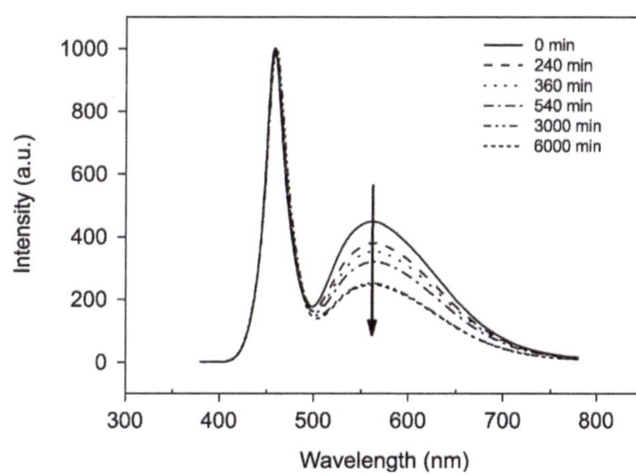

2.8 LED Chip-Level Device Structures

Electrical, optical, and thermal properties of LEDs are affected by device structure and fabrication processes. The typical mesa structure nitride-based LED is shown in Fig. 2.20. The LED epitaxial structure is grown on c-plane sapphire substrate by *metal-organic chemical vapor deposition* (MOCVD), including n-type gallium nitride (n-GaN) layer, MQWs, p-type gallium nitride (p-GaN) layer, etc. The p-GaN

Fig. 2.18 Optical microphotograph of the emissive region of one of the analyzed samples, taken before and after stress [29]

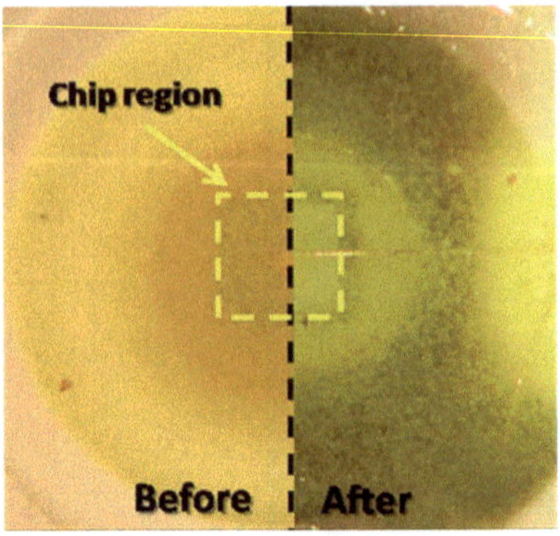

layer and MQWs active layer are selectively etched to expose the n-type layer using an inductively coupled plasma system. A transparent contact layer, most commonly indium-tin oxide (ITO) is then deposited onto the surface of p-GaN layer as a transparent ohmic contact and to provide current spreading on the surface. A metal layer is then deposited as pads used for package bonding (p-pad and n-pad).

The light extraction efficiency of GaN-based LEDs is quite low because of total internal reflection, Fresnel reflection, and absorption caused by the epitaxial layer and electrodes. In order to increase the light extraction efficiency, surface texturing is applied on the p-GaN layer, ITO layer, sapphire substrate, and LED side walls. By cutting the chips into a truncated inverted pyramid (TIP), high-performance p-electrode materials, antireflective (AR) coatings, high reflective (HR) coatings on sapphire are effective approaches to improve light extraction.

This typical mesa structure is widely used in gallium-nitride LED industries for its simple processing and low cost. However, when used for high power and high injection LEDs, its thermal performance is poor due to the relatively low thermal conductivity of sapphire (35 W/mK). Hence, a grinding process of the sapphire

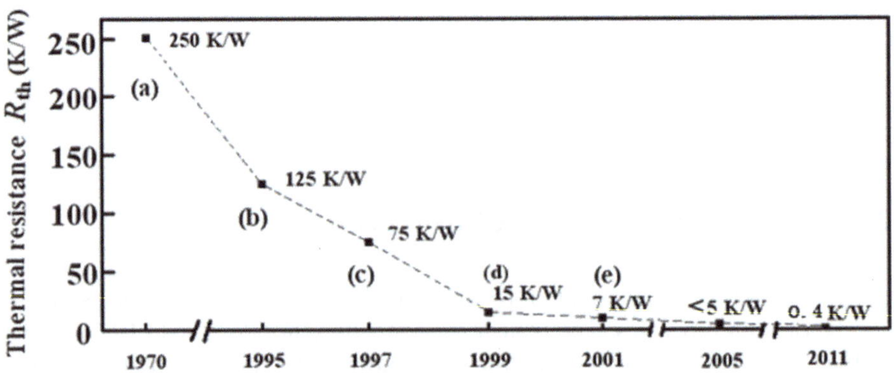

Fig. 2.19 Thermal resistance of LED package **a** 5 mm, **b** low profile, **c** low profile with extended lead frame, **d** heatsink slug, **e** heatsink slug mounted on print-circuit board [13, 30]

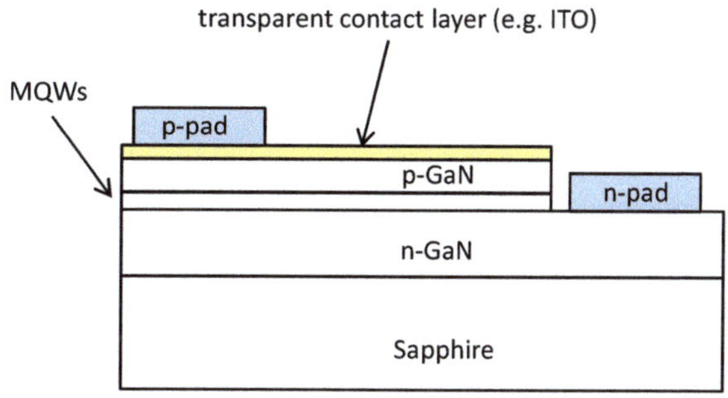

Fig. 2.20 Typical gallium-nitride LED structure

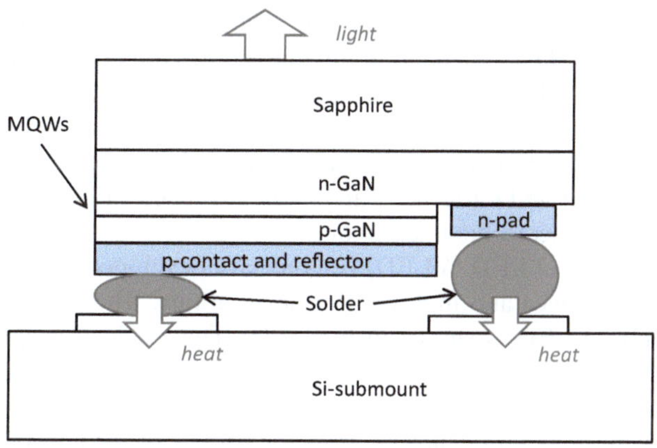

Fig. 2.21 Flip chip GaN LED

substrate is needed to lower the thermal resistance of the LED chip. Another problem is that the sapphire substrate is an electrical insulator and the current flow is lateral in mesa structure LEDs. Current crowding effects cannot be avoided because of the lateral current path. The efficiency of an LED drops greatly and the reliability decreases when the device is operating at a higher current density. Better geometry designs, current spreading techniques, and current block techniques are adopted to obtain uniform current distribution.

A flip chip power LED design has been proposed for better thermal performance. A flip chip GaN LED is shown in Fig. 2.20. The flip chip LED is "flipped-over" or inverted compared to the conventional gallium-nitride LEDs. The inverted chip is soldered onto a silicon submount and light is extracted from the transparent sapphire substrate. Fabrication processing of flip chip LEDs is as follows: the p-GaN layer and MQWs active layer are selectively etched. A thick reflective p-contact (conductive layer that joint p-type GaN and reflector) is deposited on top of the p-GaN layer. No transparent contact layer is needed. Then the n-pad is defined on the exposed n-GaN layer. The flip chip LED is then soldered onto a specially designed silicon submount via the interconnecting metallization.

Compared to the conventional top-emitting structure, flip chip assembled LEDs have some advantages. First, flip chip LEDs with thick p-contact layer have better current spreading ability. At the same time, the thermal resistance of flip chip LEDs

Fig. 2.22 Example of a vertical GaN-LED

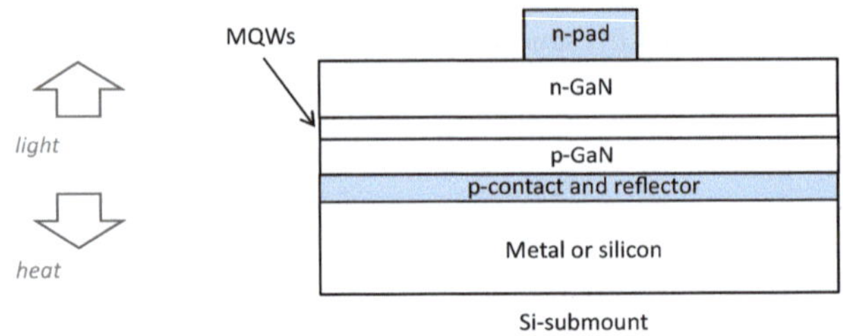

can be reduced by special design. The thermal conductivity of silicon is at least 145 W/mK at room temperature while at operating temperatures this reduces to 110 W/mK. Heat generated in flip chips flows directly through the thinner epitaxial material, solder and silicon submount (shown in Fig. 2.21). This avoids the much higher thermal resistance caused by the sapphire substrate. The most attractive property of flip-chip LED is its ability to integrate electronics. A wide range of electronics can be integrated into silicon and more advanced LED products can be developed based on optoelectronic integration.

Vertical LED (VLED) is a recently developed gallium-nitride LED structure for high power and high current applications. The cross-section of vertical gallium-nitride LED is shown in Fig. 2.21. The vertical design of LED chip is based on electrically conductive substrate, such as silicon, silicon carbon, metal alloy, etc. Current flows directly from the top n-contact to the bottom p-contact in a straight path, preventing current crowding effects in the conventional mesa structure LED. This improves LED efficiency at higher current injection and allows flexible large chip size scaling with maintained performance levels. At the same time, thermal management of VLEDs based on substrates with lower thermal resistance can be improved greatly. For example, the thermal conductivity of copper is 398 W/mK, much higher than that of sapphire and silicon. Efficient heat dissipation maximizes LED operating current and output power. Furthermore with p-side down configuration, no transparent contact layer is needed in VLEDs and maximizes the current injected into the active layer of MQW. Surface engineering (patterning, texturing, current block) on the thick n-GaN layer is easier than that on a p-GaN layer.

An example of a vertically grown LED on sapphire substrate is given in Fig. 2.22. First a high reflective p-contact layer (p-reflector) is deposited on p-GaN layer [8]. Then LED epitaxial material is transferred to a metal or silicon substrate by a wafer bonding or an electroplating process. The sapphire substrate is then removed using laser lift-off (LLO), wet etching, or Chemical-Mechanical Polishing (CMP) techniques. In the end, the n-pad (cathode contact) is deposited. In order to get higher extraction efficiency, the n-GaN layer can be patterned by wet etching. Only one bonding wire is needed in VLED packages.

The VLEDs offer the great advantages of uniform current spreading, smaller thermal resistance, improved light extraction efficiency, and smaller series electrical resistance (resulting lower additional forward voltage drop) compared to the conventional lateral LEDs. However, the fabrication processing is much more complicated. Nowadays, the cost of VLEDs is still higher and the fabrication techniques are still under development. To some degree, VLEDs based on GaN substrates may be an ideal structure for gallium-nitride LEDs (blue-green family and phosphor-converted white LED).

2.9 Encapsulation, Packaging

Commonly speaking, packaging technologies of microelectronics are also used in LED packaging: enclosing, providing electrical connections, cooling, and protecting chips from the ambient environment as well as protecting the ambient environment from the chips. On top of this, special process steps are involved to add wavelength converting phosphor layers (to obtain white light form the light emitted by a blue LED chip) and to equip LED chips with the necessary optics. It must be concluded that signal and power transmission, thermal dissipation, electromagnetic interference shielding as well as physical and environmental protection act as the bridge that interconnects the chips and other components with electronic systems. Some different level in electronic packaging can be defined as bare chip, packaged chip, printed circuit board assembly, electronic subassembly, and electronic assembly to systems. Typical trend in system integration is to leave out certain levels. This way, for example, dies can be directly mounted on printed circuit boards, allowing higher system-level integration density and reduce costs. Similar levels and design trends in LED packaging and building LED assemblies can be seen in the solid-state lighting industry: bare LED dies, LED chips directly attached to heat sinks, packaged LEDs with built in cooling tabs (using different packaging styles), packaged LEDs assembled to a metal core printed circuit board (MCPCB), etc. Each level of packaging requires different interconnection devices, optical devices, and processes.

As already mentioned, an LED package should include optic elements besides those usual in conventional discrete semiconductor device packages. The LED packages in the early days were very simple structures, such as shown in Fig. 2.23. It includes two pins, a reflector cup, wire bond, and a resin lens. It is simple, reliable, high intensity, and low cost. This bullet-shaped structure is still the most widely used in the world. Unfortunately, thermal management of this structure is too poor to enable high-power applications.

With continuous demand for higher performance, low cost and miniaturized LEDs (especially after white LEDs were realized in the 1990s), the LED packaging is continuously creating new sets of changes that require advanced packaging technology including innovative thermal management solutions. The first power LED package that included a heat sink slug was soldered onto a board for heat dissipation, a novel design from Lumileds in 2000. Compared to the first LED packages shown in Fig. 2.22, power LEDs are packaged such that the thermal resistance is reduced at least by an order of magnitude. As an example from this class, Fig. 2.24 presents a packaging structure common for many high power top emitting LEDs, such as Osram's Dragon LEDs.

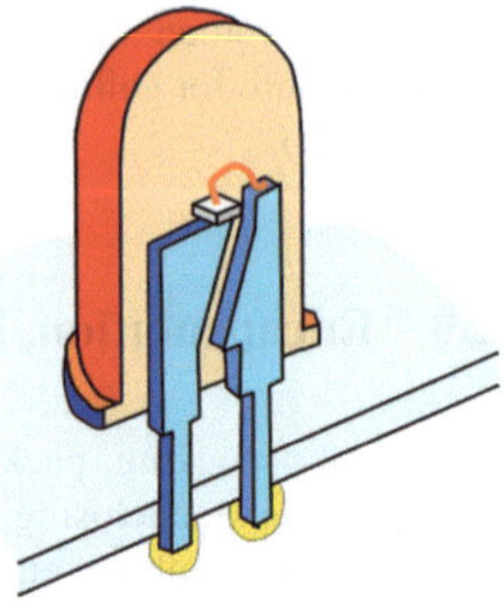

Fig. 2.23 Packaging of early low power (radial) LEDs. Typical thermal resistance of such LEDs is roughly 200–300 K/W, attached to a printed circuit board with through-hole assembly

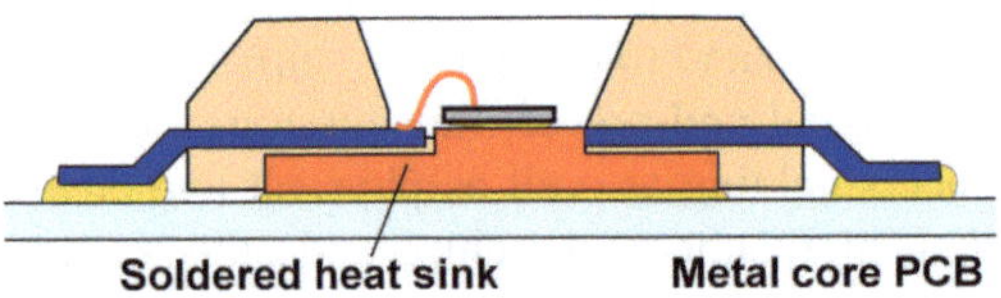

Fig. 2.24 Typical package structure of a top emitting LED (attached to a metal core printed circuit board)

Fig. 2.25 Example for ceramic substrate assembled LED soldered to a board with surface mount techniques (Cree XLamp XPG LEDs [9]) Thermal resistance with such packaging techniques is reduced to a few K/W

From a thermal management perspective, LED packaging could be classified as chip-level packaging, frame-level packaging, board-level packaging, and system-level packaging. Each level of packaging requires different interconnection devices, optical devices, and processes.

Following the miniaturization trends of microelectronics in general, LEDs with surface mounted technology (SMT) have also been employed recently. It has a much smaller volume, less parasitics, and a small thermal resistance (in the order of magnitude of a few K/W only). Such SMT LEDs are widely used in backlights and displays. Ceramic substrates and surface mount technology together further reduce thermal resistance of LED packages as ceramic provides a higher thermal conductivity. Recently, most of the leading LED vendors turn their designs to the ceramic packages—such an example is Cree's XLamp XPG LEDs (as shown in Fig. 2.25).

Beyond this, new methods in microelectronics packaging have been transplanted to LED packaging, such as chip on board (COB) technology. It usually includes more than one chip directly mounted on printed circuit boards or large ceramic substrates in order to provide much higher luminous flux with the lowest possible thermal resistance.

In LEDs packaging, encapsulation is commonly shaped as a hemisphere—forming a lens—such that the light angle of incidence at the lens-air interface is usually normal. Hence, there is less total internal reflection at the interface. However, there are many types of LEDs that do not have a hemispherical shape. For example, there are rectangular, cylindrical, straw-hat, or double hemisphere shapes. Those shapes are frequently used for different light distribution requirements.

The thermal resistance of LED packages is related to the maximum temperature of operation which influences lifetime. Early LED packages had the highest thermal resistance of about 250 K/W (junction-to-ambient). The current Chip on Metal Board technology allows thermal resistances much below 10 K/W, in the range of a few K/W only (junction-to-case). This is just one component of the total junction-to-ambient thermal resistance of such LEDs being used in solid-state lighting applications. Heat sinks and different active cooling technologies are used to transfer the heat from the LED package to the further environment, finally, to the ambient. Chapter 2 provides basics on thermal design for LED applications while Chaps. 7, 8, and 9 discuss details of different thermal management solutions and provide insight into their design.

In the future, some progress in mineral diffusers or nanoparticle-assisted high refractive materials may improve the efficiency of LEDs. However, from a general purpose point of view, light weight, integration, intelligence, etc. are key trends in LED packaging structures. Some applied optics developments such as photon crystal, 2D optical elements, and MOMES, etc. will play key roles in wafer-level LED packaging technologies.

2.10 Summary

GaN-based LEDs have attracted much attention due to their significant application in solid-state lighting. Generally, an LED is grown by an MOCVD process on a sapphire substrate. The epitaxial structure usually includes a low-temperature buffer layer, an n-GaN layer, GaN/InGaN MQWs, an AlGaN EBL, and a p-GaN layer. The MQWs are composed of a wide bandgap material and a narrow bandgap material, which are grown periodically.

One challenge in developing LEDs is their thermal behavior, because higher temperatures will reduce the lifetime and the light output of the devices. Much research has been performed to investigate the mechanisms that control the thermal behavior of LEDs.

When an LED is powered, electrons and holes are injected into the MQWs, where they will recombine and emit photons or generate heat since only part of the injected carriers could reach the MQWs. After electrons and holes are injected into the active region, they can recombine either radiatively or nonradiatively. The radiative recombination will emit photons, which is the ultimate goal of using light-emitting diodes.

The nonradiative recombination usually includes the defect-related recombination and Auger recombination, which will not result in photon emission. Nonradiative recombination refers to the carrier-lattice vibration or carrier-phonon interaction. Especially, the relaxation of the Auger carrier from the high-energy excited state to the bottom of energy band generally includes a large number of emitted phonons, which results in self-heating of LEDs. This way the nonradiative recombination process not only wastes energy but also leads to self-heating and degradation of LEDs. Furthermore, if temperature increases, the nonradiative recombination loss will increase as well, leading to further increases in temperature. This is a vicious circle, ultimately leading to thermal runaway. Hence, it is important to decrease the defect density.

Current spread is another major factor which will impact the thermal behavior of LEDs. Typical GaN-based LEDs grown on an electrically insulating sapphire substrate have an etched mesa shape and a side-by-side contact configuration due to the insulating substrate. In these diodes, there is both a lateral and a vertical current flow. In particular, the lateral current flow can lead to nonuniform current spreading, which locally produces light emission and heat generation. It was reported that the nonuniform current spreading, the so-called current crowding effect, was strongly related to chip reliability in addition to local light emission. The design and optimization of p-pad and n-pad electrodes have been employed to reduce the current crowding effect. The location and size of ohmic contacts are relevant to light extraction, because a portion of the injected carriers will still be confined underneath the thick metal bonding pad, even with the transparent current spreading layer ITO.

In conclusion, LED thermal behavior has serious influence on LED performance. Because the thermal behavior will consume power, the wall-plug efficiency of LEDs will be decreased. Furthermore, self-heating of LEDs will also lead to the degradation of the material quality, which will result in decreased efficacy. Finally, as the junction temperature increases, the lifetime of the LEDs will be shortened.

A large part of the remainder of this book is therefore devoted to ways to reduce the junction temperature and the temperature differences.

Acknowledgments Thanks are due to the following researchers from the Institute of Semiconductors, Chinese Academy of Sciences: Dr. Lixia Zhao, Dr. Ping Ma, Dr. Hua Yang, Dr. Xiaoyan Yi, Dr. Xiaoli Ji, Dr. Tongbo Wei, Dr. Jianchang Yan, and MPhil Bin Xue.

References

1. Capasso F, Ripamonti G, Tsing WT, Hutchinson AL, Chu SNG (1989) New photoconductive gain mechanism by electric field modulation in multiquantum-well heterostructures. J Appl Phys 65:388
2. Casey HC, Panish M (1978) Heterostructure lasers. Academic press, New York
3. Digle R, Wiegmann W, Henry CH (1974) Quantum states of confined carriers in very thin AlxGa1-xAs-GaAs-AlxGa1-xAs heterostructures. Phys Rev Lett 33:827
4. Miller RC, Kleinman DA, Gossand AC (1984) Energy-gap discontinuities and effective masses for GaAs-AlxGa1-xAs quantum wells. Phys Rev B 29:7085

5. Nakamura S, Sench M, Iwasa N, et al (1995) High brightness InGaN blue, green and yellow light-emitting diodes with quantum well structures. Jpn J Appl Phys Lett 34:L797–L799
6. Krames MR, Ochiai-Holcomb M (1999) High-power truncated inverted-pyramid (AlxGa1–x)0.5In0.5P/GaP light-emitting diodes exhibiting >50 external quantum efficiency. Appl Phys Lett 75(16):2365–2367
7. Wierer JJ, Steigerwald DA, Krames MR, O'Shea JJ, Ludowise MJ, Christenson G, Shen Y-C, Lowery C, Martin PS, Subramanya S, Götz W, Gardner NF, Kern RS, Stockman SA (2001) High-power AlGaInN flip-chip light-emitting diodes. Appl Phys Lett 78(22):3379–3381
8. Wuu DS, Hsu SC, Huang SH, Horng RH (2004) Vertical-conducting p-side-up GaN/mirror/Si light-emitting diodes by laser lift-off and wafer-transfer techniques. Phys Status Solidi (A) 201(12):2699–2703
9. Cree product family sheet (2010) Cree XLamp XP-E High efficiency White LEDs Cree official website. http://www.cree.com/products/pdf/XLampXP-E-HEW.pdf. Accessed 18 July 2011
10. Colinge JP, Colinge CA (2002) Physics of semiconductor devices. Kluwer Academic Publishers (Chaps. 4 and 9)
11. Nakamura S, Chichibu SF (2000) Introduction to nitride semiconductor blue lasers and light emitting diodes. CRC Press (Chaps. 1 and 3)
12. Sze SM, Ng KK (2006) Physics of semiconductor devices. Wiley-Interscience (Chap. 2.3.1)
13. Schubert EF (2006) Light-emitting diodes, 2nd edn. Cambridge University Press (Chap. 4)
14. Hall RN (1960) Recombination processes in semiconductors. Proc Inst Electr Eng 106B(suppl 17):983
15. Garbuzov DZ (1982) Radiation effects, lifetimes and probabilities of band-to-band transitions in direct A3B3 compounds of GaAs type. J Lumin 27:109
16. Hall RN (1952) Electron-hole recombination in germanium. Phys Rev 87:387
17. Shockley W, Read WT (1952) Statistics of the recombinations of holes and electrons. Phys Rev 87:835
18. Bhattacharya P (1996) Semiconductor optoelectronic devices, 2nd edn. Prentice Hall, Englewood Cliffs
19. Hader J, Moloney JV, Koch SW (2005) Suppression of carrier recombination in semiconductor lasers by phase-space filling. Appl Phys Lett 87:201112
20. Guo X, Schubert EF (2001) Current crowding and optical saturation effects in GaInN/GaN light-emitting diodes grown on insulating substrates. Appl Phys Lett 78:3337
21. Guo X, Schubert EF (2001) Current crowding and optical saturation effects in GaN/InGaNligh emitting diodes on insulation substrates. J Appl Phys 90:4191
22. Kim H, Cho J, Lee JW, Yoon S, Kim H, Sone C, Park Y, Seong T (2007) Consideration of the actual current-spreading length of GaN-based light-emitting diodes for high-efficiency design. IEEE J Quantum Electron 43:625
23. Chang SJ, Shen CD, Chen WS, Ko TK, Kuo CT, Yu KH, Shei SC, Chiou YZ (2007) Nitride-based LEDs with an insulating SiO2 layer underneath p-pad electrodes. Electrochem Solid-state Lett 10:H175
24. Leung KK, Fong WK, et al (2010) Physical mechanisms for hot-electron degradation in GaN light-emitting diodes. J Appl Phys 7:107
25. Eunjin J, Hyoung RJ, et al (2011) Optical degradation of phosphor-converted white GaN-based light-emitting diodes under electro-thermalstress. J Electrochem Soc 2:158
26. Hsu YC, Lin YK, et al (2007) Failure mechanisms associated with lens shape of high-power LED modules in aging test, LEOS 20th Annual Meeting of the IEEE Lasers and Electro-Optics Society
27. Lin Y-H, You JP, et al (2010) Development of High-performance optical silicone for the packaging of high-power LEDs. IEEE Trans Compon Packag Technol 33(4):761–766
28. Xi Y, et al (2005) Junction and carrier temperature measurements in deep ultraviolet light emitting diodes using three different methods. Appl Phy Lett 86:031907
29. Meneghini M, Meneghesso G, Trevisanello L, Zanoni E (2008) A review on the reliability of GaN-based LEDs. IEEE Trans Device Mater Reliab 8:1530–4388

30. Cree product family sheet (2011 Cree XLamp XM-L high efficiency white LEDs cree official website http://www.cree.com/led-components-and-modules/products/xlamp/arrays-nondirectional/xlamp-cxa2011. Accessed 30 May 2013
31. IESNA LM-80 Lumen maintenance: measuring lumen maintenance of LED light sources. http://apps1.eere.energy.gov/buildings/publications/pdfs/ssl/led_luminaire-lifetime-guide.pdf (for additional considerations and guidance page 4). Accessed 30 May 2013

Chapter 3
Basics of Thermal Design for LEDs

Cathy Biber

Abstract This chapter addresses the key thermal mechanisms active in light-emitting diode (LED) packaging and describes a big-picture approach to thermal design of the LED package. It also highlights differences in thermal phenomena between LED chips and integrated circuit (IC) semiconductor chips. Best-practice thermal design activities are identified.

3.1 Overview

Thermal design of a light-emitting diode (LED) system is becoming essential as LEDs become more powerful and find wider applications. Although they are known to be energy-saving and efficient, these terms are used in comparison to previous lighting technologies (mainly incandescent) and are therefore relative. Previous lighting technologies aside, LED efficiencies—light power per unit input electrical power—are increasing all the time. While it seems that this increase should be a good thing for thermal design, it does not necessarily mean that heat loads are decreasing; rather, light output can be increased for a given heat load.

There are several reasons to manage the temperature of the LED. The light output decreases as the temperature increases (see, e.g., Fig. 5.1 in Chap. 5) so a hotter device is a dimmer device. The color of the light shifts with temperature change; this could be desirable or undesirable, depending on your target. The lifetime of the device decreases at elevated temperature; lifetime (or rather "end of lifetime") for LEDs is usually defined in terms of relative luminous flux. This relative flux is typically a percentage, say 70 %, of the initial light output. So, the LEDs do not usually suffer "fatal failure" in the conventional semiconductor sense of causing errors or stopping altogether; they just get dimmer. And finally, there is an absolute maximum temperature, at which there is a risk that the device suffers irreversible damage. The foregoing three temperature effects (light output, color shift, and lifetime) very often drive an operational temperature requirement that is significantly lower than the absolute maximum rating.

C. Biber (✉)
Biber Thermal Design Ltd., Portland, OR, USA
e-mail: biberthermal@gmail.com

C. J. M. Lasance, A. Poppe (eds.), *Thermal Management for LED Applications*, Solid State Lighting Technology and Application Series 2,
DOI 10.1007/978-1-4614-5091-7_3,

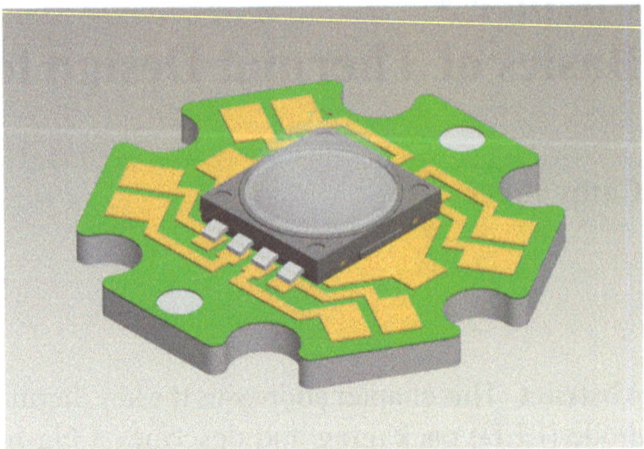

Fig. 3.1 Schematic representation of thermal substrates [9, 10]

Since there are many factors that influence the device temperature, let us look at some general categories that cause temperature rise as heat flows from the die to the ambient. In general (there are exceptions), these temperature rises are additive.

First is the temperature rise due to the die itself. The heat in the active layer is generated by light that is produced but that does not make it out of the front face of the die and is reabsorbed, as well as by charge carriers which during the recombination process interact with the semiconductor crystal lattice (see Chap. 2 on LED physical basics). This heat must conduct through the thickness of the die and exit the back of the semiconductor; often, a thermal pad is provided as part of the package to make thermal but not electrical contact to the mounting surface. Choice of materials and geometry influence the temperature rise values obtained here and are decided by the manufacturer.

Next is the chip connection to the carrier or board. The anode and cathode are electrical connections, and all electrical connections also carry heat, although relative amounts may be small because of the small dimensions. If a thermal pad is provided, this should be the major heat path into the board, and would be part of the mechanical attachment. The solder used for this purpose forms a thermal interface. The combined effect of the LED chip itself and the connection to the carrier form the thermal resistance of the package. Look for this value on the LED datasheet. The most common one is "junction-to-case" as for other power components, but it may simply be called "thermal resistance."

The packaged LED chips when part of an assembly are attached to a "macroscopic" carrier or board. The thermal features of the board significantly affect the thermal path, and these are typically a choice made by the product design team. The board material (metal core or insulating), thermal via pattern if used, and surface metallization design all play a part in the temperature rise of the LED chip, and should be a subject of close cooperation between thermal and electrical designers. LED manufacturers often publish application notes containing recommended designs, and sometimes even quantify the differences between design options. Lacking this type of recommendation, thermal experts should be consulted. Figure 3.1 shows a sample schematic drawing of a metal core board, sometimes known as a thermal substrate. The board or carrier serves as a "first heat sink"—air flows around the carrier can absorb some of the heat. If this proves insufficient because of limited area, limited air flow, or both, further measures are needed.

Removing heat from the board for higher power LEDs will require contact to a larger surface area, and this contact is managed by using a thermal interface material. There are many choices for this, including some that further spread the heat, and others that mainly displace air from the small spaces between the carrier and the extended surface. The heat spreaders include solid materials with high in-plane conductivity like graphite or diamond. The air-displacers are materials like thermal paste, phase change material, or gap fillers. Although many of these have seemingly low thermal conductivity, they are two orders of magnitude better than air and so can offer measurable (junction) temperature reductions if properly applied. Critical to know about these materials is that their performance on datasheets is described by the common (erroneous) term "thermal impedance," by which is actually meant the unit area thermal resistance, or temperature rise per unit heat flux density. This value actually affects the extent to which the heat spreads in the carrier or heat spreader. However, the data sheet value is not necessarily the value achieved in practice, so prediction of a temperature rise based on the data sheet value is not straightforward. Complexities of heat spreading are discussed further below. See also Chap. 8 on thermal interface materials (TIMs) in LED applications.

The heat sink or heat pipe base serves to further spread the heat, if necessary, and in the heat pipe case, serves to concentrate it into the heat pipe. The base also serves a mechanical function. In general, this temperature rise is minor compared with the ones previously mentioned.

Lastly, the heat must be dissipated to the ambient fluid medium—almost invariably air—which is a lousy heat transfer fluid, thus requiring large surface area. This extended area is in the form of fins or housing. The fins can be on the housing or luminaire, or exposed to air flowing through the luminaire. The large surface area may also be just the housing if it is much larger than the LEDs. With both finned and unfinned surfaces, the flow plays an obvious role, whether buoyancy-induced, fan-induced, or synthetic jet-induced. In many cases, thermal radiation is also an important contributor to dissipating the heat. The radiative path is in parallel with the convective path, adding heat dissipation capability without increasing the surface temperature conversely reducing the surface temperature when the dissipation is held constant. The surface characteristics determine the radiative emission; the wavelengths important for heat dissipation near room temperature are in the far infrared range, 8–10 μm. Thus, the optical color is of no importance whatsoever to heat dissipation by radiation, as optically perceived wavelengths are more than an order of magnitude shorter than the far infrared wavelengths. More important is the surface texture and emissivity; powder coatings and paints perform better than bare aluminum or shiny chrome surfaces. Anodize finishes are in the intermediate range.

Fig. 3.2 Multi-LED array [9, 10]

3.2 LED Unique Features

LED cooling differs from traditional IC semiconductor cooling in several important ways. This is not surprising since they are used in completely different ways.

3.2.1 Die Size and Arrays

Typical LED dies are smaller (today) than many high performance IC chips, although they may be tightly arrayed into similar form factors. Figure 3.2 illustrates a multichip LED package. The tight arrays of chips in the multichip case are possible with LEDs because there are only very few electrical connections, instead of hundreds or thousands as for an IC chip.

Heat densities found on LEDs are similar to the maximum heat densities on integrated circuit (IC) chips—but this heat density is nearly evenly distributed across the entire die. Heat densities up to 10 W/mm^2 have been reported for special applications like projection light sources [13]. On an IC, the high heat density areas frequently have adjacent low-density areas into which the heat can diffuse through the relatively high conductivity silicon substrate. These low-density areas are not present on an LED chip.

Since the function of an LED is to produce light, and this occurs at the outward-facing large surface, it is not possible to remove heat by conduction at this face. This provides only one cooling path leaving the LED chip, unlike most high-power ICs, which in theory at least, can be cooled both from the active side (if flip chip) and the substrate side.

Arrays of LED chips can be more closely spaced than ICs because the small number of electrical connections requires little space on the substrate. The array begins to approximate a single source as the gap between the chips approaches zero. A smaller gap increases the thermal challenge, as the average heat density over the array increases as the array outline shrinks (if the total array dissipation is kept constant). Increasing the gap will reduce the temperature. If your design allows it, use multiple LEDs instead of a single one, given a certain area and power.

3.2.2 System Housings

LEDs for illumination also face an additional challenge not shared by ICs. The housing or luminaire presents a number of constraints.

The luminaire is supposed to look decorative, and it is mounted on a wall or ceiling. ICs on the other hand can be housed in an ordinary enclosure that may be on the floor, where hot air can rise up and away from it. The good news is that LEDs are typically not hidden away in a closet; the light is being produced for people to see, so the ambient air temperatures are ambients that humans are used to. A big exception to this is recessed can lighting, where according to present construction standards building insulation may surround the light fixture and severely reduce heat dissipation capability. As LED lighting becomes more common, architects could design direct conductive cooling for the lighting system into the structure of the ceiling. In addition, there may be electrical safety standards that require electrical insulation between a metal PCB and the luminaire enclosure. If so, the thermal contact to dissipate the heat must take this requirement into account.

Another housing design constraint is that frequently it is expected to look like a reflector, although the reflector function is not needed with the directional lighting capability of LEDs. The reflector shape may make it more difficult to achieve good thermal contact to the housing.

Illumination housings also are subject to safety requirements for electrical isolation. The LED substrate may be electrically active, but should be insulated from the housing, while at the same time mounted with a thermally conductive path.

Housings that are made of sheet metal for previous lighting technologies also differentiate LEDs from ICs; electronics cooling has enough history of overcoming thermal challenges that sheet metal electronics housings are not expected to conduct heat. If an IC is in a sheet metal housing, there are vents to allow air to flow through. But, vents would have decreased the reflectivity needed for older lighting technologies, so are a new design element for luminaires. Sheet metal housings for LEDs also typically have poor conduction characteristics like thin metal and imperfect (high contact resistance) joints formed by fasteners.

The usual approach to cooling electronics—applying a heat sink—often adds a lot of weight to a system that must be mounted up high, increasing the risk associated with mechanical failure. Electronics heat sinks, being typically hidden inside an outer enclosure, do not require a decorative look. By contrast, an LED housing acting as a heat sink should be aesthetically pleasing. Figure 3.3 shows an example of a finned luminaire.

Fig. 3.3 Decorative finned structure on luminaire. (Image courtesy of Journee Lighting. Used by permission)

3.2.3 Heat Density and Spreading

Finally, for LEDs, heat spreading in the packaging plays a far more significant role than in IC packaging, where the die is the first heat spreader. In both ICs and LED chips, the heat generation is in the active layer, which is continuous with the bulk of the die. However, in ICs, there are localized regions of high heat flux density with adjacent regions of low heat flux density. These low-density regions are typically absent in an LED chip, so that the bulk of the die cannot act as a heat spreader. Thus, the LED packaging is the first heat spreader. As a consequence there are thermal interfaces with very high heat flux density compared with the IC packages. The small size of the LED renders the interface performance susceptible to imperfections in the interface. These can have devastating effect on the LED temperature.

It is well known [12] that the spreading resistance is a function of the heat transfer coefficient on the far side of the spreader. For the same reason, the spreading resistance is a function of the interfaces on the far boundary of the heat spreader. The heat will "puddle"— spread farther away from the source—where there is a "downstream" high resistance, so that the larger area of the heat path reduces the interface resistance somewhat. This reduction is a good thing for thermal design if a well-designed heat spreader is included in the packaging.

3.3 Considerations for Conductive Path

Since the primary path adjacent to the heat source is conductive, it is important to estimate the dominant contributors to temperature rise in the conductive path. The three main areas are the packaging, the board or carrier, and the interfaces. The contributions of these areas are readily identifiable by the transient experimental method using structure functions. This method shows the main thermal resistances and capacitances in the physical system (Fig. 3.4).

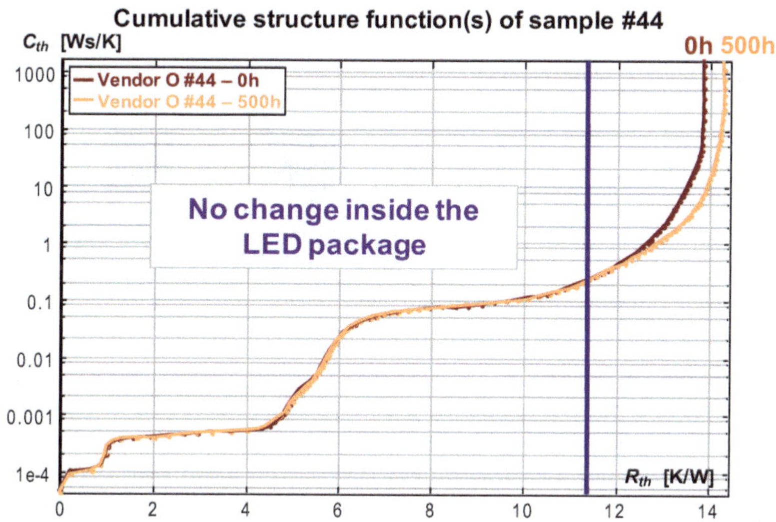

Fig. 3.4 Structure function plot showing details of thermal features of different parts of packaging [9]

3.3.1 *Packaging*

The packaging conductive path is mainly determined by the manufacturer, since most of the package construction details are not given to customers. The datasheets normally give a value for thermal resistance, usually meaning junction to "case" (or mounting surface, which for power components like LEDs amounts to the same thing). These values vary widely, depending on whether the die is mounted directly to a board ("Chip on Board," or COB), how many dice are in a package (one or more), the type of board, and the configuration of the electrical and thermal contacts provided to the next layer. Increasingly, manufacturers are providing a "thermal pad" with area much larger than the electrical contacts, which need only a small cross-section to pass sufficient current. All electrical conductors carry heat, but not all thermal conductors carry electrical current. The thermal pad, an example of package cross-section, which is illustrated in Fig. 3.5, would normally be isolated from the electrical contacts, and may be either metal or ceramic, or a combination for solderability. This thermal pad provides a dedicated heat path to minimize thermal resistance. Applications requiring high power density would do well to choose components that come in packages using this feature.

3.3.2 *Circuit Board*

The circuit board conductive path is decided by the customer. Some manufacturers give one or more recommended circuit board configurations and performance values

Fig. 3.5 Package cross-section showing thermal pad [8]

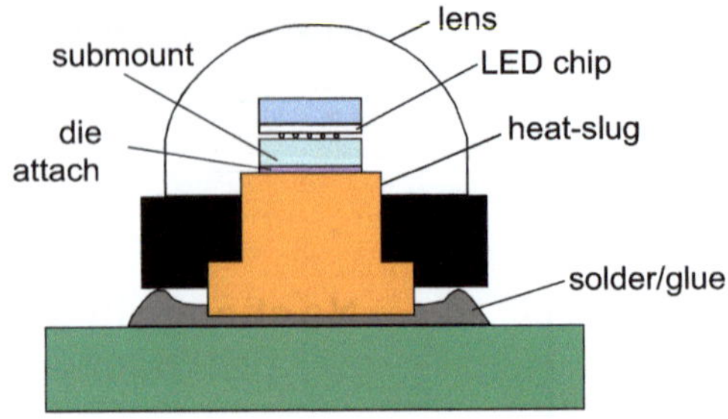

in an application note. Others provide less guidance. The first decision to make concerns the type of board.

Several board types are commonly used for LEDs: traditional insulating boards with metal conductors, more conductive boards like metal core printed circuit boards (MCPCB) or insulating metal substrates (IMS). The metal core boards are very similar to the insulating metal substrates, and consist of a sheet of aluminum or other metal covered with a thin dielectric and then surface metal. The basis for the decision about insulating versus metal core boards will be not only cost and thermal performance, but also comfort level with manufacturing and assembly. Expect some differences in soldering parameters between the two types of boards.

The conductive thermal paths differ widely between the two types of boards; insulating boards will need thick surface metal to spread heat and an array of vias to conduct heat through the board for each LED, while the body of the metal boards will do both the spreading and through-conducting jobs. See, for example, Lasance [3]. Some of the LED manufacturer application notes address a comparison between the two, although they seldom include the effect of the far-side thermal interface on the heat spreading. Because the same board can perform differently depending on the circumstances, board supplier data sheet values cannot be used to evaluate the conductive path for the LED application. Physical or numerical simulation must be used to obtain actual values.

When deciding on the number of thermal vias to use in an insulating board, it is a good idea to take into account not only the LED manufacturer's recommendation, but also the thermal solution that will be used. For natural convection cooling and lower performance TIMs, plan for a larger array of vias, as the heat will spread more than with higher performance cooling and interfaces.

For a metal core board, through resistance will be small, although not necessarily negligible; the main source of resistance is the thin dielectric layer, but its resistance starts only to become significant with high-power LEDs and/or liquid cooling. The cross-sectional area of the dielectric to use for calculation is approximately the same as the source area due to its small thickness and low thermal conductivity. For the metal core we face a different situation. On the LED mounting surface, it is reasonable to assume this area to be the size of thermal pad; on the opposite side, the heat will have spread out and the area will be larger. But, how much larger is unknown a priori, and depends on the performance of both the second dielectric layer and the adjacent

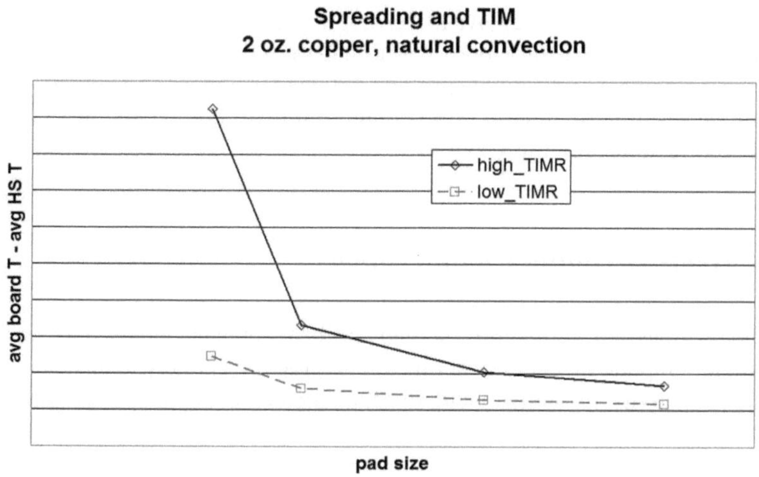

Fig. 3.6 Effect of thermal interface on heat-spreading performance. (Unpublished work of the author)

thermal interface. As general advice, consider that thermal interface performance on the side opposite the LED will be less critical than for the insulating board, because the metal core does such a good job of spreading the heat in the radial direction. It may even be the case that the resistance of the second dielectric layer could be negligible (say, an order of magnitude smaller) compared with the first dielectric layer, which has a much smaller area of heat conduction.

3.3.3 Thermal Interfaces

The interface conductive path seems straightforward enough, except that the cross-sectional area is unknown as it was for the board, and depends on the extent of heat spreading in the board. Conversely, the extent of heat spreading in the board depends on the interface conductive path. This situation, as in the preceding discussion, comes up again and again. For example, let us examine the case of an insulating board with thick (70 μ, or "2 oz." copper) surface metal and a nominal array of 14 small thermal vias. The board attaches to a cast aluminum heat sink cooled by natural convection and thermal radiation. Two TIMs are considered, their data sheet values differing by a factor of four. The results for interface temperature rise are shown in Fig. 3.6. With a pad of surface metal slightly larger than the LED and the better (lower datasheet unit-area thermal resistance) TIM, the performance is about the same as with a very large area of surface metal—nearly covering the entire width of the board—and the worse TIM with higher unit-area thermal resistance. With a small surface metal area, the difference between the two TIMs is quite large. With a large surface metal area, the difference is small. So, a large metal area on the board can give some design leeway on TIM selection. On the other hand, surface metal might seem expensive, and a board might be less expensive to make if it were to use smaller surface metal—but then the design will require more attention to the TIM. Understanding this design tradeoff in the context of the whole product will lead to a more cost-effective choice, although the result may initially seem counterintuitive. For details on TIMs aimed for LED applications, refer to Chap. 8.

3.3.4 Heat-Spreading Enhancements

If the board does not sufficiently spread the heat, enhancements may be added, for example, graphite layers or heat pipes. The graphite is likely to have anisotropic thermal conductivity, meaning that its structure dictates differences between in-plane and through-plane conductivity. A diamond heat spreader could be used instead of a small circuit board. The size of the spreader that is most effective will depend on both the cooling method and the thermal interface material. At present the authors do not know of any reliable analytical methods for predicting heat spreading enhancements in LED applications for these or any other materials; physical or analytical experiments are needed.

A heat pipe, on the other hand, is a thermal superconductor. There is minimal temperature gradient within the pipe itself, but in the typical case, the heat density entering the heat pipe from the source is quite high—this is why a heat pipe is needed. Therefore, the interface at that location is critical. On the other hand, the heat pipe is able to transport the heat quite a distance, and allows it to be dissipated into a much larger volume of air remote from the heat source. Thus, a heat pipe is suited to a situation in which the dissipation region is not adjacent to the heat source. If the product can be designed such that moving air volume is close to the back of the die, the thermal solution (maybe a simple piece of copper) will be simpler, cheaper, and more straightforward than a heat pipe.

To estimate heat pipe performance, focus on the interface at the source (see discussion of interface materials, both previous and next sections) as the largest thermal resistance in the design.

3.4 Modeling Strategies

3.4.1 Conductive Path-Modeling Strategies

Given the difficulty of estimating the through-board conductive paths using traditional thermal network analysis, it is appropriate to consider what analysis method can be useful. For example, although it is not possible to calculate the path values analytically because of variable heat cross-sectional area, minimum and maximum possible values of this area can be used to bound the problem. In some cases, there might be an order of magnitude spread between the minimum and maximum areas.

Then, using a geometric average $A_{mean} = \sqrt{A_{\min} \cdot A_{\max}}$ of the two values will minimize the likely error in most estimates. This neither implies accuracy, nor precision! It is just a way to estimate whether the design is somewhat likely to meet the criteria, or to decide whether a given material has any chance of being suitable. The idea behind using a thermal network, even with such imprecision, is to identify the big roadblocks. These are fundamental limitations in the design—elements whose thermal resistances are an order of magnitude larger than others. The highest resistances should be identified, or ideally, resolved before proceeding to more detailed analysis. Thermal networks can also be used to quantify the effect of various proposed solutions in the earlier design phases. They are a great communication tool in discussions with electrical engineers; most are familiar with the electrical-thermal analogy. For all of these situations a first-order estimate is valuable. Thermal networks for this type of activity should be constructed by approximating the conduction paths. But, if physical samples are available, the network could also be constructed from transient test results, as illustrated in [7].

3.4.2 Spreading Resistance-Modeling Strategies

To estimate spreading resistance, it might be useful to employ analytical relationships for heat spreading (see, for example, [1, 2, 4–6, and 12]) with an artificially low heat transfer coefficient at the far side of the spreader to obtain maximum spreading, and a somewhat reasonable heat transfer coefficient to obtain an estimate for the minimum spreading. The geometric average of these two bounds provides a first-order estimate.

More accurate estimates would be obtained by using 3D conduction analysis. Even within this category, there are several key points to consider for a good estimate for spreading resistance. The simplest version of 3D conduction analysis is a conduction model. Because no equations of motion have to be solved—just the diffusion equation—solve times are short. To make the solution realistic, it is important to include the base of the heat sink or heat pipe and its associated interface material (or small air gap, if there will be one instead). These features will determine the extent of heat spreading in the board, whether insulating or metal. On the dissipation side of the base, there should be a heat transfer coefficient, either an estimated coefficient to ambient applied to the surfaces of any fins plus the base exposed to air, or an effective coefficient applied to the base. The effective coefficient increases over the actual coefficient by the ratio of total area to model base area. Of course, the actual coefficients are not known, but if they are in the correct range, then the spreading resistance will be somewhat insensitive to their exact value. The overall level of temperature, however, might not be correct. But, this method can be used to make the best-possible design decisions about the board construction even if the estimate is a little off on the absolute level.

The most representative version of 3D conduction analysis is a conjugate model that includes all three heat transfer modes: conduction, convection, and radiation.

This type of model is especially well suited to problems where there are large variations in heat transfer coefficient or local ambient temperature. The variations in heat transfer coefficient come from the system level flow pattern, for example, locally strong air flow from a fan, or locally weak air flow due to upstream blockage. The variations in local ambient temperature come from air flow that is previously heated by upstream dissipation—usually the lower part of the dissipation surface in natural convection, or upstream devices in the case of a multi-LED board. They may also come from actually having two separate ambients, as, for example, a recessed lighting fixture, where the recessed portion dissipates into the ceiling space and the exposed portion dissipates into the room, although this consideration applies more to the system housing than to the circuit board.

3.4.3 Heat Pipe Conduction-Modeling Strategies

The dominant conduction resistance in a heat pipe is the interface resistance at the hot end. Here the heat density is highest, resulting in the largest temperature difference. An interface plate can help assure good contact and spread the heat to the perimeter of the heat pipe. In this case, the area is easy enough to estimate, except that the interface resistance affects the area, as before. Since the heat pipe is an excellent conductor of heat, if properly engineered and sized its internal temperature differences will be small compared with those at the hot-end interface. Interfaces at the cold end to extended surfaces are important too. If fins do not make very good contact, the fins will stay cold and the heat pipe temperature will be higher than expected. For estimation purposes, if the fins are attached directly to the heat pipe the interface resistance may be neglected (although in practice this could vary from unit to unit, and would lead to a difference in heat pipe temperature from one part to another). If a heat sink is used, interface resistance should be treated as if the heat pipe were a uniform-temperature source.

3.5 Estimating Convective and Radiative Path

The convective path is the interaction between solid surfaces and the surrounding medium (air). Strictly speaking, convection is just conduction to a moving, as opposed to stationary, medium. If temperature gradients in the medium cause density gradients that then induce motion, it is called buoyancy-induced or "natural" convection. The strength of the convection is a function of the temperature gradient, and also of motion induced above or below (with respect to gravity) the area of interest. For forced convection, there may be upstream blockage reducing airflow on both a local and a global basis. But for estimates, it is usually sufficient to assume a typical value of heat transfer coefficient over the whole geometry. Values in the range 5–7 W/m^2 K for natural convection, and 25–50 W/m^2 K for forced, are the right order of magnitude. In some cases, for example, downward facing hot surfaces, the air will stratify with the hot air at the top, and then there is no motion, so that these surfaces should not be counted on to do a lot of the convective cooling job. They may still dissipate heat by radiation, however.

3.5.1 Luminaire Convection

If the luminaire is acting as the heat dissipation element, it will probably have a nonuniform heat transfer coefficient. It may have a nonuniform ambient temperature, and even a nonuniform temperature within its field of view to which it will radiate. It could even have nonuniform emissivity. In the face of all these variations, averages or typical values can be used as a starting point for a sensitivity analysis. For large variations, treating them as active in separate zones might be a workable approach. Each zone gets its own path in the thermal network.

3.5.2 Fin Array Convection

Analysis of heat sink or fin array performance proceeds the same way as for any other heat sink; the finned element senses no difference in the heat source being an LED and not a semiconductor chip. Even in pulsed applications, the frequencies are so high that the heat sink itself will average out the power fluctuations with time, if the package has not done so already. Temperature gradients within the fins may be neglected for natural convection, as they are much smaller than the gradients within the adjacent air. For forced convection, fin efficiency treatments are a practical approach and are described in heat transfer textbooks.

3.5.3 Ambient Conditions

There are several categories of ambient conditions to consider: temperature, air flow, and radiative flux.

For open systems, the local air temperature around the product should be used.[1] This may differ from the general air temperature due to upstream heating from other components, due to stratification (hot air rises and stays against the ceiling), due to natural convection boundary layers (along the wall of a building, for example, where preexisting convection may produce higher local temperatures, albeit at nonzero air

[1] This is akin to the principle of the adiabatic heat transfer coefficient. The interested reader may consult Moffat, R., "The Benefits of Using Hadiabatic in Thinking About Electronics Cooling," *Electronics Cooling*, Vol. 10, No. 1, 2004.

speed), or due to a portion of the product being in an insulated ceiling space. It may be effective to divide multiple temperature boundary conditions into separate zones and treat each as a separate thermal path. In strict thermal terms, the reference temperature should be the "adiabatic temperature"—the temperature attained by a device when it is the only element in an array that is not dissipating. The adiabatic temperature includes the effects of the surrounding air as well as heat conducting to it from neighboring components via the board. This complexity makes it relatively difficult to estimate the adiabatic temperature a priori.

Air flow conditions around sealed enclosures indirectly affect LED designs, in that the heat must be conducted to the exterior housing. Higher air flow produces lower housing temperatures, but in general do not exist all the time. Outdoor applications like indoor or outdoor general lighting, architectural or specialty lighting, and traffic signals need to work whether the wind is blowing or not. The need to protect the light source from dirt, moisture, and animals usually dictates sealing the enclosure.

Radiative flux to consider is most often incoming radiation, either from a nearby warmer object (e.g., engine block or solar-heated ground) or from solar load, also called insolation. The solar load can be doubly difficult, as high solar load also means a bright environment, making the LED appear dimmer to the eye, as well as increasing its temperature and decreasing its light output. This is a consideration for displays, signage, and traffic signals. The effect on light output can be estimated by applying vendor data for the light decrease for temperature increase. These data are often given in chart form, so for first-order estimates some curve fitting will be helpful.

3.6 Estimating Thermal Load

As LED efficiencies increase, the difference between thermal load and electrical load increases also. The thermal load is the electrical power minus the optical power. Neglecting the light power leads to overestimating temperature, thereby overdesigning the thermal solution; this is great for cooling the LED better than expected, but it potentially involves higher cost and complexity than needed.

A secondary effect is the dependence of light output on temperature. There is a dependency of both forward voltage and relative luminous flux on temperature, so that a given current setting does not directly specify the light output. Vendor spec sheet give curves of chromaticity coordinate and relative luminous flux as a function of forward current, with the curves varying with either ambient or board temperature. As the temperature increases, the light output decreases, indicating decreasing efficiency. Figure 3.7 shows measurements of dimensionless optical efficiency as a function of the current for various reference temperatures.

The difficulty of estimating thermal load is exacerbated by the data usually presented by the vendors, where the effective thermal "resistance" defined as the temperature rise divided by the total electrical power. A true thermal parameter would be the efficiency, which is the light power (preferably in watts) divided by the

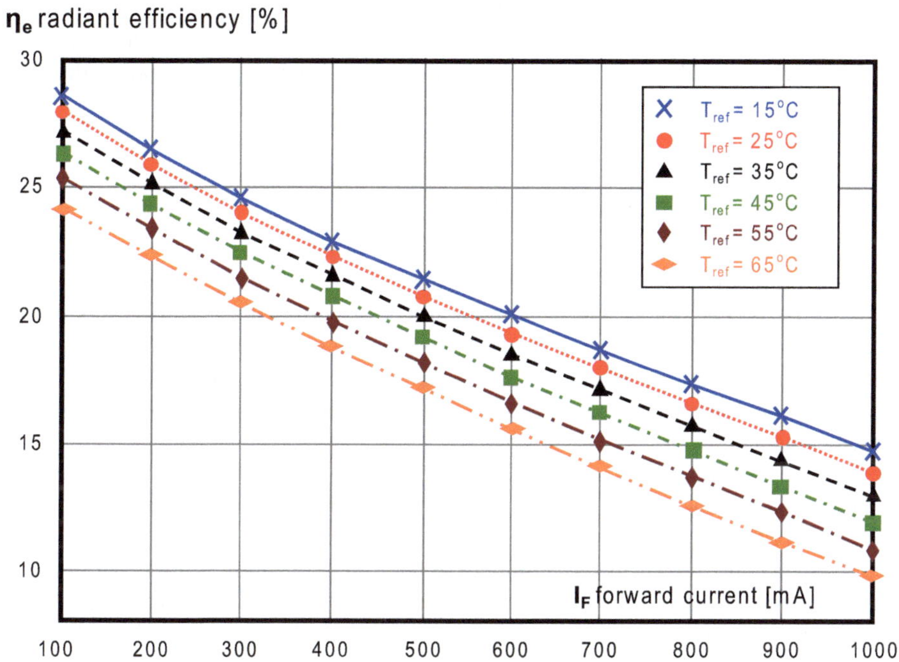

Fig. 3.7 Variation of LED efficiency with temperature [8]

electrical power. In that case, the real thermal resistance could be obtained as the temperature rise divided by the real electrical (= total − optical) power. The necessary measurements can be made using advanced test equipment, but the test standard is in development and not yet fully adopted. Ideally, a light measurement in an integrating sphere (whose output is watts) is made simultaneously with an electrical power measurement.

Another piece of information that is missing from most data sheets is the decrease in luminous efficacy over time; some vendors define "failure" as the relative light output falling below 70 %. This implies that the thermal power increases as the LED ages, which raises the temperature, which accelerates the aging. For more on LED aging, see [11].

3.6.1 Maximum Temperature

The data sheet value for maximum temperature does not represent a recommendation for actual operating temperature. In practice, the system designer targets a lower temperature, but it is not always clear what it should be. A lighting designer is interested in the light output and color appearance; energy efficiency certification may dictate that the smallest possible driving power should be used to achieve it.

Reliability targets may not be in the forefront of the design requirements, but if present also play into the choice of operating temperature. Therefore, a thermal design for an LED device needs to aim for a much lower target operating temperature than the standard for silicon IC devices.

3.7 Best-Practice Design Advice

First, estimate the heat load accurately for the temperature range at which the device is likely to run (i.e., not 25 °C). Vendors should be providing these data to users. There are emerging test standards relating to this issue (see [9]), and the U. S. Department of Energy is addressing this in the Solid-State Lighting Reliability Working Group; however, at the time of writing these data are not generally available. Actual test data may be needed.

Second, use the smallest TIM resistivity you can afford—to minimize need for heat spreading. The closer the TIM is to the LED, the more important it is to have low resistivity.

Third, use the most heat spreading possible close to the chip. A high conductivity, "thick" layer (either surface copper more than 35 μ (1 oz.) thick, or metal core board) at least three times the chip dimension in extent will do a good job spreading the heat before it encounters higher resistivity. With detailed thermal modeling or careful physical experiments to look at what is happening in the spreading region, it is fairly straightforward to quantify the benefit obtained by incurring the cost of the spreader, and reduces the risk of unexpected high temperature, or an overdesigned spreader. First-order analysis is insufficient!

Beyond the first heat spreader, the usual good thermal design practices apply as for other semiconductors. Fill gaps with TIMs; the smaller the cross-section to heat flow, the more importance should be attached to this. Then, engineer sufficient heat dissipation surface area and flow to remove heat with small enough temperature rise. At a lower junction temperature, the chip can run more efficiently and thus emit more light.

References

1. Lasance C (2008) Heat spreading: not a trivial problem. Elect Cooling 14(2). http://www.electronics-cooling.com/2008/05/heat-spreading-not-a-trivial-problem
2. Lasance C (2010) How to estimate heat spreading effects in practice. J Electr Packag 132:031004
3. Lasance C (2011) Basics of (PCB) thermal management for LED applications. IPC/APEX expo Las Vegas (updated paper from the author through lasance@onsnet.nu)
4. Lee S, Song S, Au V, Moran KP (1995) Constriction/spreading resistance model for electronics packaging. ASME/JSME thermal engineering joint conference, vol 4, pp. 199–206
5. Muzychka YS, Yovanovich MM, Culham JR (2001) Application of thermal spreading resistance in compound and orthotropic systems. AIAA2001-0366, AIAA 39th aerospace sciences meeting and exhibit, Reno, NV

6. Muzychka YS, Yovanovich MM, Culham JR (2004) Thermal spreading resistance in compound and orthotropic systems. AIAA J Thermophys Heat Transf 18(1):45–51
7. Poppe A, Farkas G, Székely V, Horváth Gy, Rencz M (2006) Multi-domain simulation and measurement of power LED-s and power LED assemblies. Proceedings of the 22nd IEEE Semiconductor Thermal Measurement and Management Symposium (SEMI-THERM'06), Dallas, USA, 14–16 March 2006, pp 191–198
8. Lasance C, Poppe A (2009) On the standardization of thermal characterization of LEDs. Proceedings of the 25th IEEE semiconductor thermal measurement and management symposium (SEMI-THERM'09), 15–19 March 2009, San Jose, USA, pp 151–158
9. Poppe A, Farkas G, Molnár G, Katona B, Temesvölgyi T, He JW (2010) Emerging standard for thermal testing of power LEDs and its possible implementation. In: SPIE Proceedings 7784, Tenth International Conference on Solid State Lighting (Eds: Ian Ferguson; Matthew H. Kane; Nadarajah Narendran; Tsunemasa Taguchi), 1 August 2010, San Diego, CA, USA, p 778414
10. Poppe A, Molnár G, Temesvölgyi T Temperature dependent thermal re-sistance in power LED assemblies and a way to cope with it. In: Proceedings of the 26th IEEE Semiconductor Thermal Measurement and Management Symposium (SEMI-THERM' 10), 21–25 February 2010, Santa Clara, USA, pp. 283–288. (ISBN: 978-1-4244-6458-1)
11. Poppe A, Molnár G, Csuti P, Szabó F, Schanda J. Aging of LEDs: a comprehensive study based on the LM80 standard and thermal transient measurements, CIE 27th Session-Proceedings, CIE 197:2011: (Vol 1, Part 1–2). Sun City, South-Africa, 10–15 July 2011. pp 467–477. Paper OP57. (ISBN: 978 3 901906 99 2)
12. Song S, Lee S, Au V (1994) Closed-form equation for thermal constriction/spreading resistances with variable resistance boundary condition. Proceedings of the 1994 international electronics packaging conference, Atlanta, Georgia, pp 111–121
13. Wilcoxon R, Cornelius D (2006) Thermal management of an LED light engine for airborne applications. Proceedings of the 22nd IEEE Semiconductor Thermal Measurement and Management Symposium (SEMI-THERM'06), Dallas, USA, 14–16 March 2006, 22: 179–185

Part II
Testing and Standardization

Chapter 4
Thermal Testing of LEDs

Gábor Farkas and András Poppe

Abstract In this chapter, after a generic discussion of thermal testing techniques used to characterize packaged semiconductor devices; the latest practical test methods widespread in thermal testing of LED components and SSL luminaires are discussed. Thus, the focus is on the latest, power semiconductor and LED-specific test procedures, environments and thermal metrics—all derived from the classical JEDEC JESD51 family of testing standards. Detailed discussion is devoted to the transient extension of the so-called static test method and the differential measurement principle in its practical realization.

Different representations of the thermal impedance are presented starting from the classical $Z_{th}(t)$ functions ending with the so-called *structure functions*. These are discussed in depth because they became the de facto standard in laboratory testing of thermal properties of LED components, in reliability analysis and in quality assurance at leading LED manufacturers. The basic concepts are introduced through practical examples.

4.1 Introduction

A common feature in present engineering tasks is struggling with growing *power level* in electronic systems. Solid-state lighting luminaires now operate at dozens of watts; the driving electronics has to produce many amperes. In other fields of electronics, processors now run at aggressive clock frequencies, electric cars have commutators working at kilowatts and hundred amperes. These are challenges not only for thermal management but also need clever solutions in thermal testing.

What is more important, the *power density* also increases. Video projectors, which were formerly of suitcase size, now resemble a pocketbook, mobile phones produce although a few watts only but in an extremely tight case with no ventilation at all.

Furthermore, many systems work in extremely *harsh environments*. Automotive electronics, for example, has to operate in the − 30 to + 80 °C external temperature range; this is similar for LED-based automotive lighting solutions or for street lighting

G. Farkas (✉) · A. Poppe
Mentor Graphics MAD MicReD unit, Infopark D, Gábor Dénes utca 2,
Budapest 1117, Hungary
e-mail: gabor_farkas@mentor.com; andras_poppe@mentor.com

C. J. M. Lasance, A. Poppe (eds.), *Thermal Management for LED Applications*,
Solid State Lighting Technology and Application Series 2,
DOI 10.1007/978-1-4614-5091-7_4,

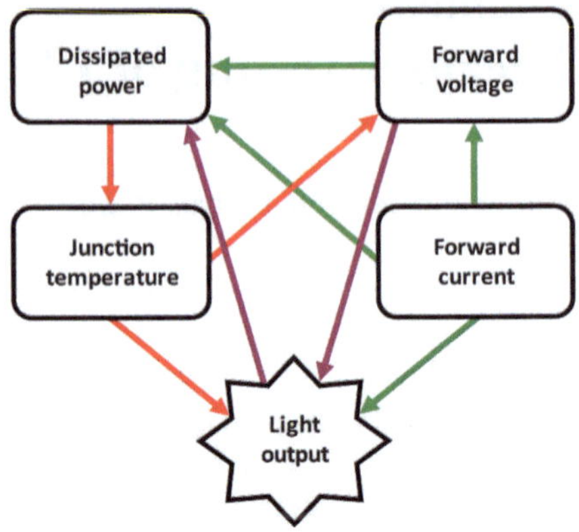

Fig. 4.1 Light output depends on "everything"

luminaires. Of course, at the hottest points of such a system, semiconductor junctions can reach 150 °C or even more, near their operation limits.

In conventional electronics as well as in solid-state lighting, the junction temperature (T_J) is a primary quantity influencing system reliability and lifetime. The junction temperature of an LED is not just a performance indicator of the thermal design but also plays a major role in lighting design since many properties of the light output of an LED depend on the absolute junction temperature (Fig. 4.1). This means that thermal management should be an integral part of the system design of an LED-based lighting solution, resulting in changing roles of different engineering disciplines in the overall design process.

The spectacular technological development described above has been achieved by *thermally aware system design*.

As it is known, thermally aware design has two pillars, *measurement* and *simulation*. However, both of them implicitly rely on a third one, modeling the thermal behavior of the system.

This mandatory modeling step is very evident in case of simulations where you create a solid model first, then do some meshing, etc. However, many limitations and misunderstandings originate from improperly handling the inherent model creation in case of measurements.

At first sight, one could think that the precision of thermal testing depends on the *resolution*, *accuracy*, *repeatability*, and *reproducibility* of the measurement.[1] A simple example can prove that this is not generally true.

If we ask, for example, how much is the distance between London and Moscow, the correct answer is 2,500 km. This statement already contains a lot of implicit modeling factors such as: we measure the length of the arc on the globe and not the chord; we do not talk about driving distance, etc.

At present, we have length-measuring instruments with an accuracy of centimeters, resolution of millimeters, and we also have satellites. With these tools, we can

[1] Some of these topics will be discussed later in this chapter and many more in Chap. 11, but their detailed treatment is beyond the scope of this book.

give a much more precise number for the distance of two towers in the cities, for example, but this was not the question.[2] We cannot improve measuring the distance of the two cities, as they are not point-like objects, and modeling them as such is an obvious mistake. The achievable accuracy of this measurement is 20 km considering the real objects to be measured and the relevant models to be used for them. These factors manifest as inherent *uncertainty* of the measurement.

Even in special literature, we find papers where some quantities with limited relevance like chip temperature or package case temperature are measured at four-digit accuracy and then derived quantities are calculated at six-digit accuracy. In the following sections, we will clarify possibilities and limits of measurements.

Before the advent of new thermal design methods, the major cause of system breakdown was the overheating of critical components. Failure analysis (FA) shows that nowadays systems are correctly designed in this respect. The typical component breakdown is caused by repeated *thermal transients*. Heating and cooling induces shear stress at the material interfaces in the structure (die attach, solder joint) resulting in delamination, tear off, etc. The poorer heat removal through a diminished surface can then cause thermal runaway.

This is one of the reasons for introducing thermal transient measurements and simulation techniques. As we can see later, these methods also unfold internal structural details that cannot be identified by static measurements.

4.2 Modeling of Devices for Measurement and Simulation

When we measure parameters of semiconductor devices, we inherently have an underlying *model* in mind. The model parameters like single junction temperature, single lumped thermal resistance, or thermal capacitance are highly simplified.

In reality, there is always a given distribution of the temperature over the active area of a device; and thermal systems are infinite distributed RC systems. The question is how good our implicit models are reduced to single (average) temperature value or to a few thermal network elements.

For a long time, packaged electronic components were represented by a single thermal resistance in the data sheet. Power devices were usually encapsulated in packages with a dedicated cooling surface, called "case."

At discrete semiconductors (diodes, transistors), the hottest portion of the device was the pn-junction. As first estimation, the engineer accepted that applying P_H heating power at the junction of the packaged device mounted on a surface of T_C

[2] The mentioned cities span over a large surface while distance is defined between two points. The question is: are there some points in these cities which are characteristic for the real target of the distance measurement and where are they located? We face similar questions in the thermal testing of semiconductor devices. To which point of the active semiconductor we assign the "junction temperature"? Resistance (strictly speaking) is defined between two nodes of a graph. Which are the distinct points between which we define the thermal resistance? See Chap. 6 for further details on these issues.

temperature; the device will reach maximum junction temperature in a longer run.

$$T_J = P_H \cdot R_{thJC} + T_C \tag{4.1}$$

At present, the range of LED devices is very wide. There are small devices for which the usual approximations and assumptions of the classical thermal testing are still more or less valid, but nowadays we face a couple of challenges, which are not yet followed up by standards:

- Instead of LED packages containing a small chip, we have to characterize large modules with multiple devices, chip arrays, etc. causing various dissipation patterns on the chip surfaces.
- Packages have become large and very flat, the idea of a homogenous temperature on a case surface (and also on the chip surface) is no more acceptable.
- For many devices (LEDs and driving electronics), the heat flow through package pins became a major, often dominant factor.
- There are direct mounting (chip on board) solutions.
- With surface mount technologies the heat leaves in very complex and manifold ways through pins, exposed tabs, mould surface towards different layers of the printed board, to the air, sometimes to heat sinks, fans, etc.

Single thermal resistance values such as the junction-to-case R_{thJC} are still part of a data sheet facilitating component selection at early design phases. However, the complex, three-dimensional reality can only be handled with sophisticated simulation tools along with thermal measurements.

Not only the conventional concept of a single thermal resistance is insufficient for adequate design in solid-state lighting, but we have also serious doubts about the nature of the innocent P_H heating power in Eq. (4.1). This will be treated more in detail in Sect. 4.3.7.

The most exciting question is the temperature of the semiconductor in a specific environment. With more measurements at various boundaries, we can get relevant data about this. A simple and effective use of the measured values is if we order them into a consistent structure of a few model elements (compact model).

Most accurate results, such as temperature and heat flux at many points of the chip, package, and environment can be gained using *detailed models* consisting of many hundreds of thousand elements, e.g., with finite element (FEM) or finite difference (FDM) methods (see Fig. 4.2). If we want to see the effect of convection ("still" or moving air around the component), we can rely on computational fluid dynamics (CFD) methods.

This kind of simulation needs special expertise and this level of accuracy can be achieved at the manufacturer of the component only, where all geometrical and material details are known.

Building compact models of a few elements gives a much faster and simpler approach, describing the component with reasonable accuracy for the application engineer. Some compact models are behavioral ones while others also reveal essential structural information.

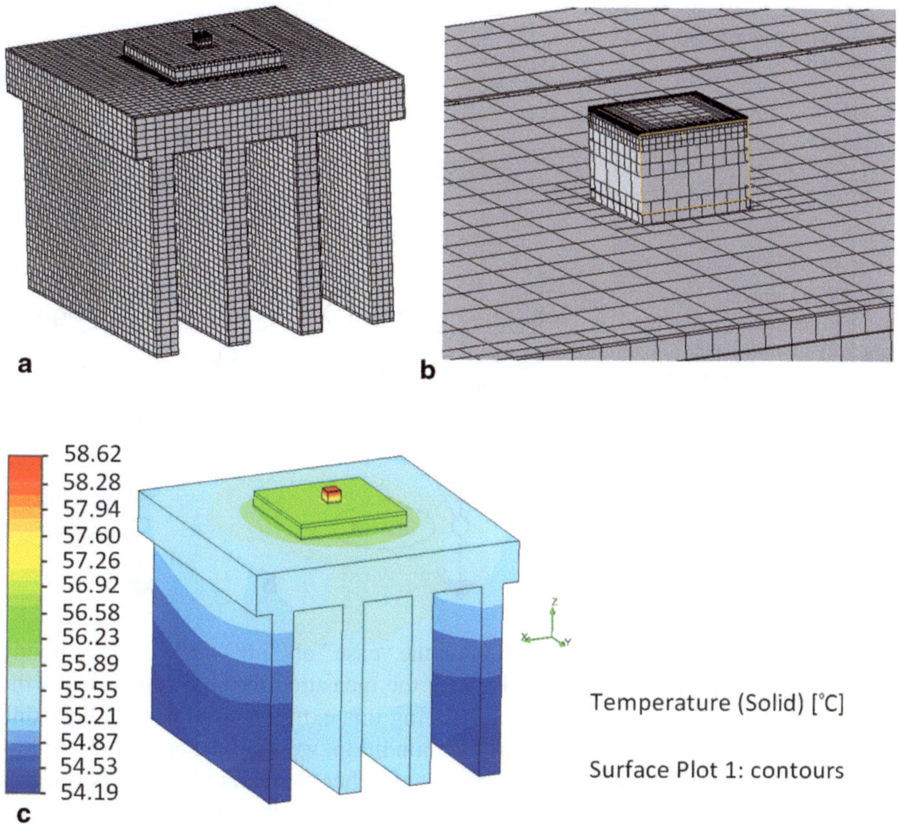

Fig. 4.2 An LED device on heat sink: **a** detailed thermal model, meshed for a finite difference solver (FloTHERM), **b** close-up view of the model at the LED chip, **c** results of the CFD simulation using this model

Creating compact models is not too simple either. In classical electronics, thermal management and modeling techniques were developed in the last two decades and became formally standardized recently. In case of LEDs, however, compact modeling is still evolving; it is far from a mature state that one could call standard. A survey on LED compact modeling status is provided in Chap. 6 on standardization.

Figure 4.3 shows an LED device packaged and a heat sink to be attached to the "case" surface. Fig. 4.4 shows different possible model topologies—which are in practical cases nodes assigned to surfaces of more or less uniform temperature and thermal resistances between these.

A compact model approach helps in interpreting measurement results, and we shall mainly use it for this purpose in this chapter. It also enables understanding the interaction of the device (represented by RL, RR, RP) and the environment (represented by R1, R2, R3 in Fig. 4.5).

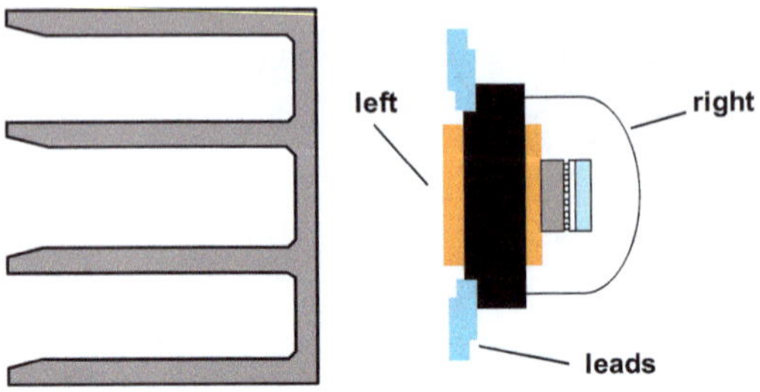

Fig. 4.3 A real package: LED packaged and a heat sink to be attached

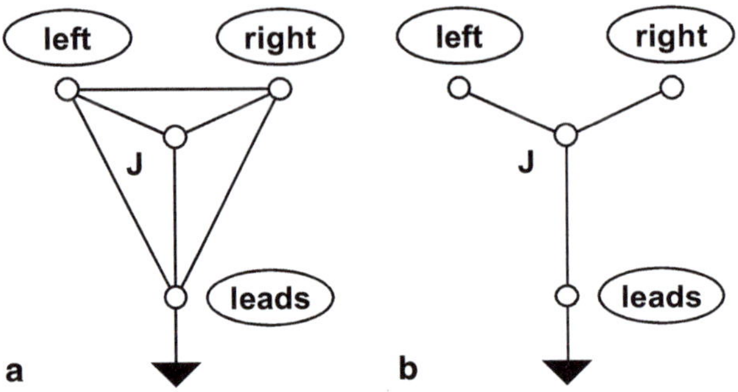

Fig. 4.4 **a** General, **b** simplified model network topology of a packaged device. (*Lines* represent graph edges, *small circles* represent graph nodes)

Applying an extremely good cooling on the "case" or "left" surface and applying known power on the LED, we can identify the measured thermal resistance as the classic junction-to-case value (RL). Replacing this extreme cooling by a heat sink for which a catalogue claims R1 at still air condition, we shall measure more or less RL + R1 thermal resistance. Rather less, as all other elements of the model (if it is correct) will be "active" in this new situation, the air cannot be so "still"—and we shall see later on that even the package section represented by RL will behave in a very different way.

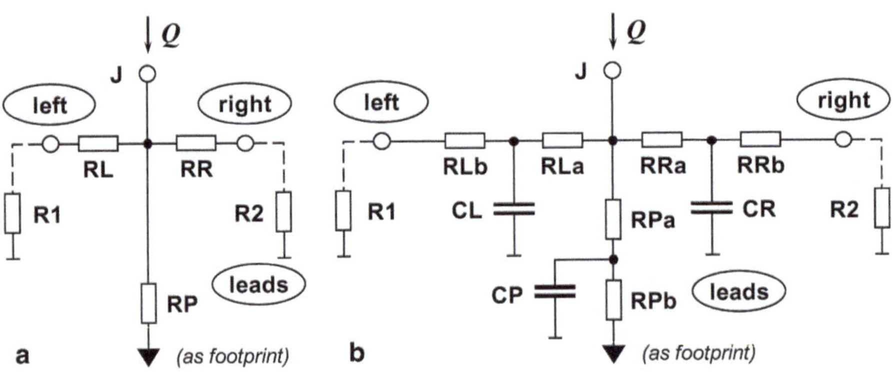

Fig. 4.5 **a** Static, **b** dynamic compact model of a packaged device according to the model topology shown in Fig. 4.4b

Usually, model values derived for a given topology strongly depend on the actual environment. Models containing more elements such as in Fig. 4.5a show less dependence on actual boundary conditions.

Instead of single thermal resistance values, major semiconductor companies nowadays offer compact models that can be used in design tools like board or system-level thermal simulators. It is expected that a thorough optimization step yields acceptable values for a broad range of actual boundary conditions (*boundary condition-independent* models), see also [1].

Besides packaged devices, other system elements such as heat sinks, fans, heat pipes, etc. may also have compact models derived from measurements and simulations [2].

4.3 Thermal Measurements

We can summarize the consequences of the previous section as follows:

- Measurements can be done for component or for system characterization.
- They can be steady-state or transient measurements.

In this chapter, we mainly focus on *transient measurements*.

As we will see in later sections, besides the primary information of the temperature at selected points; the heat propagation starting from the heaters in the system also reveals structural details of the assembly, comparable to ultrasonic or X-ray investigation. In such a way, transient measurements and their evaluation serve *multiple purposes*. They:

- Yield essential data to application engineers in format of data sheets or compact model libraries.
- Characterize the thermal behavior of components at package, board, and system level.
- Help production engineers to perform quality checks on packaging, die-attachment.
- Give feedback to designers of packages or luminaires on the actual performance of constructions.
- Help design engineers to evaluate thermal properties of package, thermal interface, and board materials.

As we can see, some measurement tasks serve research and development (R&D), or can help perform structural FA and production monitoring as part of a quality assurance (QA) process.

Applying a steady low power level to the device and then instantly switching to a higher level (Fig. 4.6), we can observe a heating transient. Similarly, switching from a higher level to a lower one, we can monitor the cooling of the device.

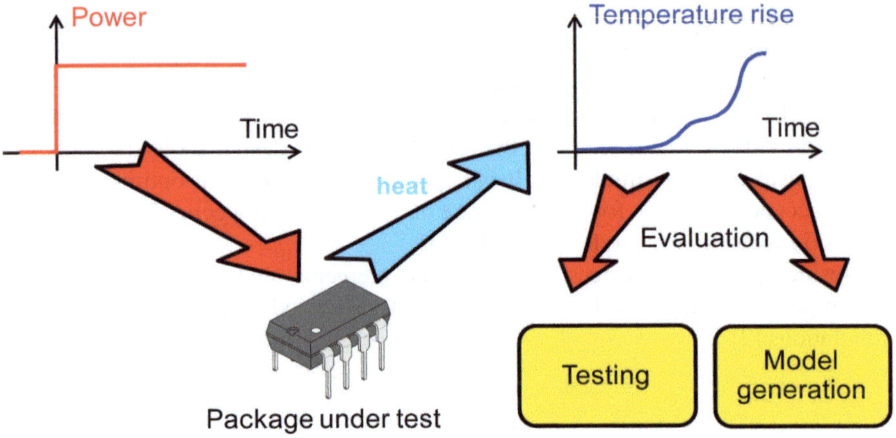

Fig. 4.6 Thermal transient testing of electronics parts (with heat generation only)

At R&D and FA, the time spent for each measurement doing transient analysis is not critical. At these measurements, we are in the fortunate situation that we can completely capture the transients up to steady-state level. This practice can also be followed at QA if we restrict investigations to a few selected samples.

When aiming at a broader characterization performed on all manufactured samples, the time per measurement is much more limited; i.e., we have to carry out short transients and search for correlation between short time and complete transient behaviors. Some results related to this problem in case of LEDs are presented in a couple of conference papers [3, 4]. Figure 4.6 illustrates testing of semiconductor devices with a single-energy transport mechanism, all electric energy flowing towards the device will be transformed at the end to heat.

In Fig. 4.7, we see a case where a portion of the input electric energy is converted to some other sort of energy: in case of LEDs, laser diodes, etc. this is typically emitted light. While recording the temperature transient, this other quantity can also be recorded. Further consequences of this type of measurement will be discussed in Sect. 4.3.7.

4.3.1 Device Categories and Their Features

Semiconductor devices can act as *heaters* by applying appropriate voltage and current at some of their internal structures. They can also act as *sensors* using their *temperature-sensitive parameters* (further referenced as TSP, for more details see Sect. 4.3.8 on the TSP calibration of devices).

Dedicated thermal test chips have separate heater (dissipator) and sensor structures (Fig. 4.8d).

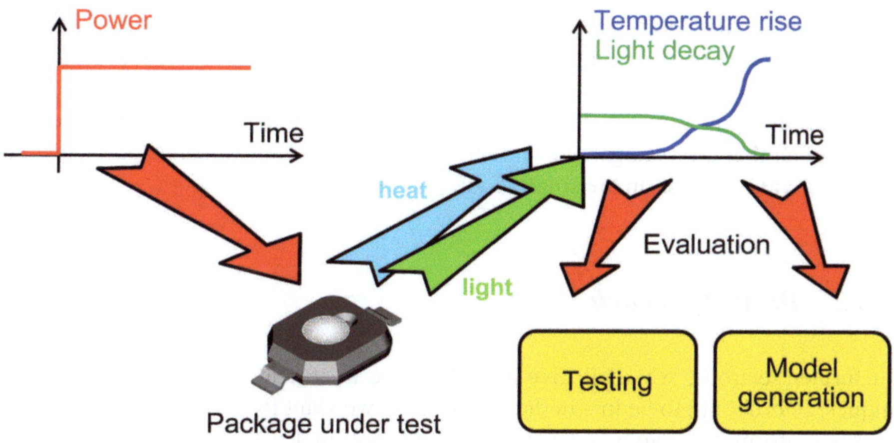

Fig. 4.7 Thermal transient testing of LEDs (with heat generation and light emission)

Fig. 4.8 Heater/sensor structures used in thermal testing: **a** MOS transistors, **b** bipolar transistors, **c** pn-junction or diodes, **d** separate heaters and sensors

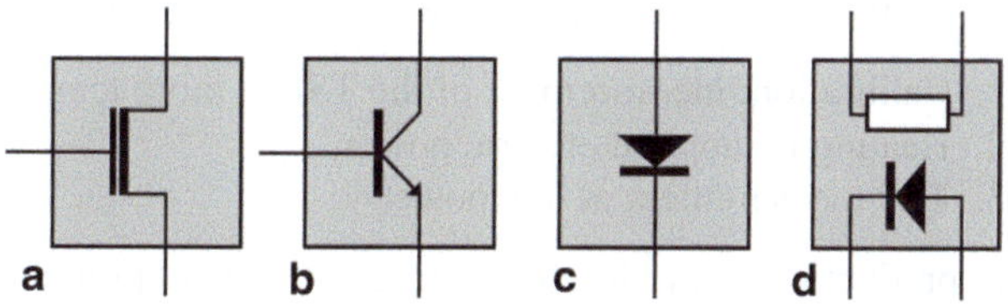

The most common three-pin semiconductor devices like MOS field effect transistors (MOSFET) and bipolar junction transistors (BJT; Fig. 4.8a, b) have an inherent feature of sensing temperature—parameters influenced by the temperature such as V_{BE} or V_{GS} voltages can be measured. These three-pin devices have at least partially separated ports for heating and sensing, e.g., heating can be applied across collector and base electrodes, while the emitter-base junction can be used for sensing purposes. The power step can be induced by a sudden change in the collector-base voltage V_{CB} (voltage jump) or in the emitter current I_E (current jump).

Diode-type discrete devices (Fig. 4.8c) have only one pair of pins, the port used for heating and sensing is the same. The typical powering mode is "current jump," switching between a larger heating current and a smaller sensing or measurement current. At heating, the voltage on the diode changes significantly during the transient (which is the effect we use for measurement actually). Preferred measurement mode is cooling, at a small measurement current. This ensures a suitable V_F forward voltage, which can be used for sensing purposes. In case the power level at cooling is very small, it can be considered to be constant zero. (See considerations of proper choice of the measurement current later.)

Standard silicon integrated circuits can be measured in a "dull" way with their inherently present substrate diodes; such falling into category c. MOSFET devices also can be measured using their reverse diode structures.

A more sophisticated way of measuring integrated circuits is powering heating structures, which dissipate at normal operation. The same structure can act as sensor in many cases. Also, input protection diodes or output parasitic diodes can be used for multiple point sensing. In this case, selecting a proper low sensor current is essential so that the voltage on eventual series resistors can be neglected compared to the forward voltage on the small diode.

4.3.2 Basic Approach

In further sections, we shall give an exhaustive treatment of standards and techniques. In order not to be lost in the details here, we want to emphasize that relevant thermal information can be gained even with a multimeter, thermocouple, and other simple devices. Of course, advanced methods give a much deeper insight into device structure and failure locations.

All methods have to comprise the following steps:

1. Calibration: measurement of the TSP at more temperature values.
2. Heating by applied electric power.
3. TSP measurement at low power.

In production lines, thermal testing is often simplified to a *static method* recording one or a few characteristic temperatures.

In a measurement targeting steady-state *thermal metrics* such as junction-to-ambient *thermal resistance*, step 3 has to be carried out twice, once in hot state, just after switching off the heating power, and once near cold steady state. The measured TSP values yield the "hot" and "cold" temperatures using the calibration values from step 1. This calibration is sometimes carried out just for a few samples of a manufacturing lot.

In a correct static measurement, the cold steady state is reached after a longer equalization time,

- Either just before switching on the power, or
- Using the same long equalization time at the end of a cooling transient (which is not measured).

The critical question is how the time point "just after switching off" is found.

Production line testers often arbitrate just based on one or few "hot" temperature values in go-no go tests.

With "*continuous transient*" *methods* (sometimes, as a result of ill-formed terminology of an old thermal testing standard [5] mistakenly referred as static), step 2 has to be applied for an equalization time and then step 3 has to be carried out more times at a certain sampling rate. The first and the last sample yields the "hot" and "cold" temperatures, the temperature change in between provides device structure-related information.

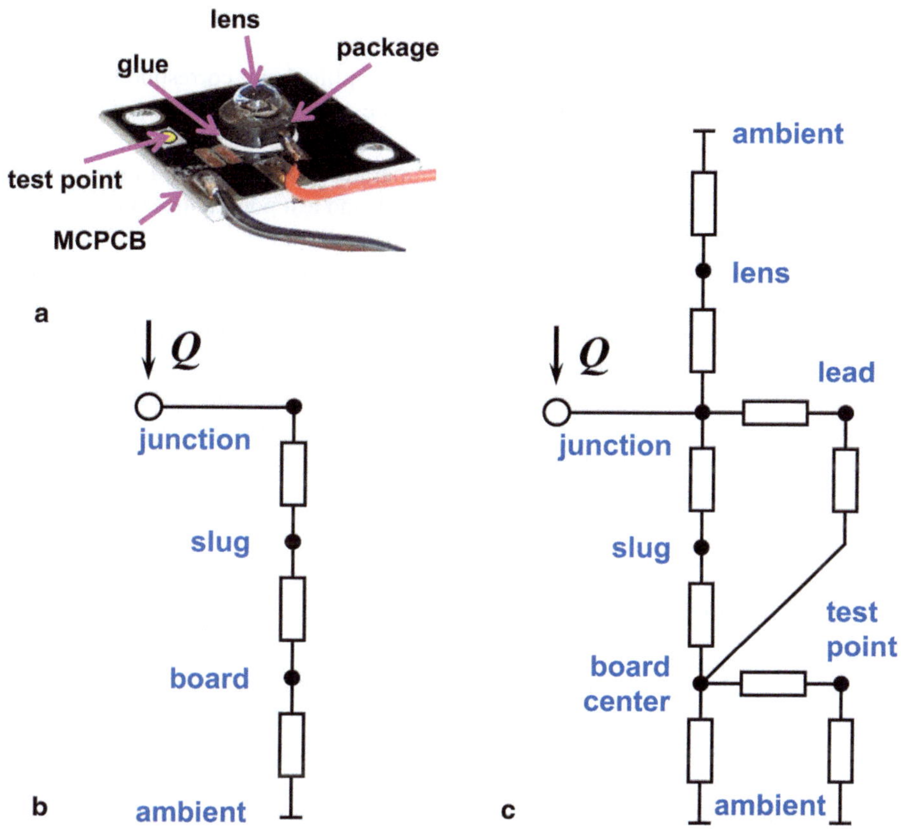

Fig. 4.9 LED device on metal core PCB with thermal testing point, **a** photograph, **b**, **c** compact models of different accuracy and complexity

While at simple static and "continuous transient" methods step 2 and step 3 last for a time long enough for reaching temperature equalization, *pulsing methods* repeat step 2 and step 3 for each sampling. Here, actually step 3 has to be carried out before and after each heating pulse.[3]

For large-scale testing, often simple comparison measurements are used.

In Fig. 4.9, we see a power LED mounted on a metal core printed circuit board (PCB) used as heat spreader having a thermal test point. LED manufacturers often specify the safe operating area of the LED in terms of test point temperature—allowing their end-users to perform a very simple temperature measurement at this point, often considered as the final thermal testing of their LED application.[4]

[3] Measurement waveforms for the static test methods are depicted in Fig. 6.20 of Chap. 6; (see also [6] and [7]). A concise summary of thermal measurement of diodes is given in [7, 8, 9]. A more detailed analysis is provided in Sect. 4.3.6.

[4] The *junction-to-test point* thermal characterization parameter is a thermal metric, which can help relate junction temperatures to test point temperatures. Discussion of such simple tests is not the main target of this chapter.

For LEDs of this construction (Fig. 4.9a), application notes suggest a compact model as presented in Fig. 6.16 of Chap. 6.

The model in the application note is rather simplified and corresponds approximately to Fig. 4.9b. Although it can be used well despite its simplicity, partial thermal resistance values and internal node temperatures cannot be measured.

Instead, at certain powering, the chip temperature can be derived using one of the methods described before. The package, lead, and test point temperatures can be measured by attaching a thermocouple there.

The more detailed model suggested in Fig. 4.9c hints that such simple static methods yield information on junction, lead, and test point temperatures. However, test point temperature can significantly differ from board temperature under the LED, depending on the cooling mount attached.

In comparative tests, the measurement restricted on leads and test point already provides basic information like "some failure occurred in the chip–submount–heat slug–glue–distributor plate" heat conduction path. Temperature of internal elements in the device and structural information can be gained by transient measurements.

4.3.3 Thermal Testing Standards

Repeatability of a measurement on the same device, moreover the reproducibility among systems at different sites serving the same purpose are key issues in all measurements. In case of thermal characterization, this can be ensured provided the *measurement technique*, *boundary conditions (test environments)*, in some cases, *test devices* along with the *measured quantities (thermal metrics)* are defined by widely accepted standards.

Since 1990, the Joint Electron Device Engineering Council (JEDEC), under the Electronic Industries Association (EIA), has been creating a set of thermal measurement standards for semiconductor device packages. The JEDEC JC-15 committee (Thermal Characterization Techniques for Semiconductor Packages) has been formed by over 40 member companies, among them semiconductor, packaging, and software companies.

The JEDEC thermal testing standards recommend specific environmental conditions, measurement techniques, fixturing, heating power, and data reporting guidelines. Most recently, compact thermal modeling guidelines and LED thermal testing guidelines have been developed and published. (For details on these latter, refer to Chap. 6 on standardization.)

Further in this chapter, we provide details on the basic concepts of thermal testing of packaged semiconductor devices (keeping in mind what is important for LED thermal testing). The basics of thermal testing are described in the JEDEC JESD51 document [10], which is the so-called overview document of the thermal characterization of packaged semiconductor devices. The family of pertinent JESD-51 standards is shown in Fig. 4.10. This figure is an updated version [11] of the chart published in the original JESD51 document [10]. (The most recent update on the

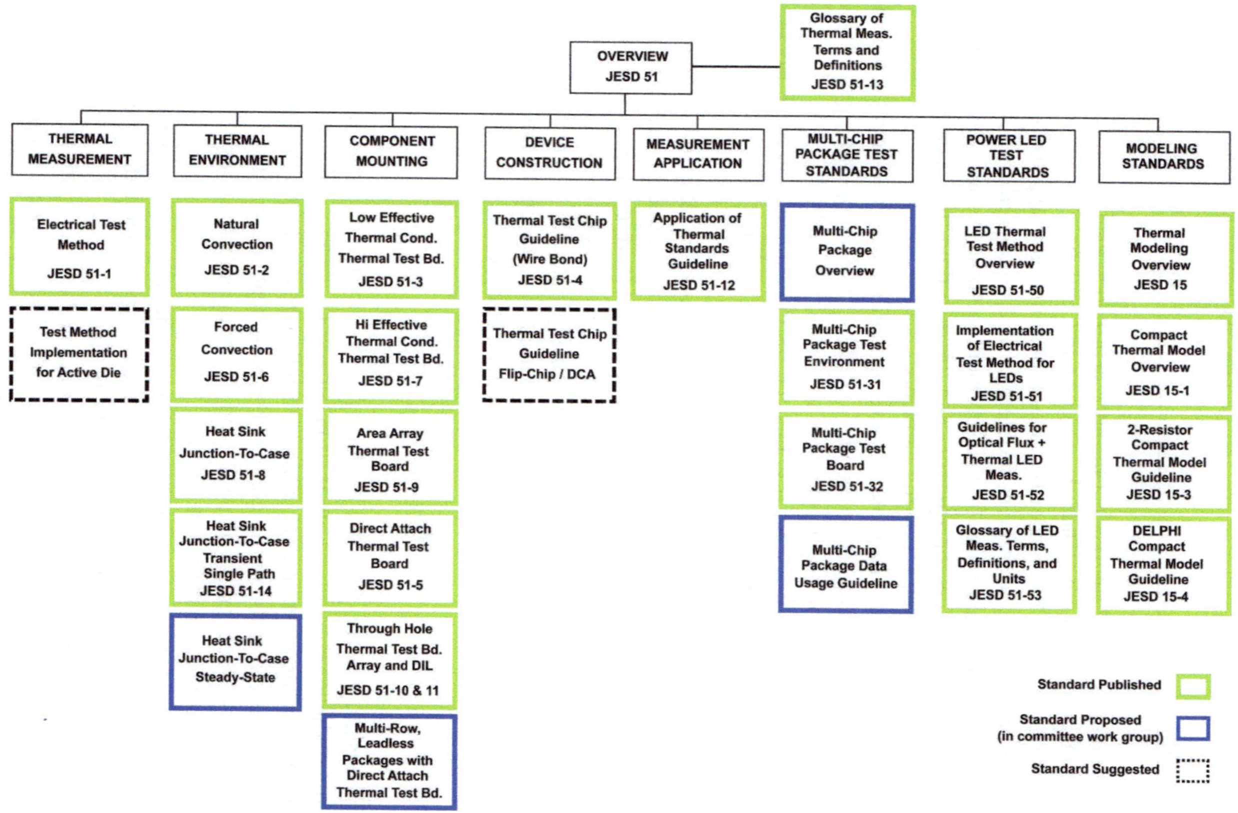

Fig. 4.10 An updated version of the JEDEC thermal testing standards overview chart of the JESD51 document [10]. (Chart provided in private communications by Bruce Guenin, chairman of the JEDEC JC15 committee. Reproduced with permission of Bruce Guenin)

Temperature measurement technique:	Electrical test method
Boundary: composed of	
Test environment:	Natural convection
Component mounting:	Low conductivity test board for surface-mount leaded package
Device construction:	Wire bond thermal test chip

standardization activities of the JC15 committee is provided in [12]. More details on JEDEC's LED thermal testing standards are given in Chap. 6).

The series of the JESD 51 standards is based on a modular approach. A testing protocol for a particular measurement problem can be composed by selecting the appropriate documents describing the measurement method, the test environment, component mounting, device construction, and the data reporting. For example, a process can be built up on the following elements:

The measurement process yields one or more standardized *thermal metrics*, which need to be reported with information recommended by the data reporting guidelines and according to the specific requirements of the given test method.

The modular approach of the JEDEC thermal testing standards is explicitly described in the JESD51 document by mentioning possible addition of new documents as the need arises.

JEDEC thermal metrics types aim to express power-induced temperature change. The measurement occurs between the point where the powering occurs (driving point, "*junction*") and some reference point or reference surface or reference environment.

The first type of thermal metrics is based on a stricter definition and is called *thermal resistance*, denoted as R_{thJX} or Θ_{JX}. Supposing that all heat generated by P_H heating power at the J driving point (usually referred to as *junction*) flows through an isothermal surface in the measurement arrangement called X; we define R_{thJX} as

$$R_{thJX} = \frac{T_J - T_X}{P_H} \tag{4.2}$$

In thermal design, we still often use the term *junction* for the thin region on the top of the semiconductor chip where dissipation occurs. The term is inherited from the time of discrete components having a large pn-junction as source of dissipation. In case of LEDs, it is really the pn-junction where most heat is generated.

Usually, the X surface is not a plane. T_J and T_X are the temperatures of the driving point and the reference environment.

In many cases, it is not possible to find an isothermal surface where all heat passes. In such cases, a less rigorous thermal metric type is defined, called *thermal characterization parameter*, which is denoted as Ψ_{JX}.

Supposing the heat generated by P_H power at the J driving point flows through more alternative heat conducting path sections, and there is a well-defined *point* in the measurement arrangement called X, Ψ_{JX} is defined as

$$\Psi_{JX} = \frac{T_J - T_X}{P_H} \tag{4.3}$$

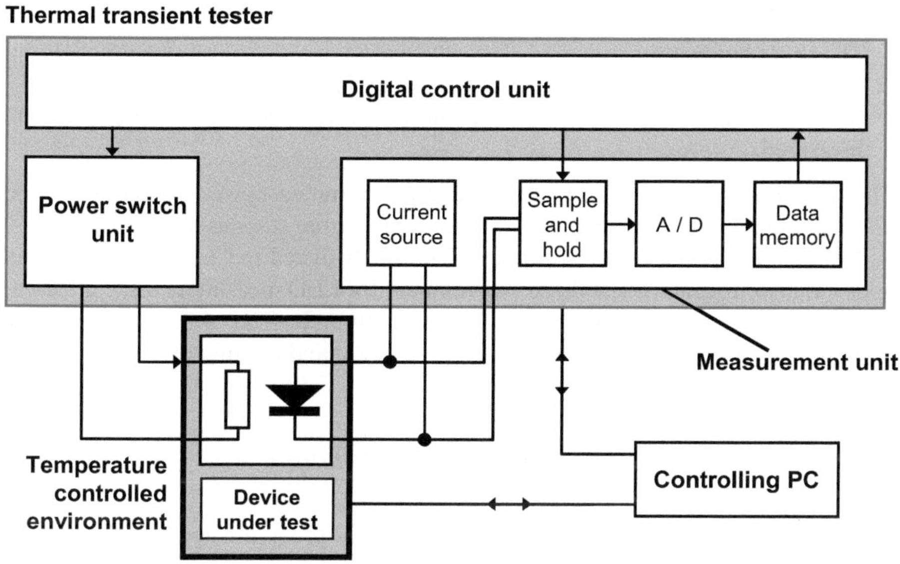

Fig. 4.11 The general scheme of the tester and test environment

T_J and T_X are the temperatures of the driving point and of the reference point. The test point in Fig. 4.9 can serve for producing such a thermal characterization parameter in comparison measurements.

For further details on standardization of thermal characterization of LEDs, consult Chap. 6 of this book. Some new thermal testing standards are presented for power semiconductor devices in [13] and for power LEDs in [6, 14].

4.3.4 Measurement Instruments and Environments

As shown before, we have to use controlled voltage and current sources to apply the appropriate power levels. For measuring the *junction temperature* by *electrical test method*, we also need fast and accurate amplifiers and data acquisition, at least in transient testing. Powering, sampling, and data acquisition is integrated in *thermal transient testers*.

Another task related to the thermal testing of semiconductor devices is the realization of *repeatable boundary conditions*. The JEDEC JESD 51 series of standards define *convective* and *conductive* environments.

Convective boundaries are typically realized by still air chambers, wind tunnels, and liquid baths. Conductive environment is realized by different cold plate structures.

Plates and baths can be programmed to force different temperature values to the device under test. Using them with the tester equipment, these can be also applied to calibrate the temperature-sensitive parameter of the devices (Fig. 4.11).

The thermal transient testing system consists of:

- The tester hardware,
- Software elements for measurement control and data acquisition,
- Software elements for the subsequent postprocessing and display of data measured.

The measurement control software applies programmed power excitations to the dissipator element and records the complex temperature responses.

Relevant radiometric and photometric standards (listed in Chap. 6) prescribe an appropriate temperature-stabilized environment for LED measurements, e.g., an integrating sphere with temperature-stabilized fixture. In Sect. 4.3.7, we present a combined methodology for radiometric/photometric and thermal testing.

4.3.5 The Differential Measurement Principle

Formerly, when steady-state thermal measurements predominated, usually the *absolute measurement technique* was used. This technique takes the meaning of Eq. (4.2) literally.

One has to apply a P_H heating power and then has to measure the absolute T_J and T_X values separately. The advantage of this measurement is that it can be carried out by very simple means, like a multimeter. On the other hand, this approach unleashes a lot of problems in measurement and calibration (referred later in this section).

One can get rid of most measurement errors using a *differential technique* for measuring quantities, which are by nature differences.

For example, at junction-to-ambient measurements, we can apply two different level of the heating power P_{H1} and P_{H2}, and measure the temperature after equalization in each case. We shall get:

$$T_{J1} = P_{H1} \cdot R_{thJA} + T_A \tag{4.4}$$

$$T_{J2} = P_{H2} \cdot R_{thJA} + T_A \tag{4.5}$$

$$(P_{H1} - P_{H2}) \cdot R_{thJA} = T_{J1} - T_{J2} \tag{4.6}$$

$$R_{thJA} = \frac{(T_{J1} - T_{J2})}{(P_{H1} - P_{H2})} \tag{4.7}$$

As shown before, the temperature-sensitive element is in most cases a pn-junction (or a resistor), needing some bias (usually a constant current) to produce a temperature-dependent voltage. The temperature sensitivity of these structures has typically good linearity over a large temperature interval, thus, if the V_{F0} reference value of the forward voltage is known for the T_{J0} reference value of the junction temperature, the forward voltage at any junction temperature can be expressed with a linear relationship:

$$V_F(T_J) = V_{F0} + S_{VF} \cdot (T_J - T_{J0}) \tag{4.8}$$

where the S_{VF} sensitivity slightly depends on the I_M bias current, too. With this, for R_{thJA}, we get

$$R_{thJA} = \frac{V_F(T_{J1}) - V_F(T_{J2})}{S_{VF} \cdot (P_{H1} - P_{H2})} = \frac{\Delta V_F}{S_{VF} \cdot \Delta P_H} \quad (4.9)$$

The differential principle offers a lot of advantages. There is no need to directly measure T_A, just T_J. Even when doing the calibration for getting the S_{VF} sensitivity for Eq. (4.8), only a differential measurement is needed. All offset problems at measurement and calibration cancel out.

In the literature, often the $K = 1/S_{VF}$ reciprocal sensitivity factor is used. Therefore, the calibration process for finding the value of S_{VF} is also called *K-factor calibration*. With the K-factor, Eq. (4.9) reads as

$$R_{thJA} = K \cdot \frac{\Delta V_F}{\Delta P_H} \quad (4.10)$$

The results apply to transient measurements, too. Here, a full capture of $\Delta T_J(t)$ time function (i.e., using the voltage of the temperature-sensitive structure) is needed between the equilibrium at P_{H1} and that at P_{H2}. It is often thought that one of the power levels, P_{H2} for example, has to be kept low in order to avoid self-heating of the device. Using the differential technique, this effect is only of minor importance; even the effect on the K-factor calibration can be neglected.

P_{H1} and P_{H2} can be applied in many ways, such as changing V_{CB} on a transistor, or changing the clock frequency of a processor.

If in Eq. (4.10) we consider the $\Delta V_F(t)$ time function of the change of the forward voltage, we obtain $R_{thJA}(t)$—a time dependent thermal resistance value referred to as $Z_{thJX}(t)$ thermal impedance, see also Sect. 6.1.4.1 of Chap. 6.

$Z_{thJX}(t)$ impedances, i.e., thermal impedances towards a point different than the ambient can be derived similarly, by measuring T_J and T_X at two power levels. Here, all offset problems cancel out again.

4.3.6 *Current Jump Measurements on Two–Pin Devices*

At diode-like devices (including LEDs), the usual way to apply the P_{H1} and P_{H2} power levels is to change the current bias. While maintaining a constant I_{sense} sensor current, a sudden jump in the power level can be generated by switching on and off an I_{drive} heating current. Throughout the JESD51 series of thermal testing standards the I_{sense} sensor current is also known as I_M *measurement current* and the sum of I_{sense} and I_{drive} currents is called I_H *heating current*.

4.3.6.1 Continuous Cooling Measurements

With this measurement technique, the switching from the I_H *heating current* to a smaller I_M *measurement current* occurs only once, as shown in Fig. 4.12.

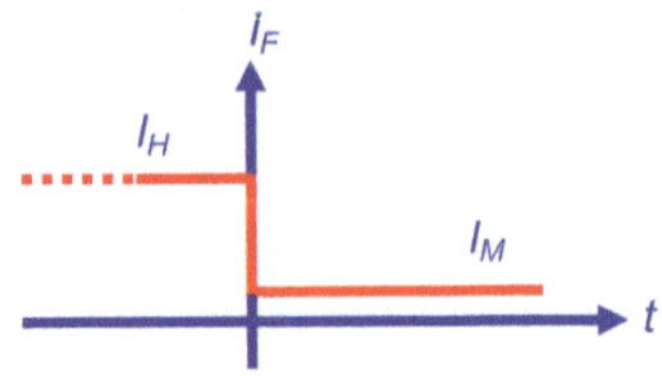

Fig. 4.12 Stepwise change of the forward current on the diode under test

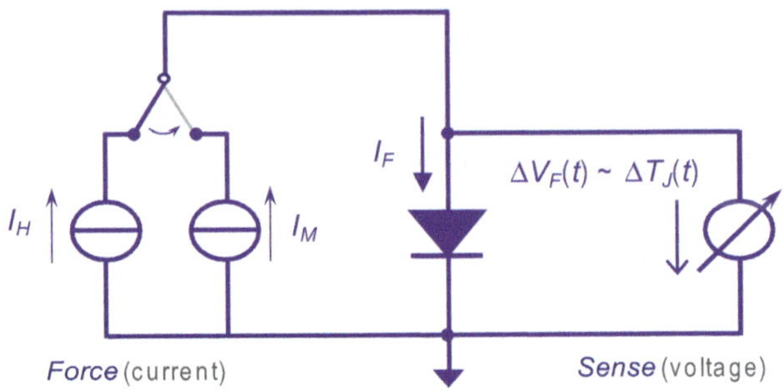

Fig. 4.13 Diode measurement scheme as defined by JESD51-1 and JESD51-51

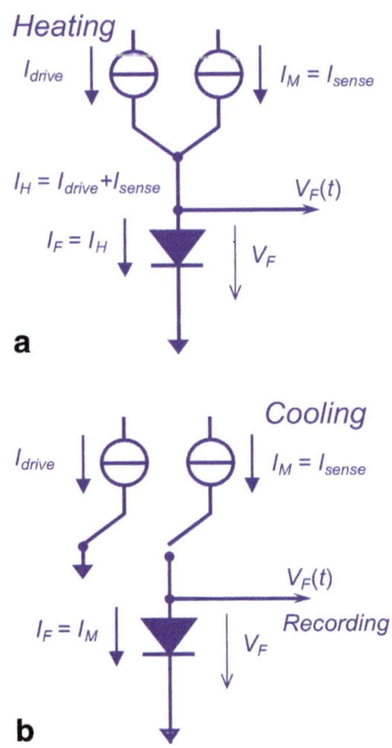

Fig. 4.14 Practical realization of the diode measurement scheme: **a** applying heating current, **b** applying measurement current

Figure 4.13 shows the theoretical measurement arrangement as recommended by the classical technical literature [7] and the JEDEC thermal testing standards [5, 10]. In Fig. 4.14a, b, the practical realization of the switching is shown.

The I_H heating current is provided as a sum of the currents of the I_{drive} and I_{sense} current sources, thus, in the heating period (Fig. 4.14a) both currents are applied on

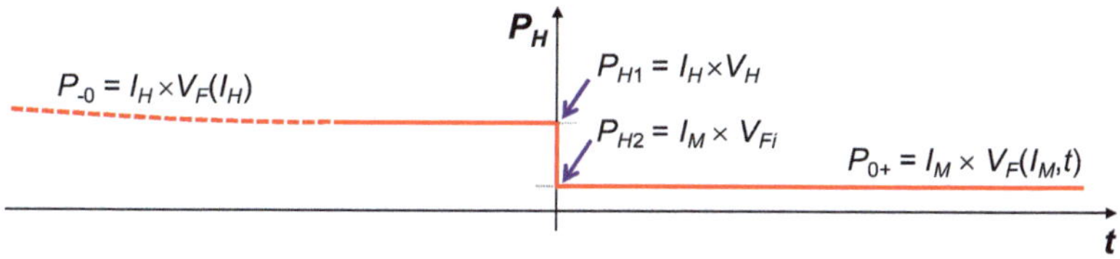

Fig. 4.15 Power change before and during the cooling measurement on a diode

Fig. 4.16 Thermal (cooling) transient of the diode at constant measurement current shown in its I–V characteristics

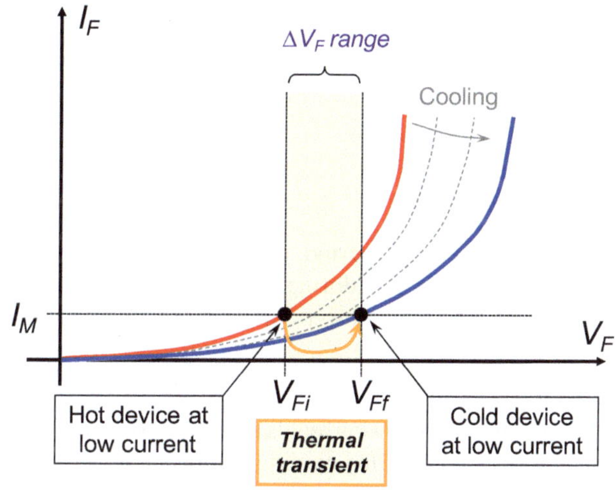

the device for an equalization time of appropriate duration. When the device reached "hot" steady state, the drive current will be switched off and the transient recording of the cooling will start (Fig. 4.14b) at the I_M measurement current provided by the I_{sense} current source only.

Diodes have negative temperature sensitivity ($S_{VF} \approx -1$ to -2.5 mV/°C), so during the heating period the temperature increases and the V_F forward voltage diminishes. As the current is constant, P_{H1} also diminishes during the heating (Fig. 4.15), it has to be measured just before switching out. We can state that

$$P_{H1} = I_H \cdot V_H, \ P_{H2} = I_M \cdot V_{Fi} \tag{4.11}$$

and

$$\Delta P_H = P_{H1} - P_{H2} = I_H \cdot V_H - I_M \cdot V_{Fi} \tag{4.12}$$

where V_H is the forward voltage of the hot diode biased with the I_H heating current (heating voltage) at the time instance of switching, V_{Fi} is the initial value of the forward voltage at the beginning of the cooling transient when the diode is supplied with the small I_M measurement current only (for notations see Fig. 4.16).

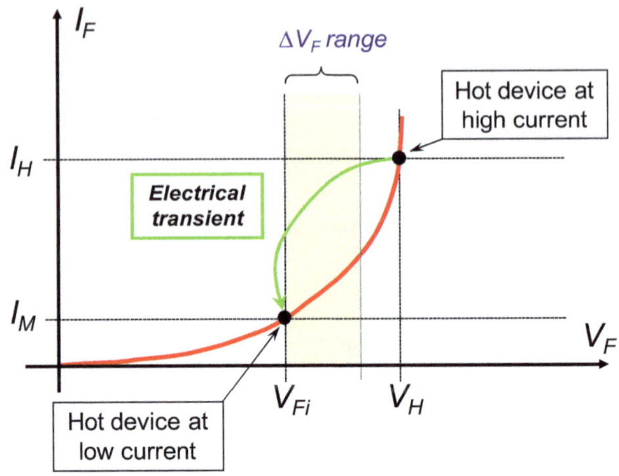

Fig. 4.17 Electrical transient of the hot diode shown in its I–V characteristics, switching from the heating current to the measurement current

During the cooling period, the V_F forward voltage will grow, it nearly repeats the change that took place during the equalization, but now in the opposite direction. The slightly changing power during cooling can be expressed as

$$P_{H2} = I_M \cdot V_{Fi} + I_M \cdot \Delta V_F(t) \tag{4.13}$$

If the I_M measurement current is significantly lower than the I_H heating current, then the *power change* during cooling can be neglected.

For a simple estimate on an LED measurement, let us program $I_H = 350$ mA and $I_M = 5$ mA. Assuming a forward voltage of about 3 V at I_H and 2.9 V at I_M, the total power step will be (350 mA × 3 V) − (5 mA × 2.9 V) = 1.035 W.

If the temperature sensitivity is −2 mV/°C at the measurement current and the temperature change during the transient is 50 °C, then the temperature-induced forward voltage change is 100 mV. The error term (power instability during the cooling) will be $I_M \cdot \Delta V_F(t) = 0.5$ mW, which is about 0.5 % of the total power change. Thus, at low ΔV_F and low I_F the change of the heating power is a secondary order effect only. In such a way, at cooling, we could realize a nearly perfect *power step*.

After switching off first, we can observe a sudden large voltage jump on the diode—the V_F forward voltage sinks from the V_H value belonging to the "hot" diode characteristics at $I_H = I_{drive} + I_{sense}$ to the lower V_{Fi} value belonging to the "hot" diode characteristics at the smaller $I_M = I_{sense}$ only (Fig. 4.17).

This change, referred to as electrical transient, can be many hundred millivolts for diodes, for diodes with higher series resistance[5] even more than 1 V. The voltage change can be slow (10–100 μs), as large amount of stored diffusion charge has to be removed from the forward biased pn-junction (see also [7]).

After this electric transient, we can capture the thermal transient, as the V_F forward voltage slowly increases from its V_{Fi} initial value when the operating point moves

[5] The effect of the series resistance is discussed in detail in Chaps. 2 and 6 and in papers [15, 16].

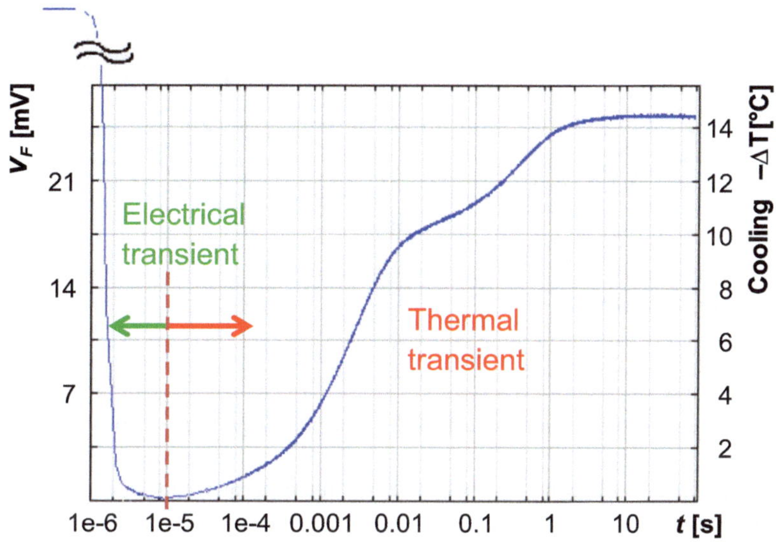

Fig. 4.18 Recorded transient of an actual diode, scaled in voltage (on the *left*) and in temperature (on the *right*)

from the "hot" to the "cold" diode characteristics, reaching its final, V_{Ff} value, always at I_M bias (Fig. 4.16).[6]

Selecting a measurement range of a few millivolts (shown in Fig. 4.16 as ΔV_F *range*), we always capture first the end of the electric transient at early times (intersection of ΔV_F *range* and the electric transient), and then the thermal transient.

In Fig. 4.18, a recorded transient of a diode is presented; the electric transient finishes at approximately 10 μs.

4.3.6.2 Selecting the Heating Power

As thermal testers are typically sensitive, for just analyzing the structure integrity, we need a power level that can ensure a few centigrade temperature elevation. For reliability testing, normal operational power has to be ensured. For accelerated reliability testing, higher than normal power has to be applied unto the device.

4.3.6.3 Selecting the Measurement Current

A common mistake in the literature is that the sensor current should be kept low in order to avoid self-heating of the component. In reality, the self-heating has only some minor influence in the *absolute* measurement technique, and can be nearly neglected

[6] Voltage and current notations correspond to the notations used the JESD51-5x series of standards, see also Chap. 6.

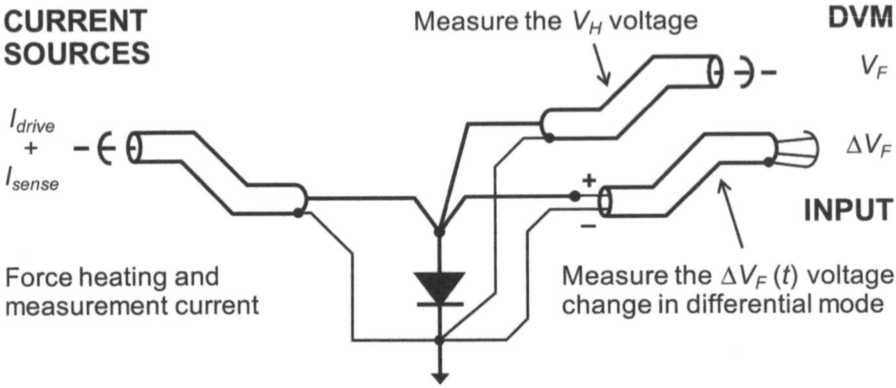

Fig. 4.19 Force/sense measurement of a diode. (Connection diagram of a thermal transient tester)

in the *differential* technique. In this case, we have great freedom for selecting the sensor current.

One of the advantages of selecting higher sensor currents is smaller and faster electric transients.

As we will see in Sect. 4.3.8, semiconductor diodes biased at different current level obey the

$$(V_{F1} - V_{F2}) = V_T \cdot \ln(I_1/I_2) + R_S \cdot (I_2 - I_1) \tag{4.14}$$

law. The first term of the equation describes the ideal diode equation of the junction. For calculating the change of the voltage caused by this term, we can use the approximate values as $V_T = kT/q \approx 26$ mV (at room temperature) and ln (10) $\approx$ 2.3. The result is that at a measurement current ten times lower than the heating current the voltage change during the electric transient is approximately 60 mV. Similarly, at a measurement current hundred times lower than the heating current the voltage change during the electric transient is approximately 120 mV.

The second term in Eq. (4.14) is practically dependent on the heating current only; the influence of the measurement current can be neglected.

Using forward voltage of diodes as temperature-sensitive parameter, we experience lower noise at higher measurement current. The physical background of the effect will be given later in Sect. 4.3.9.

In actual current jump measurements, a *four-wire* force/sense arrangement (as shown in Fig. 4.13) is preferred for correct power calculation, not adding the dissipation on the wires to the powering of the device.

Up-to-date testers can measure separately the high V_F value on a DVM unit for calculating the power and the small temperature-induced ΔV_F change on a sensitive differential measurement channel (Fig. 4.19).

An actual thermal measurement arrangement with the tester and the equipment realizing the thermal boundary is shown in Fig. 4.20.

As continuous heating measurements and pulsed measurements suffer from serious drawbacks in LED measurements and in such a way their use has limitations, we discuss these below only briefly.

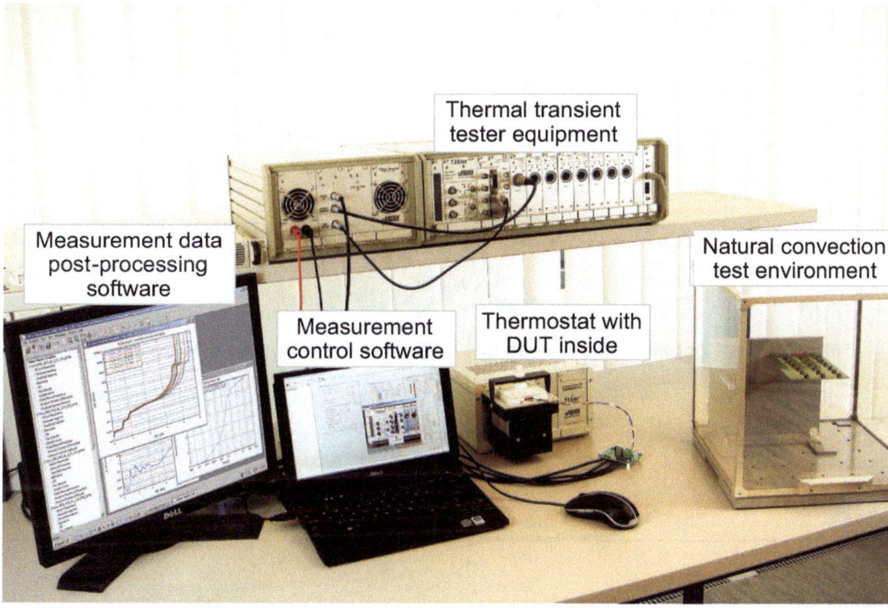

Fig. 4.20 Typical thermal measurement arrangement: test equipment, thermostat for K-factor calibration or to be used as a cold plate, natural convection test environment and measurement control, and data processing software running on a PC

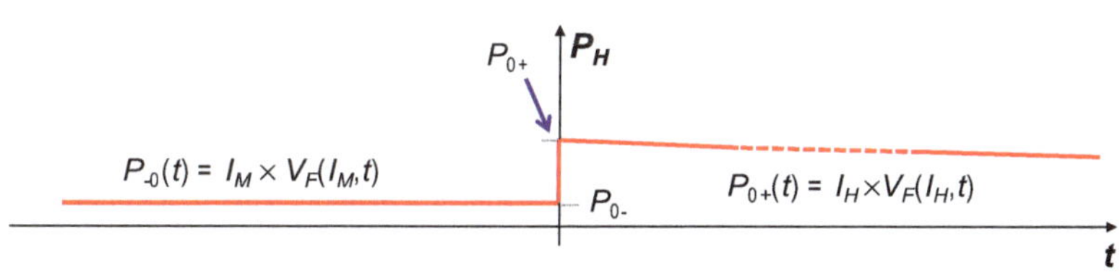

Fig. 4.21 Power change before and during the heating measurement on a diode

4.3.6.4 Continuous Heating Measurements

In these measurements, only the sensor current is applied to the device for an equalization time of appropriate duration. When the device reached "cold" steady state (the equilibrium at $P_{H1} = I_M V_{F1}$), the drive current will be switched on, too, and the transient recording of the heating will start.

Most considerations introduced for the cooling operate in the same way. However, here we experience a larger power change on the hot device during recording time, approximately the opposite forward voltage change, which we saw at cooling measurements occurs now at high current. The powering is far from an ideal step (Fig. 4.21).

This problem can be mathematically handled, but it is easier to use the cooling for the thermal measurements of two pin devices.

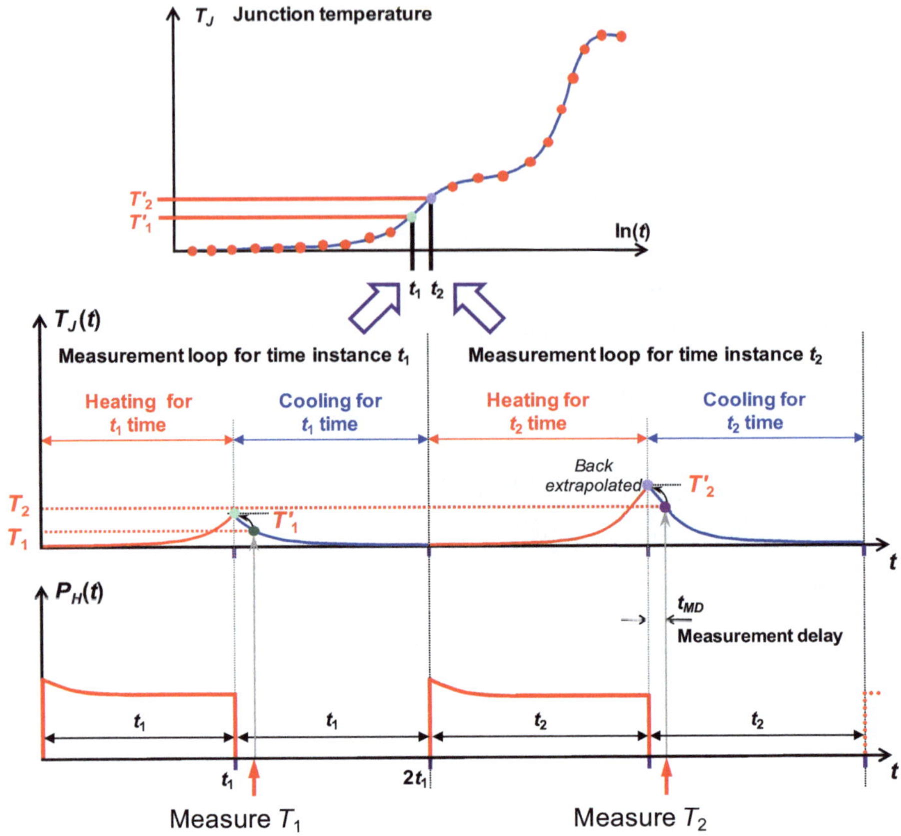

Fig. 4.22 The principle of the JEDEC JESD51-1 dynamic test method: **a** the series of heating pulses and corresponding temperature responses, **b** heating curve composed of individual temperature values measured at the end of the heating pulses

Both the cooling and the heating measurements described up to this point correspond to the transient extension of the JEDEC JESD51-1 "static" test method,[7] defined in [5] and Chap. 6.

4.3.6.5 Pulsed Measurements

The JEDEC JESD51-1 standard [5] also defines the so-called "dynamic" method. The measurement principle is illustrated in Fig. 4.22. In this method, the measurement

[7] This extension was first defined in details in the JEDEC JESD 51-14 standard [13]. This proposes a transient method for the measurement of the junction-to-case thermal resistance of power semiconductor device packages with a single heat flow path and with an exposed cooling surface—such as power LED packages. JEDEC JESD 51-14 also prefers the cooling mode transient measurement for diodes since the error introduced by the slightly changing power after switching is negligible in this case (as we also pointed out by a numerical example). The most recent LED thermal testing standard JESD51-51 [6] also recommends the cooling mode measurement if the task is to measure the real $Z_{th}(t)$ thermal impedance of an LED package.

is based on a series of high current pulses for heating and switching back to low current for temperature recording: the temperature value T_1 corresponding to t_1 time instance is measured such that a power pulse with a duration of t_1 is applied to the device under test. When t_1 time is elapsed, the power pulse is switched off (switching from I_H heating current to the I_M measurement current) and with a short delay t_{MD} (called the measurement delay) the value of the TSP (forward voltage) is measured and through the K-factor is converted to junction temperature. Then, the device under test needs to cool down, the cooling time must be at least as long as t_1. Then, the process is repeated for a longer pulse width t_2, etc.

The test result (referred to as heating curve) is composed from these responses to individual heating pulses of different length. This technique distorts each recorded point by an electric transient and the data correction problem (back extrapolation of the measured ith junction temperature at $t_i + t_{MD}$ time instance to the t_i time instance) is also present at every data point of the composed Z_{th} curve. Moreover, as we see in Sect. 6.5.1 of Chap. 6, these recorded points belong to different heating powers due to temperature-dependent LED efficiency, even if the applied electrical power is kept constant during the entire measurement process. Last but not least: the physical time needed for the measurement by the dynamic test method is by orders of magnitude longer than the length of the real cooling transient measured by the transient extension of the static test method.

As such, use of pulsing measurements is very restricted in LED applications: the JESD51-51 standard recommends this only for measuring steady-state thermal metrics such as total junction-to-ambient thermal resistance of a given test setup. The description of this pulsed method as "dynamic" is a bit misleading; all transient measurements of any style are by nature dynamic.

4.3.6.6 Measurement of Devices with Separated Heating and Sensing

Typically, complex systems or thermal test chips have separate leads for heating and sensing. They can also be measured in *current jump* mode with a sudden change in a current of the system. More typically, *voltage jump mode* is used for such devices.

Even simple devices widely used in SSL lighting as current drive stage, such as bipolar transistor or MOSFETs, have a sort of separation between heating and sensing. For example, for transistors, we can maintain a steady I_E emitter current and then switch between a high and a low V_{CB} collector-base voltage value for generating a power step. For sensing purposes, we can capture the voltage change at the emitter-base junction. Then, using different leads, we have good separation of powering and sensing. In such a way, the above mentioned problem of changing power during a heating type measurement is also eliminated.

This measurement mode offers the best resolution when mapping the fine details of the thermal structure, belonging to shortest time constants. Due to the separation, the electric transient is very fast, just a few microseconds.

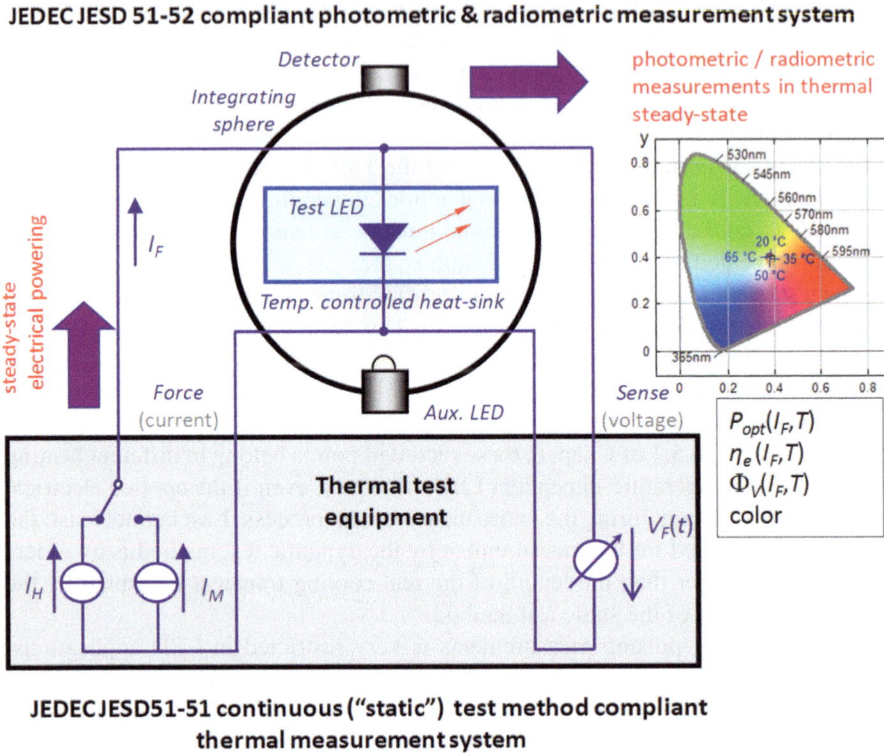

Fig. 4.23 Photometric/radiometric measurement of an LED device in a combined arrangement

4.3.7 Combined Thermal and Photometric/Radiometric Measurements

As mentioned before, thermal and photometric/radiometric measurements can be combined for faster and more reliable testing.

In these combined measurements, first the device is mounted on a temperature-stabilized plate, as anyway requested in optical measurements of high-power LEDs and then this plate is fixed to an integrating sphere equipped with detector, filters, spectrometers, etc.

The device is heated by appropriate heating current. When steady state is reached, emitted optical power, luminosity, color coordinates, etc. are provided by the optical test system (Fig. 4.23).

The forward voltage of the device at the heating current is measured by the thermal test equipment. The measured value is stored and used for power calculation. As hinted in Fig. 4.7 "power" can have more interpretations:

- First of all, we can calculate the *input electric power*, $P_{el} = I_F \cdot V_F$
- The radiometric measurement provides the *emitted optical power*, P_{opt}
- The heat remaining in the structure $P_H = P_{el} - P_{opt}$ is called *heating power*

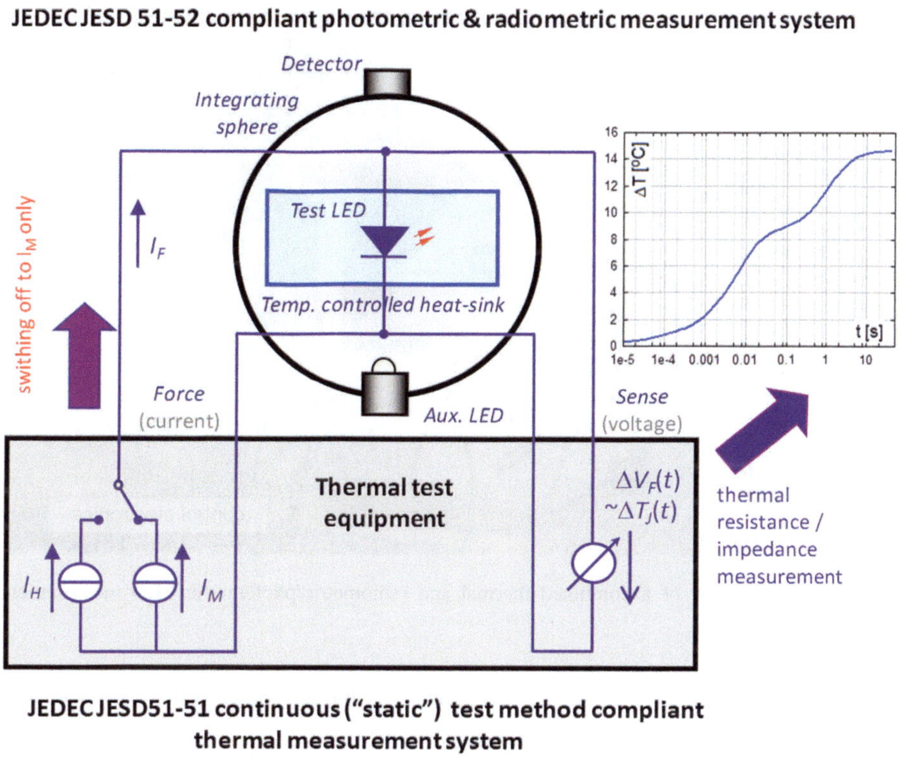

Fig. 4.24 Thermal transient measurement of an LED device in a combined arrangement

P_{opt} is also known as *total radiant flux* and is also denoted by Φ_e.

Finishing the test, we switch down the heating current to a low measurement current and capture the thermal transient (Fig. 4.24) in accordance with the continuous cooling transient measurement technique discussed in Sect. 4.3.6.1.

Now all thermal descriptive functions, which are normalized by the power, will have two different sets with different meaning. Even in the simplest steady-state case, we can define

$$R_{th-el} = \Delta T_J / P_{el} \tag{4.15}$$

as *electrical only* thermal resistance for calculating the thermal stress on the actual device at a certain electrical powering (not considering light emission), and

$$R_{th-real} = \Delta T_J / P_H = \Delta T_J / (P_{el} - P_{opt}) \tag{4.16}$$

as *real* thermal resistance for characterizing package quality, structural details, not influenced by the actual type (color, etc.) of the packaged LED device.

The JESD51-51 standard [6] prescribes measuring the real thermal resistance for package characterization.

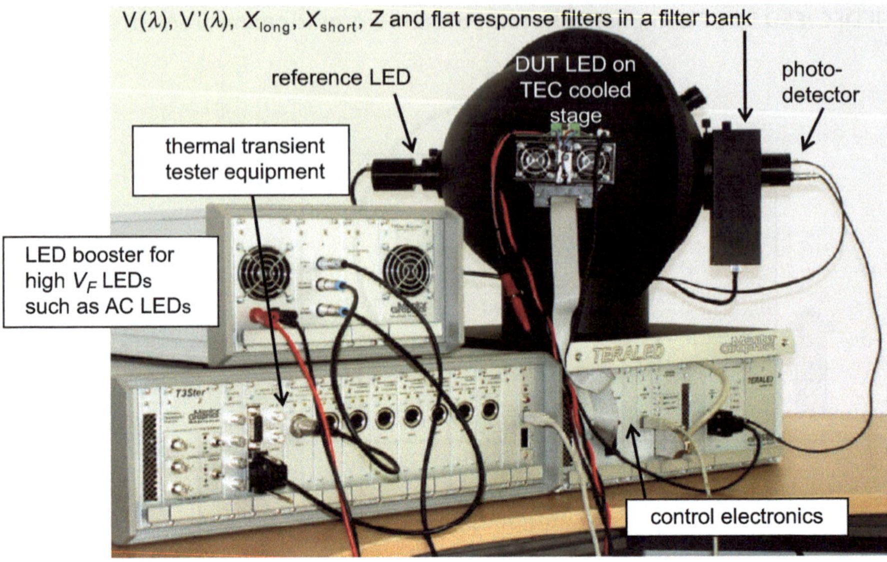

Fig. 4.25 Realization of a combined thermal and radiometric/photometric LED measurement station

Further analysis of $R_{th\text{-}el}$ and $R_{th\text{-}real}$ thermal characteristics will be provided in Sect. 4.4.

Figure 4.25 shows a practical realization of a combined thermal and optical LED measurements station. The device has to be mounted on the temperature-stabilized cold plate only once. The system automatically measures all temperature- and current level-dependent LED parameters, going through a user-defined set of temperatures and currents. After an initial step needed for the optical measurements, the automated measurement is carried out in three embedded loops. The complete measurement sequence is as follows:

1. Dark offset measurement with DUT LED off
2. Self-absorption correction with DUT LED off and reference LED on
3. New temperature is programmed (this can be changed in the slowest way) the system waits for temperature stabilization
    4. New current is programmed, the system waits for voltage and temperature stabilization
        - 5/a. radiometric and photometric parameters are measured by filters and/or a spectrometer
        - 5/b. switch from programmed forward current down to the measurement current: thermal parameters are measured by the thermal transient tester

    Next current is programmed

Next temperature is programmed

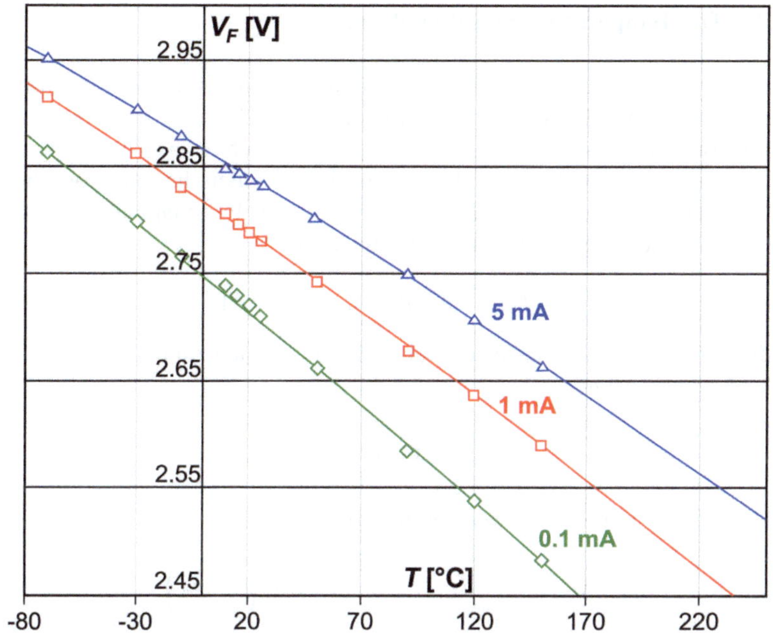

Fig. 4.26 Temperature dependence of the forward voltage, LED at different bias currents

In step 1, the DUT LED is switched completely off: the dark-level offset measurement of the photodetector (and/or spectrometer) takes place. In step 2, the measurements needed for the self-absorption correction of the DUT LED are performed using an auxiliary reference LED with known and stable parameters. (For further details on this, refer to Sect. 5.2 of Chap. 5).

In a postprocessing step, after making the measurements for every (I_F, T) operating point the real heating power is calculated from the electrical power $(I_F \cdot V_F)$ and the measured radiant flux $(P_{opt} = \Phi_e)$.

With these data, the real thermal resistance can be gained from which the real junction temperature is also calculated back: $T_J = T_{ref} + R_{th\text{-}real} \cdot P_H$, where T_{ref} is the cold plate temperature and $P_H = I_F \cdot V_F - P_{opt}$ denotes the actual heating power.

Plots of temperature- and current-dependent LED parameters are presented in Figs. 6.15 and 6.26 of Chap. 6.

4.3.8 The K-Factor Calibration Process

The precise calibration process is of high importance because this step influences the overall accuracy of the measurement. All other steps in thermal measurements are practically voltage and current measurements for which we have instruments of high precision and high time resolution. On the contrary, it is easy to perform a bad calibration and undermine the validity of thermal data.

4.3.8.1 The Temperature-Sensitive Parameter

The most often used parameter for temperature sensing is the forward voltage of a diode-like structure. Temperature sensitivity can be deduced from basic semiconductor theory. We present the case of diodes, which behave nearly ideally in a wide range of temperature and current, but the results are much the same for other device types, too. The forward current (I_F)–forward voltage (V_F) characteristics of an ideal diode follows the Shockley model of pn-junctions[8]

$$I_F = I_0[\exp(V_F/mV_T) - 1] \tag{4.17}$$

Introducing the R_S internal electrical series resistance (and neglecting the very small—I_0 term), we get

$$V_F = mV_T \ln \frac{I_F}{I_0} + I_F R_S \tag{4.18}$$

The V_T and I_0 parameters are temperature-dependent

$$V_T = \frac{kT}{q}, \quad I_0 \sim n_i^2 \sim T^3 \exp\left(\frac{-W_g}{kT}\right) \tag{4.19}$$

where n_i is the intrinsic concentration of moving electrons in the pure semiconductor, showing the above dependence on the T absolute temperature of the semiconductor material.

m is a device-specific constant called *ideality factor*. m is 1 in the normal operation mode of an ideal diode.

At very small currents, $m = 2$ because of recombination/generation effects in the depleted junction region. Theoretically, at very high currents $m = 2$, again because of ambipolar diffusion of carriers of both type. These regions generally overlap thus giving an m factor valid for a wide current range. The m factor can be calculated by more methods, the simplest is choosing two appropriate points from the characteristics:

$$V_{F2} - V_{F1} = mV_T \ln \frac{I_{F2}}{I_{F2}}, \quad m = \frac{V_T}{V_{F2} - V_{F1}} \ln \frac{I_{F2}}{I_{F1}} \tag{4.20}$$

In practical cases, Eq. (4.17) is suitable for the description of the diode behavior for a wide range of the forward current with an appropriate selection of the ideality factor m.

The temperature dependence of the forward voltage (temperature sensitivity) can be calculated from the above equations. An often-used form is

$$S_{VF} = \frac{dV_F}{dT} = \frac{\left(V_F - 3mV_T - \frac{Wg}{q}\right)}{T} \tag{4.21}$$

[8] See also Chap. 2 on the physical basics of LEDs.

Table 4.1 Major properties of some semiconductor materials

Properties	Si	Ge	GaAs
Atoms/cm^3	5.0×10^{22}	4.42×10^{22}	4.42×10^{22}
Density [g/cm^3]	2.32	5.32	5.32
Bandgap energy at 300 K [eV]	1.12	0.66	1.424
Intrinsic carrier concentration [cm^{-3}]	1.45×10^{10}	2.4×10^{13}	1.79×10^{6}
Lattice constant [nm]	0.543	0.564	0.565
Minority carrier lifetime [s]	2.5×10^{-3}	$\sim 10^{-3}$	$\sim 10^{-8}$
Optical phonon energy [eV]	0.063	0.037	0.035
Specific heat [kJ kg^{-1}K^{-1}]	0.7	0.31	0.35
Thermal conductivity at 300 K [Wm^{-1}K^{-1}]	150	60	46

Table 4.2 Bandgap energy of some color LEDs

Color	Wavelength [nm]	Bandgap energy [eV]
Deep red	700	1.77
Red	660	1.88
Orange	623	1.99
Amber	594	2.09
Green	523	2.37
Cyan (verde green)	501	2.47
Deep blue	470	2.64
Violet	410	3.02

where W_g is the bandgap of the semiconductor. Important material parameters (including bandgap energy) for some semiconductors used in electronics are listed in Table 4.1. Table 4.2 lists bandgap energies of LEDs emitting light in different spectral ranges. The tables show values at room temperature (300 K), at actual operating temperature they can be significantly different.

Figure 4.26 shows the V_F forward voltage of an LED at different bias current (0.1, 1, and 5 mA). The marks in the plot in the temperature range from 10 to 90 °C show measured V_F values of the actual diode. Using the measured values at 1 mA bias, we calculated the I_0, m and R_S values and produced analytic curves (solid lines) for other bias than 1 mA and an extrapolated temperature range (– 80 to 250 °C).

The curves, corresponding to the analytic equations above, are obviously nonlinear, but show only very small nonlinearity over a broad temperature range.

Previously, we used a symbolic form for the same temperature dependence of the temperature-sensitive parameter in Eq. (4.8)

$$V_F(T_J, I_M) \cong V_{F0}(I_M) + S_{VF} \cdot (T_J - T_{J0}) \tag{4.22}$$

Thermostats produce highly repeatable boundary conditions at various temperature levels. So they can be used for *temperature–voltage calibration*, which means *recording* V_F forward *voltage* (or other parameter) values *at different component temperatures*.

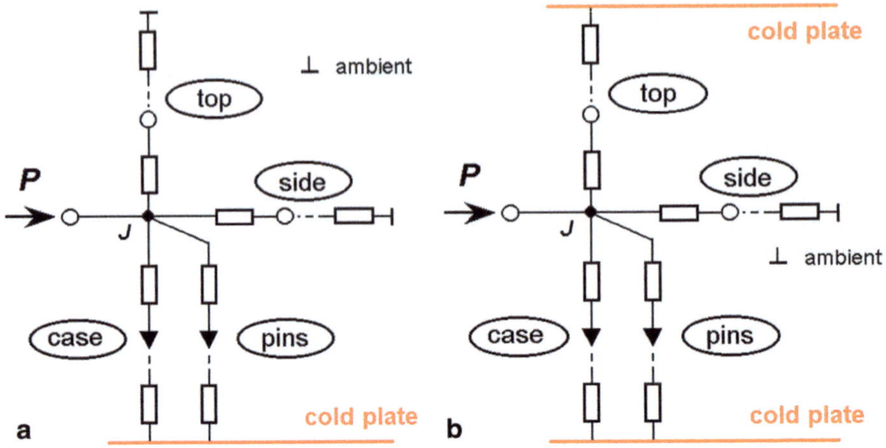

Fig. 4.27 Calibration of a component mounted on single and dual cold plate

We have to make a difference between *absolute* and *relative* calibration. The former means mapping the $V_F(T, I)$ function, the latter deriving the S_{VF} sensitivity parameter only.

Absolute calibration has theoretical limitations. For operating our component, we need some P power applied on it. Supposing all paths in the heat conduction path arrive at the same temperature-controlled surface named X, we experience a

$$T_J = P \cdot R_{thJX} + T_X \tag{4.23}$$

junction temperature. Even with the P power kept low, the junction temperature will differ from the controlled temperature.

Relative calibration has practical limitations only. As we saw in Sect. 4.3.5 for the differential measurement principle, we only need the S_{VF} sensitivity value (or its reciprocal, the K-factor).

4.3.8.2 Calibration on a Cold Plate

Components having a large cooling surface (case, tab) where most heat flows through can be measured and calibrated when mounted on a cold plate (Fig. 4.27). This is actually an easy process:

- Set the cold plate temperature to several values.
- Record the corresponding V_F forward voltage.

Despite its simplicity, we have to be aware of some important issues for doing the calibration correctly:

- As the S_{VF} sensitivity introduced in Eq. (4.8) depends on the operating point; always *apply the voltage and/or current* on the component to be calibrated, which *corresponds exactly to the transient measurement* circumstances.

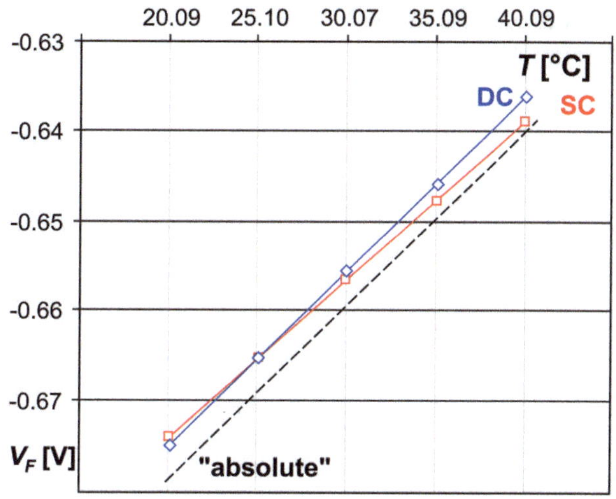

Fig. 4.28 Example on the absolute calibration error

- For cooling measurements, this should be the lower, for heating the higher powering of the two levels used in the differential method.
- For diode-type components, this practically means applying the sensor current only if cooling (or pulsed method) is the selected transient type. The same is true for other devices in current jump mode.
- If the power sinking capability of (liquid circulator-driven) cold plates is high, even the calibration at high power needed for heating can be carried out easily.

Figure 4.27 shows the simplified model of a surface mounted device (SMD) component mounted on a single and on a dual cold plate. The figure reveals that some surfaces of the package are terminated by the ambient (room temperature) rather than by the temperature-controlled plate. The junction temperature is "downscaled" by the appropriate thermal resistances in the thermal circuit; it does not follow exactly the setpoint of the plate.

Figure 4.28 shows the consequences. The dashed line labeled as "absolute" shows the $R_{thJX} = 0$ case ("chip not packaged, just attached to the cold plate"). The solid SC curve corresponds to the forward voltage of the packaged component on a single cold plate, while the solid DC curve corresponds to the dual cold plate case. The temperature axis is scaled in the measured *cold plate temperature*. V_F is negative (calibration of an anode grounded device).

If the cold plate is set to room temperature (25 °C in the figure), then the SC and DC curves coincide. The junction-to-ambient thermal resistance can be derived from the plot, using the $T_J = P \cdot R_{thJA} + T_A$ equation. At all other temperatures, the junction is between the ambient and the cold plate temperature and we underestimate the actual S_{VF} sensitivity.

Figure 4.27 also hints that if a large portion of the heat leaves through the pins (package with small tab and many pins), a good thermal contact is also needed between the wires feeding the package through the pins and the cold plate.

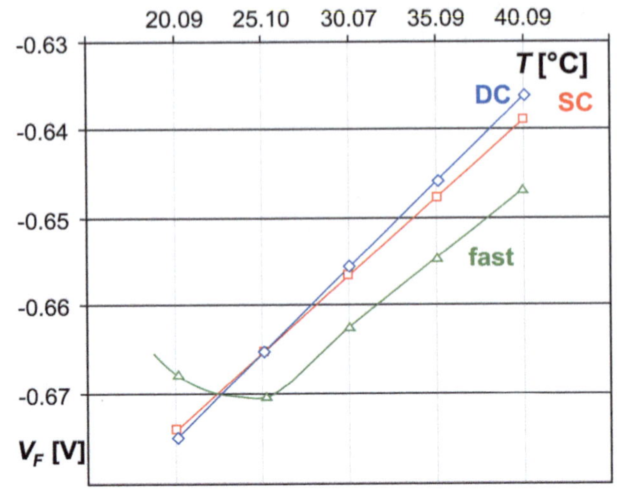

Fig. 4.29 Example on the dynamic calibration error

The calibration process can be manual or automatic, using calibration software. In both cases, the following steps have to be carried out:

- Select four or five temperature setpoints, spanning the whole temperature range of the future measurement.
- Apply the appropriate power on the device.
- Program the lowest temperature value and wait for t_1 time until the cold plate temperature stabilizes.
- Check the component voltage, wait for t_2 time until the voltage stabilizes.

The t_1 and t_2 waiting times are needed because the "thermal resistance" elements shown in Fig. 4.27 are complex impedances and their capacitive part expresses heat storage in different material sections (see Sect. 4.4.4). If t_2 is too short, we face the problem shown in Fig. 4.29. The component voltage (dynamic curve labeled "fast" in the figure) follows the cold plate with some delay, at the lowest temperature point, we also see the previous cooling from room temperature.

Even waiting for long t_2 times, we cannot completely get rid of the dynamic effect. On the other hand, even with long t_1 times, we experience small changes in the cold plate temperature also due to the control loop of the liquid circulator. The best practice to minimize these effects is:

- Record the actual temperature of the cold plate after a t_1 stabilization time instead of the programed setpoint temperature (as in Fig. 4.29).
- Omit the first voltage reading at lowest temperature.
- Try to read all voltage values at equidistant $t_1 + t_2$ times (in such a way the dynamic curve runs parallel to the SC or DC curve).

Example 1—K-factor calibration
A flat SMD package with exposed cooling surface (tab, heat slug) is calibrated in a single-side cold plate setup.

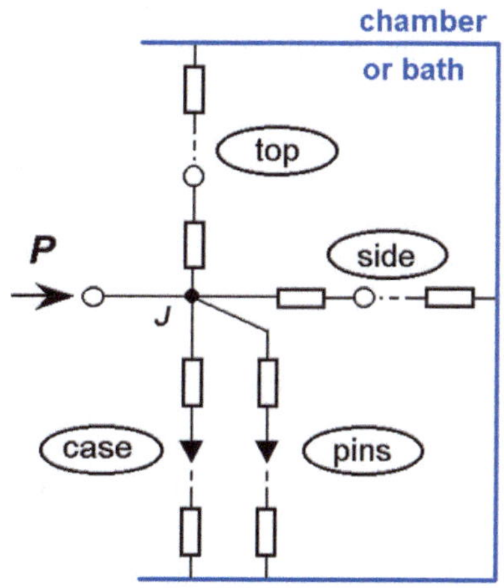

Fig. 4.30 Calibration of a component in closed chamber or bath

Suppose that in the model of Fig. 4.27 the junction-to-top resistance is approximately 10 K/W and the still air cooling on the top surface corresponds to approximately 20 K/W.

The parallel path composed by the junction-to-case and junction-to-pin resistance is 2 K/W, the spreading in the cold plate is below 0.1 K/W. The wires towards the pins are in good thermal contact with the cold plate. The junction-to-side path is approximately 200 K/W.

Without detailed calculations, we can see that nearly 10 % of the heat leaves towards the ambient, we underestimate the sensitivity by almost 10 % in this setup.

In the dual cold plate setup, we have some heat loss towards the sides only. However, as there is a thin air gap between the two temperature-stabilized metal plates, the air is practically of the same temperature as the cold plate. The side resistance is connected to the cold plate temperature rather than to room temperature.

4.3.8.3 Calibration in a Closed Chamber or Bath

In case all branches of the heat conduction path end at the same temperature, many problems of the previous section are automatically solved. This is the case when using a closed chamber with thermoelectric heating and cooling, or a liquid bath (Fig. 4.30).

Otherwise, the way of programming temperatures and reading voltages is much the same as in case of cold plate calibrations.

The t_1 and t_2 equalization periods are not fixed values. Instead, we should accept the state as thermal equilibrium if the changes of the chamber temperature remain below a given limit in a given time window, and after reaching this, also the measured voltage does not change more than a predefined limit for a similar time window.

For cold plates and chambers, the actual liquid or plate temperature can be different from that of the sensor used for controlling the system. In real life, 3–5 % repeatability can be expected for TSP calibration.

As mentioned before; all other steps in thermal tests need some sort of voltage and current measurement instruments, which are of high accuracy and stability. Recalibration of these is also only very rarely needed. Calibration thermostats need regular recalibration using stable reference devices.

4.3.9 *Electrical Noise Calculation*

The electrical noise in thermal measurements has two distinct sources: internal noise (shot noise in case of diodes) and external noise from the sensor current source.

For the latter, the noise calculation is very simple. From Eq. (4.17), the differential resistance of the diode is simply inversely proportional to the measurement current, $r_d = V_T/I_M$. As a consequence, the more shunted the external noise is, the higher is the measurement current.

The internal shot noise comes from the thermal equations, at a certain Δf bandwidth of the measurement it is:

$$i = \sqrt{4kT\Delta f/r_d} = \sqrt{4kT \cdot (qI_M/kT) \cdot \Delta f} = \sqrt{4qI_M\Delta f} \tag{4.24}$$

where k is Boltzmann's constant in J/K, T is the absolute temperature in K, and r_d is the dynamic resistance. The Δf is the bandwidth of the measurement, q is the charge of the electron, and I_M is the steady current (measurement) through the diode.

However, testers measure the device voltage, along with the noise voltage. As $u = i \cdot r_d$, the noise voltage shows square root decrease with higher measurement current.

Electrical noise calculation plays an important role in tester construction, but also has importance for a broader audience when selecting the measurement current for thermal transient measurements.

In practical cases, we select the lowest measurement current with already acceptable noise. For a broad category of silicon devices, 1 mA and for LEDs 10 mA is an acceptable value.

4.4 Evaluation of the Thermal Transients: Theory and Practice

The world of thermal transient measurements could be explored in a fully theoretical way. Now, we rather select some practical examples to show the highlights of the evaluation of thermal measurements characterizing power LEDs.

Example 2—Thermal impedance of a high-performance 1 W white LED
An Osram Dragon LED was selected for measurement. For characterizing the package, we selected a set of measurements at two different boundaries.

Fig. 4.31 Sketch of the two boundaries: with TIM layer ("wet" condition) and without TIM layer ("dry" condition)

Table 4.3 Measurement settings and powering data of LED of Example 2

Programed parameters	Measured parameters	Calculated parameters
$I_H = 0.700$ A	$V_H = 3.33$ V	$P_2 = 2.30$ W
$I_M = 0.010$ A	$V_{Fi} = 2.68$ V	$P_1 = 0.03$ W
$t_{max} = 20.0$ s	$t_{corr} = 20$ µs	$P_{el} = P_2 - P_1 = 2.27$ W
$T_{coldplate} = 25$°C	$P_{opt} = 0.55$ W	$\eta_e = P_{opt}/P_{el} = 0.24$
Sampling rate: 200/octave		$P_H = 1.75$ W

The difference in boundaries was realized by different thermal interface at the exposed cooling surface of the package as suggested in the JEDEC JESD 51-14 standard [13].

The measurement was carried out in an integrating sphere, in a combined thermal and radiometric/photometric arrangement (Fig. 4.25). During the measurement, effective cooling was assured by the cold plate mounted on the side of the sphere. The two boundary conditions were realized as (Fig. 4.31.):

- Mounting the LED on the dry plate surface (which always implies a thin insulating air gap between the faces); and then
- Mounting the same LED on the plate wetted by thermal grease.

For calibrating the *temperature-sensitive parameter,* we put the device in a closed thermostated chamber and recorded the forward voltage at different temperatures with $I_M = 10$ mA applied (Fig. 4.31).

The measured temperature sensitivity of the forward voltage was approximately $S_{VF} = -1.21$ mV/°C.

Then, we placed the LED on the cold plate of the integrating sphere and connected the device to the appropriate connectors of the *thermal transient tester.*

A few trial measurements were carried out to find the measurement time needed to reach steady state and the proper sampling rate of the tester. The power level was chosen to achieve several degrees Celsius of temperature change. The selected parameters are listed in the first column of Table 4.3.

Cooling measurements were carried out. The power was applied on the diode by switching on a higher I_{drive} current, as an addition to the I_M measurement current, which had been already used in the calibration step. The LED was left to reach its hot thermal steady state powered by the $I_H = I_{drive} + I_M$ heating current.

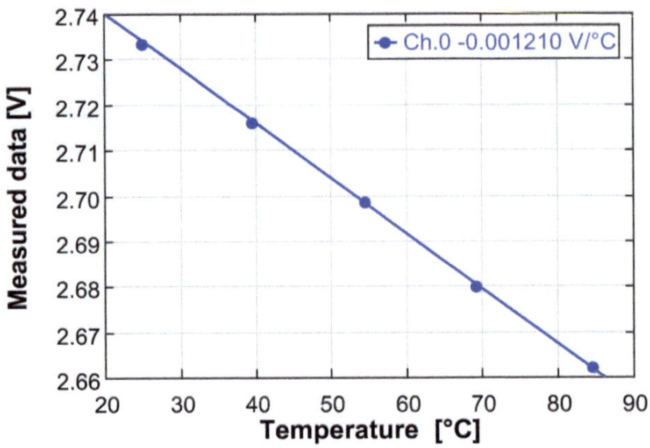

Fig. 4.32 Calibration curve of a power LED at 10 mA

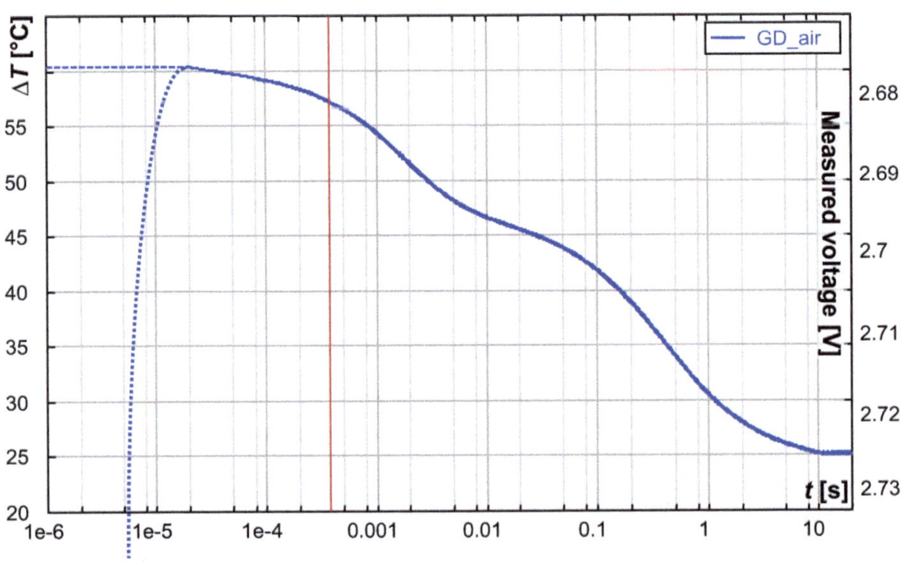

Fig. 4.33 Raw cooling transient: 1 μs to 20s with initial correction

At the start of the cooling, the current was switched from I_H to I_M only. Before switching off, $V_{FH} = 3.33$ V was measured on the hot diode at 700 mA bias. Immediately after switching off, the forward voltage dropped to 2.68 V at 10 mA bias. So the *electric* power step was $P_{el} = (3.33 \cdot 0.7)\text{W} - (2.68 \cdot 0.01)\text{W} = 2.27$ W.

The radiometric measurement yielded $P_{opt} = 0.55$ W emitted optical power resulting in a radiant efficiency of $\eta_e = P_{opt}/P_{el} = 0.24$ and $P_H = 1.75$ W heating power.

Figures 4.32 and 4.33 show the recorded transient as change of voltage and temperature.

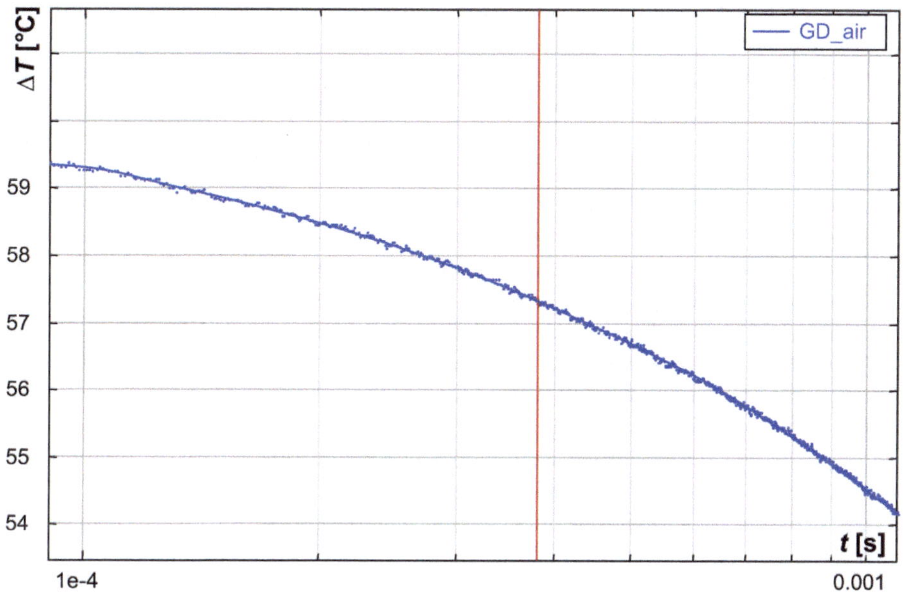

Fig. 4.34 Excerpt of the raw cooling transient, 100 μs to 1 ms

In the 1 to 20 μs interval, we see the electric transient of the diode as the voltage sinks from 3.33 towards 2.68 V. (Reference directions of the picture have been selected in a way that moving up in the plot represents diminishing voltage—so cooling is shown in the usual way).

The electric transient of the device finishes at 20 μs, after this point, we see the temperature-induced voltage change, the real thermal transient.

The excerpt in Fig. 4.33 demonstrates that for subsequent postprocessing a good-quality transient record is needed. The precise sampling at as high as 200 samples/octave rate produces many thousand points. (An octave is a 1:2 time span, such as 100 to 200 μs, etc.). Moving average over the samples serves good noise suppression.

4.4.1 Temperature Change Plots

Figure 4.34 presents the relative temperature change of our Dragon LED at the two selected boundary conditions. Here, and in all subsequent plots, GD_air denotes the curve belonging to a dry condition with the inherent air gap when no thermal interface material (TIM) is applied and GD_grease denotes the curve belonging to a wet condition with applied grease on the package cooling surface.

A cooling (or heating) plot already yields a lot of useful information. Usually, the *time axis is logarithmic*, this helps analyze the thermal behavior of the packaged device, we see the *early time details* that *characterize the chip, package, and pin region* and later we can identify the cooling mount.

However, this information is very specific, it describes the component behavior on cold plate, at $P_{el} = 2.27$ W power step only. We want to find descriptive functions that predict component behavior at *different boundary conditions, different power waveform and magnitude*, etc.

Fitting the cooling curves at their hottest point in Fig. 4.34, we find that the cooling is *not influenced by the actual boundary condition* until 20 ms and all curves coincide perfectly. This can be easily explained stating that until 20 ms the thermal changes occur inside the package; we still did not reach the outer thermal interface.

A beautiful world of powerful description tools can be entered assuming that *the behavior of our thermal systems is linear*. Luckily, the material parameters of the components show only a small dependence on temperature. However, e.g. LEDs' radiant efficiency shows a stronger change.

In silicon devices, linearity assumption is typically valid over a 50 °C temperature rise. A study on handling nonlinearity is presented in [17]. As shown in Fig. 6.32 of Chap. 6, assuming temperature-independent material properties for LEDs in a temperature range of 60 °C is also valid for power LEDs. Similar experimental evidence for LEDs can be found in the technical literature too (see, e.g., Fig. 11 of [18]).

4.4.2 Z_{th} Curves

The first evident step of generalizing our temperature measurement result can be done by *normalizing* it by the applied power. This normalized temperature transient is the Z_{th} curve also known as *thermal impedance curve.*[9]

In the world of silicon (single-energy transport) devices, all input electric energy is converted into heating power: $\Delta P_H = \Delta P_{el}$, thus: $Z_{th}(t) = \Delta T_J(t)/\Delta P_{el}$.

The power step can be negative (cooling) resulting in a negative change of the junction temperature as well or the power step and the junction temperature change can be positive (heating); in both cases, the thermal impedance curve will be a monotonically increasing function, therefore we usually consider the direction of the heating as positive in Z_{th} curves.

The curves shown in Figs. 4.36 and 4.37 come from dividing the measured cooling curves (Fig. 4.35) by $P_{el} = -2.27$ W.

In silicon devices, the approximate junction temperature transient for any power step can be easily gained from the Z_{th} curve, simply multiplying each time point by the actual power.

This process has a small error as linearity is not perfect. At higher temperature, the cooling is generally better, turbulent convection is more effective, radiation grows quickly. If we apply on our system an actual power, which is higher than the one used

[9] In electronics, the impedance is interpreted in the frequency domain, not in the time domain as a step-response function. In Sect. 6.1.4.1 of Chap. 6, the thermal resistance concept is generalized to thermal impedances.

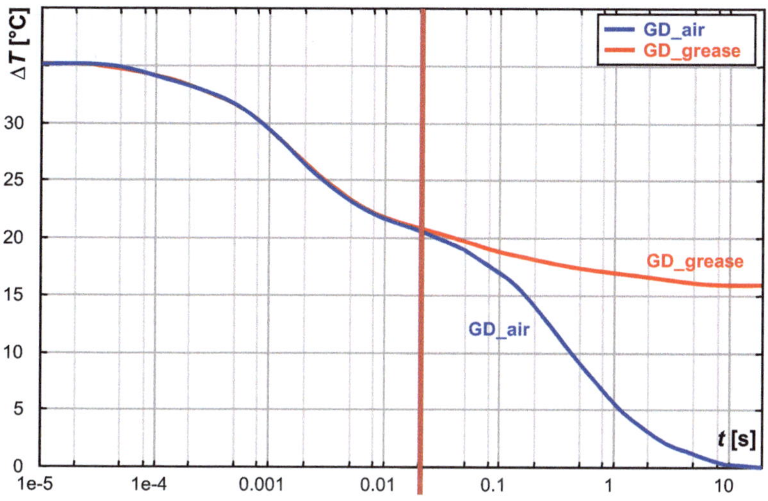

Fig. 4.35 Smoothed response curves matched at hot point

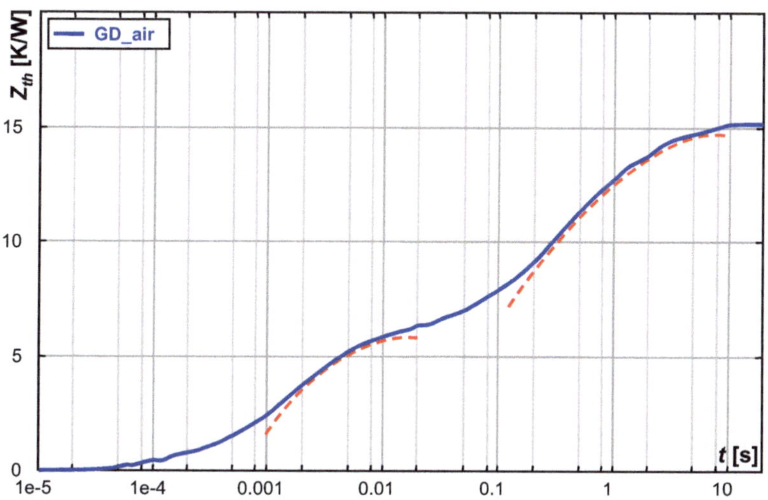

Fig. 4.36 Z_{th} curve (without correction of emitted optical power), LED component on cold plate, dry surface. Two possible exponential components shown

during the Z_{th} measurement, the actual temperature elevation will be lower than the calculated. In such a way, the error is made on the safe side.

From the beginning, the Z_{th} curve was used for analyzing device structure. As we can see in Fig. 4.36 the Z_{th} curve is "bumpy," we see the heating of structural elements (chip, submount, heat spreader) superposed. The height and position of a bump can be used to check the structural health of the device or for identifying failures.

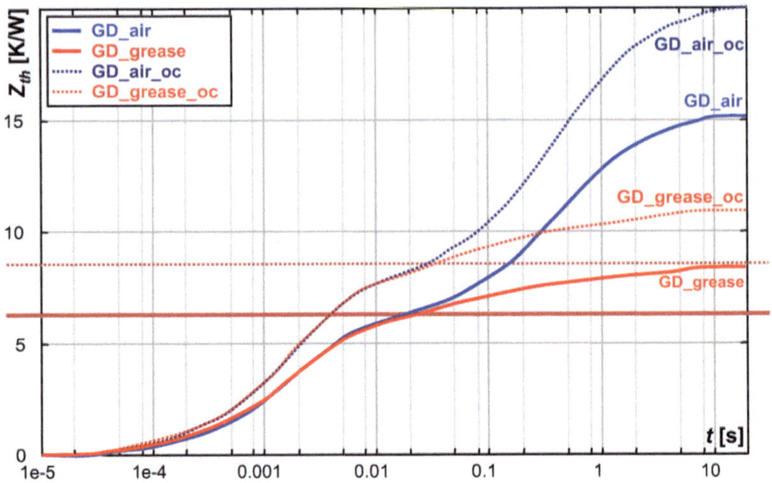

Fig. 4.37 Z_{th} curves with and without correction for the emitted optical power, LED measured twice with different case-to-cold plate thermal interfaces

In Fig. 4.37, we compare more Z_{th} curves constructed from the cooling curves of Fig. 4.35. Dividing the cooling curves by P_{el}, we get the GD_air and GD_grease "electrical only" $Z_{th\text{-}el}$ curves of Fig. 4.37.

Using $P_H = P_{el} - P_{opt}$, we get the GD_air_oc and GD_grease_oc "real" $Z_{th\text{-}real}$ curves, where the postfix _oc stands for "optical power corrected".

Putting the question which one is the "good" Z_{th} curve, the answer is that all of them are correct but they have different information content and may serve different applications.

Repeating the Z_{th} measurements at different power level or ambient temperature, we will see that the real $Z_{th\text{-}real}$ curves fit nearly perfectly; they reflect only the possible temperature dependence of material parameters and the cooling mount characteristics.

As exposed many times in this book, the $\eta_e = P_{opt}/P_{el}$ radiant efficiency depends on the LED's forward current and junction temperature. In the range of interest, the efficiency is higher at lower currents and lower temperatures.

According to the above, $Z_{th\text{-}el}$ is valid only for one current and one specific ambient temperature. Therefore, publishing it in product data sheets is correct only along with these conditions.

Both curve types can be relevant in proper context. In the data sheet of an LED module or luminaire where practically one optimal driving current is specified, and all other use is limited to dimming at smaller current; $Z_{th\text{-}el}$ gives answer to questions like "after switching on or off the unit what will be the junction temperature at a certain point in time."

However, in an R&D project where LED packages or LED light output properties have to be compared, $Z_{th\text{-}real}$ can be used for adequate comparison. Also, this is the thermal metric to be used in thermal simulation—allowing modeling of application scenarios with any electrical and thermal conditions.[10] As presented later in this section, postprocessing $Z_{th\text{-}real}$ can reveal structural details and can identify failure locations in a correct way.

Looking at the GD_air and GD_grease "electrical only" $Z_{th\text{-}el}$ curves in Fig. 4.36, we can observe in the plot that until the point we proceed in the internal structures of the heat conduction path a bit more than 6.3 K/W (reached at 20 ms), the Z_{th} curves coincide. Leaving the cooling slug, we step into the thermal interface region. The measured curves show 15 K/W total change with the LED on dry surface and 8.2 K/W on the surface with thermal grease applied.

This way, we identified the "electrical only" junction-to-case thermal resistance as $R_{thJC\text{-}el} = 6.3$ K/W.

The "effective" junction-to-ambient thermal resistance is 8.2 and 15 K/W for the two boundaries, interpreting "ambient" as the end of the cold plate–liquid-based thermostat system.

The corresponding "real" quantities are 8.3, 11, and 20 K/W, respectively. These bigger numbers for the real impedance or resistance values underline the importance of proper standardization. The requirement of reporting these values on product data sheet does not leave any room for possible misleading of inexperienced end-users.

Figure 4.36 already proves that a drastic change in the structure provokes a visible change in the transient thermal behavior, but quantitative statements are limited on one specific point only. The Z_{th} curves can be used as starting point for more "views" of the same measurement, which provide much clearer picture of the device and its environment. The most refined way for structure identification will be presented in Sect. 4.4.4.

A basic statement of linear system theory is that knowing the system response to a short pulse (Dirac-δ pulse) or to unit step (Fig. 4.36 or 4.37) we know all possible transient responses. The transient change caused by any excitation of any waveform can be easily calculated using the so-called *convolution integrals* [20].

A closely related problem is studying system response on periodic excitations at different frequencies. The result is that time domain transients can be converted to frequency domain response using the *Fourier transformation*.

Descriptive functions of such nature are now common in data sheets regarding electric parameters, and are more and more important for the thermal parameters.

4.4.3 Thermal Time Constants

A further way of representing the thermal system is highlighted in Fig. 4.36. The Z_{th} curves are of "bumpy" nature. This is natural, at heating we can observe how we first heat up the chip, then internal package elements, followed by the package body, the board, etc.

---

[10] See Sect. 6.5.2.2 of Chap. 6 for an overview of LED multidomain modeling.

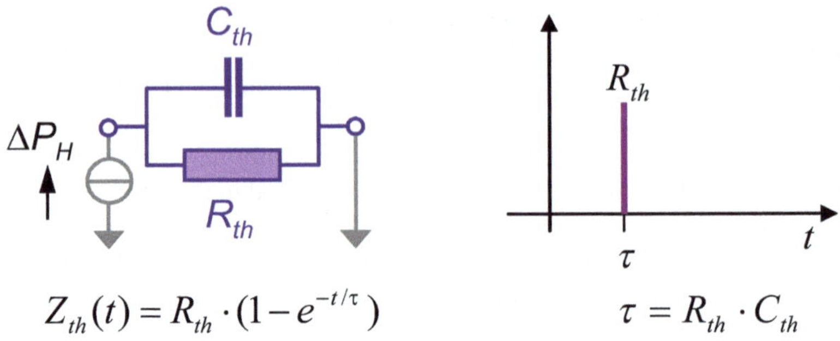

Fig. 4.38 The simplest dynamic thermal model: a parallel thermal resistance and thermal capacitance and its discrete time constant representation. (Source: [13] © JEDEC, reproduced with permission)

Fig. 4.39 Time response of a single RC stage to a step function excitation with its magnitude and time constant shown

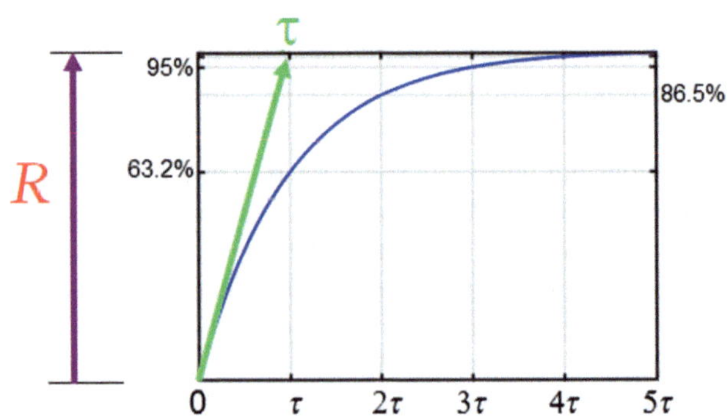

Such a curve can always be interpreted as a sum of exponential components. This exponential composition automatically yields a simple one-dimensional dynamic compact model, a series or chain of parallel connected thermal resistance–capacitance pairs.

In the simplest case, our system can be represented by a single thermal resistance expressing heat conduction and a parallel thermal capacitance expressing energy storage (Fig. 4.38).

Applying a step-wise power change to this equivalent network, the temperature quickly grows until $t = R_{th} \cdot C_{th}$ time, then gradually stabilizes at the $T = P_H \cdot R_{th}$ value following the $T(t) = P_H \cdot R_{th} \cdot (1 - e^{-t/\tau})$ time function (Fig. 4.39). (In the analogous electric network, power is replaced by current, temperature by voltage.) If 1 W power is applied, we get the $Z_{th}(t)$ curve.

Composing now a Z_{th} curve like the one in Fig. 4.36, we have to sum up such exponential heating curves:

$$T(t) = \sum_{i=1}^{n} P_H \cdot R_{thi} \cdot (1 - e^{-t/\tau_i}) \tag{4.25}$$

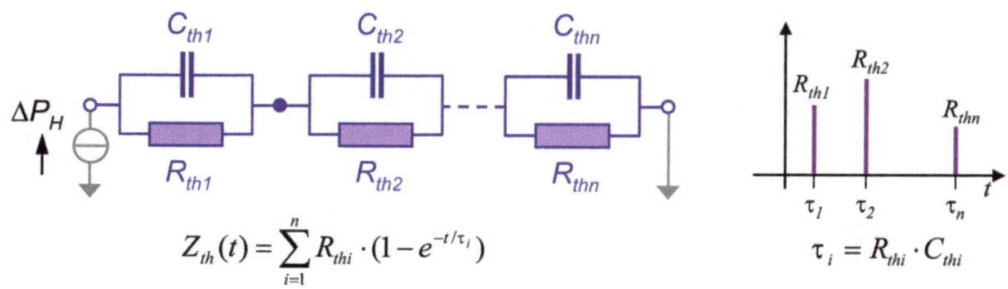

Fig. 4.40 A popular behavioral dynamic thermal model: a chain of parallel thermal resistance and capacitance stages its discrete time constant representation. (Source: [13] © reproduced with permission)

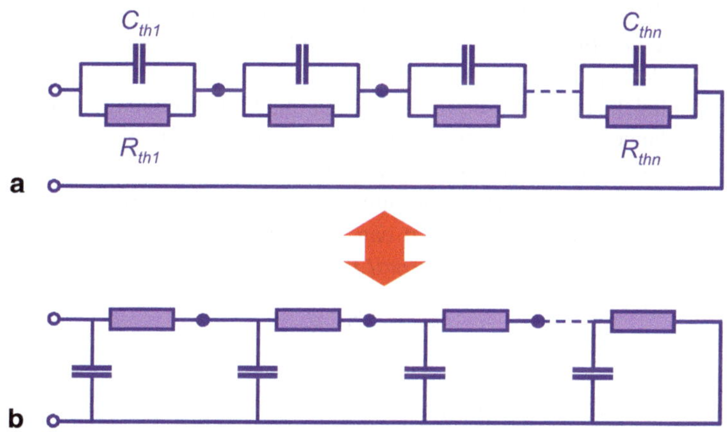

Fig. 4.41 **a** *FOSTER* type and **b** *CAUER* type representations of a driving point[12] thermal impedance. (Source: [13] © JEDEC, reproduced with permission)

The addition of temperatures corresponds to model network composed as a chain of parallel RC stages as shown in Fig. 4.40: the same power ("current") flows along the chain, and the total temperature ("voltage") is calculated as the sum of the components. At 1 W power, we get the $Z_{th}(t)$ curve again.

We could quantitatively describe the chain model with a large table of R_i and $C_i$pairs. For the visual representation, it is practical to give the R_{thi} and $\tau_i = R_{thi} \cdot C_{thi}$ values instead (see on the right in Fig. 4.40), because R_{thi}-s give direct information on the magnitude of the given component, and τ_i-s on the place of the "bump" along the time axis.

The network model shown in Fig. 4.40 is called the *FOSTER model* of the impedance. As this model perfectly describes the time response of the thermal impedance, it can be considered as a black box model of the thermal impedance.

It would be misleading to associate these thermal resistances and capacitances to the different physical regions of a heat conduction path structure. The FOSTER model is unsuitable for this since it contains node-to-node thermal capacitances representing no physical reality. An equivalent model exists as well for the RC one-ports: the so-called *CAUER network*. This model is a ladder network, shown in Fig. 4.41b.

This model is excellently matched to the idea of associate the circuit elements with physical regions. This behavior will be the base of the heat flow path identification by means of *structure functions* as will be shown later.

4.4.4 Structure Functions

In the late 1990s, the way of interpreting thermal measurement results radically changed. A plausible modification of the time constant representation caused a real revolution that enabled direct investigation of the physical structures, reverse engineering, failure analysis (FA)—instead of just viewing the thermal changes as time behavior of a "black box" system. This change was due to the introduction of the so-called *structure functions*, introduced in thermal testing of packaged semiconductor devices by Székely [21].

We concluded the previous section by stating that CAUER equivalent models of the thermal impedances can be generated from the time constant spectra. The practical problem with such a network model is that 150..200 individual thermal resistance and thermal capacitance values cannot be interpreted. Therefore, after the introduction of further two simple definitions, we can represent the CAUER equivalent model graphically. Thus, the *cumulative thermal resistance* is defined as

$$R_{th\Sigma} = \sum_i R_{thi} \tag{4.26}$$

and the cumulative thermal capacitance as[11]

$$C_{th\Sigma} = \sum_i C_{thi} \tag{4.27}$$

In other words: starting from the driving point (the junction), we cumulate (sum) the partial thermal resistance and thermal capacitance values for all subsequent heat flow path sections. If we interpret the cumulative thermal capacitance as function of the cumulative thermal resistance, we obtain the so-called *cumulative structure function*, often abbreviated as *CSF*

$$CSF = C_{th\Sigma}(R_{th\Sigma}) \tag{4.28}$$

This formal definition is illustrated in Fig. 4.42. The origin of the function corresponds to the *junction*. As all thermal capacitance values are positive, the cumulative

[11] The concept of *cumulative resistance* and *cumulative capacitance* and the concept of the $C_\Sigma(R_\Sigma)$ function were first introduced by Protonotarios and Wing in their fundamental papers about the theory of nonuniform (electrical) RC lines. In the first part, Protonotarios, E. N. [22] they used this function for simplifying the telegraphists' equations when used in the synthesis of nonuniform RC lines.

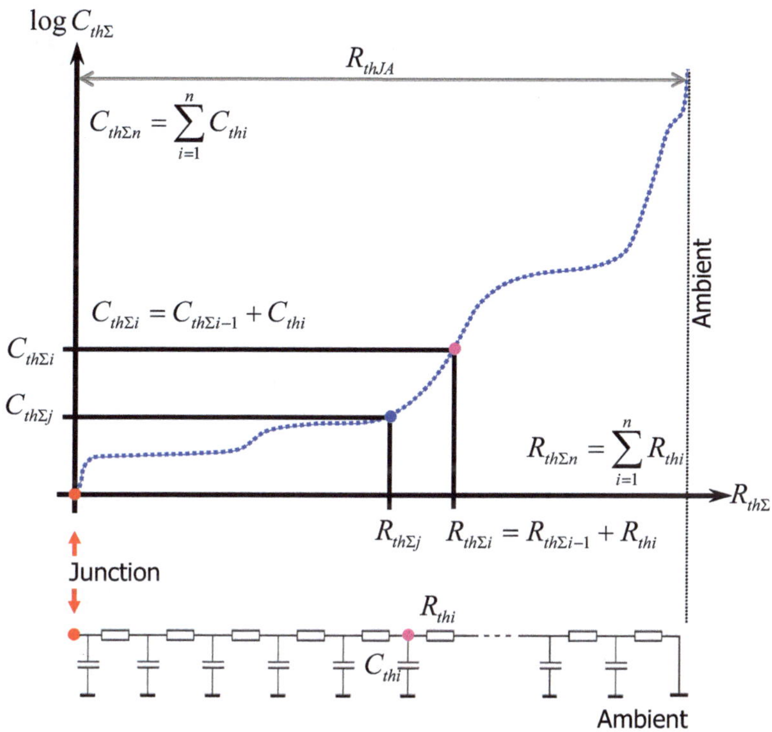

Fig. 4.42 Cumulative structure function: the graphic representation of the thermal RC equivalent of the system

structure function should be a strictly monotonically increasing function. The heat conduction path ends in the *ambient*, which is characterized by infinite heat sinking capacity, therefore the cumulative thermal capacitance must tend to infinity. This means that the cumulative structure function should end with a singularity (at the location corresponding to the ambient). As a further consequence, the distance of the singularity and the origin is equal to R_{thJA}—the junction-to-ambient thermal resistance.

A simple physical model helps understand the meaning of the cumulative structure function. If we have a heat flow through a small portion of material, we shall experience two effects. As shown in Fig. 4.43, there will be a temperature drop *between* two (isothermal) surfaces of the material (assuming an adiabatic condition at the other four faces of the cuboid).

If the material has λ thermal conductivity and P power flows through the a and b surfaces, they will have T_a and T_b temperatures, measured from the ambient. We can say that if the slice has a small dx length and a surface with A cross-sectional area then the temperature drop between the two sides can be expressed as

$$T_a - T_b = P\left(\frac{1}{\lambda}\frac{dx}{A}\right) \tag{4.29}$$

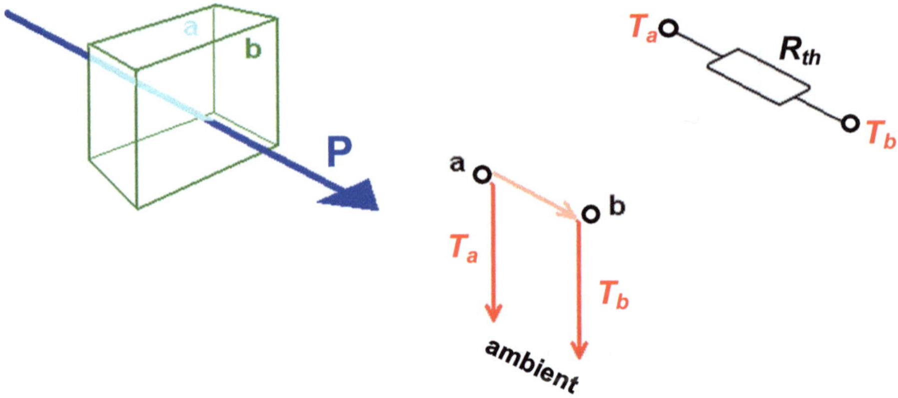

Fig. 4.43 Heat flow through a material slice

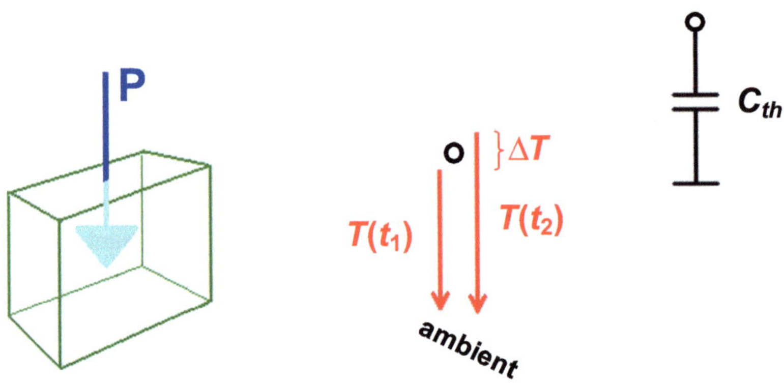

Fig. 4.44 Heat flow into a material slice

where the expression in the bracket on the right-hand side is the R_{th} thermal resistance between the a and b points corresponding to the two surfaces:

$$R_{th} = \left(\frac{1}{\lambda}\frac{dx}{A}\right) \tag{4.30}$$

On the other hand, the same material slice can store thermal energy (see Fig. 4.44). If we have a heat flow *into* the material, then in a short $dt = t_2 - t_1$ time interval the energy change is

$$dQ = Pdt = C_{th}(T_2 - T_1) \tag{4.31}$$

if $T_1 = T(t_1)$ is the temperature of the material at t_1 time and $T_2 = T(t_2)$ is the temperature of the material at t_2 time. We can represent the slice by a single point now

---

[12] Driving point means that heating and measuring the temperature response takes place at the same physical location. See also Sect. 6.1.4.2 of Chap. 6.

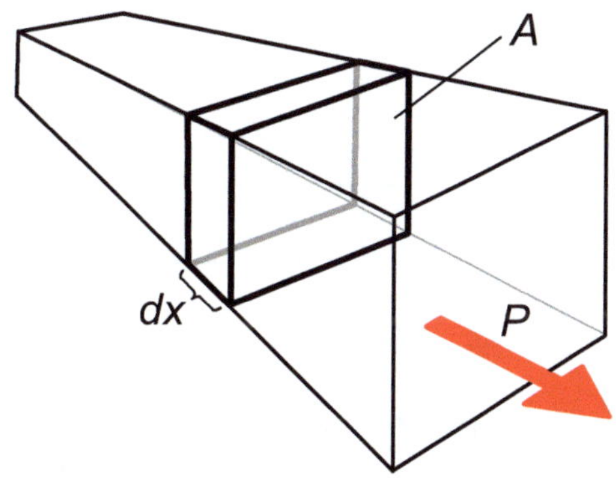

Fig. 4.45 A section of the heat conduction path

for simplest approach. As T_1 and T_2 temperatures are again measured from the ambient, Eq. (4.31) defines a C_{th} thermal capacitance between a point representing the material portion and the ambient. The value of this C_{th} thermal capacitance can also be expressed trough material parameters

$$C_{th} = c \cdot m = c \cdot \rho \cdot dx \cdot A \ or \ C_{th} = c_V \cdot V = c_V \cdot dx \cdot A \tag{4.32}$$

where c denotes specific heat, m denotes mass, ρ is the density, c_V denotes volumetric (specific) heat capacitance, and V denotes volume.

The cumulative structure function is an excellent graphic tool to analyze the physical structure.

In *low-gradient sections*, a small amount of material having low capacitance causes large change in thermal resistance. These regions have *low thermal conductivity* or *small cross-sectional area. Steep sections* correspond to material regions of *high thermal conductivity* or *large cross-sectional area.* Sudden breaks of the slope belong to material or geometry changes.

In such a way, thermal resistance and capacitance values, geometrical dimensions, heat transfer coefficients, and material parameters can be directly read from cumulative structure functions.

In a realistic environment, the heat flow can have various shapes—longitudinal along a beam, radial in a board, conical in a heat sink holding a smaller package. In most cases, we can make a "proper" slicing of the material, on the isothermal surfaces, perpendicular to the direction of the flow. These slices must be narrow, but not always of very small cross-sectional area (Fig. 4.45) and the surfaces are usually not planes.

It is sometimes easier to identify the interface between the sections using the derivative of the cumulative curve: the differential structure function. Here, peaks correspond to regions of high thermal conductivity like the chip or a heat sink and valleys show regions of low thermal conductivity like die attach or air. Interface surfaces are represented as inflexion points between peaks and valleys.

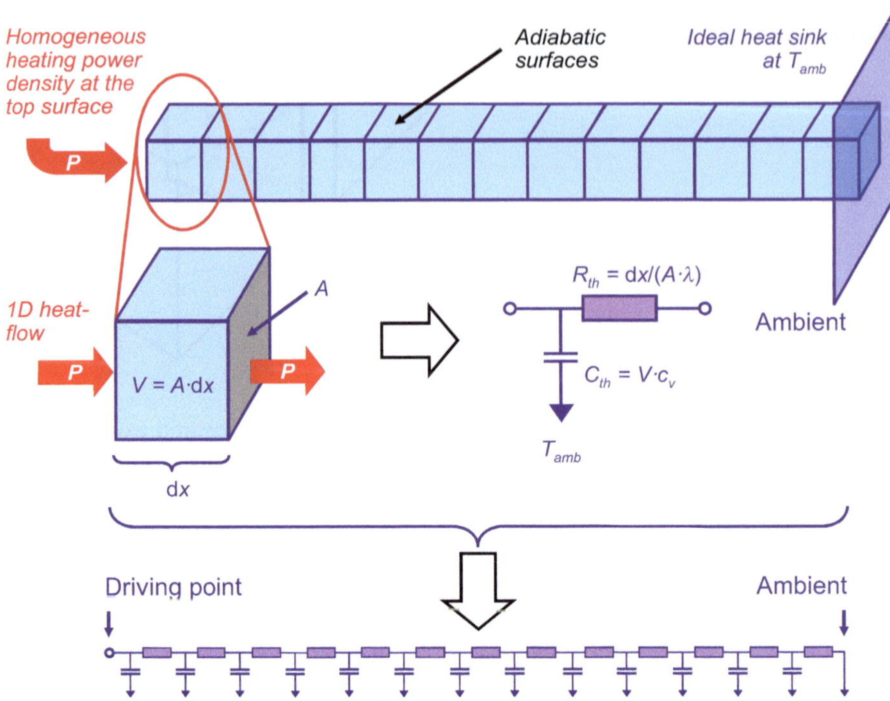

Fig. 4.46 The RC model of a narrow slice of the heat conduction path with perfect one-dimensional heat flow and the CAUER-type network model of the thermal impedance of the entire heat flow path

From Eq. (4.31) and Eq. (4.32), we can say:

$$DSF = \frac{dC_{th\Sigma}}{dR_{th\Sigma}} = c_V \cdot dx \cdot A \cdot \left(\frac{1}{\lambda}\frac{dx}{A}\right)^{-1} = c_V \cdot \lambda \cdot A^2 \tag{4.33}$$

This is called *differential structure function* (frequently abbreviated as *DSF*), which also yields information on the cross-sectional area along the heat conduction path.

As summary, let us consider a homogeneous rod with thermal boundary conditions as indicated in Fig. 4.46. This rod can be considered as series of infinitesimally small material sections as discussed above. Consequently, the network model of its thermal impedance would also be a series connection of the single RC stages as shown in Fig. 4.46. Thus, with this slicing along the heat conduction path, we create a ladder of lateral thermal resistances between two thermal nodes and thermal capacitances between a node and the ambient.

Since we assumed homogeneity, the ratio of the elementary thermal capacitances and thermal resistances in the network model shown in Fig. 4.46 would be constant. This means that the cumulative structure function of the rod would be a straight line—its slope is determined by the C_{th}/R_{th} ratio of the network model and its differential structure function would be a constant pn-junction C_{th}/R_{th} ratio of the element values, as shown in Fig. 4.47.

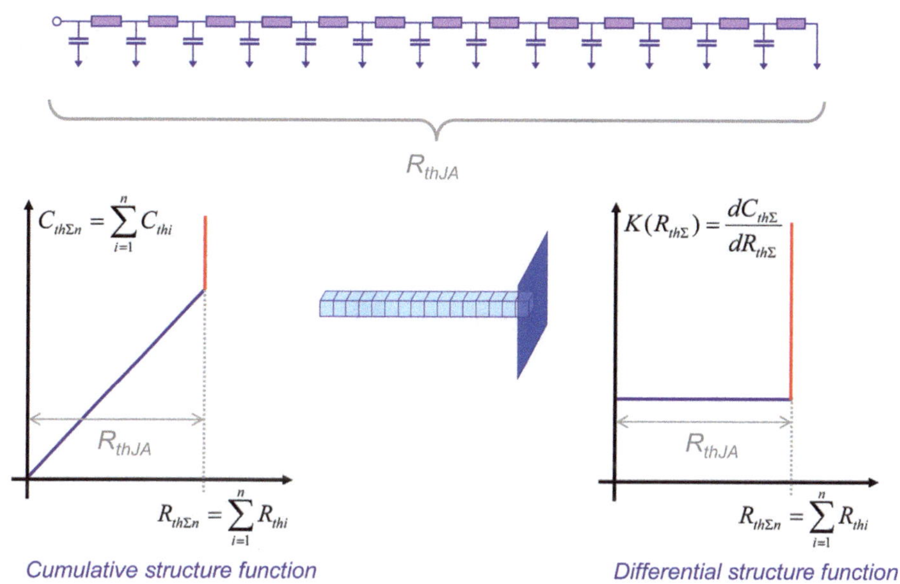

Fig. 4.47 The cumulative and differential structure functions of a homogeneous rod

With this rod example, we can demonstrate that the features of the structure functions are in a one-to-one correspondence with the properties of the heat conduction path.

Let us assume that in a given section in the middle of the rod the C_{th}/R_{th} ratio is doubled. This results in a steeper middle section in the cumulative structure function (with the slope doubled) and in a peak in the differential structure function (which is twice as high as the constant value of the other sections). This is illustrated in Fig. 4.48.

Example 3—optical correction and TIM quality in structure functions

In Fig. 4.49 and Fig. 4.50, we converted our previous Z_{th} curves (presented in Fig. 4.37) into structure functions.

Figure 4.49 presents the cumulative structure functions at two boundaries when the normalizing factor in the calculation was the P_{el} electric power and the emitted optical power was not taken into consideration. The technique used for producing the curves emphasizes small differences; we can see the divergence of the curves much clearer than in Fig. 4.37.

The curves are steep around the divergence, we can read an $R_{thJC\text{-}el} = 6.3$ K/W junction-to-case thermal resistance unambiguously. However, if the boundary change is applied on a larger surface in the structure then the divergence is less pronounced. For correct interpretation of the measured data, it is indispensable to specify the ε threshold value at which we consider the structure functions belonging to different boundaries to be separate.

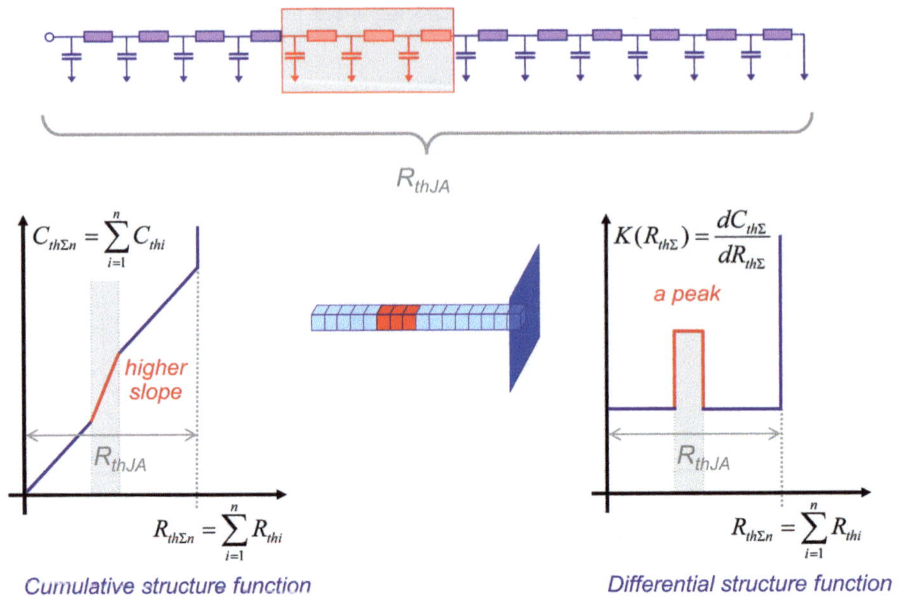

Fig. 4.48 Structure functions indicate the changes in the C_{th}/R_{th} ratio along the heat conduction path

In case we start our calculation from the $Z_{th\text{-}real}$ thermal impedances with the emitted optical power considered, we get those structure functions that are scaled in the real physical thermal resistances and capacitances (Fig. 4.50, curves with postfix _oc).

In this plot, we can read partial resistances and capacitances easily. We can clearly see the "sandwich-like structure" of the LED, again we can read the junction-to-case thermal resistance as $R_{thJC\text{-}real} = 8.3$ K/W. The curves diverge fast; the value does not change much if we vary the ε difference threshold. More reference on the standard-compliant definition of the junction-to-case thermal resistance is provided in Sect. 4.5.2.

The steep section until 8.3 K/W can be identified as the cooling slug of the LED; we can measure 18 mJ/K thermal capacitance form the start until the end of it which can be also expressed as 5 mm^3 copper calculated from Eq. (4.32).

Figure 4.51 shows the differential structure functions of the same LED measurement examples.

Example 4—locating structural elements in an LED package

In Fig. 4.52, we see the optically corrected cumulative structure function of a power LED, mounted on cold plate.

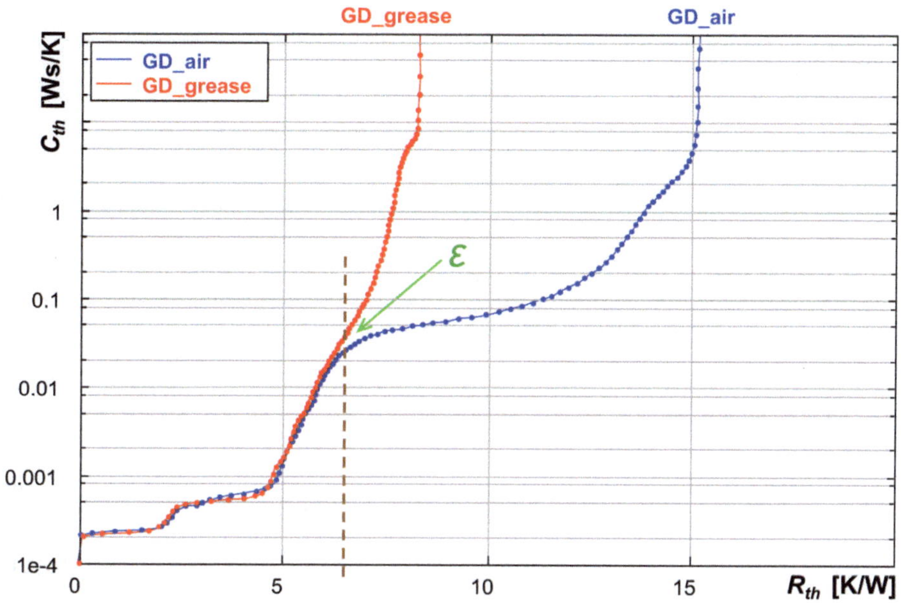

Fig. 4.49 Cumulative structure functions, LED component at two boundaries, without optical correction

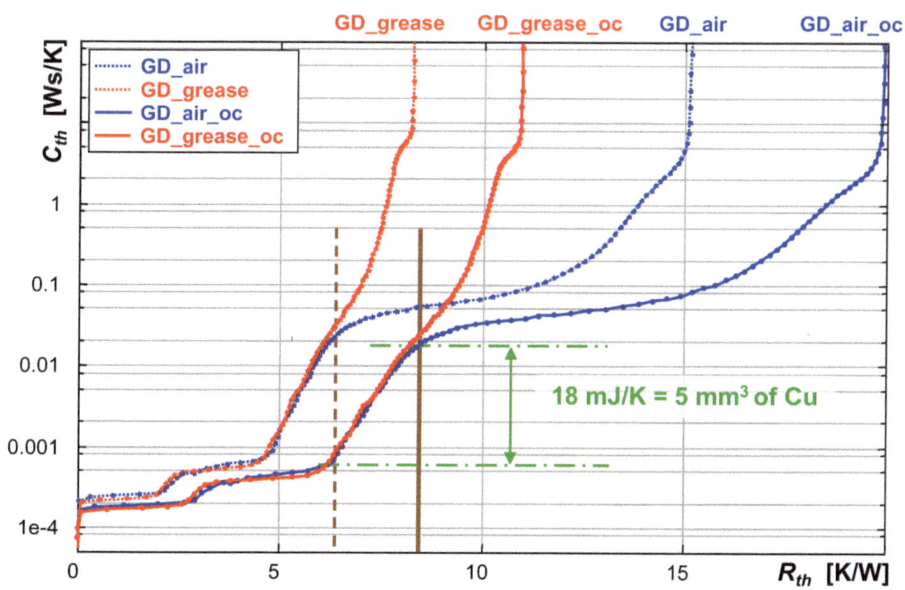

Fig. 4.50 Cumulative structure functions, LED component at two boundaries, with and without optical correction

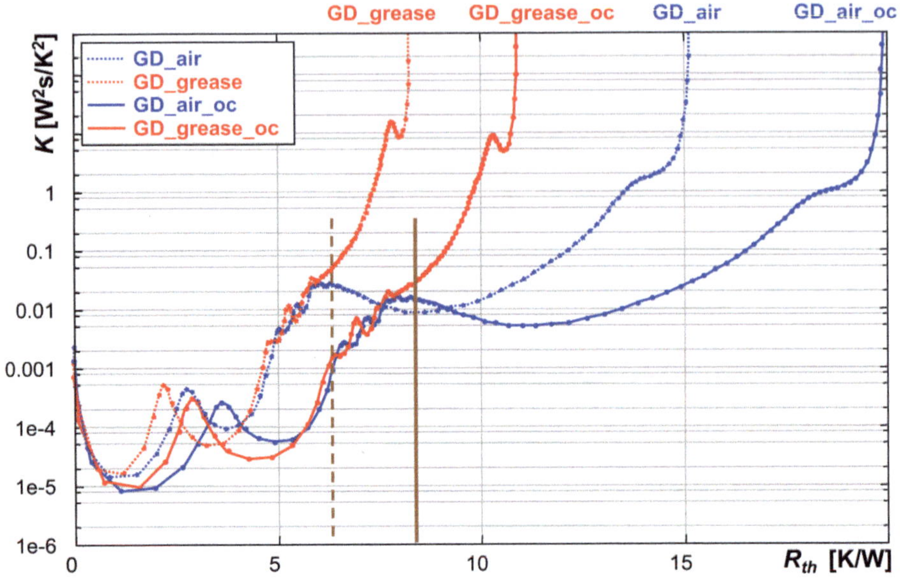

Fig. 4.51 Differential structure functions, LED component at two boundaries

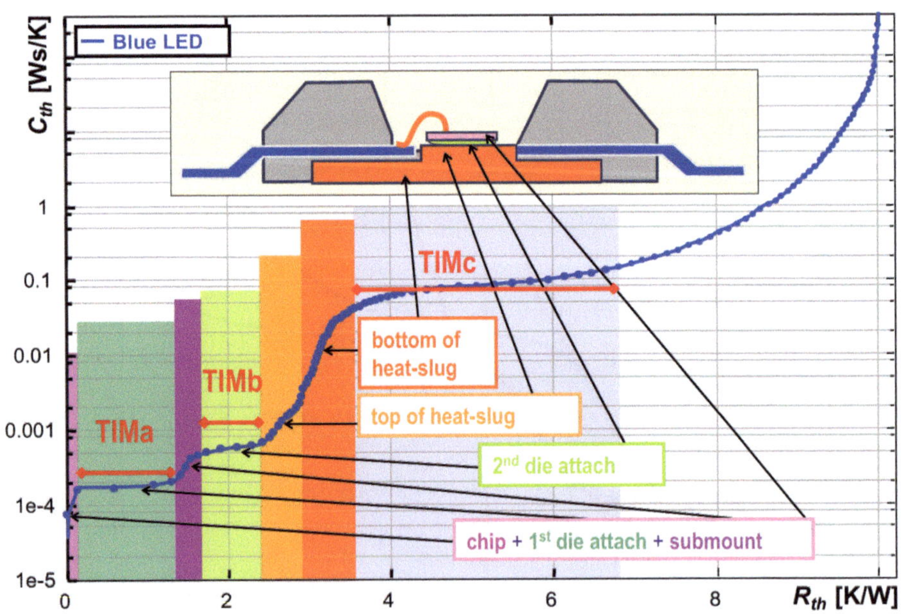

Fig. 4.52 Cumulative structure functions of a power LED with structural elements identified

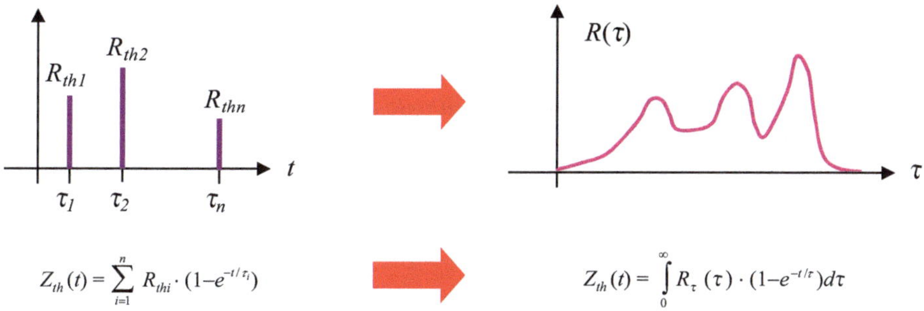

Fig. 4.53 Distributed, infinite (thermal) RC systems can be represented by the $R\tau$ (τ) time constant spectrum. (Source: [13] © JEDEC, reproduced with permission)

We can identify the sandwich-like structure corresponding to the chip, submount, and the heat slug. The flat sections represent different thermal interface layers (TIMa, TIMb, TIMc).

In case we know the material composition, we can measure the volume of an element through the thermal capacitance difference in the structure function. In case we know the exact geometry of an element, we can measure thermal conductivity and specific heat. For details of such analysis, consult, e.g., [15, 16].

4.4.5 *Theoretical Background of Thermal Time Constant Spectra, the NID Method and the Structure Functions*

Real thermal systems are infinite, distributed thermal systems, therefore real thermal impedances can be represented not by a set of discrete thermal time constants, but with a continuous spectrum of possible time constants. This representation is called thermal *time constant spectrum.* The $Z_{th}(t)$ function can be expressed in a similar way as presented in Fig. 4.40, but the sum of the exponential terms is replaced by an integral over the entire range of the possible thermal time constants

$$Z_{th}(t) = \int_{0}^{\infty} R_{\tau}(\tau) \cdot (1 - e^{-t/\tau}) d\tau \tag{4.34}$$

Thus, the discrete R_{thi} magnitude values are replaced by the $R_{\tau}(\tau)$ thermal time constant spectrum, see Fig. 4.53.

Using logarithmic time scale has advantages. A practical reason is that data acquisition with logarithmically equidistant time intervals greatly reduces the need for data storage. Using logarithmic time in graphical display of junction temperature transients allows visualizing all the details over the entire time constant range: changes corresponding to the chip + submount, heat slug, MCPCB, etc. will be visible. From data processing point of view, applying logarithmic time scale the relationships between different representations of thermal impedances can be formulated by means of convolution integrals [20, 23].

Thus, by introducing the $z = \ln(t)$ *logarithmic time* and the $\zeta = \ln(\tau)$ *logarithmic time constant* Eq. (4.34) will look as follows:

$$Z_{th}(t) = \int_{-\infty}^{\infty} R_{\zeta}(\zeta) \cdot [1 - \exp(-t/\exp(\zeta))] d\zeta \tag{4.35}$$

where the $R_{\zeta}(\zeta)$ logarithmic time constant spectrum is defined as

$$R_{\zeta}(\zeta) = R_{\tau}(\exp(\zeta)) \cdot \exp(\zeta) \tag{4.36}$$

In the following, we shall always refer to $R_{\zeta}(\zeta)$ as the time constant spectrum. Simplifying our notation by $a(z) = Z_{th}(t = \exp(z))$, we can further write

$$\frac{d}{dz} a(z) = \int_{-\infty}^{\infty} R_{\zeta}(\zeta)[\exp(z - \zeta - \exp(z - \zeta))] d\zeta \tag{4.37}$$

Further introducing the notation

$$w_z(z) = \exp[z - \exp(z)] \tag{4.38}$$

we can write

$$\frac{d}{dz} a(z) = \int_{-\infty}^{\infty} R_{\zeta}(\zeta) \cdot w_z(z - \zeta) d\zeta \tag{4.39}$$

One can easily realize that Eq. (4.39) is a convolution integral

$$\frac{d}{dz} a(z) = R_{\zeta}(z) \otimes w_z(z) \tag{4.40}$$

where the $\otimes$ symbol denotes the convolution operation. From this, we can deduce a method of obtaining the $R_{\zeta}(\zeta)$ time constant spectrum, as $a(z)$ is the measured Z_{th} thermal impedance on logarithmic time scale, and $w_z(z)$ is a fixed function:

$$R_{\zeta}(z) = \left[\frac{d}{dz} a(z)\right] \otimes^{-1} w_z(z) \tag{4.41}$$

where $\otimes^{-1}$ denotes *deconvolution*, the inverse operation of convolution.

Implementation of this calculation is not straightforward. Deconvolution can be performed in time domain, e.g., by *Bayesian iteration* or in the frequency domain by division, also known as *Fourier-domain inverse filtering*. For details of the frequency domain implementation, see, e.g., Annex B of the JESD51-14 standard [13] or refer

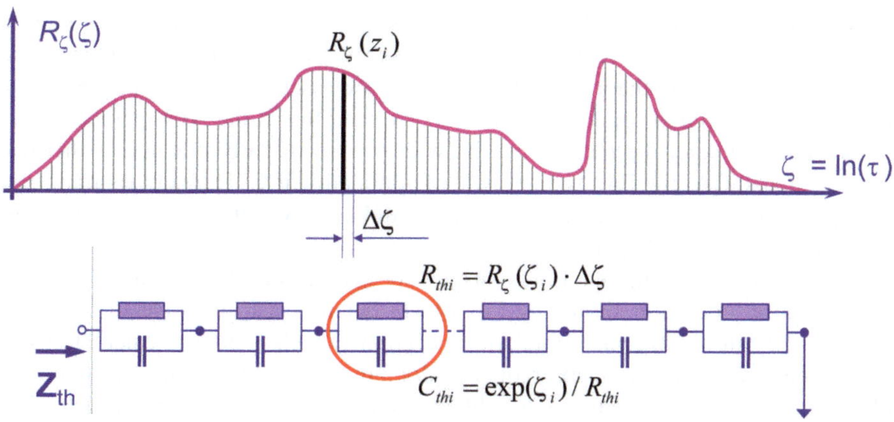

Fig. 4.54 Discretization of the $R_\zeta(\zeta)$ time constant spectrum and constructing a FOSTER model of the measured thermal impedance from this. (Source: [13] © JEDEC, reproduced with permission)

to an original paper of Székely et al. [24]. There is another practical aspect of Eq. (4.41): as both the derivation and the deconvolution operations enhance noise, if time constant spectra are to be generated from measured thermal transient curves, these Z_{th} curves must be extremely noise free. Using noisy signals, the obtained time constant spectra would include false values causing misleading artifacts in the results of further postprocessing steps.

Based on this discussion, another, more formal definition of the time constant spectrum can be given as follows [20]:

$R_\zeta(\zeta)\Delta\zeta$ is the magnitude of the components in the thermal impedance (unit-step response) that belong to the time constant range of $[\exp(\zeta), \exp(\zeta + \Delta\zeta)]$.

This formal definition implies a method for discretizing the time constant spectrum and creating a network model of the thermal impedance (Fig. 4.54). Thus, according to this definition the thermal resistance value belonging to the time constant range $[\tau_1, \tau_2]$ can be expressed as

$$R_{th}(\tau_1, \tau_2) = \int_{\ln \tau_2}^{\ln \tau_2} R_\zeta(\zeta)d\zeta \tag{4.42}$$

and the total steady-state thermal resistance of the measured structure can be calculated as

$$R_{th}(t) = \int_{-\infty}^{\infty} R_\zeta(\zeta)d\zeta \tag{4.43}$$

The procedure was first applied for the characterization of heat flow path structures of packaged semiconductor devices by Székely [21]. As the method provides network models of measured thermal impedances through a deconvolution step, it became widely known as *network identification by deconvolution* or *NID method*. The method has some theoretical limits, which are discussed in detail by Székely in his other often cited paper [23]. One needs to add that if the NID method is implemented with care, the method can be widely used in daily practice of thermal measurements, QA, and FA. Another field of application is CFD model validation or LED package compact thermal modeling as will be shown later in this chapter (see also Sect. 6.5.2.2. of Chap. 6).

It is worth mentioning that time constant spectra can be created not only by the deconvolution operation shown by Eq. (4.41) but it is also possible to calculate them directly in any thermal simulation program, which is able to calculate thermal impedances in the complex frequency domain [20]. Such calculations have the advantage that time constant spectra obtained this way are free from any distortion of measurement noise and measurement artifacts originating from the cutting and substituting the early electrical transients (see Fig. 4.16 and Fig. 4.18 in Sect. 4.3.6.1).

When converting measured thermal impedance curves to time constant spectra, small differences and noise are further enhanced by the numerical derivation and numerical deconvolution needed to postprocess data using Eq. (4.41).

Based on Eq. (4.41), time constant spectra can also be extracted from simulated thermal impedance curves. In this case, the problems are due to the quantization noise of the numerical results and the possibly too coarse time resolution[13] of such simulated transients. The direct calculation of the time constant spectra as described, e.g., in [20] is a viable alternative to the "brute force" application of the NID method to simulated Z_{th} curves. Unfortunately, commercially available program tools providing this opportunity are no longer available.

As stated already, the $Z_{th}(t)$ thermal impedance function and the $R_{\zeta}(\zeta)$ time constant spectrum are equivalent, the $R_{\zeta}(\zeta)$ function also carries all available information about the heat conduction path. Thus, it is worthwhile further transforming it into highly *detailed network models* of the thermal impedance and into other *descriptive functions* for the analysis of different aspects of the junction-to-ambient heat flow path structure. These further representations include the *cumulative* and *differential structure functions*, which were discussed in the previous sections and the *pulse thermal resistance diagrams* described later in this section.

As Fig. 4.54 shows, for further numerical processing, the continuous time constant spectrum obtained by Eq. (4.41) needs to be discretized. For a given discrete time constant value an average R_{thi} magnitude value is assigned using Eq. (4.42). From the discrete time constant and magnitude value, a thermal capacitance value describing the given discretized time constant value is calculated. This way the discretized time constant spectrum is turned into a long FOSTER equivalent model as shown in Fig. 4.54. The discretized version of the continuous time constant spectrum typically contains 150..200 R_{th}–τ pairs, thus, the FOSTER equivalent model consists

---

[13] Achieving a time resolution higher than 20 points per decade may require unaffordable simulation resources while measuring a thermal transient with about 200 or 400 points per decade resolution is not a problem with the transient extension of the JESD51-1 "static" test method.

of 150..200 stages. In the practical realization of these calculations, the stages are ordered according to the time constant values, starting with the smallest one.

The temperature response of the system described by this model can be calculated by Eq. (4.25). Due to the associativity property of the addition operation, however, the summation in Eq. (4.25) can be carried out by any order of the RC stages of the FOSTER equivalent model. In simple words, this means that the order of the RC stages of the FOSTER equivalent model can be any, it has nothing to do with the physical structure of the heat conduction path described by the model.

This underlines our prior statement that the elements of the FOSTER model are "fictitious": the entire model properly represents the thermal impedance but the individual element values of the model are not related to the physical reality.

To overcome this problem, the FOSTER equivalent model of the thermal impedance is converted into the corresponding CAUER ladder model.

The FOSTER $\leftrightarrow$ CAUER conversion is a "standard" technique in case of (passive) linear electrical circuits, but there are practical limits when one tries to implement the conversion algorithms numerically.

Fortunately, we need the FOSTER $\rightarrow$ CAUER conversion only, which works well for large models with hundreds of stages. The "only" numerical issue in this case is that the standard floating point number representations do not provide the appropriate range of orders of magnitude of values and do not provide enough numbers of decimal digits.[14] Description of the conversion algorithm is out of our scope, it can be found, e.g., in Annex C of the JESD 51-14 standard [13].

As a final result, the $R_\zeta(\zeta)$ continuous time constant spectrum discretized to about 150..200 points is converted into a detailed equivalent CAUER-type RC model of the thermal impedance. If any two RC stages of that model are swapped, the thermal response of the network model would be different, therefore the order of the RC stages of a CAUER-type ladder model is significant; the order and the element values of the RC stages are related to the physical structure of the (essentially one-dimensional) heat flow path.

As a summary regarding structure functions, we can conclude the following. Structure functions are closely related to the theory of the linear RC networks and to their different representations. Strictly speaking, structure functions are representations of impedances of distributed (thermal) RC networks characterized as follows:

- The network is linear and passive;
- Driving point behavior is considered;
- Essentially one-dimensional heat flow is assumed.

Linearity means that the thermal resistance and capacitance are independent of the temperature itself. In other words: both the thermal conductivity and heat capacity are constant values, without temperature dependence. In reality, this theoretical condition is not met exactly but in practical situations it is a reasonable approximation.

[14] This problem can be overcome by using special math libraries in the implementation, which provide any number of digits both for the exponent and for the mantissa.

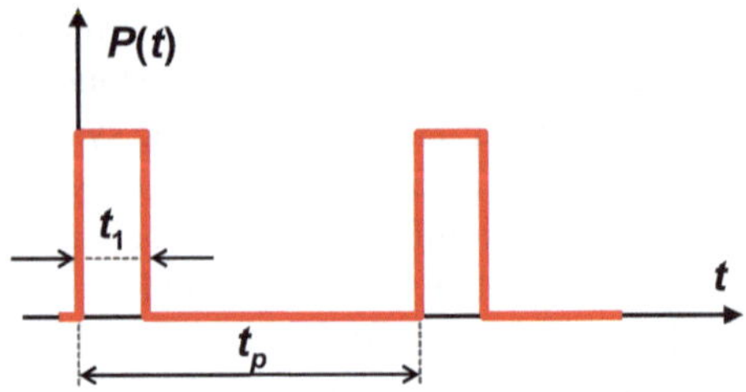

Fig. 4.55 Periodic heating pulse sequence

The subject was discussed in detail by Székely and Rencz [17] and is also touched in Sect. 6.5.2.2 of Chap. 6.

Driving point behavior means that the same location of the structure is heated (by, e.g., a power step) and measured for its temperature response; see also Sect. 6.1.4.2 of Chap. 6.

Essentially one-dimensional heat flow (besides longitudinal flow) includes more complex heat spreading, which with some coordinate system transformation can be mapped to longitudinal flow in a Cartesian coordinate system. This includes *radial spreading* in disc-like structures such as an MCPCB under a power LED or in a JEDEC standard thermal test board, *cylindrical spreading* or *conical spreading* (e.g., in a copper heat slug of a power LED package).

A heat flow path is called essentially one-dimensional path if it is formed as "series connection" of the regions with the above described heat spreading characteristics, which is often the case with real semiconductor device packages. Only the splitting of the heat flow path poses questions. When the splitting point coincides with the driving point of a main and a parasitic heat flow path and the total thermal resistance of the parasitic path is known, there is a possibility to eliminate the effect of the parasitic path from the structure function [25]. Even if such a correction is not possible because the parasitic shunting resistance is not known or the splitting point differs from the driving point, still an "equivalent" physical structure can be derived, but without direct mapping to the real physical structure in such a case.

4.4.6 Pulse Thermal Resistance Diagrams

In case we apply a square wave power profile (such as shown in Fig. 4.55) on a device for a long period of time, we can expect it to reach a stationary state: the junction temperature will also follow a periodic function and its waveform becomes stable. Examples for such stabilized junction temperature waveforms are shown in Fig. 4.56.

The most important quantity at such excitations is the peak temperature reached in stationary state. One way of getting this value is really programming different pulses of different length and amplitude and measuring the transient temperature changes (for practical reasons in the "off" state).

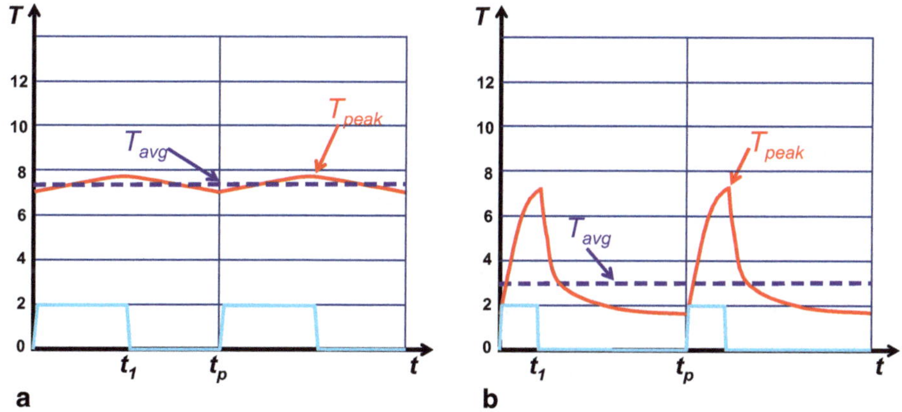

Fig. 4.56 Junction temperature waveform at different heating pulse series:
a $t_1 = 0.1$ ms, $t_p = 0.2$ ms, $\delta = 0.5$, $f = 1/t_p = 5$ kHz, $\mathrm{d}T_{peak} = 7.8$ K, $\mathrm{d}T_{avg} = 7.3$ K
b $t_1 = 1$ ms, $t_p = 5$ ms, $\delta = 0.2$, $f = 1/t_p = 200$ Hz, $\mathrm{d}T_{peak} = 6.9$ K $\mathrm{d}T_{avg} = 2.9$ K

Carrying out a series of measurements with square wave excitation at many frequencies and duty cycles is a rather tedious task. Instead, we can derive this plot from a single measurement with step excitation, and some mathematical calculation.

Actual temperature waveform can be gained by simulation, making first one of the RC models of the device defined in previous sections and then using a SPICE-like circuit simulator.

A useful plot called *pulse thermal resistance diagram* or *thermal transient impedance plot* yields peak temperature in stationary state directly; it can be easily derived from the time constant spectrum.

Let the periodic pulse load be denoted by the t_1 pulse width, t_p period, and the $\delta = t_1/t_p$ duty factor (Fig. 4.55). The curves of the pulse thermal resistance diagram can be calculated from the time constant spectrum by the following convolution operation [20]:

$$Z_{th}(z = \ln t_1, \delta) = R_\varsigma(z) \otimes \frac{1 - \exp\left[-\exp(z)\right]}{1 - \exp\left[-\exp(z)/\delta\right]} \tag{4.44}$$

Example 5—generating pulse thermal resistance diagrams
Such a diagram (Fig. 4.57) describes the behavior of the same LED used in Examples 1–3—when excited by repeated pulses of given length and duty cycle. The horizontal axis shows the pulse length; the duty cycle is the curve parameter. The vertical axis shows the peak temperature elevation in terms of an “effective thermal resistance” [K/W].

In case of a stream of very short pulses, the power source can be interpreted as an effective source with a power downscaled by the duty cycle. In case of very long pulses, the temperature has “enough time” to reach the steady-state situation.

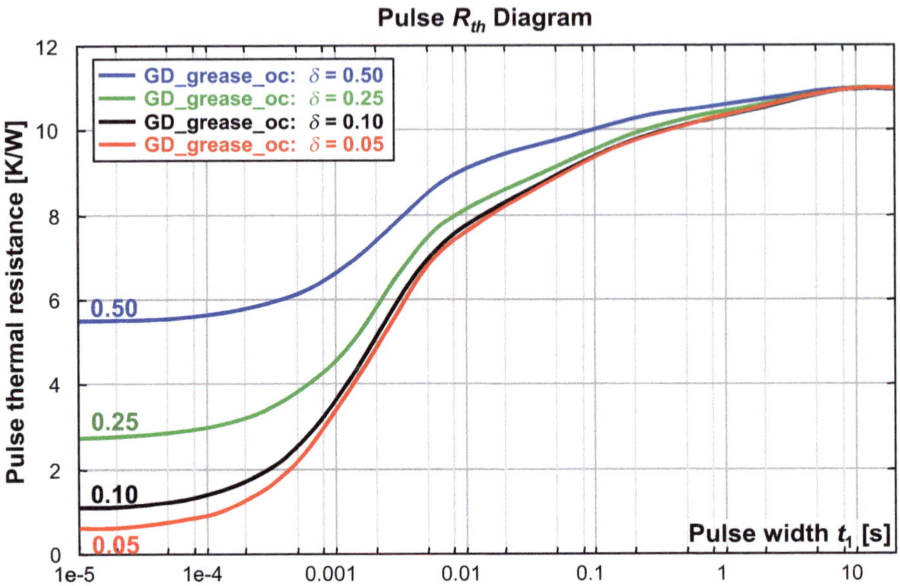

Fig. 4.57 Pulse thermal resistance diagram of the Dragon LED on cold plate, thermal grease applied, emitted optical power considered

Accordingly, in Fig. 4.57, we get back for pulses of very long t_1 again the peak temperature elevation corresponding to 11 K/W as in previous "views." At 50 % duty cycle, the peak temperature behaves like at driving a 5.5 K/W thermal system for a long time, at 10 % like driving a 0.11 K/W heat conduction path, etc. The really informative part of the figure is in between, where we have no instant guess on the peak temperature.

This curve is extremely useful for producing data sheets for devices and designing switching-type power supplies (pulse width modulation [PWM] LED drives).

The powering scheme shown in Fig. 4.55 is typical in case of dimming LEDs with pulse width modulation (PWM dimming) where the average luminous flux is controlled by the duty factor.

4.4.7 Complex Loci

The frequency domain representation of the thermal impedance can be provided in different ways. Based on the $Z_{th}(t)$ function, it can be by transformed into the $Z_{th}(\omega)$ function

$$Z_{th}(\omega) = \int_0^{\infty} Z_{th}(t) e^{-j\omega t} dt \tag{4.45}$$

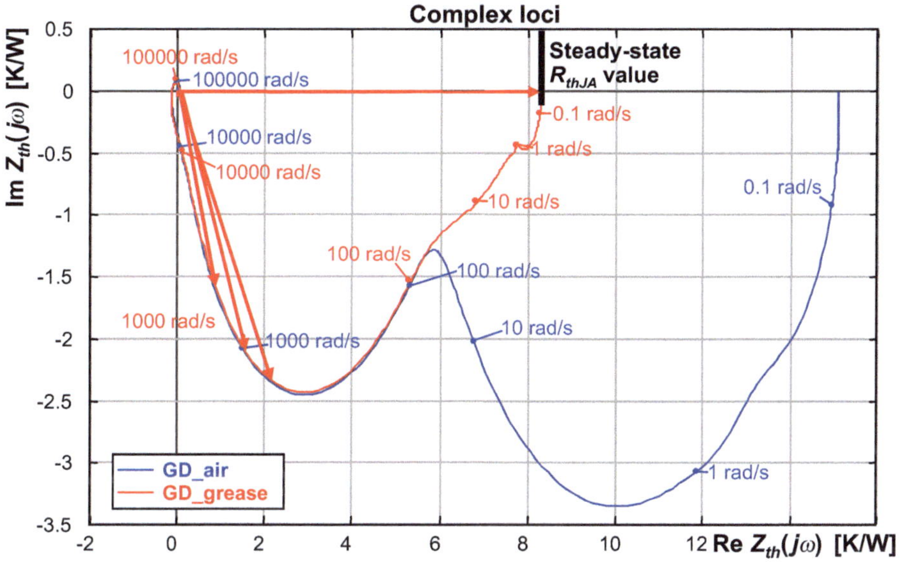

Fig. 4.58 Complex locus of the frequency domain thermal impedance of the power LED assembly. Harmonic components corresponding to 0, 100, 200, and 300 Hz (0, 628 rad/s, etc.) are shown

where ω is the angular frequency of the excitation. The resulting $Z_{th}(\omega)$ complex thermal impedance function can be visualized, e.g., by means of a *complex locus* (also known as Nyquist diagram) as shown in Fig. 4.58.

The zero frequency value is the thermal resistance defined in Sect. 4.3.3. In the actual measurement, this is again 8.2 K/W and 15 K/W for the two boundaries when measured along the horizontal axis on the "electrical only" curves.

In case the frequency of the mains *voltage*, is, e.g., 50 Hz, then the *power* will have 100 Hz base frequency and hence, harmonics of 200 Hz, 300 Hz, etc. The corresponding angular frequencies of interest will be accordingly 628 rad/s, 1,258 rad/s, 1,885 rad/s, etc.

The absolute value of the impedance at any frequency is the length of the vector pointing from the origin to the given frequency point on the $Z_{th}(\omega)$ locus (arrows in Fig. 4.58 and Fig. 4.59). The angle between the horizontal axis and the locus shows the phase shift between the sinusoid power component and the temperature change belonging to it.

The length of the vectors shows that the thermal impedance is smaller at higher frequency; their angle proves that the temperature change is delayed to the power. This is also implied by the compact network model of the thermal impedance (Fig. 4.40). The physical meaning of the smaller temperature change on the same power amplitude is that the periodically changing heat can be locally stored and released by the structures near the junction; it does not reach the ambient.

Complex loci are a very powerful representation of the component and its environment when analyzing periodic excitations. An application example is the single valued "AC thermal impedance" of LEDs [26] that will be detailed in Sect. 4.7.

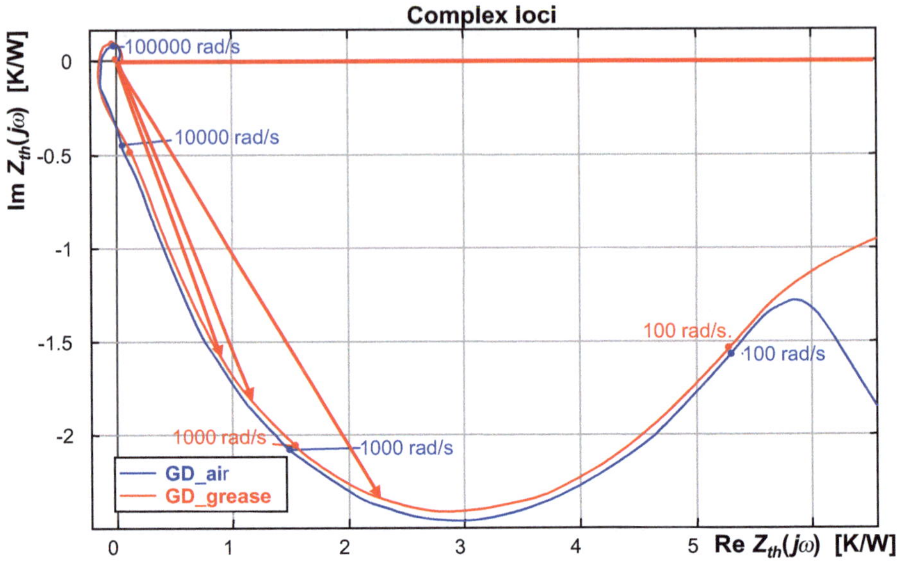

Fig. 4.59 Excerpt of the complex locus in Fig. 4.58

4.4.8 Summary About the Different Representations of Thermal Impedances

In Sect. 4.4, we described how the results of thermal transient measurements can be interpreted.

The well-known $Z_{th}(t)$ curve is obtained from the measured junction temperature change $\Delta T_J(t)$: this function is divided by the ΔP_H real heating power step applied at the device under test (in case of LEDs that is the supplied electrical power less the emitted optical power). The $Z_{th}(t)$ *time domain thermal impedance function* directly provides the junction temperature response if 1 W of heating is applied, therefore it is called *unit-step response*.

We distinguish between driving point thermal impedances (when the heating and the measurement of the dynamic temperature response takes place at the same location) and the transfer impedances (when the heating and the measurement of the temperature response are at different locations).

During the subsequent postprocessing of the measured thermal impedances, logarithmic time scale is assumed by the introduction of the $z = \ln(t)$ transformation. The unit step response in logarithmic time scale is often denoted by $a(z)$ in the technical literature.

The first concept we described was the concept of the continuous $R_\zeta(\zeta)$ thermal *time constant spectra*. The *convolution integral*, which connects the time constant spectrum and the unit-step response, was shown. This relationship provides the algorithm of extracting the time constant spectra from measured thermal impedances. Time constant spectra are interpreted both for driving point and transfer impedances. Time constant spectra of driving point impedances have only positive values, while in case of transfer impedances there are negative values also present in the time constant spectra.

Time constant spectra of driving point thermal impedances can be further processed: after discretization (as a practical step in the numerical implementation of the necessary calculations) *FOSTER equivalent models* (as black box models) and after a conversion step, detailed *CAUER equivalent network models* (with about 150..200 stages) can be obtained. These detailed network models are yet another representation of the driving point thermal impedances.

The measurement result evaluation method is named *NID* as the process is based on the series of a deconvolution and a network model identification steps.

To facilitate the further analysis of the measurement results, the *cumulative structure function*[15] (*CSF*) is defined as a map or a graphical representation of the thermal capacitance/thermal resistance distribution of the junction-to-ambient heat flow path. We have shown that for essentially one-dimensional heat flow paths the structure functions and the physical structure of the path are in a one-to-one relationship. To further enhance the heat flow path details, the derivative of the cumulative structure function, called the *differential structure function* (*DSF*) is also introduced.

Both kinds of structure functions are powerful tools in finding *partial thermal resistances* (like TIM resistances) or thermal capacitances of different structural elements of a power semiconductor device package. Knowing material properties, volumes and effective cross-sectional areas of different heat flow path sections can be identified; or, if geometrical information about the heat flow path sections is available, material properties can be identified. Some of these properties of structure functions lead to new thermal testing applications like die attach qualification or measurement of thermal conductivity of TIM. Application of structure functions in pre- and poststress analysis of LED's reliability studies was demonstrated recently [28]. This topic is further discussed in Sect. 4.5.4.

Thermal transient measurements and structure functions recently got used in measuring standard thermal metrics of packages like the measurement of the junction-to-case thermal resistance of power semiconductor device packages (see the JEDEC JESD51-14 standard [13]).

Last but not least, there are two further representations of the thermal impedances, which help study the thermal behavior of packaged semiconductor devices under periodic heat load conditions. The *pulse thermal resistance* diagrams can be derived from time constant spectra by using a convolution integral [20]. These diagrams provide information about the effective thermal resistance of the heat flow path when driven with a square wave dissipation with a given pulse duration and duty factor. This information can be useful for the thermal management design of PWM-dimmed LED applications.

Complex loci (or Nyquist diagrams) are the graphical representations of thermal impedances in the *frequency domain*. They are defined both for driving point and transfer impedances. With the help of these loci, one can identify the value of the thermal impedance at any harmonic frequency of a sinusoid power excitation. This is very useful information when one needs to characterize the thermal behavior of directly AC mains driven LEDs [26]. This topic is further discussed in Sect. 4.7.

[15] In many cases, the adjective "cumulative" is neglected and the *CSF* is simply called the structure function.

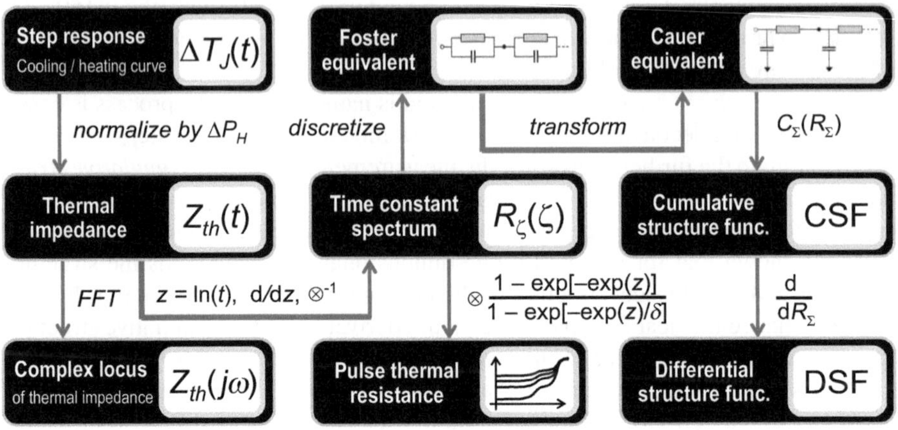

Fig. 4.60 Summary of the possible equivalent representations of a driving point thermal impedance (after Vermeersch [27])

All these equivalent representations of the thermal impedances and the conversion steps between them are summarized in Fig. 4.60.

4.5 Some Applications of Structure Functions

4.5.1 In Situ TIM Testing

As pointed out in Sect. 4.4.4, structure functions can be well used to study material property changes in junction-to-ambient heat conduction paths of different (power) semiconductor devices. As an example, interfaces between mating solid surfaces in a heat flow path are regions usually characterized by a small C_{th}/R_{th} ratio—appearing as long, flat plateaus in structure functions.[16]

Some portions of the LED structure are highly stable (chip, submount, heat slug.) On the contrary, TIM layers, which are applied to fill the small gaps between mating surfaces can show large variation even in samples from the same manufacturing lot. Testing the TIM itself, in bulk or in thin layers (see Sect. 4.5.3) gives no hint on the actual thermal resistance achieved in the fabricated TIM layer. These unavoidable variations can be best studied in structure functions.

Figure 4.61 highlights typical thermal interfaces in an LED application. Their quality can be even measured as the length along the thermal resistance axis of TIMa, TIMb, and TIMc in Fig. 4.51 (Example 4 of Sect. 4.4.4).

There could be many reasons why interfacial thermal resistance changes: temperature change, variation in the thickness of the TIM layer, aging or simply, the TIM quality is changed by purpose. In the subsequent sections, examples will be shown

[16] The analysis technique based on changing C_{th}/R_{th} ratio in the structure function was first suggested by M. Rencz et al. [29] for the study of die attach voiding and for TIM quality assessment.

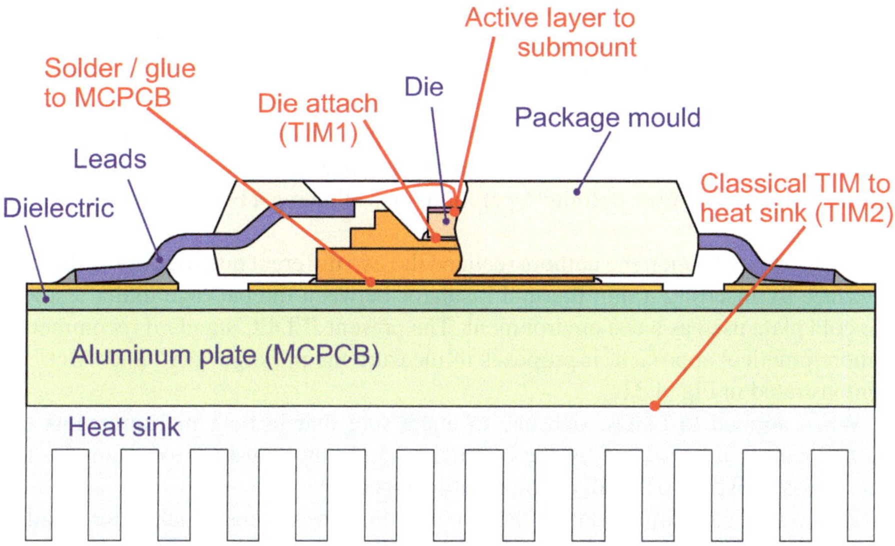

Fig. 4.61 Different thermal interfaces in the junction-to-ambient heat flow path of a typical LED application

for such cases. The overview of applications of structure functions will be concluded by an example for model calibration.

4.5.2 *The Transient Dual Interface Method for the Standard Measurement of the Junction-to-Case Thermal Resistance*

The measurement of the R_{thJC} junction-to-case thermal resistance of (power) semiconductor device packages has always been suffering from different problems. The "classical" method was putting a thermocouple into the cooling mount and making statements on measuring junction temperature and the temperature of an arbitrary point on the "case" surface. The major pain was the repeatability and another issue was accuracy—which became rather severe when small R_{thJC} values had to be measured.

Even small errors in the measurement or its repeatability may cause enormous problems. On one hand, publishing junction-to-case thermal resistance values on the product data sheet underestimated by 10 % may result in liability issues of the semiconductor vendors. On the other hand, the same thermal metric overestimated by 10 % may result in competitive disadvantage if end-users select power semiconductor parts based on the thermal performance measured by this metric.

Application of structure functions in a smart way was a workaround to this measurement problem. Today's new JEDEC standard JESD51-14 [13] is based on the original idea first published by Steffens et al. in 2005 [30], which is the following.

Since structure functions reflect structural changes of an essentially one-dimensional heat conduction path, let us introduce such a change by purpose: let us change the quality of the thermal interface at the "case" surface of a power package. That change must be reflected to some extent by the Z_{th} curves (Fig. 4.37) and much clearer by the structure functions (Fig. 4.50, Fig. 4.51). There are cases when the deviation of the first derivative of the measured Z_{th} thermal impedance function provides a more definite separation, thus allowing a better definition for the R_{thJC} value.

In the original paper, the authors realized the two different qualities of the thermal interface by inserting a thin thermal insulator between the package under test and the cold plate used as a test environment. The present JEDEC standard recommends a more practical approach: it proposes to measure the package "dry" and "wet," as demonstrated in Fig. 4.31.

When applied to LEDs, one has to make sure that in both measurements the heating power is corrected by the emitted optical power, using, for example, the curves with postfix _oc in all characteristic plots.

Finding the separation point of the structure functions is not trivial. After detailed numerical simulation studies and practical experiments, [31, 32] the method got standardized with the requirement of defining a small positive ε threshold number (Fig. 4.49, Fig. 4.50). This ε either specifies the *relative difference* of two derivative functions or the *difference* of the cumulative structure functions (depending on the method chosen) where we say, the corresponding Z_{th} value or the corresponding cumulative thermal resistance is equal to the junction-to-case thermal resistance. This ε value must be included along with the identified R_{thJC} value on the test report. Precise details are presented in the JESD 51-14 standard [13] and in papers [33] or [34].

The required data processing method is implemented in a publicly available software tool (forming an online annex of the standard) as well as in a commercially available software.

Note that the standard was worked out for silicon power semiconductor devices. Before standardization, different aspects of the applicability of the method have been analyzed by Schweitzer et al. [32, 34, 35].

One of the first studies on the applicability of this method to certain high-power LEDs was published by Müller et al. [36].

The great advantage of this transient method is that instead of measuring a *spatial difference* of two temperature values (T_J, T_{Case}) with two thermometers (the junction itself and a thermocouple attached to the "case" somehow), it takes the *temporal difference* of the junction temperature only (junction temperature transient). This way most of the uncertainty and repeatability issues of the classical steady-state R_{thJC} measurements are inherently eliminated. The completely differential approach used in the $\Delta T_J(t)$ measurement automatically cancels out offset errors and one needs to calibrate the TSP of the device under test only.

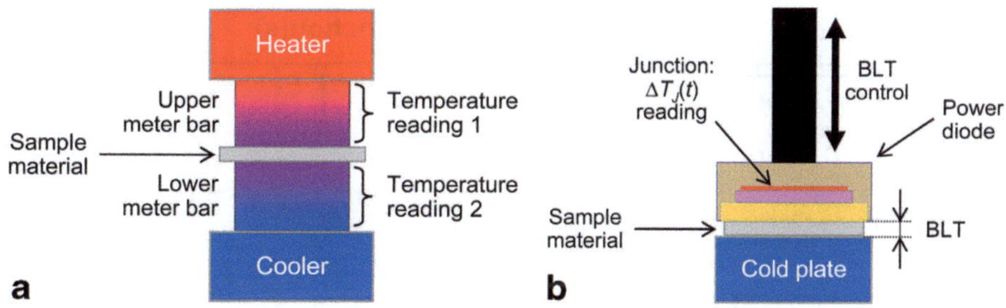

Fig. 4.62 Possible setups for TIM testing: **a** classical ASTM-like steady-state measurement setup, **b** Dynamic TIM testing setup based on junction temperature transient measurements

4.5.3 Structure Functions Used in TIM Testing

In classical (steady-state) thermal interface material testing (ASTM D-5470-01, see [37]) heat bars with known heat flux are used and the temperature drop difference due to the introduction of the TIM to be tested is measured. Based on this temperature difference, the thermal conductivity of the material can be calculated (Fig. 4.62a).

Uncertainties emerge due to the calibration issues of measuring spatial difference of two temperatures, possible errors in temperature readings of thermocouples used, etc.

A better way of testing thin TIM layers can be built on the transient method used for measuring junction-to-case thermal resistance. The material sample is placed in a dedicated test fixture where one face is the "case" surface of a power diode and the other face is the surface of a cold plate. We measure the $\Delta T_J(t)$ junction temperature transient of the diode at a well specified, preset thickness of the material (BLT—bondline thickness), see Fig. 4.62b. As we change the sample thickness with a precise, dedicated mechanics, the total measured thermal impedance of the complete test setup will change.

The change of the total junction-to-ambient resistance of the test setup is solely due to the changes in material sample thickness, as presented in the structure functions of Fig. 4.63. Therefore, the λ thermal conductivity of the sample can be calculated as follows:

$$\lambda = \frac{1}{A} \cdot \frac{\Delta L}{\Delta R_{th}} \tag{4.46}$$

where A is the cross-sectional area of the heat flow path across the sample, ΔL is the bondline thickness change, and ΔR_{th} is the corresponding change in the total thermal resistance of the test setup.

According to Eq. (4.46), we can state that the thermal conductivity of the TIM sample tested is directly proportional to the slope of the R_{th}–BLT diagram obtained for the sample. Such diagrams are shown in Fig. 4.64.

There are some advantages of this TIM testing method compared to other techniques. For example, there is reduced measurement uncertainty due to the differential

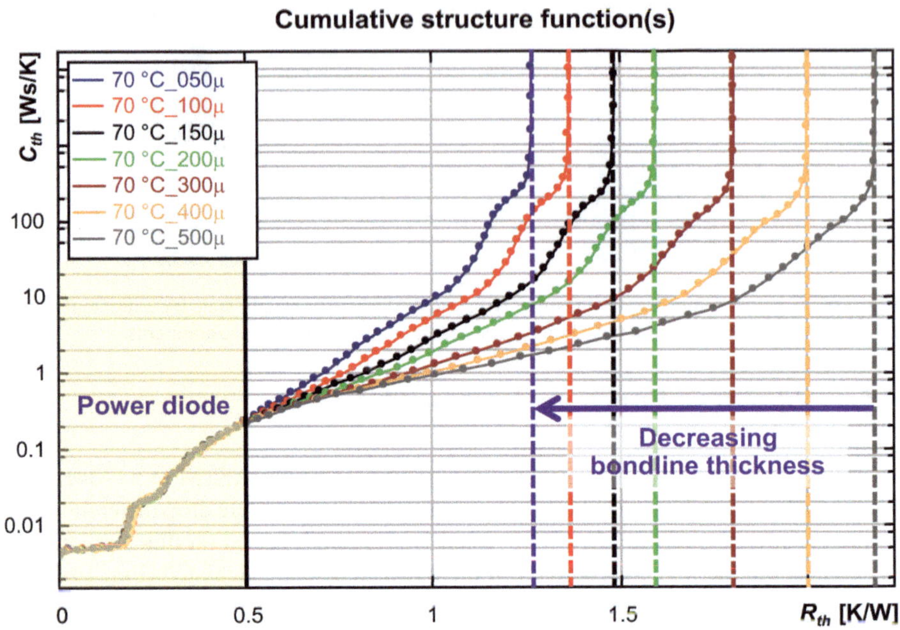

Fig. 4.63 Structure functions of the dynamic TIM testing setup measured at different preset bondline thicknesses of the material tested

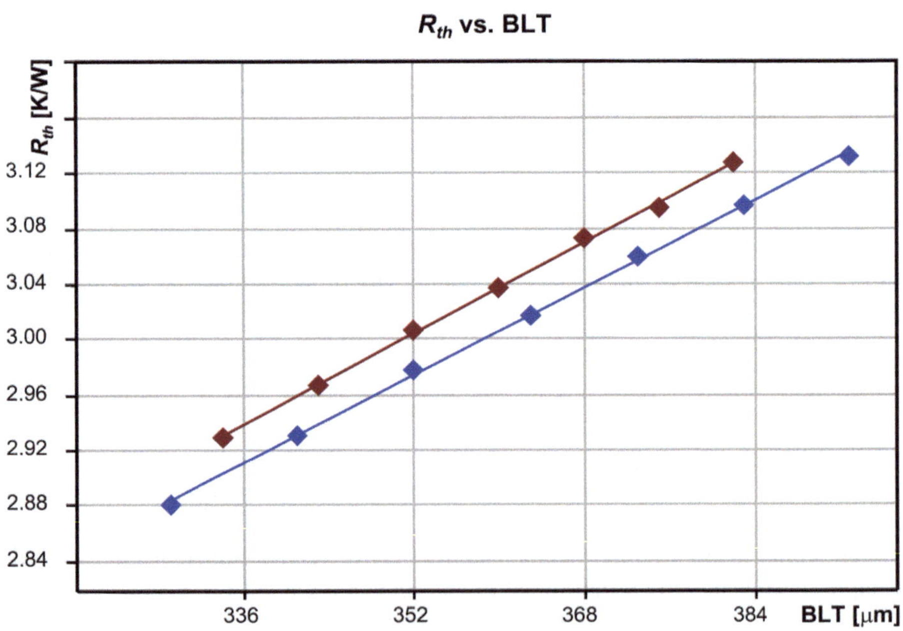

Fig. 4.64 Thermal resistance versus bondline thickness plots obtained for a given material type measured in a dynamic TIM testing setup outlined in Fig. 4.62b

approach. The test method can be considered as a quasi in situ technique as the test fixture resembles the real-life application conditions of TIM materials. Last but not least, every measurement includes an inherent self-test of the measuring system; as based on the obtained structure functions, the structural integrity of the test fixture can always be checked. Such a system is available as an add-on item [38] of a commercial thermal transient tester [39].

A general overview on different TIM testing techniques is given in Chap. 8 on TIMs.

Details of the measurement principle and its realization are presented in papers of Vass-Várnai et al. [40, 47].

4.5.4 Structure Functions in Reliability Analysis

If aging of TIMs manifests in thermal conductivity changes, these changes can thus be well measured by structure functions. Therefore, thermal transient measurements completed by structure function analysis are becoming a popular tool in reliability investigations of complete thermal management solutions both in electronics cooling and in solid-state lighting applications.

The general methodology is as follows.

First, a test heat conduction path setup is created: typically a power semiconductor device is assembled to a cold plate. The purpose of the test could be to test the semiconductor device itself or to test the TIM layer applied between the semiconductor device and the cold plate.

As second step, thermal transient measurement of the test setup is performed before any stress condition of the given reliability/or life time test is applied.

Then, the stress condition is applied. In reliability analysis, typically temperature cycling and/or power cycling are used as a thermal stress condition. Another often applied stress condition is relative humidity (RH) cycling. In case of LED life time testing (LM80 standard [42] compliant tests), the stress condition is realized by operating the LEDs at nominal forward current at elevated environmental temperature (e.g., 85 °C).

Thermal transient measurements of the test structure are usually performed regularly while the stress conditions are being applied: e.g., at 500 h, 1,000 h, 1,500 h etc. in LM80 life time measurements [29].

The first such measurements on power LEDs were published by the team of M.W. Shin [43, 44]. A power LED package was subject to high temperature and high RH conditions, which caused die attach delamination. The delamination process was "monitored" by structure functions taken after 3, 6 and 24 h of stressing. At the end of the experiment, the die attach delamination was confirmed by acoustic microscopy.

Example 6—structural changes during LM80 tests

Degradation of these thermal interfaces such as delamination or material aging results in increased thermal resistance, thus in higher junction temperature, which is among the reasons that lead to luminous flux reduction of LEDs during their life time or even to fatal failures.

End-users of LEDs have no control over the quality of most of these thermal interfaces except the ones introduced at the assembly of the LED-based product. These comprise the LEDs' attachment to a substrate like a metal core printed circuit board (MCPCB) or the so-called TIM2 layer—the TIM used to reduce the interfacial thermal resistance between the MCPCB and the heat sink (luminaire body).

In a series of experiments on aging LEDs performed by the team of Schanda et al. at the University of Pannonia (Veszprém, Hungary), the usual electrical and photometric/spectroscopic measurements during LM80 life time tests have been completed with thermal transient measurements and subsequent structure function analysis [29].

The obtained structure functions indicate all the essential elements of the junction-to-cold plate heat flow path, like the LED chip itself, die attach, the solder/glue between the primary LED package and the MCPCB used for assembly, and the TIM between the MCPCB and the temperature-controlled solid surface of the test chamber (cold plate).

With the regular thermal transient measurements during this experiment, one could observe structural degradation both within and outside the LED package. The most characteristic changes were delamination of the LED package from the MCPCB and the degradation of the applied TIM2 material.

Figure 4.65 provides snapshots for LED samples, which are considered "best" performing devices from the point of view of light output stability (samples A) and for the worst devices (samples B). In Fig. 4.65a, only a slight thermal resistance increase can be observed on samples A, the main contributor to this increase is at the end of the junction-to-cold plate heat flow path. This can be identified as the TIM (conventional thermal grease in this case) between the MCPCB and the test chamber. The increase of 0.9 K/W with respect to the original value of 15.1 K/W means about 6 % change of the total thermal resistance. In Fig. 4.65a, one can also see the high stability of the relative luminous flux of the investigated LEDs.

Figure 4.65b demonstrates that samples B show huge increase in the total thermal resistance as indicated by the structure functions. This increase took place within 500 h of aging time. Since the initial sections of the structure functions in Fig. 4.65b coincide (corresponding to the LED chip and the die attach), probably the increase reveals a delamination of the LED package from the MCPCB. Between 500 and 2,000 h, the same TIM aging took place as seen in Fig. 4.65a.

As one can see, there is a strong correlation between the maintenance of the emitted luminous flux of the LEDs and the structural stability of the junction-to-ambient heat flow path. In case of poorly performing LEDs, the significant increase of the thermal resistance is not the direct reason of the significant drop of the luminous flux.[17] The increased junction temperature resulting from the increased thermal resistance has accelerated those chip-level failure mechanisms, which cause light output reduction. Section 5.5.2 of Chap. 5 presents further details on light output measurement results obtained in this experiment.

[17] In this particular case, about 1.5 % loss can be attributed directly to the increased thermal resistance.

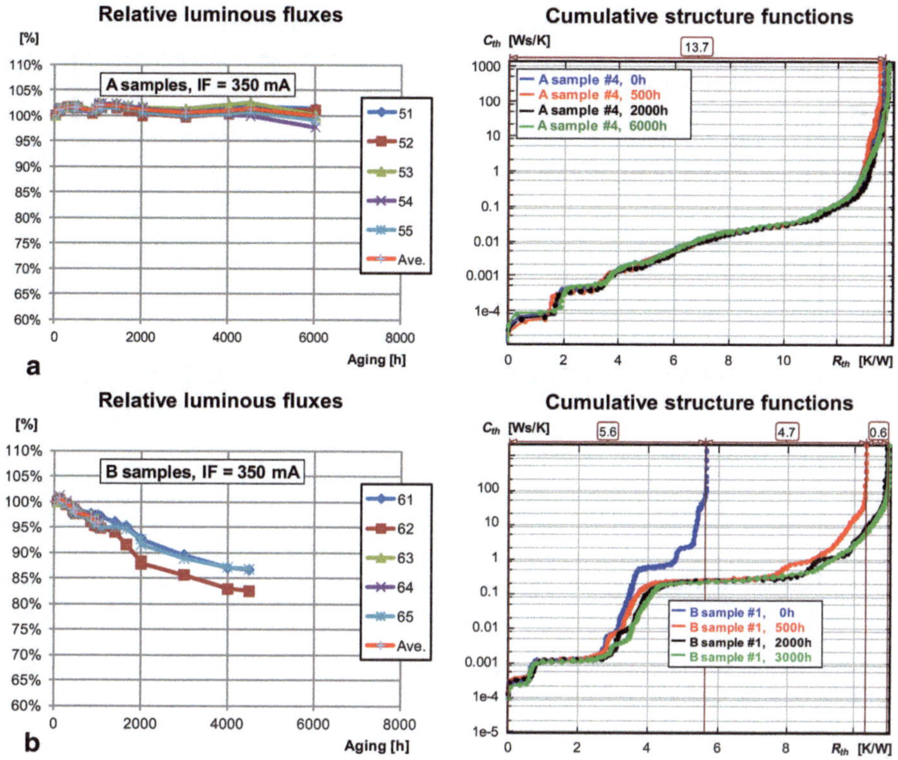

Fig. 4.65 Change of relative luminous flux and thermal properties of **a** a very stable LED and **b** a poor quality LED during LM80-compliant life time testing experiment [29]

Example 7—TIM property changes during temperature cycling

In this case study, the changes of TIM performance were monitored in situ, after repeated reliability tests. In the test arrangement, the TIM material was put between the external cooling surface of a power semiconductor device and a liquid cooled heat sink. The device was fixed to the surface of the heat sink by constant force, maintained by a special fixture.

The test was aimed at the study of the durability of TIM materials for temperature cycling, induced by power cycles applied to the device. After reaching the hot thermal steady state, the high power applied at the device was switched off and the cooling transient was measured by the transient tester. Such test conditions resemble the normal operation of the device: if the quality of the TIM gets worse, the same powering results in higher junction temperatures in each cycle.

The device was trained by approximately 20 W power steps, in 20 s periods of 50 % duty cycle. The surface temperature of the heat sink was set to 40 °C during all measurements. This allowed the junction temperature to rise up to approximately 120 °C in each cycle.

In the original study of Vass-Várnai et al. [45] the thermal behavior of eight different TIM samples was investigated; here we present the case of a nanoparticle filled composite.

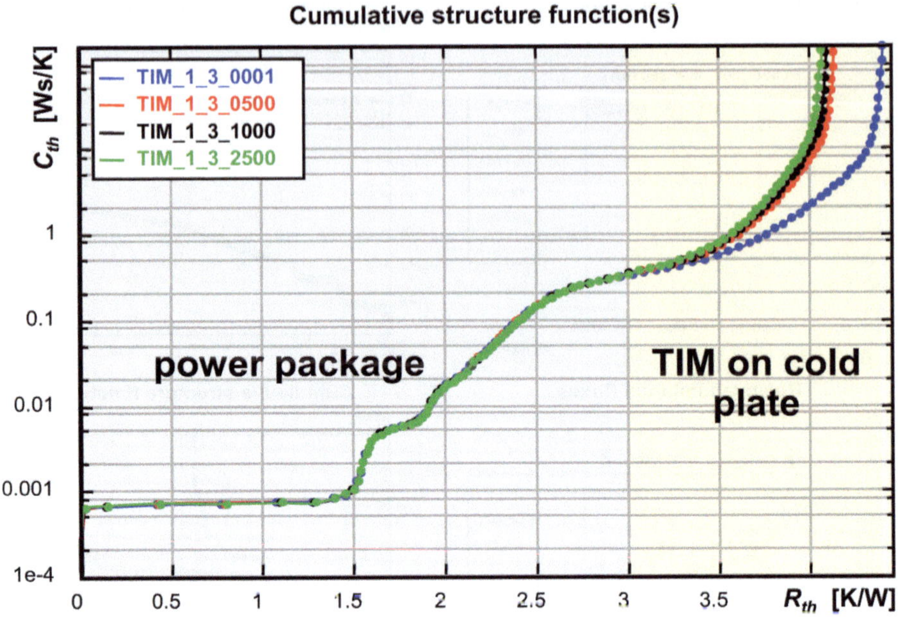

Fig. 4.66 Sample structure functions obtained from the power/temperature cycling test setup aimed at the study of the long-term behavior of different TIMs

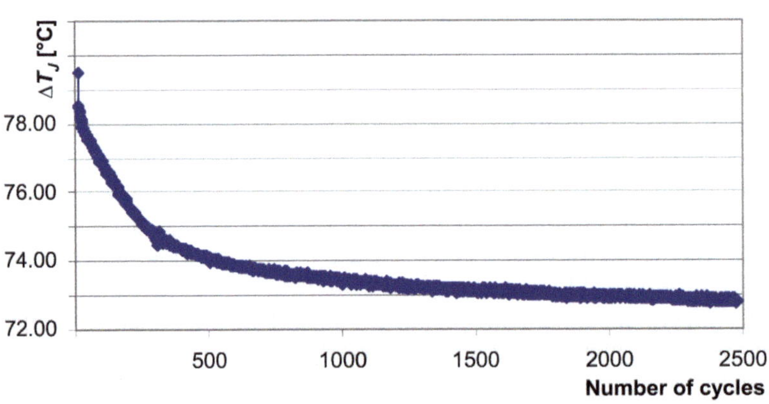

Fig. 4.67 TIM performance change during power cycling indicated by junction temperature elevation, derived from the structure functions of Fig. 4.66

During the tests of the investigated TIM materials, 2,500 cycles were performed and the cooling transients for every cycle were captured. Structure functions were generated from the junction temperature transients measured in every cycle. As seen in Fig. 4.66, as long as the heat flows in the same structure (the BJT package),

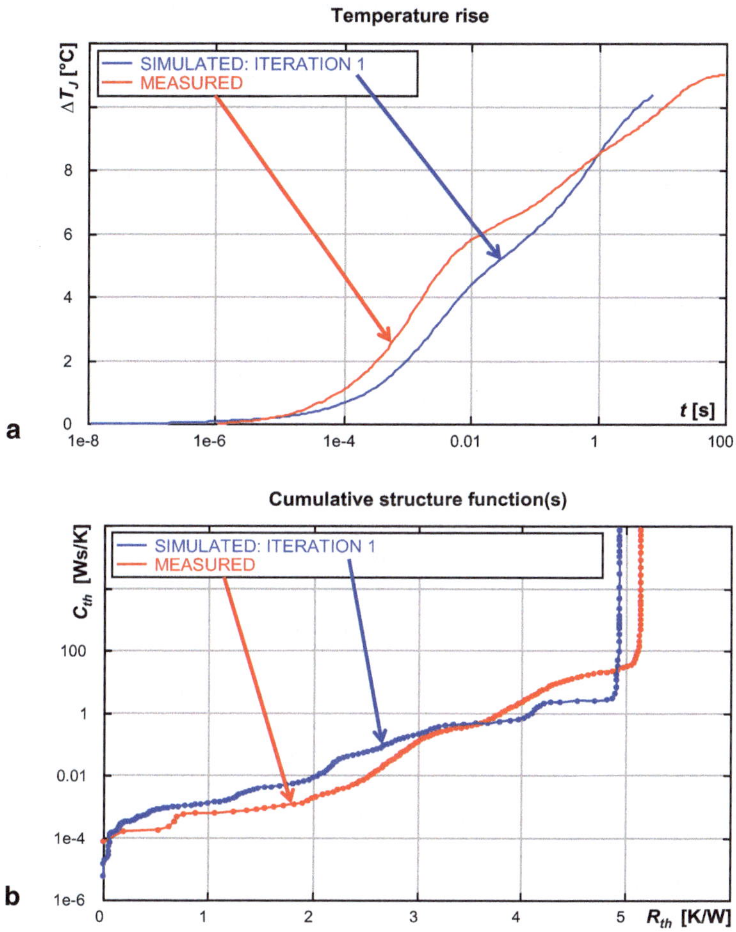

Fig. 4.68 Measurement and simulation results of a power transistor package: **a** junction temperature transients, **b** cumulative structure functions. Simulation was performed with an initial, uncalibrated detailed model

the structure functions are exactly the same. If the structure of the heat flow path changes, the difference immediately appears on these functions, showing the exact location (thermal resistance value) and magnitude (thermal capacitance value) of the change. In this example, the changes appear in the last section—corresponding to the investigated TIMs.

The reference structure function is the one obtained after the first cycle (the rightmost curve in Fig. 4.66). Further structure functions obtained after 500, 1,000, and 2,500 cycles are also shown in the figure. The observed junction temperature elevations for all test cycles of a selected TIM are shown in Fig. 4.68. In this particular example, the TIM performance has improved during the test.

The original study [45] includes examples in which TIM quality degradation was observed during the temperature cycling.

4.5.5 Structure Functions and Thermal Modeling

A simulation model is as good as the available input data: device geometry and material properties. When creating detailed models for CFD simulation tools, this is always an issue, even for semiconductor vendors who in principle must be aware of at least the device geometry. But many times, material parameters and effective volumes or area raise questions. One possible source of uncertainty in detailed thermal simulation models is the interface thermal resistance, both at TIM1 (die attach) and TIM2 (e.g., thermal grease).

Thermal transient testing was already used in the PROFIT project for the calibration of detailed simulation models further used in DELPHI-type compact model generation [46]. Though structure functions were already known as tools of heat flow path structural analysis, in this work they were not yet used: only measured and simulated junction temperature transients obtained for different DELPHI dual cold plate conditions were compared.

The idea behind using structure functions for detailed model calibration/validation is as follows. If both the geometry/material properties in the simulation model and the boundary conditions correspond to the real-life situation, then there should be no differences between the measured and simulated thermal impedance curves. Any small difference due to geometry or material mismatch should be visible in the corresponding structure functions. A paper by Bornoff and Vass-Várnai [47] provides a case study on how a power semiconductor device package model could be fine-tuned with the help of structure functions. We take the example from this paper[18]: creating the calibrated detailed model of BD-242-type transistor housed in a TO-220 package.

In the early iteration steps, the die size and the area of the active (dissipating) chip surface region were matched. With this modification, the first section of the heat flow path model became correct: as seen in Fig. 4.69a. With this, the match between the simulated thermal impedance and the measured thermal impedance became perfect up to a cumulative thermal resistance value of about 2.5 K/W.

After the die attach layer's properties also got modified (interface resistance set to the proper value by adjusting the thermal conductivity of the TIM1 material), the matching of the structure functions became perfect up to roughly 4.4 K/W, see Fig. 4.69b.

The last step in the model calibration was to properly set the thermal resistance of the applied TIM2 layer, see Fig. 4.69c. With this, the model calibration was completed. The remaining difference is attributed to the modeling of the cold plate used.

Trade-offs between simulation and measurement depend on many factors. Paper [48] discusses some of these. A recent approach is to create (dynamic) compact models of packages directly from measurement results [27], leaving out the creation of calibrated detailed CFD models used in standard compact modeling methodologies. For more on this topic, refer to Sect. 6.5.2.2 of Chap. 6.

[18] Measurement and simulation results by courtesy of Vass-Várnai and Bornoff.

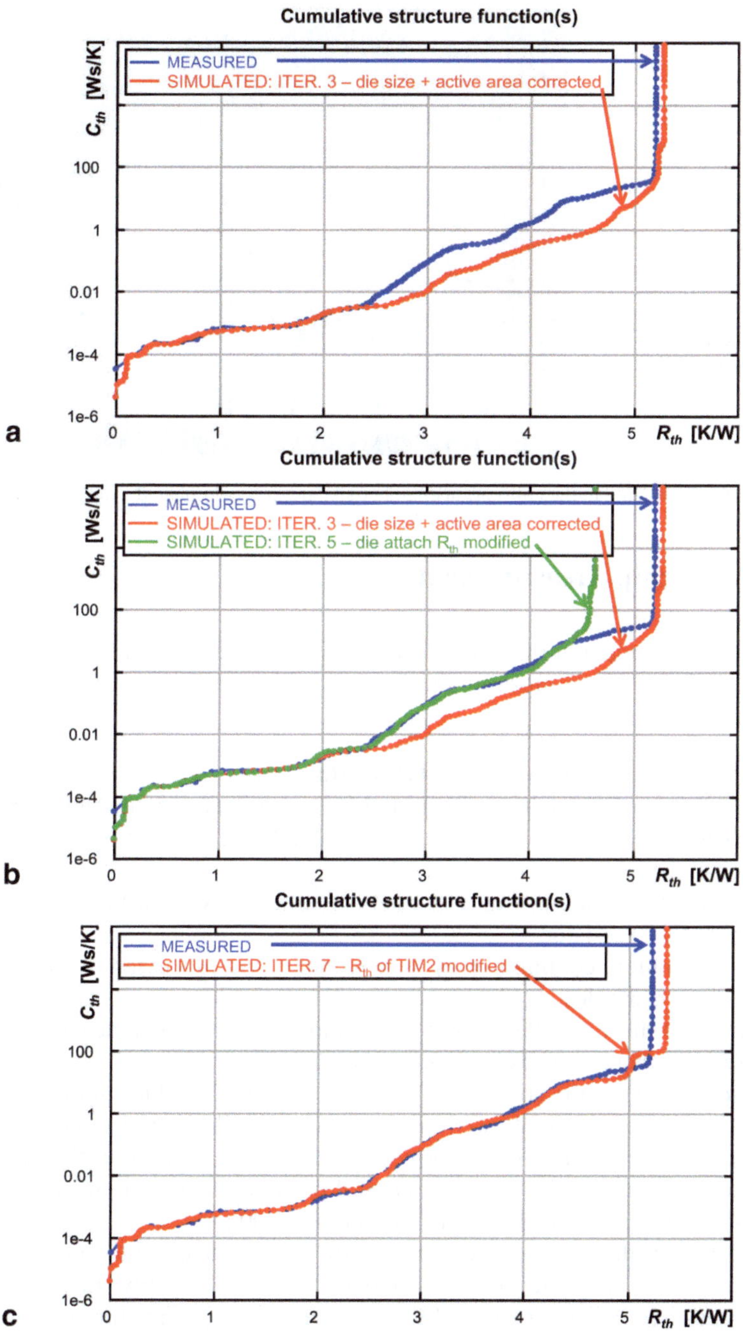

Fig. 4.69 Major iteration steps of adjusting the simulation model: **a** correction of the die size and the area of the active (dissipating) region, **b** die attach thermal resistance also corrected, **c** final calibrated model with adjusted TIM2 thermal resistance

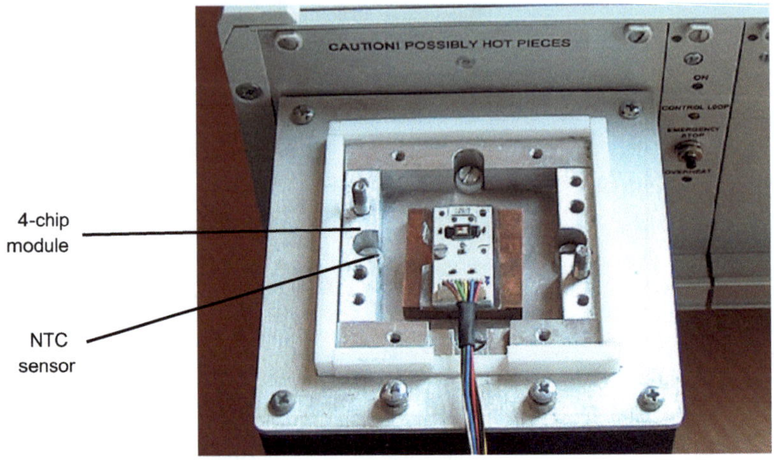

Fig. 4.70 Ostar module in the calibration chamber also used as cold plate during the measurement

4.6 Multichip Measurements

Thermal measurement of multichip modules is discussed systematically in Chap. 6, Sect. 6.1.4.2. Here, we illustrate the features of such measurement in a practical example.

Example 8—measurement of an RGB module
We measured a commercially available 4-chip Osram Ostar module built on a surface mount distributor plate. The module contained a 2 × 2 array of a red, a blue, and two green chips.

The plate temperature was separately measured by a negative temperature coefficient (NTC) resistor.

After calibration, cooling curves were recorded and processed in the same way as in previous examples.

Figure 4.70 shows the thermostat with removed lid in which the calibration and the subsequent thermal measurement was carried out.

Figure 4.71 shows the cooling curves of the system. The blue chip was driven by 350 mA, much below the allowed maximum current. We can observe that the run time from the driven blue to the nearby green chips was approximately 3 ms and to the red chip, which was in opposite diagonal position in the array it was 5 ms. The temperature change reached the NTC sensor after 400 ms.

Table 4.4 lists the measured "electrical only" $R_{thJA\text{-}el}$ junction-to-ambient thermal resistance values calculated from the input electric power only. Each row represents a measurement in which one LED was driven by heating current and then the cooling of four LEDs and the NTC sensor were recorded. In the diagonal of the table with bold numbers, we see the self-resistance (temperature change on the LED

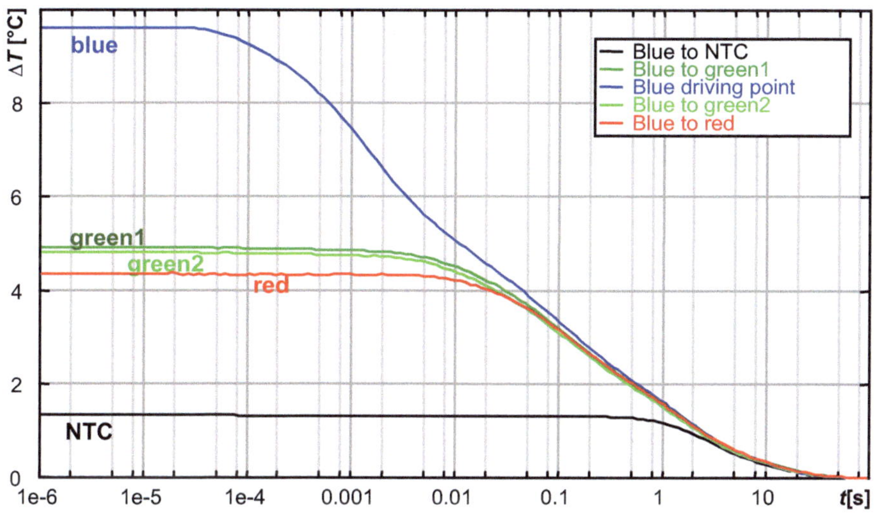

Fig. 4.71 Cooling curves of the Ostar module, blue chip-driven

Table 4.4 Measurement results for the RGB module

LED driven	$R_{ij\text{-}el}$ [K/W]				
	Channel		sensed		
	G1	B	G2	R	NTC
G1	10.44	4.67	4.03	4.53	1.15
B	4.21	9.24	4.13	3.73	1.14
G2	4.00	4.54	**9.95**	4.48	1.14
R	4.05	3.74	3.97	9.65	1.05

heated by electric power, also displayed in Fig. 4.72); other cells represent transfer resistance also known as *thermal characterization parameter* (temperature change at an unheated point divided by the input power of the heated LED).

The small asymmetry of the array is caused by the different energy conversion efficiency of the chips of different color.

The next two figures compare the $Z_{th\text{-}el}$ and $Z_{th\text{-}real}$ self-impedances of the four chips. The $Z_{th\text{-}real}$ curves are shifted such that they coincide on the longest section.

Cumulative structure functions composed from the self-impedance also show poor coincidence without optical power correction. Structure functions with correction show difference only in the chip region.

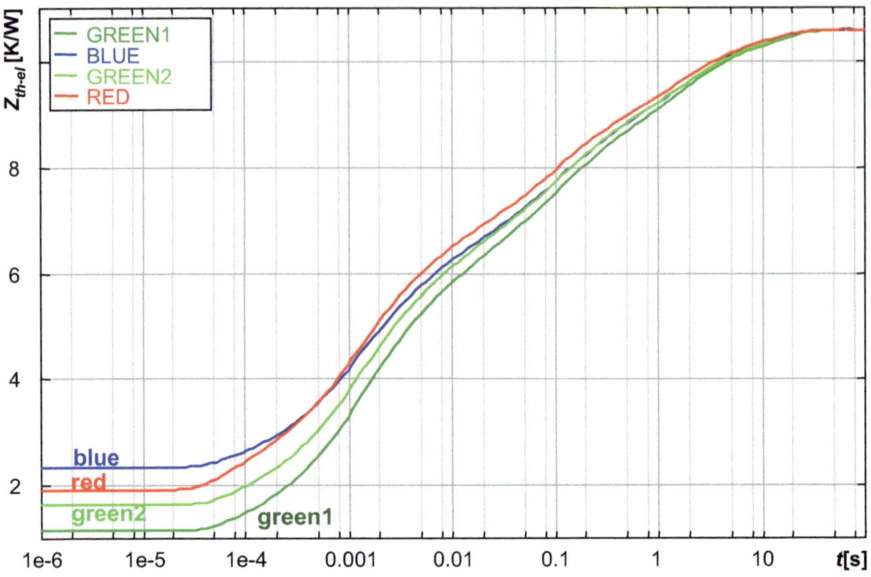

Fig. 4.72 The Z_{th-el} curves belonging to the self-heating of each LED chip

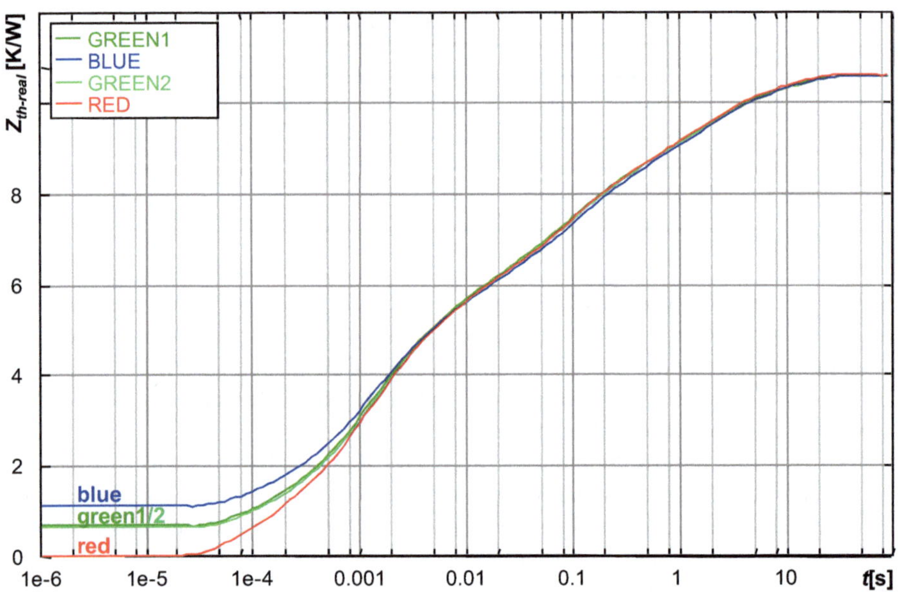

Fig. 4.73 The $Z_{th-real}$ curves belonging to the self-heating of each chip, magnified along y-axis so as to be scaled to Fig. 4.72

4.7 Issues of Thermal Testing of AC-Driven LEDs

4.7.1 Introduction

LED lighting applications can be powered either by a constant current LED driver circuit or directly from the AC mains. Such AC-powered LEDs may have different configurations [49, 50, 51]. In directly AC-driven LEDs or LED lamps, the driver electronics is simplified to a small integrated ballast, sometimes a serial resistor only, sometimes a component of reactive nature.[19]

At constant DC, powering LEDs dissipate steady power. This results in a steady junction temperature after a relatively short heating transient. Due to basically one-dimensionalheat spreading in LED components, a single *thermal resistance* value for further system (luminaire)-level thermal design is usually a sufficient model of the steady-state thermal behavior.

The first publications about techniques aimed at measuring the junction temperature of AC-driven LEDs appeared in 2010 [53, 54]. These methods are based on the idea of Zong and Ohno, first suggested for optical testing of DC-driven LEDs at a specified junction temperature [55]. In the following, we present how a single-valued thermal metric of AC-driven LEDs can be identified, based on concepts and testing methods shown for DC-driven LEDs.

As none of these methods provide any thermal metric for packages and cooling solutions of AC-driven LEDs, in the following, we try to fill this gap.

In case of AC-driven LEDs, the sinusoidal mains voltage results in a periodic waveform of the actual heating power, after an initial, transitional period while heating up the LED junctions. Once the shapes of the waveforms of the heating power and the junction temperature do not change any more, we can say that the directly AC-driven LED is in a *stationary state*. As our systems are nearly linear in the thermal domain, the thermally stationary situation can be treated similar to the small signal AC operation of linear electronic circuits.

In AC conditions, the frequency-dependent temperature response is of interest. Using the frequency domain description of the thermal impedance, the temperature can be calculated as:

$$T_J(\omega) = Z_{th}(\omega) \cdot P_{dissAC}(\omega) \tag{4.47}$$

where ω is (angular) frequency and P_{dissAC} is the AC power dissipation (as we show later, obtaining P_{dissAC} is not trivial). If P_{dissAC} were a purely sinusoidal signal in the form of $P_{dissAC}(t) = P_{\max} \cdot \sin(\omega t)$, the time domain representation of the AC junction temperature response would also be a sinusoidal function:

$$T_{JAC}(t) = T_{J\max} \cdot \sin(\omega t - \varphi),\ T_{J\max} = |Z_{th}(\omega)| \cdot P_{\max} \tag{4.48}$$

where φ is the phase shift between the power and the temperature waveforms.

[19] These ballasts usually hamper thermal testing and thus need to be bypassed for the sake of this test.

For fairly good approximate calculations, the dynamic compact network model representation of the thermal impedance (Fig. 4.74c) can be used in formulas (4.47) and (4.48) as usual in the simple small signal calculations of linear electrical circuits:

$$Z_{th}(\omega) = \frac{1}{j\omega C_{th1}} \times \left\{ R_{th1} + \frac{1}{j\omega C_{th2}} \times \left(R_{th2} + \left[\frac{1}{j\omega C_{th3}} \times (R_{th3} + Z_{hs}) \right] \right) \right\} \tag{4.49}$$

For the element values in Eq. (4.49), notation of Fig. 4.74c was used. Z_{hs} denotes the thermal impedance of the TIM and the heat sink (modeled by R_{th4} and C_{th4} in case of the present example) and the × symbol represents the "re-plus" operation (i.e., resulting in the net impedance/resistance of two parallel branches).

This means assuming constant amplitude of the power the amplitude of the junction temperature changes with the frequency since the value of $|Z_{th}(\omega)|$ vanishes with increasing frequency as it already concludes from Eq. (4.49). This also suggests that the *AC thermal impedance* (yet to be exactly defined in a subsequent section) of thesame AC LED lamp is of different value when operated from a 50 Hz mains or from a 60 Hz mains (see also Fig. 4.74b). Figure 4.74c also helps understand this: the dynamic heat storage in capacity of the different heat flow path sections described by the dynamic compact model act as "shunting branches," which become more effective at higher frequencies.

4.7.2 The Harmonic Components of the Periodic Heating Power

So far, the simple model of a purely sinusoidal dissipated power waveform helps understand the frequency-dependent nature of LEDs' AC thermal impedance. But this simplistic approach does not consider that reality: sinusoid power change does not occur. However, periodic power waveforms have a Fourier decomposition of harmonic changes around a mean (DC) power level. Terms in Eq. (4.50) describe any of these harmonics. The relationship between a periodic dissipation function and its harmonic components is

$$P_{dissAC}(t) = \sum_{n=0}^{\pm\infty} P_n \cdot e^{jn\omega_0 t} \tag{4.50}$$

where the complex number P_n denotes the Fourier coefficient of the nth harmonic of the $P_{dissAC}(t)$ "signal." Introducing the notation of $Z_{th\_n} = Z_{th}(n \cdot \omega_0)$—where ω_0 refers to the frequency of the base harmonic—the Fourier series expansion of real $T_{JAC}(t)$ time functions can be expressed as

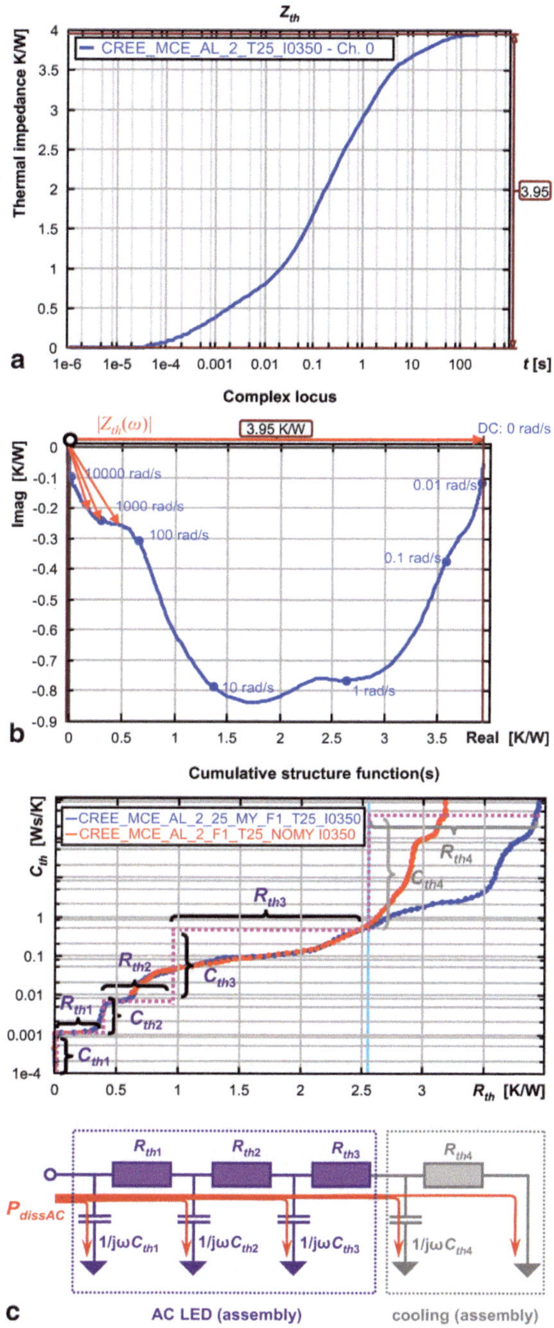

Fig. 4.74 Different representations of the thermal impedance of a 10 W LED: **a** the usual Z_{th} curve (time domain representation), **b** the complex locus of the $Z_{th}(\omega)$ function (frequency domain representation), **c** cumulative structure function and dynamic compact model representation of the same thermal impedance. (Source: [26] © IEEE, with permission)

$$T_{JAC}(t) = \sum_{n=0}^{\pm\infty} Z_{th\_n} \cdot P_n \cdot e^{jn\omega_0 t} \tag{4.51}$$

As introduced by Eq. (4.45) in Sect. 4.4.1.7, the $Z_{th}(\omega)$ function is known once the $Z_{th}(t)$ thermal impedance is measured, thus, the $Z_{th\_n} = Z_{th}(n \cdot \omega_0)$ values in Eq. (4.51) can be easily calculated.

4.7.3 The AC Heating Power of LEDs

To have a better understanding of the nature of the real AC heating power of LEDs, let us consider the model of LEDs' I–V characteristics in which we separate the forward current component responsible for light output and heat loss. As discussed in Chap. 2, light output is associated to the so-called radiative recombination processes and heat dissipation is due to the so-called nonradiative recombination processes, as illustrated by Fig. 6.34 of Sect. 6.5.2.2 in Chap. 6. With this in mind, we can write

$$\begin{aligned} I_F = I_{dis}(V_F) + I_{rad}(V_F) = I_{0\_dis} \cdot [\exp(V_F/mV_T) - 1] \\ + I_{0\_rad} \cdot [\exp(V_F/nV_T) - 1] \end{aligned} \tag{4.52}$$

where I_{0_dis} and I_{0_rad} are the current constants of the different current components associated with the nonradiative and radiative recombination processes; *m* and *n* are the corresponding ideality factors (introduced in Sect. 2.1.3.2 of Chap. 2).

Equation (4.52) is true for the forward characteristics of the LED only. However, most AC LEDs utilize both halves of the sine wave with appropriate circuitry. This way Eq. (4.52) can be used for AC voltage, although during the two different half-waves I_F may flow in different elementary diodes of an AC-driven LED chain.

With $V_{AC} = V_{MAX} \cdot \sin(\omega t)$ voltage change the current of the LED can be described as:

$$I_F(t) = I_{0\_dis} \cdot [\exp(V_{MAX} \cdot \sin(\omega t)/mV_T) - 1] + I_{rad}(V_{MAX} \cdot \sin(\omega t)) \tag{4.53}$$

The AC generator voltage is also substituted into the $I_{rad}(V)$ function representing the light emission. The dissipated AC power of the LED is determined by the AC voltage drop across the LED (the AC generator voltage) and the current *not* associated with light generation:

$$P_{dissAC}(t) = I_{0\_dis}[\exp(V_{MAX} \cdot \sin(\omega t)/mV_T) - 1] \cdot V_{MAX} \cdot \sin(\omega t) \tag{4.54}$$

Figure 4.75 shows $P_{dissAC}(t)$, the time function and the relative amplitudes of its first eight harmonic components.

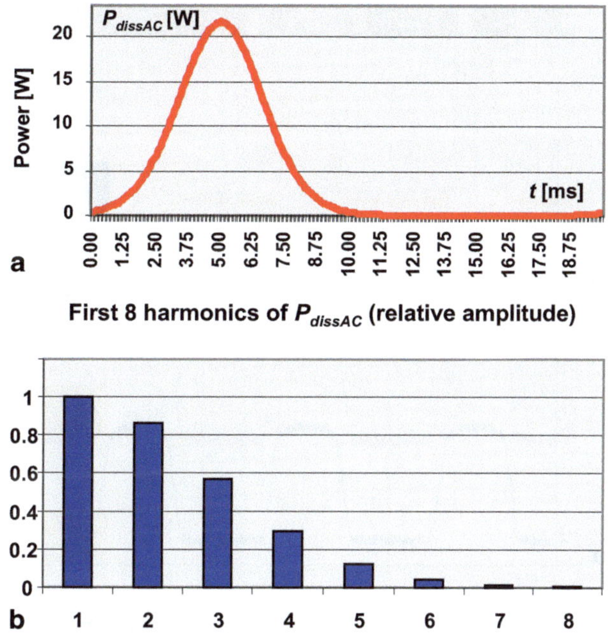

Fig. 4.75 **a** AC dissipation waveform (1 period) for a purely AC voltage generator-driven LED as per Eq. (4.54) and **b** the relative amplitudes of its first eight harmonics—see Eq. (4.50). (Source: [26] © IEEE, with permission)

Substituting the $\exp(V_{MAX} \cdot \sin(\omega t)/mV_T)$ term with the Taylor-series approximation of the exp() function and performing the multiplication with $V_{MAX} \cdot \sin(\omega t)$ in Eq. (4.54), we obtain:

$$\begin{aligned} P_{dissAC}(t) = V_{MAX}^2 \cdot \frac{I_0}{m V_T} \cdot \frac{1}{1!} \sin^2(\omega t) + \\ + V_{MAX}^3 \cdot \frac{I_0}{(m V_T)^2} \cdot \frac{1}{2!} \sin^3(\omega t) + \\ + V_{MAX}^4 \cdot \frac{I_0}{(m V_T)^3} \cdot \frac{1}{3!} \sin^4(\omega t) + \ldots \end{aligned} \quad (4.55)$$

We can see that due to the nonlinearity of the LEDs' I–V characteristics, higher harmonics of the base frequency of the AC voltage generator will appear in the dissipation. The even harmonics add to the DC level of the dissipated power, since $\sin^2(\alpha) = [1 - \cos(2\alpha)]/2$, $\sin^4(\alpha) = [3 - 4 \cdot \cos(2\alpha) + \cos(4\alpha)]/2$, etc.

The first harmonic represents a ½, the fourth $a^3/_8 \times$ ½ part of the DC level and so on. Since $\sin^3(\alpha) = [3 \cdot \sin(\alpha) - \sin(3\alpha)]/4$, the odd harmonics cause the base harmonics to appear in the power change but at a lower factor only.

Another extreme case is the AC current driven situation when a sinusoidal AC current is forced through the AC LED. The resulting voltage waveform and the amplitudes of the harmonic components the P_{dissAC} can also be analytically expressed for this situation—for details of this calculation, refer to [55].

In Fig. 4.76, measured waveforms of an AC voltage-driven LED + ballast resistor configuration are shown together with the harmonics of its AC dissipation.

a

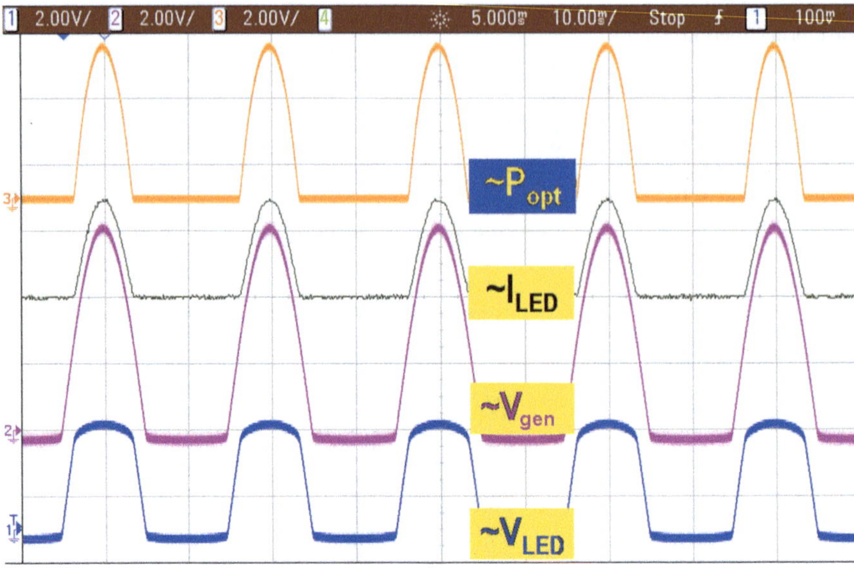

b

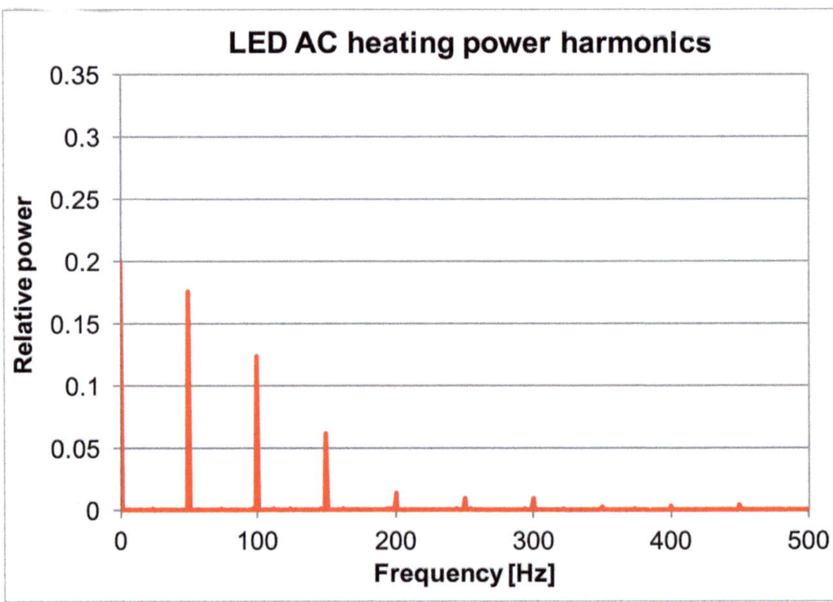

Fig. 4.76 **a** Different measured AC waveforms of power LED + ballast resistor, driven by an AC voltage generator **b** the relative amplitudes of its AC heating power

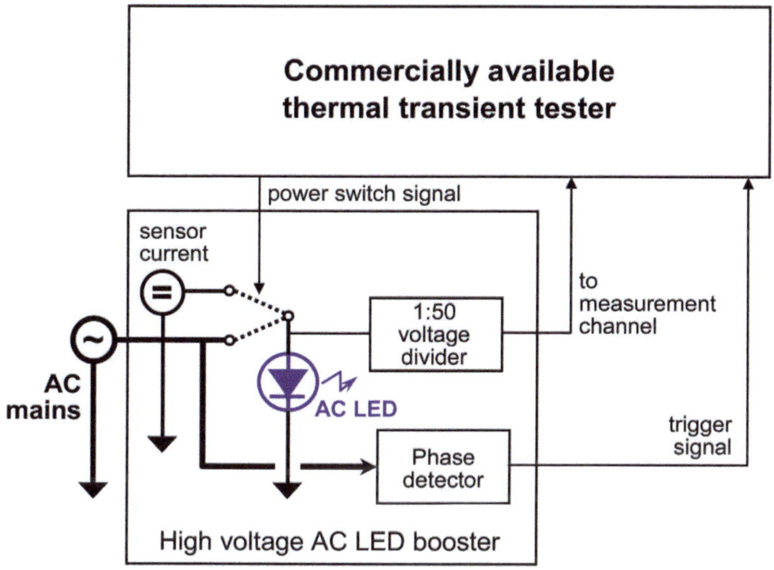

Fig. 4.77 Schematic diagram of a possible thermal test setup for the measurement of the AC thermal impedance of mains-driven AC LEDs. (Source: [53] © EDA Publishing)

4.7.4 *Possible Test Methods and Definitions of the Thermal Impedance of AC LEDs*

4.7.4.1 Direct Measurement

Direct measurement of the real AC thermal impedance of LEDs should be performed with care. Depending on the mains frequency different values are expected: the thermal impedance at 50 Hz is always larger than that at 60 Hz. The period of the mains voltage (20 ms/16.66 ms) is much longer than the smallest thermal time constants of usual LEDs—about 20 % of the thermal change of an LED occurs in this time interval.

Adapting the principles of the JEDEC JESD51-1 standard-compliant thermal transient measurements, the AC LED under test must be heated by the AC mains power source. Once reaching stationary current/voltage waveforms of a constant shape, the AC power can be cut off and the junction temperature drop can be measured by the usual electrical test method. The schematic of a possible test setup is shown in Fig. 4.77.

To assure repeatability, the power has to be switched down at a well-defined phase of the AC signal, since the measured impedance curves strongly depend on this phase. Figure 4.78 illustrates this for a low AC voltage-driven single LED + ballast resistor arrangement. Besides the dependence on the phase, another difficulty of such a measurement approach is the realization of the sharp switching between a 120 or 230 V AC power source (with no harmonic distortion) and a DC measurement current.

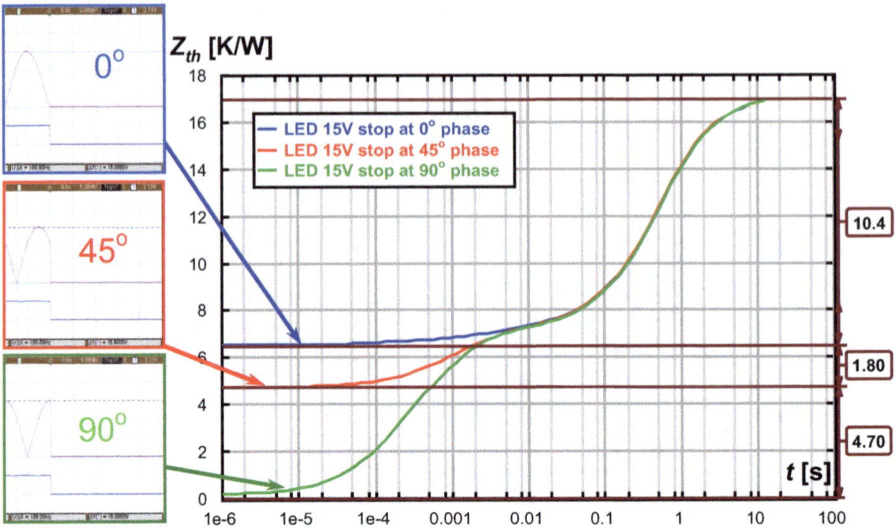

Fig. 4.78 Measurement results obtained by a low voltage version of the test setup outlined in Fig. 4.77

4.7.4.2 JESD51-1-Compliant Thermal Tests and Calculus

The most straightforward way is using the junction-to-ambient *thermal impedance*, which is a *unique property* of a semiconductor package. There are several ways of representing it, with well-defined transformation methods among the different representations as summarized in Sect. 4.1.8.

Powering an AC LED by appropriately high DC voltage (up to 150..280 V), one can measure the Z_{th} thermal impedance as defined in the JEDEC JESD51-51 electrical test method for LEDs. As recommended by the JESD51-52 standard, the measurement of the total radiant flux should also be performed. Such measurements can be performed with usual, commercially available test equipment.

The measured time domain thermal impedance curve can be transformed into frequency domain—this way one can obtain the $Z_{th}(\omega)$ function. At this point, we have the frequency domain model of the LED suitable for being driven by arbitrary waveforms.

As a next step, the AC voltage needed (120 V / 60 Hz or 230 V / 50 Hz) has to be applied. Once the AC LED has reached its stationary state, the waveforms of its AC voltage, current, and its periodic radiant flux can be measured (see, e.g., Fig. 4.76a in Sect. 4.7.3). From these waveforms, the actual $P_{dissAC}(t)$ heating power can be calculated. With Fourier analysis, the P_n harmonic components of the LED's heating power can be calculated.

Taking the values of the $Z_{th}(\omega)$ at the harmonic frequencies of the dissipation, the waveform of the AC junction temperature can be calculated with the help of Eq. (4.54).

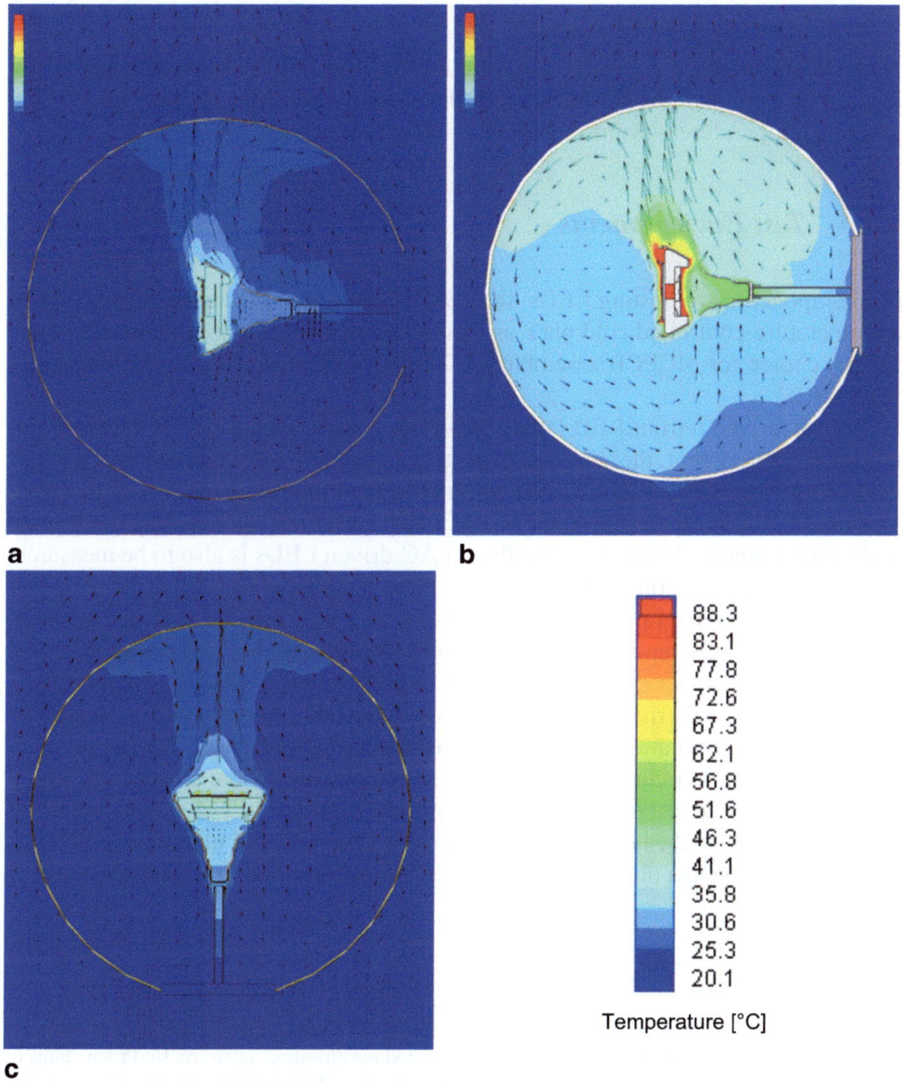

Fig. 4.79 An integrating sphere as thermal test environment for an LED lamp. Sphere properties (size, material) and lamp orientation influence the maximal junction temperature

Once the $T_{JAC}(t)$ and $P_{dissAC}(t)$ functions are known, the *effective AC thermal impedance* as a *thermal metric* can be derived. There are several possibilities to define such a thermal metric, such as the ratio of (some proper) mean values. Equation (4.56) and (4.57) provide two different examples for a possible calculation of such a value

$$Z_{thAC} - Z_{thAC\_mean} = T_{JAC} - Z_{thAC\_mean}/P_{dissAC\_RMS} \qquad (4.56)$$

$$Z_{thAC} - Z_{thAC\_mean} = T_{JAC} - Z_{thAC\_mean}/P_{dissAC-RMS} \quad (4.57)$$

where in both cases, e.g., P_{dissAC_RMS} is the RMS value of the periodic heating power. In the first case also the RMS value, in the second case the maximal value of the junction temperature waveform is used.

4.7.5 Test Environments

For component-level testing LEDs, the JESD51-50 series of standards recommend a temperature-controlled cold plate as thermal test environment. This would be the natural choice for MCPCB assembled AC-driven LEDs such as the Acriche series from SSC.

For AC-driven retrofit LED lamps, however, *natural convection* is the usual operational environment, therefore, it seems that a standardized natural convection test environment would be required and another thermal metric, some sort of junction-to-ambient AC thermal impedance would make sense. As we are interested in the real heating power, the total radiant flux of AC-driven LEDs is also to be measured, typically in an integrating sphere.

Actually, the integrating sphere can also serve as a thermal test environment, providing hopefully repeatable test conditions. Again, care has to be paid, since the lamp orientation in the sphere, the sphere size, the thermal conductivity of the sphere wall, and even the thermal conditions of the testing laboratory may have significant influence on the ultimate stationary junction temperature response. This can be easily revealed by CFD simulation. In Fig. 4.79, an integrating sphere serves as a natural convection thermal test environment for a retrofit LED lamp. The sphere properties (size, material) and the lamp orientation influence the maximal junction temperature (CFD analysis results obtained by the FloEFD tool from Mentor Graphics).

References

1. Lasance C J M, den Hertog D., Stehouwer P. (1999) Creation and evaluation of compact models for thermal characterisation using dedicated optimisation software. In: Proceedings of the 15th IEEE Semiconductor Thermal Meas-urement and Management Symposium (SEMI-THERM'99), San Diego, USA, 9-11 March , pp 189-200
2. Farkas G, Poppe A, Kollár E, Stehouwer P (2003) Dynamic Compact Models of Cooling Mounts for Fast Board Level Design. In: Proceedings of the 19th IEEE Semiconductor Thermal Measurement and Management Symposium (SEMI-THERM'03). San Jose, 11-13 March 2003. pp. 255-262.
3. Szabó P, Poppe A, Rencz M Studies on the possibilities of in-line die at-tach characterization of semiconductor devices. In: Proceedings of the 9th Electronics Packaging Technology Conference (EPTC'07), 10-12 December 2007, Singapore, pp. 779-784 (ISBN: 978-1-4244-1324-9)
4. Szabó P, Rencz M, Farkas G, Poppe A. Short time die attach characteri-zation of LEDs for in-line testing application. In: Proceedings of the 8th Electronics Packaging Technology Conference (EPTC'06): Volume One. Singapore, 6-8 December 2008, pp. 360-366 (ISBN: 1-4244-0664-1)

5. JEDEC Standard JESD51-1. Integrated Circuits Thermal Measurement Method - Electrical Test Method (Single Semiconductor Device). www.jedec.org/sites/default/files/docs/jesd51-1.pdf (December 1995)
6. JEDEC Standard JESD51-51 Implementation of the Electrical Test Method for the Measurement of the Real Thermal Resistance and Impedance of Light-emitting Diodes with Exposed Cooling Surface. www.jedec.org/sites/default/files/docs/JESD51-51.pdf. (April 2012)
7. Siegal B. An_Introduction_to_Diode_Thermal_Measurements, www.thermengr.net/An_Introduction_to_Diode_Thermal_Measurements6.pdf
8. Siegal B (1978) Measuring thermal resistance is the key to a cool semiconductor. Electronics 51:121-126
9. Sofia J W (1995) Analysis of thermal transient data with synthesized dynamic models for semi-conductor devices. IEEE Trans Comp Pack Manuf 18(1):39-47
10. JEDEC Standard JESD51. Methodology for the Thermal Measurement of Component Packages (Single Semiconductor Devices). This is the overview document for the JESD51- series of specifications. www.jedec.org/sites/default/files/docs/Jesd51.pdf. (December 1995)
11. Guenin B Update on Thermal Standards Work by JC15.1. www.ewh.ieee.org/soc/cpmt/presentations/cpmt0303a.pdf
12. Guenin B Update on JEDEC Thermal Standards. Electronics Cooling Vol.18. www.electronics-cooling.com/2012/09/update-on-jedec-thermal-standards. 17 September 2012
13. JEDEC Standard JESD51-14 Transient Dual Interface Test Method for the Measurement of Thermal Resistance Junction-to-Case of Semiconductor De-vices with Heat Flow through a Single Path. www.jedec.org/download/search/jesd51-14.pdf. (November 2010)
14. JEDEC Standard JESD51-50 Overview of Methodologies for the Thermal Measurement of Single- and Multi-Chip, Single- and Multi-PN Junction Light-Emitting Diodes (LEDs). www.jedec.org/sites/default/files/docs/jesd51-50.pdf. (April 2012)
15. Farkas G, van Voorst Vader Q, Poppe A, Bognár Gy (2005) Thermal Investigation of High Power Optical Devices by Transient Testing. IEEE Trans Comp Pack Techn 28(1):45-50
16. Farkas G et al Electric and Thermal Transient Effects in High Power Optical Devices" Proceedings of the 20th IEEE Semiconductor Thermal Measurement and Management Symposium (SEMI-THERM'04). San Jose, March 2004, pp. 168-176.
17. Rencz M, Székely V (2004) Studies on the nonlinearity effects in dynamic compact model generation of packages. IEEE Trans Comp Pack Techn 27(1):124-130
18. Poppe A, Molnár G, Temesvölgyi T Temperature dependent thermal re-sistance in power LED assemblies and a way to cope with it. In: Proceedings of the 26th IEEE Semiconductor Thermal Measurement and Management Symposium (SEMI-THERM'10), 21-25 February 2010, Santa Clara, USA, pp. 283-288. (ISBN: 978-1-4244-6458-1)
19. Székely V (1991) On the representation of infinite-length distributed RC one-ports. IEEE Trans Circuit Sys 38(7):711-719
20. Székely V, Rencz M (2000) Thermal Dynamics and the Time Constant Domain. IEEE Trans Comp and Pack Techn 23(3):687-594
21. Székely V, Bien T V (1988) Fine structure of heat flow path in semiconductor devices: a measurement and identification method. Solid-State Electronics 31(9):1363-1368
22. Protonotarios E N, Wing O (1967) Theory of nonuniform RC Lines, part I: analytic properties and realizability conditions in the frequency domain. IEEE Trans Circuit Theory 14(1):2-12
23. Székely V (1998) Identification of RC Networks by deconvolution: Chances and Limits. IEEE Trans Circuits and Sys I - Fundamental Theory and App 45(3):244-258
24. Székely V, Rencz M, Poppe A, Courtois B (2001) THERMODEL: a tool for thermal model generation, and application for MEMS. Analog Integrat Circuit Signal Process 29(1-2): 49-59
25. Rencz M, Poppe A, Kollár E, Ress S, Székely V (2005) Increasing the Accuracy of Structure Function Based Thermal Material Parameter Measurements. IEEE Trans Comp and Pack Techn 28(1):51-57
26. Poppe A, Siegal B, Farkas G Issues of Thermal Testing of AC LEDs. Proceedings of the 27th IEEE Semiconductor Thermal Measurement and Management Symposium (SEMI-THERM'11), San Jose, USA, 20-24 March 2011, pp 297-304

27. Vermeersch B (2009) Thermal AC modelling, simulation and experimental analysis of microelectronic structures including nanoscale and high-speed effects. PhD Tesis, D/2009/10.500/24, University of Ghent, Ghent, Bel-gium. http://lib.ugent.be/fulltxt/RUG01/001/330/941/RUG01-001330941_2010_0001_AC.pdf
28. Poppe A, Molnár G, Csuti P, Szabó F, Schanda J Ageing of LEDs: A Comprehensive Study Based on the LM80 Standard and Thermal Transient Measurements. In: CIE 27th Session-Proceedings, CIE 197:2011: (Volume1, Part 1-2). 10-15 July 2011, Sun City, South Africa, Vienna: CIE, pp. 467-477
29. Rencz M, Székely V, Morelli A, Villa C (March 2002) Determining Partial Thermal Resistances with Transient Measurements, and Using the Method to Detect Die Attach Discontinuities. In: Proceedings of the 18th IEEE Semiconductor Thermal Measurement and Management Symposium (SEMI-THERM'02), March 12-14, 2002, San Jose, USA, pp. 15-20
30. Steffens O, Szabó P, Lenz M, Farkas G (March 2005) Thermal transient characterization methodology for single-chip and stacked structures. In: Proceedings of the 21st IEEE Semiconductor Thermal Measurement and Management Sym-posium (SEMI-THERM'05), 15-17 March 2005, San Jose, USA, pp 313-321
31. Schweitzer D, Pape H, Chen L, Kutscherauer R, Walder M (March 2011) Transient dual interface measurement-A new JEDEC standard for the measurement of the junction-to-case thermal resistance. In: Proceedings of the 27th IEEE Semiconductor Thermal Measurement and Management Symposium (SEMI-THERM'11), 20-24 March 2011, San Jose, USA, pp 222-229
32. Pape H, Schweitzer D, Chen L, Kutscherauer R, Walder M (2011) Development of a standard for transient measurement of junction-to-case thermal re-sistance. In: Proceedings of the 12th International Conference on Thermal, Mechanical and Multi-Physics Simulation and Experiments in Microelec-tronics and Microsystems (EuroSimE'11), 2011, pp 1/8- 8/8
33. Schweitzer D, Pape H, Chen L (2008) Transient Measurement of the Junction-to-Case thermal resistance using structure functions: chances and limits. In: Proceedings of the 24th IEEE Semiconductor Thermal Measurement and Management Symposium (SEMI-THERM'08), 16-20 March 2008, San Jose, USA, pp 193-199
34. Schweitzer D (2008) Transient dual interface measurement of the Rth-JC of power packages. In: Proceedings of the 14th International Workshop on THERMal INvestigation of ICs and Systems (THERMINIC'08), 24-26 September 2008, Rome, Italy, pp 14-19
35. Schweitzer D, Pape H, Kutscherauer R, Walder M (2009) How to evaluate transient dual interface measurements of the Rth-JC of power packages. In: Proceedings of the 25th IEEE Semiconductor Thermal Measurement and Management Symposium (SEMI-THERM'09), 15-19 March 2009, San Jose, USA, pp 172-179
36. Müller S, Zahner T, Singer F, Schrag G, Wachutka G (2011) Evaluation of thermal transient characterization methodologies for HighüPower LED applications. In: Proceedings of the 17th International Workshop on THERMal INvestigations of ICs and Systems (THERMINIC'11), 27-29 September 2011, Paris, France, pp 104-109
37. ASTM Standard D 5470, 2012, Standard Test Method for Thermal Trans-mission Properties of Thin Thermally Conductive Solid Electrical Insulation Materials. ASTM International, West Conshohocken PA USA, 2012, DOI: 10.1520/D5470-12; www.astm.org
38. Mentor Graphics' DynTIM tester webpage: www.mentor.com/products/mechanical/products/dyntim
39. Mentor Graphics' T3Ster thermal transient tester webpage: www.mentor.com/products/mechanical/products/t3ster
40. Vass-Várnai A, Székely V, Sárkány Z, Rencz M (2011) New level of accuracy in TIM measurements. In: Proceedings of the 27th IEEE Semiconductor Thermal Measurement and Management Symposium (SEMI-THERM'11), 20-24 March 2011, San Jose, USA, pp 317-324
41. Vass-Várnai A, Sárkány Z, Rencz M (2012) Characterization method for thermal interface materials imitating an in-situ environment. Microelectronics J 43(9):661-668
42. ANSI/IESNA IES-LM-80-08 Standard (2009) Approved Method for Measuring Lumen Maintenance of LED Light Sources. ISBN 978-0-87995-227-3
43. Hu J, Yang L, Shin M W (2005) Mechanism and Thermal Effect of Delamination in Light Emitting Diode Packages. In: Proceedings of the 11th International Workshop on THERMal

INvestigations of ICs and Systems (THERMINIC'05), 28-30 September 2005, Belgirate, Italy, pp 249-254
44. Hu J, Yang L, Shin M W (2007) Mechanism and Thermal Effect of Delamination in Light-Emitting Diode Packages. Microelectronics Journal 38:157-163
45. Vass-Várnai A, Sárkány Z, Rencz M (2009) Reliability Testing of TIM Materials with thermal transient measurements. In: Proceedings of the 11th Electronics Packaging Technology Conference (EPTC'09), 9-11 December 2009, Singapore, pp 823-827
46. Pape H, Schweitzer D, Janssen J H J, Morelli A,Villa C M (2004) Thermal transient modeling and experimental validation in the European project PROFIT. IEEE Trans Comp and Pack Techn 27(3):530-538
47. Vass Várnai A, Bornoff R, Sárkány Z, Ress S, Rencz M (2011) Measurement Based Compact Thermal Model Creation—Accurate Approach to Neglect Inaccurate TIM Conductivity Data. In: Proceedings of the 13th Electronics Packaging Technology Conference (EPCT'11), 7-9 December 2011, Singapore, pp 67-72
48. Poppe A, Vass-Varnai A, Farkas G, Rencz M (2008) Package characterization: simulations or measurements?. In: Proceedings of the 10th Electronics Pack-aging Technology Conference (EPTC'08). 9-12 December 2008, Singapore, pp 155-160. Paper A6.4.
49. Yen H-H, Kuo H-C, Yeh W-Y (2008) Characteristics of Single-Chip GaN-Based Alternating Current Light-Emitting Diode. Japanese Journal of Applied Physics 47: 8808-8810
50. Chou P-T, Yeh W-Y, Lin M-T, Tai S-C, Yen H-H (2009) Development of On-chip AC LED Lighting Technology at ITRI. Proc. of the 2009 CIE Midterm Conference, Light and Lighting, Budapest, Hungary
51. Cho J, Jung J, Chae J H, Kim H, Kim H, Lee J W, Yoon S, Sone C, Jang T, Park Y, Yoon E (2007) Alternating-current light emit-ting diodes with a diode bridge circuitry. Japanese J Appl Phys 46:1194-1196
52. Zong Y, Chou P -T, Lin M -T, Ohno Y (2009) Practical method for measure-ment of AC-driven LEDs at a given junction temperature by using active heat sinks. In: SPIE Proceedings 7422, Ninth International Conference on Solid State Lighting (eds.: Tsunemasa Taguchi, Ian T. Ferguson, Christoph Hoelen, Jianzhong Jiao), 2 August 2009, San Diego, CA, USA, p 742208
53. Y. Liu Y, Jayawardena A, Klein T R, Narendran N (2010) Estimating the junction temperature of AC LEDs. In: SPIE Proceedings 7784, Tenth International Conference on Solid State Lighting (eds.: I. Ferguson; M. H. Kane; N. Nar-endran; T. Taguchi), 1 August 2010, San Diego, CA, USA, p 778409
54. Zong Y, Ohno Y (2008) New practical method for measurement of high-power LEDs. In: Proceedings of the CIE Expert Symposium 2008 on Ad-vances in Photometry and Colorimetry, CIE x033:2008, pp 102-106
55. Poppe A, Farkas G, Temesvölgyi T, Katona B, Molnár G (2010) Thermal Impedance of AC LED-s. In: Proceedings of the 16th International Workshop on THERMal INvestigation of ICs and Systems (THERMINIC'10). 6-8 October 2010, Barcelona, Spain, pp 127-132

Chapter 5
Laboratory Measurement of Optical Properties of LEDs

János Schanda, Péter Csuti and Ferenc Szabó

Abstract Light-emitting diodes (LEDs) are solid-state sources, where the emitted light depends on the characteristics of the semiconductor material. It is usual to call an LED a light-emitting diode prepared from inorganic semiconductors, in contrast to organic light-emitting diode (OLEDs), prepared from organic compounds. In this chapter, devoted to the measurement of optical properties, we will deal only with inorganic semiconductor LEDs.

5.1 Photometric and Colorimetric Properties of Light-Emitting Diodes

Two classes of semiconductors are used nowadays to manufacture LEDs: aluminum gallium indium phosphide (AlGaInP) for producing red to yellow LEDs and indium gallium nitride (InGaN) for building green to blue LEDs. The two classes of LEDs have slightly different photometric and colorimetric properties.

For example, the emission wavelength and intensity of AlGaInP (or shortly GaP-based) diodes depend more strongly on temperature than that of GaN-based diodes. Figure 5.1 shows the temperature dependence of a red and of a blue LED. As can be seen, both the intensity of the radiation diminishes, and the spectrum changes to longer wavelength in both cases, but the changes are much larger in case of the red LED. Due to such temperature dependence, all photometric and colorimetric measurements have to be performed at well-defined constant temperature, and if white light is produced for example in a Red-Green-Blue (RGB) LED module by mixing the light of red, green, and blue LEDs, keeping the color constant needs some active control.

J. Schanda (✉) · P. Csuti · F. Szabó
Department of Electrical Engineering and Information Systems,
University of Pannonia, Egyetem u. 10, 8200 Veszprém, Hungary
e-mail: schanda@vision.vein.hu; schanda.janos@virt.uni-pannon.hu

P. Csuti
e-mail: csutip@vision.uni-pannon.hu; csuti.peter@virt.uni-pannon.hu

F. Szabó
e-mail: szabof@vision.uni-pannon.hu; szabo.ferenc@virt.uni-pannon.hu

C. J. M. Lasance, A. Poppe (eds.), *Thermal Management for LED Applications*,
Solid State Lighting Technology and Application Series 2,
DOI 10.1007/978-1-4614-5091-7_5,

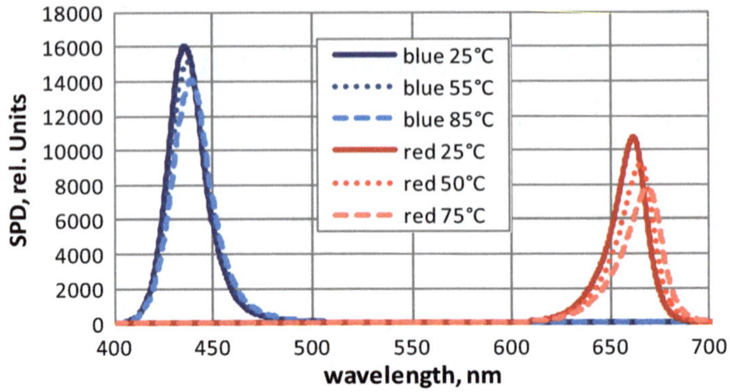

Fig. 5.1 Temperature dependence of the spectrum of a *blue* gallium nitride (GaN)-based (curves on the *left*) and a *red* gallium phosphide (GaP)-based (curves on the *right*) light-emitting diode (LED) (temperature was measured at the case)

LEDs were used first as signal lamps with little care about their photometric and colorimetric characteristics, partly due to the very unstable manufacturing conditions. It became necessary to define some measurement procedures for LEDs only at the end of the 1980s. The International Commission on Illumination (CIE: Commission Internationale de l'Eclairage) is the international commission that produces light measuring standards. The most basic concept in light measurement, the spectral sensitivity of the human eye, or as it is officially called the spectral luminous efficiency function, abbreviated as $V(\lambda)$ was published in 1924 [17]. LED photometric measurements are still done using this function (for more details see also Sect. 5.5). LED measurement recommendations were published in a technical report first in 1997, followed by a second edition in 2007 [1], where the most important measurement conditions for LEDs are described. In the following, we will summarize the most important photometric quantities.

5.1.1 Light-Emitting Diode Luminous Intensity

The fundamental photometric quantity is *luminous intensity*, as the unit of this quantity is part of the SI system[1], and couples the psychophysics of photometry to the physical quantities. The first obstacle to newcomers of photometry is the complicated definition of luminous intensity, using a second quantity, the luminous flux:

"Luminous intensity" (of a source, in a given direction) is the quotient of the luminous flux, $d\Phi_v$, leaving the source and propagated in the element of solid angle,

[1] The unit of luminous intensity is the candela. Its definition is: candela is the SI base unit for photometry: luminous intensity, in a given direction, of a source that emits monochromatic radiation of frequency 540 × 10^{12} Hz and that has a radiant intensity in that direction of 1/683 $W\cdot sr^{-1}$ Symbol: cd = $lm\cdot sr^{-1}$.

$\mathrm{d}\Omega$, containing the given direction, by the element of solid angle

$$I_{\mathrm{v}} = \frac{\mathrm{d}\Phi_{\mathrm{v}}}{\mathrm{d}\Omega} \tag{5.1}$$

Unit: cd = lm·sr^{-1}

with a note: "The definition holds strictly only for a point source."

Thus, one needs first the definition of luminous flux to understand this definition:

"**Luminous flux** is a quantity derived from the radiant flux, Φ_{e}, by evaluating the radiation according to its action upon the CIE standard photometric observer.

$$\Phi_{\mathrm{v}} = K_{\mathrm{m}} \int_0^{\infty} \frac{\mathrm{d}\Phi_{\mathrm{e}}(\lambda)}{\mathrm{d}\lambda} V(\lambda)\mathrm{d}\lambda \tag{5.2}$$

where, K_{m} is the maximum luminous efficacy (683 lm/W for photopic vision), $\Phi_{\mathrm{e}}(\lambda)$ is the spectral distribution of the radiant flux, and $V(\lambda)$ is the spectral luminous efficiency function.

Unit: lm. Here—on one side—we can be glad that the definition relates to a physical quantity, radiant power (in other chapters of this book also referred to as emitted optical power, denoted by P_{opt}), measured in watts, but we have immediately three further open questions: it says "for photopic vision": practically all photometric measurements are performed for photopic vision, as this covers vision under normal daylight conditions. For this situation, $K_{\mathrm{m}} = 683$ lm/W. The CIE standard photometric observer is described by the $V(\lambda)$ function, and is intended to represent average human eye spectral sensitivity (see Sect. 5.5).

With this in mind, the next obstacle is in the note for luminous intensity: valid "for a point source." To conform to this note is not simple in LED measurements, as it means that the measuring detector, measuring in reality illuminance, has to be so far from the source that the so called inverse square law of photometry should be valid, i.e., with increasing distance between the source and detector the illuminance should diminish with the square of the distance:

$$E_{\mathrm{v}} = \frac{I_{\mathrm{v}}}{d^2} \tag{5.3}$$

where, I_{v} is the luminous intensity (the v index refers to luminous = *visual* quantities, in the following we will not use this index if it is self-explanatory that the quantity is a photometric quantity), d is the distance between the source and the detector, and E_{v} is the illuminance.

Most early LEDs had a very collimated light beam to increase visibility in the preferred direction, and therefore the d distance had to be high. At the same time the luminous intensity was still low, and LEDs being small sources compared to the traditional ones it was difficult to adjust them properly along their optical or geometric axis. All this led to increased measuring uncertainty. Therefore, the CIE coined a new term: *Averaged LED* (*luminous*) *Intensity* (*ALI*) and defined its measurement geometry. Actually two geometries have been defined. CIE Publication 127-2007

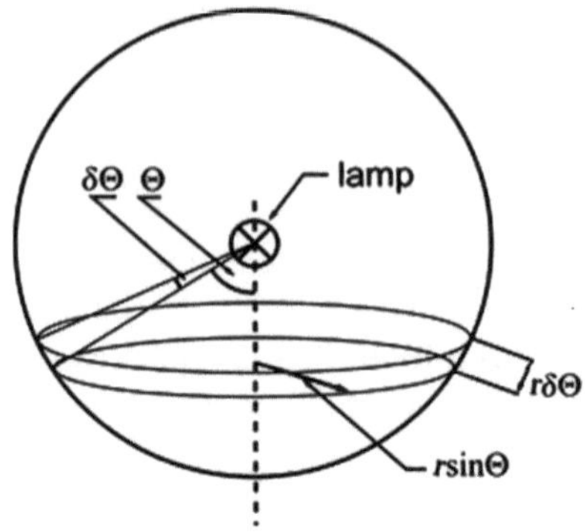

Fig. 5.2 Coordinate system around a light source to measure the total luminous flux

defines these geometries and the quantity ALI [1]. With modern high-intensity LEDs the ALI concept seems to be already outdated, we will not deal with it here, but refer to the original literature [1].

5.1.2 *Total Luminous Flux*

For traditional light sources and for nondirectional emission LEDs the most important photometric quantity is the total luminous flux of the device. This is again a quantity that is often mixed-up with luminous flux, as defined in the previous section. Luminous flux itself can be defined anywhere in the space, if the total amount of light emitted by a source is questioned one should always use the term *total luminous flux*, unfortunately often the adjective "total" is not included and this can lead to confusion.

The total luminous flux is the luminous flux emitted by the device in 4π geometry, i.e., the total sphere. If a point source lamp would emit in all directions evenly and would have a 1 cd luminous intensity (in all directions) its total luminous flux would be 4π lumens.

Luminous flux is either measured by summing up the partial luminous fluxes emitted in the different directions around the source, this is the fundamental method used by most national standards laboratories by deriving their luminous flux standards from luminous intensity standard lamps.

In practice, one usually takes luminous intensity measurements along a ring on the surface of a sphere drawn around the lamp to be measured, see Fig. 5.2. The surface of the ring is

$$A = r \cdot \mathrm{d}\Theta \cdot 2\pi r \sin(\Theta) \tag{5.4}$$

The solid angle Z represented by this ring is given by

$$Z = 2\pi r(\cos\Theta_1 - \cos\mathrm{d}\Theta) \tag{5.5}$$

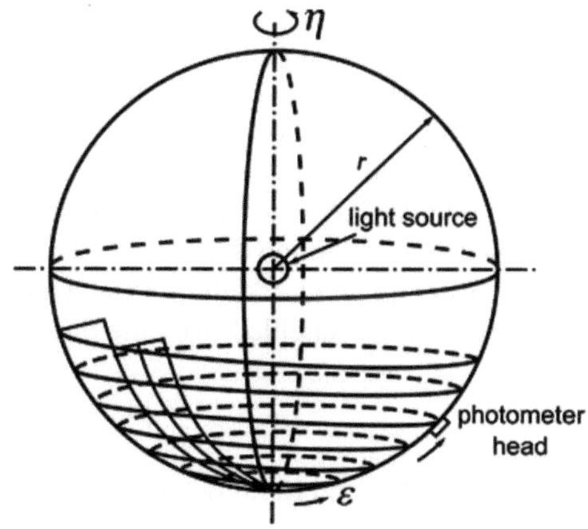

Fig. 5.3 Luminous flux measurement is based on illuminance measurement. (The figure has been taken from the International Commission on Illumination (CIE) publication 84–1989 and the authors are indebted to the CIE Central Bureau for their kind permission to include this figure here)

The luminous flux emitted into this ring (zone) is

$$\Phi_n = \int I_n \mathrm{d}\Omega = \int_{\Theta_1}^{\Theta_2} I_n \cdot 2\pi \sin\Theta \mathrm{d}\Theta = 2\pi I_n (\cos\Theta_1 - \cos\Theta_2) \tag{5.6}$$

where I_n is the average luminous intensity in the given zone and $2\pi(\cos\Theta_1 - \cos\Theta_2)$ is called the zone factor, which is in reality the solid angle of the given zone.

Critical here is how the zones are selected. Classical literature recommends to take 10° wide zones, and to measure the average luminous intensity in the middle of such zones, in that case the total luminous flux is

$$\Phi = \sum_{n=1}^{18} I_n Z_n \tag{5.7}$$

and I_1 is the average luminous intensity at 5°, ... I_{18} is the average luminous intensity measured at 175°. The zones have to be calculated in the following form:

$$\begin{aligned} Z_1 &= 2\pi(\cos 0^\circ - \cos 10^\circ), \ldots, \\ Z_{18} &= 2\pi(\cos 170^\circ - \cos 180^\circ) \end{aligned} \tag{5.8}$$

One can determine the total luminous flux also simply by illuminance measurement, see Fig. 5.3.

$$\Phi = \int_{(A)} E \mathrm{d}A = r^2 \int_{\varepsilon=0^\circ}^{180^\circ} \int_{\eta=0^\circ}^{360^\circ} E(\varepsilon, \eta) \sin\varepsilon \mathrm{d}\eta \mathrm{d}\varepsilon \tag{5.9}$$

here, E is the illuminance, r the radius of the virtual sphere, along which the detector takes the measurements, ε is the polar angle, and η is the azimuth angle.

The other possibility is to measure the total luminous flux in a photometer sphere (Ulbricht sphere, integrating sphere). This is usually a substitution measurement, where the total luminous flux of the test lamp is compared to that of a standard lamp (see [2]), although absolute methods have also been developed [3].

The theory of the integrating sphere tells us that if we have a hollow sphere, the inside of which is covered with a diffuse neutral paint of high reflection factor (with $\rho(\lambda)$ spectral distribution of reflectivity), and the light source is placed into the middle of the sphere, the indirect illuminance (E) reaching an area of the wall (i.e., this wall element has to be shielded from the direct light of the lamp) is

$$E = \frac{\rho}{4\pi(1-\rho)}\Phi \tag{5.10}$$

where, Φ is the total luminous flux of the lamp.

In practice, as mentioned, one uses a substitution method, as the corrections that have to be taken for the nonideal characteristics of the sphere, the loss due to openings, etc. would make the direct use of Eq. 5.10 difficult.

5.1.3 Luminance

Luminance is the most important photometric quantity, as the human observer sees luminance, actually luminance contrast, the contrast between the target and its immediate surroundings.

Luminance (in a given direction, at a given point of a real or imaginary surface) is defined by the following equation:

$$L_{\mathrm{v}} = \frac{\mathrm{d}^2\Phi_{\mathrm{v}}}{\mathrm{d}A\cos\theta\,\mathrm{d}\Omega} \tag{5.11}$$

where

- $\mathrm{d}^2\Phi_{\mathrm{v}}$ is the luminous flux transmitted by an elementary beam passing through the given point and propagating in the solid angle $\mathrm{d}\Omega$, containing the given direction;
- $\mathrm{d}A$ is the area of a section of that beam containing the given point;
- $\mathrm{d}\Omega$ is the elementary solid angle in which the beam is passing;
- θ is the angle between the normal to that section and the direction of the beam.

Looking at an extended source or seeing an illuminated surface, the luminance of the viewed surface will determine the visual impression. Luminance is the quantity that cannot be increased by optics. If the light comes from an extended source (such as a white LED module consisting of several blue chips with a single extended surface phosphor layer), the illuminance produced by this source can be increased by a proper optical system, but the luminance in the image surface can never be higher than the luminance of the source itself. In fact the luminance can be only smaller due to losses by reflection and absorption by the optical media.

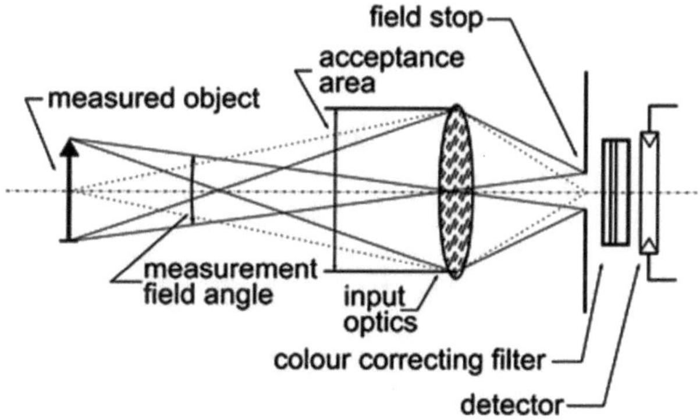

Fig. 5.4 Entrance optics of a luminance meter

Luminance measurement is usually performed by using a collimating optics that focuses the light from a small surface area of the source, see Fig. 5.4. With this arrangement it will be possible to measure always only the average luminance in this field of view. Critical for a luminance meter are the acceptance area (area where the instrument is sensitive) and the field of view, as within this the radiation should be homogeneous.

Luminance measurements (and radiance measurements with some different spectral weighting functions) are important in LED radiometry for the determination whether an LED is an eye-safe source or not. With very high luminance sources—and LEDs tend to get into this category—the blue light radiation might be so high that even a short time look into the direct light of an LED can cause retinal damage. (For details on photobiological hazards of light sources see [4].)

The detector of luminance meters is often a small charge-coupled device (CCD) array detector spectrometer that provides spectral data as well as the required colorimetric, photometric or radiometric quantities (as provided by embedded software of the spectrometer).

5.1.4 Colorimetry

The four most important colorimetric descriptors of LEDs are: chromaticity, dominant wavelength (for "monochromatic" LEDs), correlated color temperature (CCT) and color rendering index (CRI) (for white LEDs).

5.1.4.1 Chromaticity, Dominant Wavelength

For any of the CIE colorimetric characteristics first three quantities, the *X*, *Y*, *Z* *tristimulus values*, have to be determined. These are provided from the spectral power distribution (SPD) of the LED by folding the SPD with the so called color

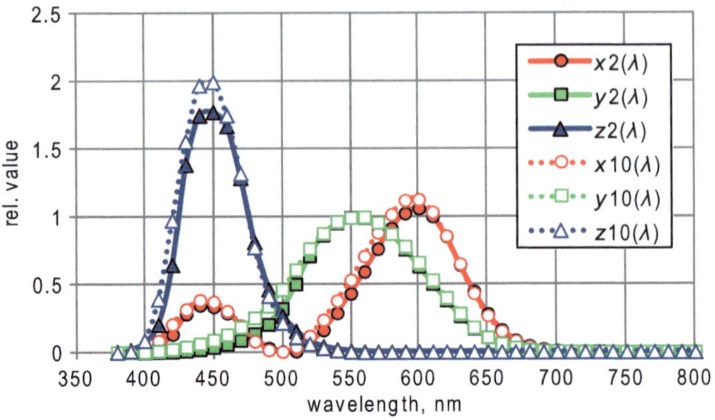

Fig. 5.5 CIE color matching functions (CMFs) for the 2° system (*solid lines*) and the 10° system (*broken lines*)

matching functions (CMF) over the entire visible range:

$$X = k \int_{380nm,}^{780nm} \Phi_e(\lambda)\bar{x}(\lambda)\mathrm{d}\lambda$$
$$Y = k \int_{380nm,}^{780nm} \Phi_e(\lambda)\bar{y}(\lambda)\mathrm{d}\lambda \qquad (5.12)$$
$$Z = k \int_{380nm,}^{780nm} \Phi_e(\lambda)\bar{z}(\lambda)\mathrm{d}\lambda$$

where $k = 683$ lm/W, $\Phi_e(\lambda)$ has to be measured as radiant quantity to get to the equivalent photometric quantity, thus, e.g., to get to luminous flux it has to be the radiant power density, and $\bar{x}(\lambda)$, $\bar{y}(\lambda)$, and $\bar{z}(\lambda)$ are the CMFs, see Fig. 5.5.

Figure 5.5 shows the CMFs for the two systems CIE has recommended, for LED color description usually the 2° system is used, but for some applications such as wall-washer lights the 10° system would be more appropriate. The 2° and the 10° refer to the visual size of the object seen.

From the X, Y, Z tristimulus values one usually calculates the x, y chromaticity coordinates:

$$x = \frac{X}{X+Y+Z}, \ y = \frac{Y}{X+Y+Z} \qquad (5.13)$$

Chromaticities are two dimensional (2D), real colors can be described only with three numbers, to characterize the color one needs beside the chromaticity coordinates the photometric description of the stimulus, thus, e.g., the luminance of the LED, or its total luminous flux.

Figure 5.6 shows the CIE x, y diagram, with the approximate representation of chromaticities (the chromaticities of highly saturated colors (near to the boundary of the graph) cannot be realized on a display or in print, we show the entire

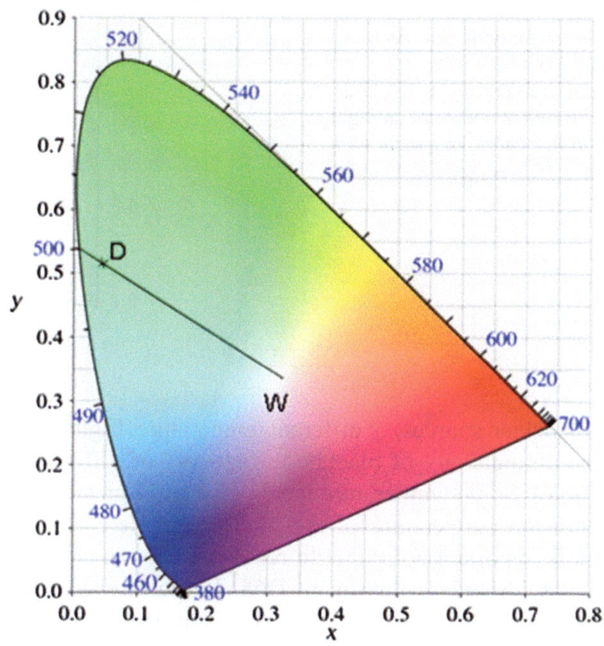

Fig. 5.6 CIE chromaticity diagram with approximate color representation

graph filled with color only to approximately show in which part of the diagram what chromaticities are found. The equienergetic white point is at the $x = 0.3333$, $y = 0.3333$ point (W). At the boundaries of the diagram the monochromatic (spectral) chromaticites are located from about 380 nm (deep purplish blue) to 700 (780) nm (deep red). Along the straight line connecting the two ends of the spectrum locus one finds the purple hues.

Drawing a line from the white point towards a monochromatic hue one finds chromaticities with roughly similar hue. If a chromaticity lies along such a line and is near to the spectrum locus the short term for that chromaticity is the *dominant wavelength*, the wavelength of that monochromatic hue that lies at the end of the line. In Fig. 5.6 the dominant wavelength of chromaticity D is 500 nm. The dominant wavelength is typically used to characterize monochromatic LEDs.

5.1.4.2 Correlated Color Temperature

As can be seen from Fig. 5.6 in the middle part of the chromaticity diagram there is a larger region where the light is perceived as white. This is even more so if this light produces the illumination in the room, and the visual system adapts to this light.

As mankind accustomed itself to changes in daylight spectrum—and the light of incandescent materials—it turned out that different phases of near white light can be well described by comparing their color to that of blackbody radiators

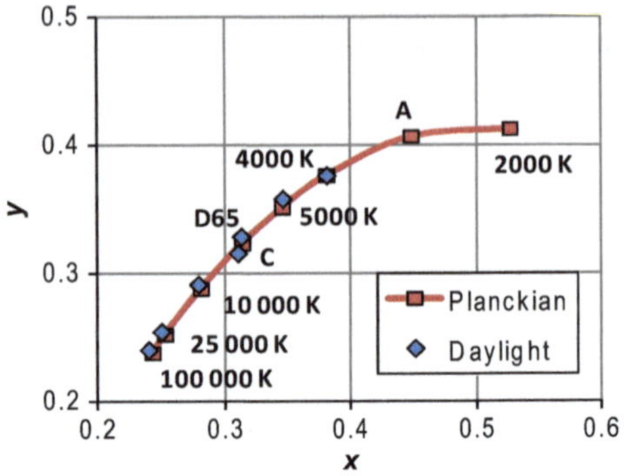

Fig. 5.7 Loci of the chromaticity of blackbody radiators of different temperature (*red* curve), and of different phases of daylight (*orange* curve) in a section of the CIE chromaticity diagram

(Planckian sources) of different temperature. Figure 5.7 shows the curve along which the chromaticity of blackbody radiators of different temperature are located (red curve). Some temperature labels give an impression how the chromaticity changes with the temperature of the blackbody. Here also the chromaticity of the three most important colorimetric reference illuminants (Standard illuminant A, labeled as "A," Standard illuminant D65, labeled as "D65" and illuminant C, labeled as "C") is shown. Blue diamonds show chromaticities of daylight illuminants of different phases of daylight. For higher temperatures (above 5,000 K) these are used as reference.

If the chromaticity of a light source lies exactly on the blackbody curve, we speak of its color temperature; if the chromaticity is above or below the blackbody curve one refers to that blackbody radiation the color of which resembles that of the test source most closely, and one speaks about *correlated color temperature* (CCT). (To be precise for the determination of CCT CIE has defined some rules, to discuss these here would go too far, see instead [5]). CCT gives a good first approximation on the chromaticity of the white light sources; it is used to characterize white LEDs.

5.1.4.3 Color Rendering Index

White light sources are usually used to illuminate objects, thus it becomes important how these objects look under the illumination of different sources. CIE decided to describe this by determining the color difference for a number of objects (eight for a general CRI R_a, and further six for some supplementary information) illuminated by the test lamp and a reference illuminant. As reference illuminant one has to choose a blackbody radiation of equal CCT if the CCT is below 5,000 K, and a phase of daylight if it is above 5,000 K.

The detailed description of determining the CRI would lead us too far; for those who are interested should read either the CIE publication [6] or a text that gives also some further information [7].

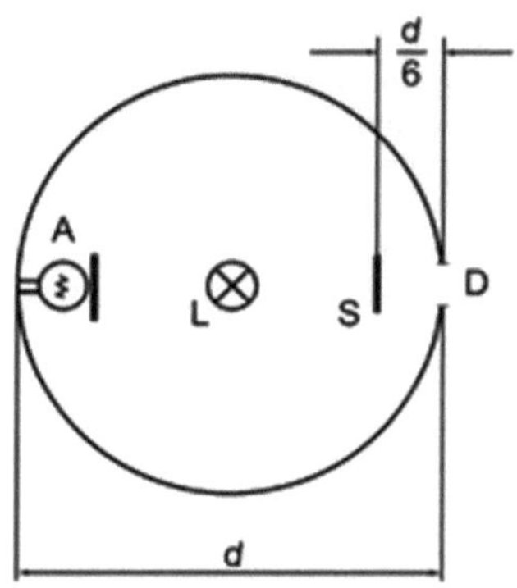

Fig. 5.8 Arrangement for measuring total luminous flux in an integrating sphere, where *A* refers to an auxiliary light source, *S* is a shutter, *D* the detector port, and *L* the lamp to be measured

5.2 Photometric Measurements

5.2.1 Total Luminous Flux Measurement in Photometer Sphere

The most common measurement in an LED manufacturer's or user's laboratory is the determination of the total luminous flux of an LED by using a photometer sphere and comparing the flux of the test LED to that of a standard LED. A photometer sphere is an integrating sphere (as introduced in Sect. 5.1.2) equipped with an appropriate detector called *photometric detector*. Thus, a photometer sphere is a hollow sphere painted internally with a flat diffuse white paint of high reflectance. An ideal sphere would have no openings and no objects inside the sphere. In this case, with a 100% reflecting white paint, the illumination measured at one point of the sphere wall and originating from all other point of the sphere wall (i.e., no direct light form the source) would be proportional to the total luminous flux of an infinitesimally small source inside the sphere. Naturally that would not be a practical solution, thus some compromise has to be found. The traditional form of a photometer sphere for luminous flux measurement has the form as shown in Fig. 5.8 [2]. Here L represents the light sources (test and standard LED placed alternatively) in the middle of the sphere. The diameter of the sphere (d) should be for LEDs smaller than 3 W electrical power consumption not less than 20 cm (for larger LED modules spheres of 1–2 m diameter are also customary).

A cosine corrected detector (i.e., a detector, the luminance sensitivity of which is independent of the angle of incidence) is placed at the opening[2] D. If the spectral distribution of the responsivity of the detector corresponds to the CIE $V(\lambda)$ function (the luminous efficiency function for photopic vision), the detector is called *photometric detector*. (In case of colorimetric measurements photo detectors with other spectral responsivity are used as well.) The detectors are usually based on silicon photocells as sensors, providing a photocurrent directly proportional to the incident light.

[2] Openings on spheres aimed at attaching external devices such as detectors are called ports.

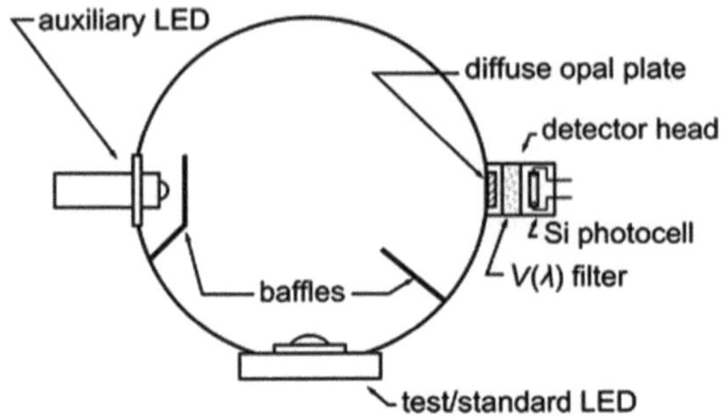

Fig. 5.9 Photometer sphere arrangement with the light-emitting diode (LED) at the sphere wall, corresponding to the arrangement of Fig. 9b of reference [1]

It is important that the detector measures the illuminance at the port of all light not emanating directly from the source L. To prevent direct light reaching the detector a shutter (denoted by S) is placed between the source and the detector opening.

The test and calibrating source distort the light distribution in the sphere, absorbing part of the light reflected from the wall. As the test and calibrating sources might have different absorption, one has to correct for this difference. This is done by placing permanently a further, auxiliary source (A) in the sphere in such a form that no direct light from this can reach the exit port. In practice four photocurrent measurements are made:

- i_t Photocurrent of detector with *t* test source in sphere and switched on.
- i_s Photocurrent of detector with *s* standard source in sphere and switched on.
- i_{At} Photocurrent of detector with *t* test source in sphere, but not lit and A auxiliary source switched on.
- i_{As} Photocurrent of detector with *s* standard source in sphere, but not lit and A auxiliary source switched on.

If the total luminous flux of the standard lamp is Φ_s the total luminous flux of the test lamp Φ_t will be:

$$\Phi_t = \Phi_s \cdot \frac{i_t}{i_s} \cdot \frac{i_{As}}{i_{At}} \tag{5.14}$$

If the sample to be measured has large bulky and light absorbing components (as, e.g., larger heat sinks or cooling units), placing them into the middle of a sphere will absorb a lot of radiation. In such cases, it is not good to place the light source in the middle of the sphere, especially if the light emission is only into the hemisphere, as is the case with most high power LEDs and LED modules. In such cases it is better to put the LEDs near to the sphere wall, see Fig. 5.9. Naturally, it is important that the LED should intrude slightly into the sphere, so that all light is captured by the sphere.

Critical for proper measurement is that the spectral distribution of the responsivity of the detector head + sphere wall absorbance should be well adjusted to the $V(\lambda)$ function. As the sphere wall paint ages, this adjustment also changes in time, therefore regular re-calibration of sphere + detector system is recommended. Besides

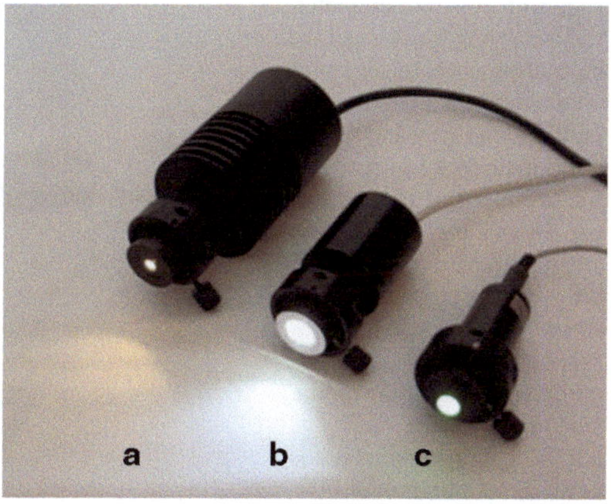

Fig. 5.10 Three standard light-emitting diodes (LEDs): **a** a traditional 20 mA one, **b** a special design for very uniform luminance, and **c** a high-intensity Peltier-cooled standard LED

maintaining the accuracy of the sphere + detector system one can also minimize the spectral mismatch error of LED measurements by using standard LEDs of the same SPD as the test LEDs. This is called *strict substitution* [1]. In this case by knowing the spectral emission properties of the standard/test LEDs and the spectral responsivity of the sphere + detector system a spectral mismatch correction factor can be applied. The correction factor, usually denoted by F can be calculated as follows [1]:

$$F = \frac{\int_{380\mathrm{nm}}^{780\mathrm{nm}} s_{\mathrm{t}}(\lambda) \cdot V(\lambda)\mathrm{d}\lambda}{\int_{380\mathrm{nm}}^{780\mathrm{nm}} s_{\mathrm{s}}(\lambda) \cdot V(\lambda)\mathrm{d}\lambda} \cdot \frac{\int_{380\mathrm{nm}}^{780\mathrm{nm}} s_{\mathrm{s}}(\lambda) \cdot s_{\mathrm{rel}}(\lambda)\mathrm{d}\lambda}{\int_{380\mathrm{nm}}^{780\mathrm{nm}} s_{\mathrm{t}}(\lambda) \cdot s_{\mathrm{rel}}(\lambda)\mathrm{d}\lambda} \tag{5.15}$$

where,

$s_{\mathrm{t}}(\lambda)$ is the SPD of the test LED,
$s_{\mathrm{s}}(\lambda)$ is the SPD of the standard LED,
$s_{\mathrm{rel}}(\lambda)$ is the spectral responsivity of the sphere + detector system, and
$V(\lambda)$ is the spectral luminous efficiency function.

The question of the standard LED is more complicated. To obtain constant luminous flux the LED has to be at a constant temperature, and fed with a constant current. However, even when meeting these conditions, not every LED supplies a constant output luminous flux, see Sect. 5.4. Figure 5.10 shows three standard LEDs, a traditional 20 mA LED standard, where an internal heating element keeps the LED's ambient temperature slightly above room temperature, a special standard LED for highly uniform light distribution and a high power LED standard where a built in Peltier-element is used to keep the LED's junction temperature at a constant value. All constructions have built in current supplies. The stability of well-aged and selected standard LEDs can have short-term stability well below 0.1%, and a long-term stability of about 0.5%/100 h.

Figure 5.11 shows a small LED measuring integrating sphere where beside of total luminous flux measurement also the chromaticity coordinates and the total radiant

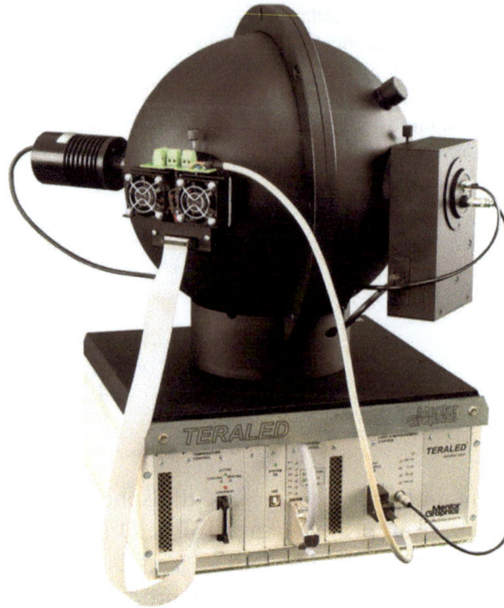

Fig. 5.11 A 30 cm diameter light-emitting diode (LED) measuring photometer sphere with a detector system including a $V(\lambda)$, and colorimetric/radiometric filters and a port for connecting, e.g., a charge-coupled device (CCD) array spectrometer to the test setup. Test LEDs in this system are to be mounted on a Peltier-cooled stage on the side of the sphere. The auxiliary light source here is realized by a high intensity white standard LED such as shown in Fig. 5.10

power of the LEDs can be determined by using specially designed filters to correct the spectral responsivity of the detector to the respective standard functions.

5.2.2 *Goniophotometric Measurements*

Goniophotometers have the advantage that with them the total luminous flux can be directly determined from luminous intensity measurements, as discusses in Sect. 5.1.2. Disadvantage is that goniophotometers need a large dark space, thus this is an alternative only for a laboratory where light distribution measurements have to be performed on a daily basis.

Officially, LED modules should be rotated in the goniophotometer only around a vertical axis, as otherwise the temperature of the chip might change, causing a luminous flux change. In practice this is often corrected by doing measurements at different C angles and $\gamma = 0$, see Fig. 5.12. As the LED is set to different C values, a change of luminous intensity in the $\gamma = 0$ direction is due to the temperature change and this has to be corrected to the normal position of the LED module.

LED light distribution is needed, e.g., when on the single LEDs some optics have to be mounted and the influence of this optics has to be tested. Figure 5.13 shows the γ angle dependence for different C planes of a single LED chip with a special microoptics to produce a batwing light distribution required for street lighting luminaires.

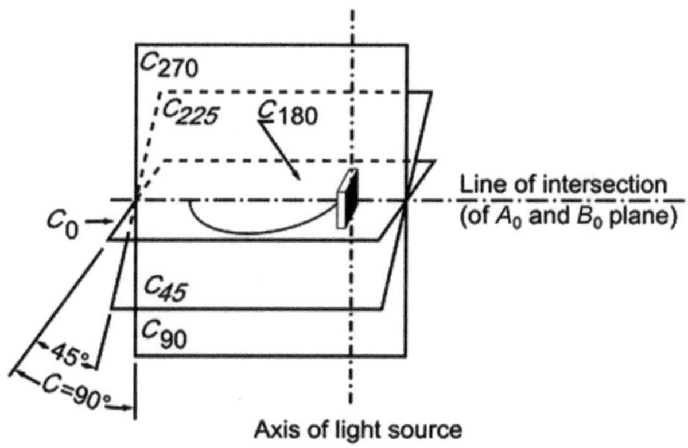

Fig. 5.12 A "C-gamma" set-up rotated so that the main emission axis of the lamp is in the horizontal plane

5.3 Spectrometric and Colorimetric Measurements

5.3.1 Spectrometric Instruments

Two types of spectrometers are used in LED spectral measurements: scanning type instruments and array type ones. Scanning instruments "scan" the spectrum, i.e., the single parts of the spectrum get consecutively recorded, and if the light intensity or spectrum changes in time an incorrect spectrum is recorded. Array spectrometers get the information in one shot, as a snapshot slice of a varying light.

5.3.1.1 Scanning Spectrometers

Scanning type spectrometers can be realized in different ways. Schematic of one possible realization, a single diffraction grating instrument is shown in Fig. 5.14. In scanning instruments the light to be measured falls via an input slit and collimating mirror or lens onto a diffraction grating (or in some older equipment onto a dispersing prism). This produces the spectrum, of which the spectrally resolved light (the picture shows one monochromatic component) is focused by a second mirror or lens onto the exit slit. By rotating the diffraction grating/prism different parts of the spectrum can be focused onto the exit slit. By slowly scanning the dispersion element the spectrum consecutively appears at the output slit, and can be recorded by a photoelectric detector placed behind the output slit.

Prism instruments have the disadvantage that the change in wavelength at the output slit is highly nonlinear with the angle of rotation, in case of grating instruments the wavelength-rotation angle calibration is much easier (although irregularities in mechanics can produce smaller deviations). The main problem with diffraction gratings is that the grating diffracts always into different orders, and e.g., the deviation angle of wavelength λ_1 of the first order will be in the same direction as of the wavelength $\lambda_1/2$ of the second order. These higher order—shorter wavelength—components have to be cut by separate higher order cut-off filters.

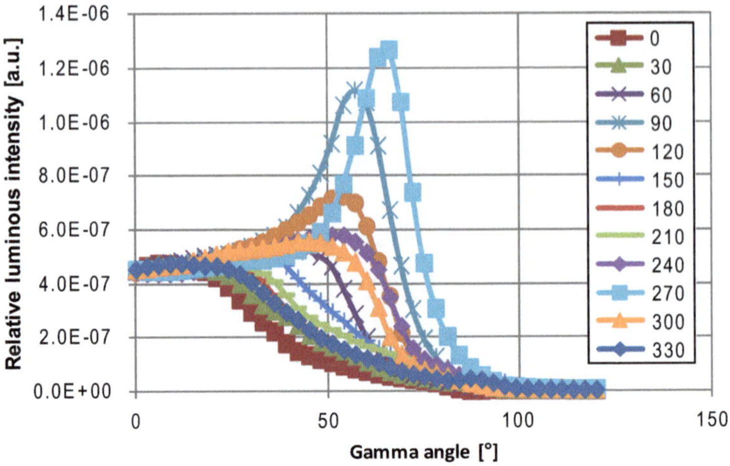

Fig. 5.13 Luminous intensity curves in different *C* planes

Fig. 5.14 Schematic view of a diffraction grating monochromator

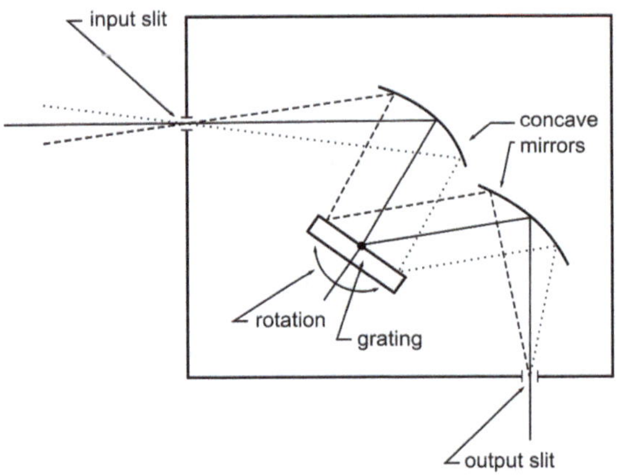

There are no optical parts that are perfect, the grating, the mirrors, etc. all have—even if only very low—diffuse reflection as well, and this produces scattered light in the monochromator. Scanning systems are of the advantage that they can be placed one behind the other, i.e., using two monochromators in tandem a double monochromator can be built with highly reduced stray light.

5.3.1.2 Array Spectrometers

The input part of the array spectrometer is similar to the scanning type, only behind the second (telescope) mirror the entire spectrum is focused in the focal plane, and a multipixel detector is placed in this focal plane. In modern array spectrometers mostly CCD arrays are used as detectors; their quality will highly influence the sensitivity, linearity, and lowest light level that can be measured. The working principle of an array spectrometer is similar to a digital camera: The dispersed light can reach the detector array for a given time, during this time period, charge carriers are collected under the single pixels, before the most heavily irradiated pixel gets overfilled the measuring cycle is finished and the accumulated photo-induced charge from each pixel is read out.

As is obvious in an array spectrometer no double monochromator system can be applied, thus the stray light of such systems is higher. To overcome this drawback some correction methods have been worked out, but these seem to work only with tunable laser sources [8].

5.3.1.3 Bandpass and Sample Interval

In a scanning instrument, besides the optical parameters of the instrument (grating line number per millimeter and total number of lines irradiated, focal length of collimator and telescope) the bandpass is set by the width of the input and exit slits. In a well-adjusted monochromator the bandpass function for equal slit widths can be approximated with an isosceles triangle.

In case of an array spectrometer the bandpass function is not that simple. Quite often, the input is not a slit but the output of a fiber, thus a round surface, and the image of this on the output—the array of the detector—might cover several pixels.

The general principle in SPD measurement is—and this can be realized in scanning instruments—that the bandwidth and the step-size with which one takes the measurements should be equal. As can be seen easily, if the step-size is larger than the bandwidth, some parts of the spectrum will not be measured, and if it is smaller, some parts will be measured twice. If the spectrum has fine structure and e.g.; a line is just left out or measured twice this will produce big errors. The remedy to overcome this error is to highly oversample the spectrum, this is the only possibility for array spectrometers.

If LED spectrometry is performed in order to calculate the tristimulus values and chromaticity coordinates from the spectrum, a further question to be answered is what the band-width of the measurement should be. Y. Ohno at NIST has made detailed calculations on the possible errors when different types of LEDs were measured with different bandwidth settings [9] and he came to the conclusion that the bandwidth error would appear only in the fourth decimal digit of the chromaticity coordinates if the bandwidth was kept below 5 nm.

5.3.2 Tristimulus Colorimetry

In Sect. 5.1.4.1 we have shown that to obtain the tristimulus values the SPD of the test source had to be folded with the CMF (see Eq. 5.12). Folding with the CMFs and integrating over the visible spectrum can be done in one step by using a detector

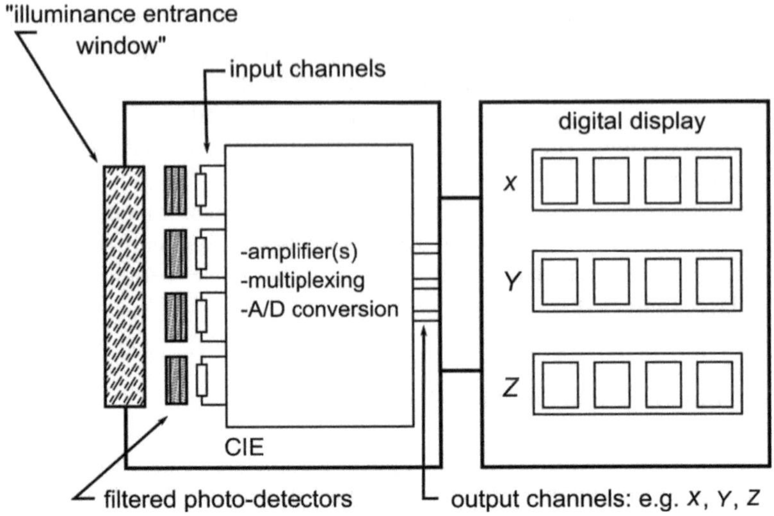

Fig. 5.15 Schematic diagram of an illuminance measuring tristimulus colorimeter for light source measurement

the spectral responsivity of which equals to the CMFs (see Fig. 5.5). This is called tristimulus colorimetry [10]. Figure 5.15 shows the schematic structure of a tristimulus colorimeter (of the illuminance mode, i.e., where the spatial responsivity approximates a cosine dependence), where the two humps of the $\bar{x}(\lambda)$ function had been realized by two independent detectors and filter sets. (The photometric test setup shown in Fig. 5.11 realizes this principle, but instead of four photo detectors it contains a single detector head and the filters realizing the CMFs are rotated in front of the detector head.)

One of the most critical points in a tristimulus colorimeter is how good the filtered detector responsivity corresponds to the CMF. There is no internationally accepted method for characterizing the goodness of fit of the spectral match of a colorimetric detector head. At present, many manufacturers use the method suggested by the CIE for radiometers and photometers [11]. This method is based on the spectral mismatch error index $f_1{}'$ that has to be provided for each channel of the colorimeter:

$$f'_{1,\mathrm{i}} = \frac{\int_0^\infty \left| s_{\mathrm{rel,i}}{}^*(\lambda) - \bar{t}_\mathrm{i}(\lambda) \right| \cdot \mathrm{d}\lambda}{\int_0^\infty \bar{t}_\mathrm{i}(\lambda) \cdot \mathrm{d}\lambda} \cdot 100\% \qquad (5.16)$$

where $\bar{t}_\mathrm{i}(\lambda)$ one of the CMFs $\bar{t}_1(\lambda) = \bar{x}_s(\lambda), \bar{t}_2(\lambda) = \bar{x}_l(\lambda)$ (index "s" describes the short-wave, and "1" the long-wave peak of the $\bar{x}(\lambda)$ function), $\bar{t}_3(\lambda) = \bar{y}(\lambda)$, and $\bar{t}_4(\lambda) = \bar{z}(\lambda)$; i has similar meaning in the following equations. $s_{\mathrm{rel,i}}{}^*(\lambda)$ is the normalized relative spectral responsivity:

$$s_{\mathrm{rel},i}{}^{*}(\lambda) = \frac{\int_{0}^{\infty} S_{\mathrm{m}}(\lambda) \cdot \bar{t}_i(\lambda) \cdot \mathrm{d}\lambda}{\int_{0}^{\infty} S_{\mathrm{m}}(\lambda) \cdot s_{\mathrm{rel},i}(\lambda) \cdot \mathrm{d}\lambda} \cdot s_{\mathrm{rel},i}(\lambda) \tag{5.17}$$

Where $S_{\mathrm{m}}(\lambda)$ is one of the standard illuminants, in photometry usually Standard Illuminant A, in colorimetry often Standard Illuminant D65, and some manufacturers use the equi-energy spectrum (illuminant E). $s_{\mathrm{rel,\,i}}(\lambda)$ is the spectral responsivity of the detector–filter combination. For a "laboratory grade" colorimeter the $f'_{1,3}$ values of the $\bar{y}(\lambda)$ function approximation can be better than 1.5%, for the other functions the spectral mismatch indices are usually 3–5 times higher.

Tristimulus colorimeters are used, in their standard form, mainly for color difference evaluation. Measurements can be done in a very short time (signal levels are much higher than in spectral measurements, hence integration times can be shorter as with an array-type instrument) thus, tristimulus colorimeters lend themselves for rapid chromaticity (dominant wavelength or CCT) determination of test samples if calibrated with a reference standard.

5.3.2.1 Image-Taking Colorimeter

A group of colorimeters that is becoming more and more popular is that of the image-taking colorimeters. The input optics of these instruments is very similar to that of the luminance meter (see Sect. 5.1.3), the main difference being that they measure the luminance not in one small field angle, but image the scene onto an image recording 2D detector array such as a CCD. This way the luminance and the colorimetric coordinates in each pixel of the detector can be determined. Such an equipment is similar to a digital camera, the main difference is that there are no color filters in front of the individual pixels, (i.e., the detector itself is not filtered) but the necessary filters needed for the realization of the CMFs are changed (e.g., rotated, like in case of the single detector LED measurement station shown in Fig. 5.11). Information from a high number of pixels has to be processed; therefore, these instruments are always coupled to a computer. Computer control provides observation of the scene on the computer screen with a subsequent image capture and processing operation.

Although tunable filters which permit the capture of the image in narrow spectral bands exist, and thus make it possible to take spectra of the entire scene, the processing of very large amount of captured data, and the still not perfect spectral selectivity of such tunable filters (considerable stray radiation), make the tristimulus approach attractive.

The fabrication of filters for tristimulus colorimeters is not a simple task, but to produce good filter sets for image-taking colorimeters is even more demanding: the filters have to have equal optical thickness to avoid refocusing for every channel, and have to be extremely free of any optical inhomogeneity, as this would be seen in the image. A further problem that limits the maximum achievable accuracy of such filter-detector combinations is that the length of the light path through the filter to a pixel in the middle of the matrix is different from the length of the path in the corners of the array. This limits the maximum spectral correction that can be achieved [12].

The flexibility of an image-taking colorimeter can be increased if different imaging optics are attached to the colorimeter. As the spectral transmission of different optics is usually different, individual sets of color correcting filters optimized for each input optic have to be supplied, or the set of optics to be used with a given colorimeter has to be corrected individually to the colorimeter.

5.3.2.2 Matrix Correction of Tristimulus Colorimeters

The idea of matrix correction of the tristimulus values is not new [13, 14] . Numerical optimization allowed only recently reaching a level by these methods that provide big advantages in LED colorimetry. If one knows that the targets are colored LEDs, an adaptive optimization method can be used to automatically select the best matrix transformation of a five-input tristimulus colorimeter to achieve proper corrections [15, 16]. While an average tristimulus colorimeter will measure colored LED with colorimetry errors up to about 10 ΔE_{ab} units, using the adaptive matrix transformation technique this can be reduced to below $\Delta E_{ab} = 0.5$, thus the error due to practical approximations of the standard CMFs will become lower than other error components of such measurements.

5.4 Standard Laboratory Lifespan Test Methods of Light-Emitting Diodes

5.4.1 Basic Questions—Lifespan Test Method

5.4.1.1 Semiconductor and Light-Emitting Diode Lifespan Tests

Semiconductors lifespan tests are well described in industrial standards. In those tests, the samples are operated at an increased and stabilized temperature for several thousand hours. Based on the in-between measured data the lifespan of the device can be extrapolated. A semiconductor's lifespan will end as soon as it loses its conductivity in forward mode or the isolating capability in reverse mode. For LED lifespan investigations similar tests can be run, but these are not enough, as for LEDs the most important characteristic is their luminous output, hence the change of this has to be investigated too.

The lifespan of an SSL product may differ considerably from that of a traditional light source, where usually the end of the life is reached when the device gives no light output anymore. In case of an SSL product the situation is not that simple because an SSL product may contain beside the LEDs also other electrical components with

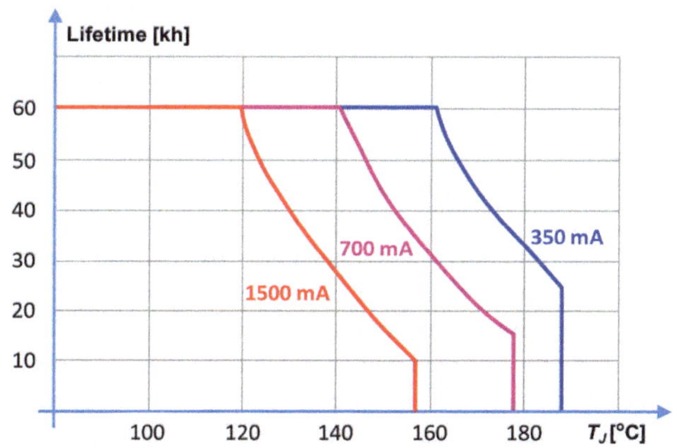

Fig. 5.16 Expected life based on drive current and junction temperature

different (shorter or longer) lifespan rating. In this book only LEDs are addressed so no luminaire systems are discussed.

5.4.1.2 Testing of Light-Emitting Diodes and Light-Emitting Diode-Based SSL Products

First LEDs emitted narrow band colored light; they were used as signal lights. For these LEDs very long lifetimes were announced, up to 100,000 h and more. As these lights were by orders of magnitude more efficient as traditional signal lights, nobody cared to stress them, and as during the normal lifespan of the equipment, where they were used, their luminous output has not changed too much, no real lifespan tests were necessary. The situation changed dramatically when white light LEDs became available, and it became obvious that these can be used for lighting. Soon it was realized that the optimistic 100,000 h lifespan cannot be reached if the LED is used at elevated temperatures and/or at high current density. The efficacy of traditional light sources decreases during their operational life but the end of the life is usually an abrupt failure of the lamp. Thus "rated life" of lamps is usually given by the test duration, during which, from a high number of test lamps 50% failed. During this time, it is usual that the luminous flux of the lamp changes only a few percent. As LEDs usually do not fail abruptly this technique cannot be used. Instead of this the working hours at which the LED's light output has decreased to 70% of initial output is stated, this is termed L70 (or L_{70}). See also Chap. 6.

One and the same LED can be used at different current levels, but this will influence its operating temperature and aging. Figure 5.16 shows how above mentioned L70 changes with junction temperature and drive current. Thus to be able to compare the quality of different products it became necessary to standardize test circumstances (LED current, temperature of aging, temperature cycling, etc.). Based on industrial and laboratory experiments a recommendation was worked out by the Subcommittee on Solid-State Lighting of the Illuminating Engineering Society (IES) Testing Procedures Committee.

The IES standard lumen maintainence (LM)-79-08 titled “Electrical and Photometric Measurements of Solid-State Lighting Products” discusses how an LED-based SSL product should be measured electrically and photometrically to state in a uniform way the product’s total luminous flux (lm), light intensity distribution, electrical power (W), efficacy (lm/W), and SPD and chromaticity (x, y or u',v'). This standard covers complete LED luminaires and integrated LED sources (e.g., retrofit sources) which need only line voltage or direct current (DC) power supply. Traditionally luminaire efficiency was determined by measuring the output of the luminaire and comparing it to the output of the light sources built into the luminaire. As in early SSL luminaires—and to a large extend this is still the situation—the light sources cannot be dismounted and energized separately (especially not under the same thermal circumstances). It should be noted that instead of relative photometry (used earlier for luminaires with traditional sources) such relative efficiency measurement was not possible, goniophotometric or other total luminous flux measurement had to be done in an absolute manner, determining directly the total luminous flux in lumens.

IES also published a standard on how life expectancy of an SSL product should be measured. This IES standard has the number LM-80-08 and is called “Measuring Lumen Maintenance of LED Light Sources” [17]. The name of the standard is in error, as lumen is a unit, it cannot be maintained, the standard deals naturally with the luminous flux maintenance of LEDs. It discusses how the light output depreciation of LED light sources, arrays, and modules should be evaluated. This standard does not cover luminaire measurements and does not provide information how the lifespan should be estimated. The standard states the temperature at which aging should take place, and what the environmental conditions should be. It provides information how frequently the photometric measurements should be done.

The IES TM-21-11 titled “Projecting Long Term Lumen Maintenance of LED Light Sources” (again a wrong term used!) discusses how to project long-term luminous flux maintenance of LED light sources when testing them per IES LM-80-08. It gives methods how to calculate different life projection numbers like L_{70} (hours). See also Chap. 6 on Standardization.

5.4.2 *Practical Lifespan Tests*

For LED lifespan measurements based on the IES LM-80-08 [28] one needs at least 10 pieces (20 pieces preferred) of the device to be tested. The aging test should run at least for 6,000 h (10,000 h preferred) on at least three case temperatures (55°C, 85°C, and a manufacturer selected case temperature). Figure 5.17 shows a test box for aging 60 LEDs in parallel. The temperature stabilizer seen on the left in the picture keeps the LEDs at predetermined temperature.

Measurements should be carried out at least every 1,000 h but it is recommended to measure more frequently in the first 1,000 h. For these time-periods, the thermostat cools the system down to 25°C, as photometric measurements should always

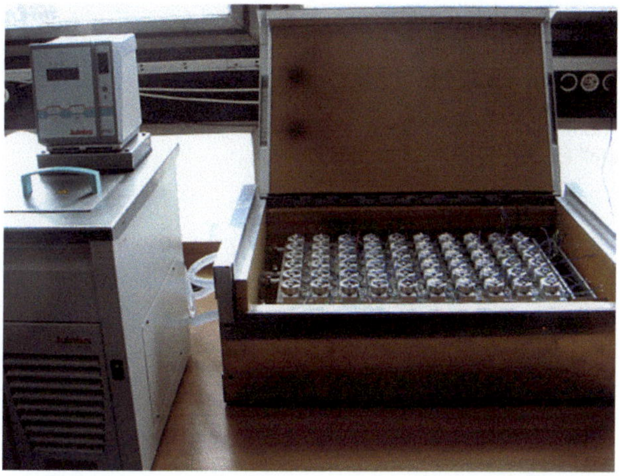

Fig. 5.17 A lumen maintainence (LM)-80 compliant test environment for aging 60 light-emitting diodes (LEDs)

be carried out at 25°C case temperature. All environmental and sample information should be reported. It is preferable to keep single devices for monitoring purposes separated (not aged) from the devices under test to keep track of the degradation of the measurement system over time. The IES LM-80-08 does not mention thermal resistance/thermal impedance measurements but it is highly recommended to collect those data also: e.g., structure functions extracted from thermal transient measurements during lifespan tests help identify structural degradations during the aging process of LEDs [29].

In Fig. 5.18 two examples of relative light output over operating time are showed. Part a shows an LED type with stable light output. Part b shows an LED type with fast degradation at 85°C aging temperature; it can be seen that after 4,000 h of aging the LED's relative light output of these LEDs dropped below 70% of the initial value at 0 h, thus these LEDs have an L70 of less than 4,000 h. As mentioned above, structure functions obtained by thermal transient tests indicate drastic structural degradation in case of type b LEDs (huge increase in thermal resistance during the test), see Fig. 5.19. For details on thermal (transient) testing and structure functions please refer to Chap. 4.

In Fig. 5.20 chromaticity shifts over aging time are showed: Part a shows an LED type which gets much yellower during the 8,000 h of aging, part b shows an LED type with relatively stable chromaticity.

From above examples it can be seen that LEDs can behave very differently, initial light output does not tell anything about aging, changes in luminous flux and/or color. The causes of degradation can be very different, thus it is difficult to extrapolate [30] from the 8,000 to 10,000 h data life expectancies of LEDs with such low degradations as shown in Fig. 5.18a. It is usual to fit an exponential decay to the data up to the 6,000 h point, but further research is needed to get better estimates of real-life expectancies of LED products.

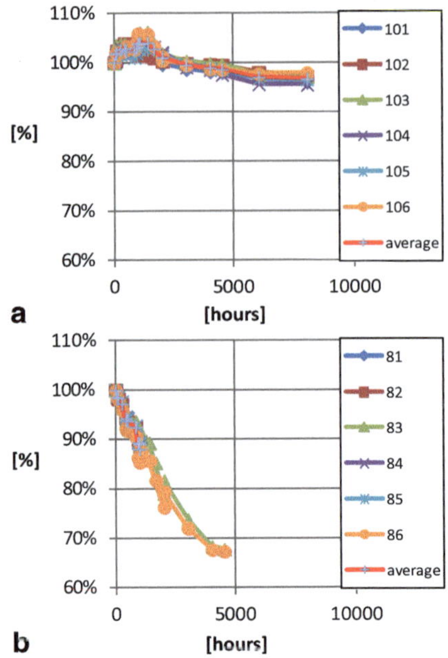

Fig. 5.18 Examples of changes of relative light output of light-emitting diodes (LEDs) with time. **a** "Long life" LEDs, **b** "short life" LEDs

5.5 Open Questions of Light-Emitting Diode Photometry and Colorimetry

5.5.1 Updating the V(λ) Function

CIE accepted the spectral luminous efficiency, or $V(\lambda)$ function in 1924 as a compromise solution of several investigations conducted using several different photometric techniques (flicker photometry, distinctness of boarder, heterochromatic brightness matching, etc.) [17]. It turned out soon that the so determined spectral responsivity is too low in the blue spectral band, compared to the average human eye sensitivity [18], however, CIE decided not to change the standard curve, as the effect on photometric calculations was negligible for the light sources used in those days. With the introduction of tri-band fluorescent lamps the difference became visible, thus in 1988 CIE decided to introduce a corrected spectral luminous efficiency function, the $V_M(\lambda)$ function [19] that corrected the discrepancy in the blue part of the spectrum, see Fig. 5.21. Difference between the 1924 $V(\lambda)$ and the modified function is only below 460 nm, resulting in difference of the measured luminous flux values for a white LED the difference is below 1%, but for a blue LED the luminous flux would be 10% higher if the $V_M(\lambda)$ function is used instead of the $V(\lambda)$ function.

Recently investigations have been performed that showed how the spectral luminous efficiency function changes with adaptation. Stockman and Sharpe published a cone fundamental-based function ($V^*(\lambda)$) that is based on the human eye long- and

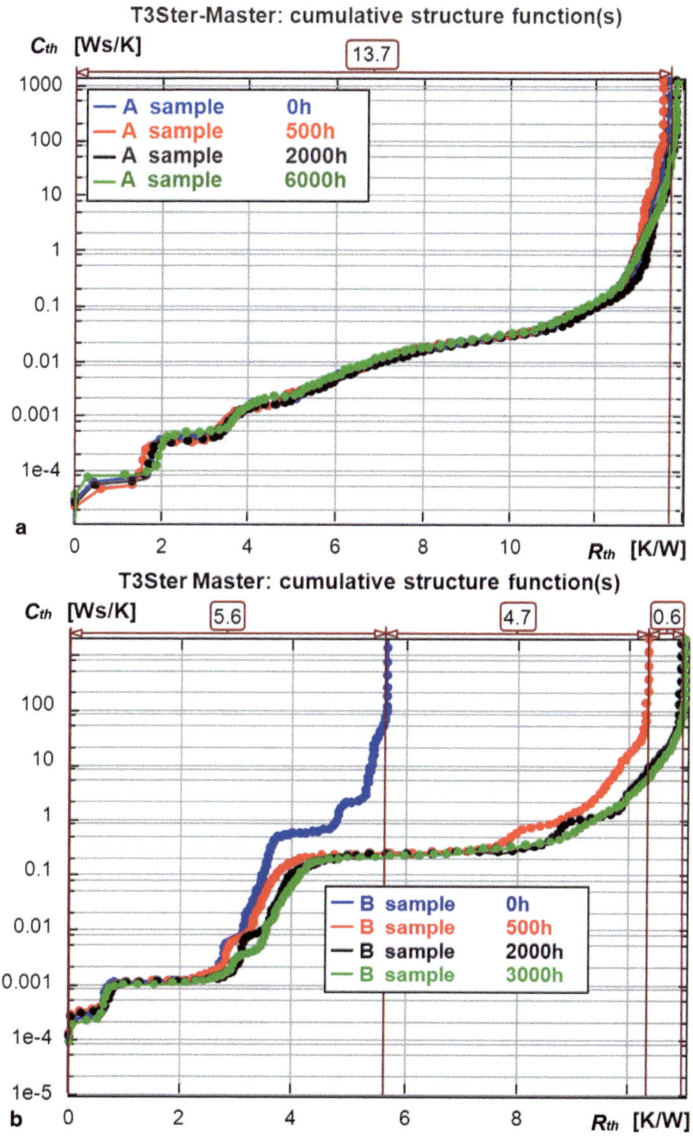

Fig. 5.19 Examples of structural changes of the junction-to-case heat-flow path of light-emitting diodes (LEDs) with time. **a** "Long life" LEDs, **b** "short life" LEDs

middle-wave sensitive cone spectral sensitivities [20]. Figure 5.22 shows this curve compared to the CIE 1924 $V(\lambda)$ function, in logarithmic form. As can be seen this curve is higher than the CIE curve not only in the blue part of the spectrum, but also in the red part. For the time being however, such a curve that would fit visual observations better has not been accepted by standardization bodies.

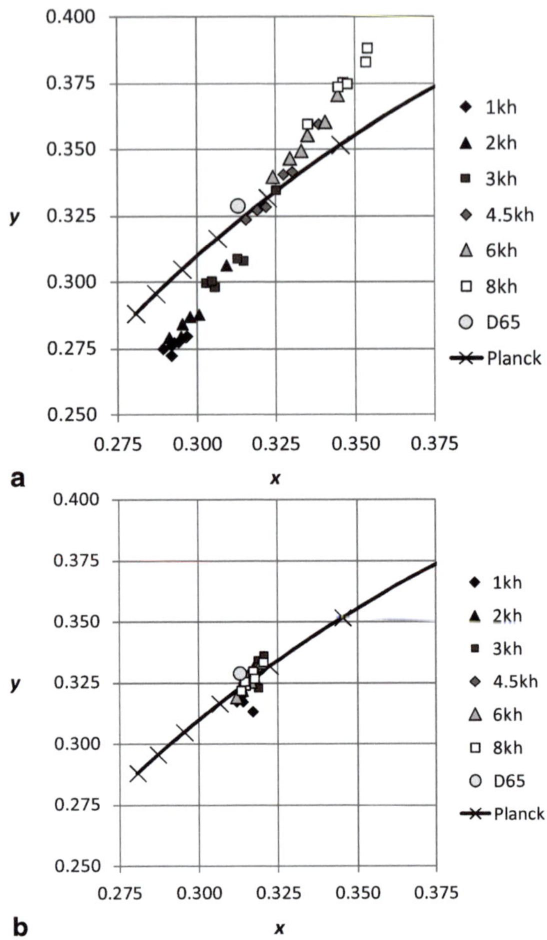

Fig. 5.20 Examples of CIE 1931 chromaticity coordinate shifts over 8,000 h aging of light-emitting diodes (LEDs) from different manufacturers at 85°C. **a** LEDs showing "yellowing" effect, **b** LEDs with high color stability

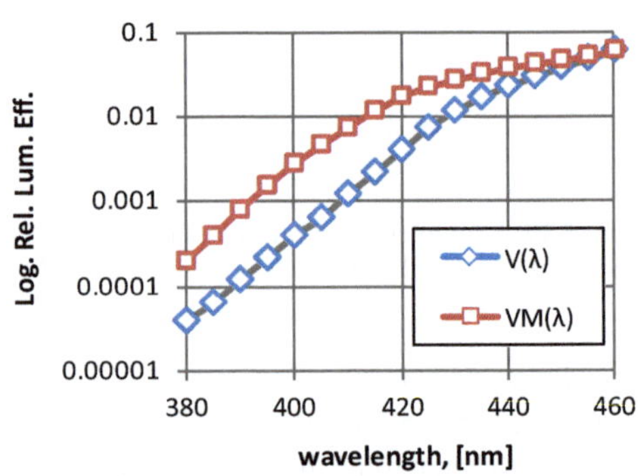

Fig. 5.21 Compare CIE $V(\lambda)$ and $V_{\mathrm{M}}(\lambda)$ functions

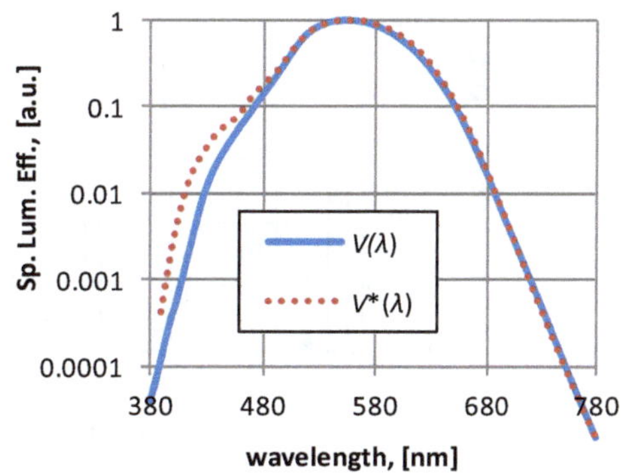

Fig. 5.22 $V^*(\lambda)$ spectral luminous efficiency function according to Stockman and Sharpe and the International Commission on Illumination (CIE) $V(\lambda)$ function

5.5.2 Updating the Colorimetric Descriptors

The CIE 1931 CMFs are transformations of CMFs based on visually determined real RGB primaries [21]. For the transformation one of the criteria was that one CMF function, the $\bar{y}(\lambda)$ function should be equal to the CIE 1924 $V(\lambda)$ function. As seen in the previous section the $V(\lambda)$ function does not describe correctly the human luminous efficiency, and this manifests itself in erroneous 2° CMFs (The CIE 10° has not been coupled to the photometric $V(\lambda)$ function and thus does not suffer from this drawback). LED and SSL colorimetry is, however, usually performed using the 2° observer system, which needs correction, as the color difference for instrumental color matches are well visible if, e.g. white light produced with RGB-LEDs is matched with an incandescent lamp light. A CIE technical committee (TC 1-36) is working on an improved colorimetric system that corresponds better to the visual assessment, and has published a tentative cone fundamental-based system of CMFs [22]. Our experiments showed that with these the errors between instrumental and visual match are halved [23]. Even better results can be obtained using a shifted S-cone sensitivity function, as seen in Fig. 5.23 [24].

Using these updated CMFs the description of LED colors could be made much more accurate, but the practical application needs still a lot of further investigations.

5.5.3 Color Fidelity and Color Preference

Light source color quality is described traditionally using the CIE CRI [25] that defines color quality by calculating the color difference of eight plus six test samples illuminated by the test light source and a reference illuminant of equal CCT. Already when tri-band fluorescent lamps became available, one noticed that the calculated General Color Rendering Index does not describe the visual impression correctly. This experience was repeated when the first white LEDs came onto the market. Several CIE technical committees worked during the decades to solve this problem, when this chapter was prepared one has the impression that a double description of

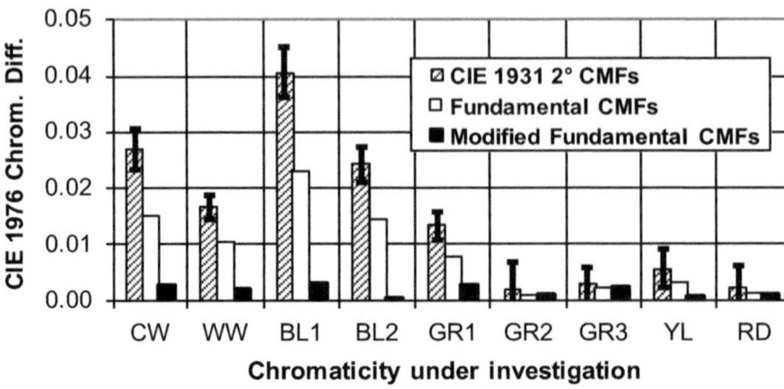

Fig. 5.23 International Commission on Illumination (CIE) 1976 u', v' chromaticity differences between visually matched stimuli of different chromaticity, using for the calculations of different color matching functions (CMFs)

color rendering will solve the problem: In some cases the requirement is that the illuminated scene should look under the test light source as it would look under a reference light source, this will be probably termed color fidelity. In some other cases, observers might accept a light source that renders the scene more pleasing as the reference light source, for that purpose a color preference index should be developed.

A candidate for color fidelity is the nCRI color fidelity index [26] that is built upon the lines of the CIE test method, only up to date colorimetric calculation algorithms are used (CIE 10° observer, CIECAM02 color space and color difference formula, some nonlinear transformation from color difference to color fidelity index, and as most important up to 200 real color test samples together with 18 very special artificial samples, that prevent gaming, i.e., to optimize color fidelity index and not real color fidelity). Just for this purpose beside a general index, that is a grand average also the characteristics of that sample that provides the poorest result is communicated.

For the color preference description the National Institute of Standards and Technology (NIST) Color Quality Scale [27] seems to be a possible candidate. Present version of this method does the calculations in CIELAB space, uses 15 test samples and has as a special feature: if under the test lamp the chroma of a sample is higher as under the reference illuminant, than this is not penalized, as authors are of the opinion that an increase in chroma makes the scene more colorful and will be preferred. Major problem with the model is that the authors of reference [27] claim that the 15 samples are critical, and truly if metameric samples would be taken quite different results would be obtained. How this will be corrected is still a question for the future.

5.6 Summary

LEDs are light sources and their photometry and colorimetry should follow general photometric principles. Due to the narrow band structure of the LED SPD and its temperature sensitivity, some precautions are needed, hence not every instrument that would provide adequate results on traditional lamps will furnish reasonable results when LEDs are measured. In the present chapter, we tried to summarize the basic photometric quantities and their measurement, pointing out where LED testing needs special care, or problems pop up that have to be solved. We expect that a number of new findings and solutions will surface within the next few years. Unfortunately, size limits do not permit us to go into further detail; the interested reader is directed to the background literature. Beside the numbered references, we draw the attention of the reader to the short list of fundamental books on colorimetry and optical properties of LEDs.

References

1. CIE (2007) Measurement of LEDs, 2nd edn. CIE Publication 127, CIE Central Bureau, Vienna
2. CIE (1989) Measurement of luminous flux, CIE Publication 84, CIE Central Bureau, Vienna
3. Ohno Y, Zong Y (1999) Detector based integrating sphere photometry. Proceedings of the 24th Session of CIE, Warsaw '99 pp 155–160. CIE Publ. 134
4. CIE (2002) Photobiological safety of lamps and lamp systems. CIE S009/E
5. CIE (2004) Colorimetry, CIE Publication 15, CIE Central Bureau, Vienna
6. CIE (1995) Method of measuring and specifying colour rendering of light sources 3rd ed. CIE Publication 13, CIE Central Bureau, Vienna
7. Schanda J (2007) Colour rendering of light sources. In Schanda J (ed) Colorimetry—understanding the CIE system. Wiley-Interscience, pp 207–217
8. Zong Y, Brown SW, Johnson BC, Lykke KR, Ohno Y (2006) Simple spectral stray light correction method for array spectroradiometers. Appl Opt 45:1111–1119
9. Ohno Y (2007) Spectral colour measurement, pp 105–133 in colorimetry—understanding the CIE system, ed. J. Schanda, Wiley-Interscience
10. Schanda J, Eppeldauer G, Sauter G (2007) Tristimulus colour measurement of self-luminous sources in colorimetry—understanding the CIE system, ed. J. Schanda, Wiley-Interscience
11. CIE (1987) Methods of characterizing illuminance meters and luminance meters, CIE Publication 69, CIE Central Bureau, Vienna
12. Fryc I (2003) Distribution of the spectral correction error of the CCD. Proc. CIE symposium temporal and spatial aspects of light and colour perception and measurement. CIE x025, pp 39–43, CIE Central Bureau, Vienna
13. Schanda J, Lux G (1973) On the electronic correction of errors in a tristimulus colorimeter. Proc AIC Colour 73, the 2nd AIC Congress. pp 466–469. Hilger
14. Ohno Y, Hardis JE (1997) Four-colour matrix method for correction of tristimulus colorimeter. pp 301–305. Proc. Fifth Color Imaging Conf. SIST
15. Kosztyán ZsT, Eppeldauer GP, Schanda JD (2010) Matrix-based color measurement corrections of tristimulus colorimeters. Appl Opt 49:2288–2301
16. Kosztyán Zs, Schanda J (2009) Using adaptive matrix transformation for decreasing colour measuring systematic error in tristimulus colorimetres. Proc. CIE Midterm Meeting—Light and Lighting Conf. Budapest 27–29
17. CIE (1926) Principales décisions de la Commission Internationale de lÉclairage, CIE 6th Session, Genève- Jul., Cambridge University Press 1926
18. Judd DB (1951) Report of U.S. secretariat committee on colorimetry artificial daylight. CIE Proceedings Stockholm, Vol 1. Part 7 Bureau Central de la C.I.E., p 11
19. CIE (1988) 1990 CIE 2° spectral luminous efficiency function for photopic vision, CIE Publication CIE 86

20. Stockman A, Sharpe LT (2000) Spectral sensitivities of the middle- and long-wavelength sensitive cones derived from measurements in observers of known genotype. Vision Res 40:1711–1737
21. CIE (2004) Colorimetry, 3rd edn. CIE 15
22. CIE fundamental chromaticity diagram with physiological axes draft Chap. 7.3: development of chromaticity diagrams based upon the principles of the CIE XYZ system. CIE TC 1–36
23. Csuti P, Schanda J, Petluri R, McGroddy K, Harbers G (2011) Improved colour matching functions for better visual matching of LED sources. CIE 27th Session & Conference. Sun City, South Africa
24. Csuti P, Schanda J (2010) A better description of metameric experience of LED clusters. Light Engng 18:44–50
25. CIE (1974) Method of measuring and specifying colour rendering of light sources, CIE Publication 13.2
26. Chou YF, Luo R, Schanda J, Csuti P, Szabó F, Sárvári G (2011) Recent developments in colour rendering indices and their impacts in viewing graphic printed materials. Color and Imaging Conf., San Jose CA, USA
27. Davis W, Ohno Y (2005) Toward an improved colour rendering metric. Proc. Fifth International Conference on Solid State Lighting SPIE Vol 5941, 59411G
28. ANSI/IESNA IES-LM-80-08 Standard, "Approved method for measuring lumen maintenance 1732 of LED light sources". ISBN 978-0-87995-227-3 (2009)
29. Poppe A, Molnár G, Csuti P, Szabó F, Schanda J. Aging of LEDs: a comprehensive study based on the LM80 standard and thermal transient measurements, CIE 27th Session-Proceedings, CIE 197:2011: (Vol 1, Part 1–2). Sun City, South-Africa, 10–15 July 2011. pp 467–477. Paper OP57. (ISBN: 978 3 901906 99 2)
30. ANSI/IESNAIES-TM-21-11 Standard, "Projecting long term lumen maintenance of LED light sources". ISBN: 978-0-87995-259–4

Further Reading

31. Colorimetry, Understanding the CIE System (2007) Schanda J (ed) Wiley
32. LEDs for lighting applications (2009) Mottier P (ed) ISTE & Wiley
33. Murdoch JB (2003) Illuminating engineering—from Edison's lamp to the LED, 2nd edn. vision Communications
34. Schubert EF (2006) Light-emitting diodes, 2nd edn. Cambridge University Press
35. Wyszecki G, Stiles WS (1982) Color science: concepts and methods, Quantitative Data and Formulae. 2nd edn. Wiley

Chapter 6
Standardization of LED Thermal Characterization

András Poppe and Clemens J. M. Lasance

Abstract The chapter discusses the increasing need for a more sophisticated thermal characterization of light emitting diodes (LEDs) and LED-based products. It goes without saying that the LED business is growing exponentially, in fact, much faster than analysts predicted 5 years ago. Unfortunately, until recently the progress in thermal characterization has not kept pace. The situation was a serious problem until the first component-level LED thermal testing standards were published. As these standards are relatively new, there are still manufacturers who are not yet aware of the new testing procedure and think they can publish whatever thermal information they want. The problem also still exists for the experienced user because the thermal data that are published are often rather useless in practice when accuracy is at stake, and accuracy is needed for an educated guess of expected performance and lifetime. This situation has much in common with the one the integrated circuit (IC) world was facing almost 20 years ago. Provided the manufacturers want to cooperate, it is relatively easy to build upon the experience gained in the IC business and their standard protocols that are in use worldwide.

In 2008, the JEDEC JC15 committee on thermal standardization of semiconductor devices decided to take action and created a task group to deal with thermal standardization issues of power LEDs. International Commission on Illumination/Commision Internationale de l'Eclirage (CIE) has also created new technical committees (e.g., TC2-63, TC2-64), which also aim to address thermal issues during measurement of high-brightness/-power LEDs. The chapter describes novel test methods,

A. Poppe (✉)
Mentor Graphics Mechanical Analysis Division, MicReD Office, Budapest, Infopark D, Gábor Dénes utca 2, 1117, Hungary
e-mail: andras_ poppe@mentor.com

Department of Electron Devices, Budapest University of Technology and Economics, Budapest, Hungary
e-mail: poppe@eet.bme.hu

C. J. M. Lasance
Philips Research Laboratories Emeritus, Consultant at SomelikeitCool, Nuenen, The Netherlands
e-mail: lasance@onsnet.nu

C. J. M. Lasance, A. Poppe (eds.), *Thermal Management for LED Applications*, Solid State Lighting Technology and Application Series 2,
DOI 10.1007/978-1-4614-5091-7_6, © Springer Science+Business Media New York 2014

which form the basis of new measurement guidelines including combined thermal and radiometric measurement of LEDs. In 2012, some new thermal testing standards were published, which are also discussed in this chapter. Initiatives to arrive at compact thermal modeling standards are also covered.

6.1 Introduction

Standardization is a wide topic: one can speak about product standards, performance standards, and testing standards. As of today, we can hardly speak about light emitting diode (LED) product standards, as it would be the case with classical light sources. Referring to, e.g., a "100 W/230 V E27/Par38 lamp" means the same product everywhere, regardless of its manufacturer. As LEDs evolve very quickly, there are no such "standard" LED products yet; LED packages exist in different forms and flavors. Despite standardized nomenclature regarding LEDs [1], there is currently no official LED classification; they can be classified by color, outline, or by luminous intensity (general, high-brightness, ultrahigh brightness).

The Zhaga consortium tries to resolve this problem by the attempt of defining different LED modules and LED light engines [2] each with clear definitions regarding their so-called electrical, mechanical, and thermal interfaces. The goal is to enable interchangeability of LED light sources made by different manufacturers. Book 1 of Zhaga specifications [3] is already available to the general public, which describes the generic concepts, terms, and definitions used by Zhaga, including thermal-related topics. Additional Zhaga specification books define particular interface details for different LED light engine types.[1] None of the Zhaga books defines how to measure thermal resistance of LEDs; the main focus is on testing the compliance of the different interfaces (electrical, mechanical, photometric, and thermal) to the Zhaga specifications.

The purpose of this chapter is to provide different concepts of thermal testing of LED components (LED packages or LED assemblies) and to show an overview of the current state of the art regarding thermal testing (characterization) standards aimed for LEDs. Information on the first, recently published LED component thermal testing standards as well as information on new characterization concepts such as test-based compact modeling of LEDs is also provided.

We start with an overview why thermal characterization is needed, we highlight the differences in objectives between manufacturers and system designers, and we shortly treat the concept of thermal resistance followed by the discussion what could be learnt from the history of standardization of thermal characterization of the integrated circuit (IC) world. Finally, we discuss the concept of the LED thermal testing standards [5–8] published by JEDEC in April/May 2012.

This chapter is partly based upon a number of earlier papers written by the authors [9–13].

---

[1] So far (as of October 2012), only Zhaga Specification Book 3 regarding *nonsocketable LED modules with 9–23 mm diameter rounded light emitting surface* is available for the general public [4].

6.1.1 The Role of Thermal Characterization

A lighting system designer wants to check if an LED's junction temperature or a solder temperature stays within prescribed limits, or, alternatively, needs a temperature for lifetime prediction. The equation that is mostly used is the following:

$$T_J = R_{thJ-ref} \cdot P + T_{ref} \tag{6.1}$$

R_{th} is a number that is supplied by the manufacturer, the power P is usually supplied by the electrical engineer, and T_{ref} is a reference temperature that depends on the definition: either it is some (unspecified) ambient temperature, or a point on the package or board in question. After T_J junction temperature has been calculated using the equation, this temperature is usually compared to a specified temperature (of which the origin is often unknown). When T_J is higher than $T_{specification}$, chances are high that a redesign is requested, with all the consequences of time-to-market, etc. Obviously, one should be convinced about the accuracy of the calculated T_J before jumping to such a decision.

Equation (6.1) looks like a very simple equation, but there are many questions that could be asked regarding its accuracy. For example:

- How is T_J defined? Can it be measured in practice? If not, how to derive it from another measurement, for example, T_{case} as suggested by many LED vendors?
- What is T_{ref}? Can it be unambiguously defined? Is it easily measurable in practice?
- How is P defined? Is it corrected for nonthermal contributions in order to obtain the real heating power?
- What is the physical meaning of the thermal resistance used in the equation?
- Is this thermal resistance really temperature-independent?
- What is the variance in the published data per manufacturer? How does it depend on LED binning class, manufacturing lot, etc.?

Because of the many uncertainties that make Eq. (6.1) *much more complex than designers think, it is possible to quote almost any value. This is one of the main drivers for the need for worldwide thermal standardization.*

6.1.2 Different Goals for Manufacturers and System Designers

Before starting the discussion on standardization, it should be recognized that the drivers for thermal characterization are quite different for manufacturers and for their customers. *The goals of manufacturers include*:

- Optimization of product thermal performance: Often, very detailed thermal/mechanical models are available, calibrated by dedicated experiments. These data are usually proprietary.
- Figure-of-Merit for comparison: These data are available for the customer in order to judge which product is better from a thermal point of view.

From the manufacturers' point of view, standardized thermal test environment and standardized thermal metrics are required, along with standardized test methods. To compare different products *relative values* of the measured thermal metrics are sufficient.

On the other hand, the goals of system designers are:

- Obtaining what-if scenarios: To investigate these, the system designer needs thermal data that can be used outside the domain in which they were measured by the manufacturer.
- Sufficiently accurate prediction of T_J during the design process: Preferably, both the absolute temperature and the temperature difference should be calculated using the data in the data sheets to have a clear relation between a temperature (T_J or T_{case}), ambient conditions, lifetime, and other performance criteria.

From the point of view of system designers, thermal metrics aimed for temperature predictions in application-specific environments are needed. To meet (thermal) design objectives, accurate *absolute* values of the applied thermal metrics are a must.

It turned out that these differences were the major cause of the "Babylonian" situation the IC world was facing about 20 years ago, and the LED world could profit from the lessons learned. However, there are differences too, as is shown in Sect. 6.2.

6.1.3 Definition of LEDs

The everyday term (high) *power LED* is somewhat ambiguous, since there is a tendency that multiple single pn-junction LED chips are packaged into a single package (sharing the same cooling assembly and optics) or multiple elementary pn-junctions on a single chip form an LED device. Also, in many cases multiple packaged LEDs are assembled to a substrate (usually a high thermal conductivity board such as a metal core PCB or MCPCB, in short) or multiple LED chips are placed onto a ceramics substrate (CoB—chip on board assembly) to form one single device. In many cases, these devices—internally formed as arrays of elementary LEDs—have only limited number of electrical connections to the outside world: this way the electrical circuitry for most high-power LED arrays does not allow for connection to an individual diode. With only two electrical contacts, the thermal measurement must treat the entire array as a single diode. The implication of this reality is that the junction temperature measurement results actually in a weighted average of multiple junction temperatures and there is also limited ability to relate the measured temperature to a spatial location within the array. Figure 6.1 illustrates the different possible electrical configurations—as an example—for a 3 $\times$ 3 physical arrangement of single LEDs.

An *LED array* is defined as *two or more individual LED chips mounted in a package or on a substrate in a manner such that any device in array can be powered through either series, parallel, or individual connections while the other devices in*

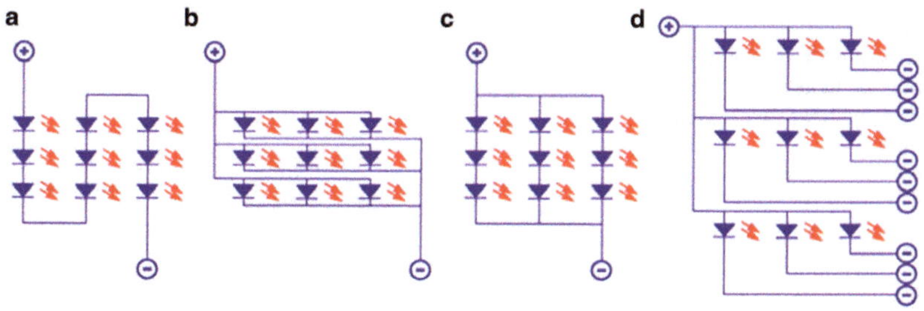

Fig. 6.1 Different LED array configurations. **a** Series. **b** Parallel. **c** Series/Parallel. **d** Individual. (Source: [6], © JEDEC, reproduced with permission)

the array may or may not be operating. The individual LED chips may also consist of an array of LED junctions on the chip as well.

Thermal measurements on an LED chip array can either be done on the array as a whole or on individual chips within the array, depending on the electrical connections available. Figure 6.1 presents schematics for different electrical connection alternatives. The *Series configuration* requires the highest power supply voltage but also requires the lowest current. Conversely, the *Parallel configuration* requires the highest current but the lowest voltage. The *Series/Parallel configuration* requires moderate current and voltage, and by modifying the series and parallel arrangement, can be more easily tailored to meet specific application requirements. The *Individual configuration* shown in Fig. 6.1d with a *common anode* connection, is most often used when the LEDs are dissimilar, i.e., different color, etc. Other individual configurations include separate contacts for each device, *X-Y* addressing and *common cathode*. Thermal measurements of any of these array configurations are necessary to ensure that each array element is operating in an acceptable junction temperature (T_J) range for that specific element.

When the array product is available only in packaged form with only the + and − leads/contacts accessible, the array can only be measured as a composite LED. All measured characteristics (thermal resistance, temperature sensitivity of the overall forward voltage, radiant flux, luminous flux, color, etc.) of such an array are *ensemble characteristics* of the array. In such cases, the array is considered as a single-chip equivalent device, which possesses the measured ensemble characteristics of the array device, i.e., the data results will assume equal power distribution and the same value of temperature-sensitive parameter (TSP) variation and the same junction temperature change (ΔT_J) for each LED in the array.

In the subsequent sections, to under the term LED (or power LED) or LED device, we mean either an individually available single LED of any LED array arrangement or an equivalent LED of an LED array where elements of the array are not accessible individually (this equivalent LED being characterized by its ensemble characteristics) with an exposed cooling surface, which is to be heat sunk during normal operation. This cooling surface could be the cooling tab (heat slug) of a single packaged LED device (usually referred to in the solid-state lighting (SSL) industry as *level0* device), or a packaged LED device attached to a substrate, see also definitions in [1].

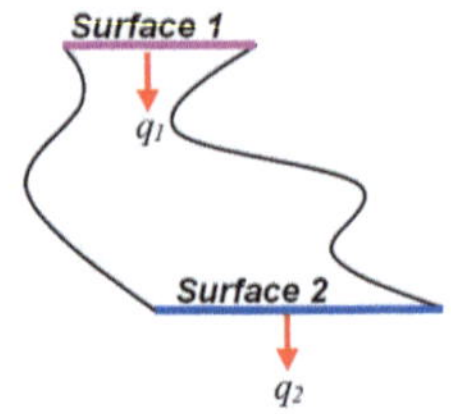

Fig. 6.2 Two isothermal surfaces connected by a heat flux tube. q_1 heat flux leaving Surface 1, q_2 heat flux leaving Surface 2

6.1.4 A General Concept of Thermal Resistance

In the early 1990s, it was found that one of the reasons for the lack of well-defined standards was a misunderstanding by electrical engineers about the physical meaning of a thermal resistance and in some cases this is still true today. Space is lacking for a serious discussion of this topic; the interested reader is referred to two papers [14, 15]. Here, we only point out why one must be careful when using a thermal resistance from a data sheet.

A formal definition of a thermal resistance is: *The temperature difference between two isothermal surfaces divided by the amount of heat that flows between them is the thermal resistance of the materials enclosed between the two isothermal surfaces and the heat flux tube originating and ending on the boundaries of the two isothermal surfaces* [16].

The essential point to understand is that a thermal resistance can never be based on measuring or calculating two *points*, *unless* the plane is isothermal. Additionally, the heat flux q_1 leaving (dissipated at) Surface 1 should be equal to the heat flux q_2 leaving (nondissipating) Surface 2 as shown in Fig. 6.2.

In reality, the above conditions can never be exactly realized but good approximations for the above conditions can be provided. This is true for standardized test environments and in many cases it is also true for real-life operating conditions, which are approximated by such test environments.

For example, in case of retrofit LED lamps (LED package attached to an MCPCB mounted onto a luminaire/heat sink like in case of, e.g., an MR16 replacement solution—see Fig. 6.3), R_{thJA}, the junction-to-ambient thermal resistance would be a good thermal performance indicator, which is to be measured in a natural convection test environment as shown in Fig. 6.4.

Though the temperature exhibits a given spatial distribution over the surface of any active semiconductor die, measuring the junction temperature by the electrical test method defined in [17, 18] results in a well-repeatable average temperature value, which is called T_J junction temperature in the JESD51 series of standards, in other words, the "junction" is considered as an isothermal source. It is an approximate model that works fairly well in everyday practice (the same kind of approximation is applied in case of LED arrays where ensemble characteristics are identified from measurements (see Sect. 6.1.3). This way the R_{thJA}, junction-to-ambient thermal resistance (measured in natural convection environment), can be considered such that the junction is surrounded by the ambient being forced to the constant T_{amb}, ambient temperature (considered ideally to possess infinite heat sinking capability).

Fig. 6.3 An MR16 halogen replacement LED lamp in a natural convection test environment. (As it will be shown later, for the correct measurement of the thermal resistance the light output needs to be considered in the calculation of the heating power. An integrating sphere could serve both as a thermal test environment for natural convection conditions and for measuring the emitted optical power (total radiant flux) of the lamp)

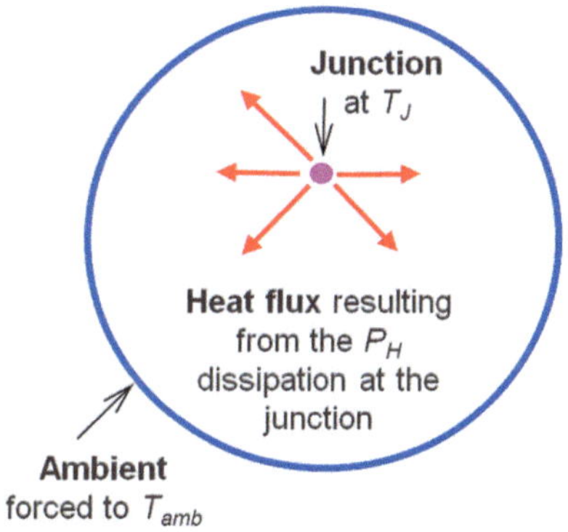

Fig. 6.4 Illustration of the junction-to-ambient thermal resistance (R_{thJA}) according to the formal definition of the thermal resistance

In this case, we have two isothermal locations between which there is no heat loss, so conditions of the formal definition of the thermal resistance are met (see Fig. 6.4).

Looking at LED thermal characterization on package level, one can generally state that packages of power LEDs have been designed to be cooled through a dedicated cooling surface and there is typically a single heat-flow path from the LED's junction toward this cooling surface. In the relatively new JEDEC standard for thermal characterization of such packages (JESD51-14 [19]), this is called the *case*. Therefore, for thermal characterization of power LEDs *junction-to-case thermal resistance* is the recommended thermal metric and the cold plate is the recommended test environment. In such a setup, most of the heat dissipated in the junction leaves the package toward the cold plate. In our approximation, the cold plate is considered isothermal as shown in Fig. 6.5. An actual realization of such a test condition is shown in Fig. 6.6. This setup again is a good approximation of the conditions of the formal definition of the thermal resistance.

Let us consider a real product. Figure 6.7 shows the most important features of a typical LED-based product. According to the above definition, it is easily shown that it is not possible to define a thermal resistance between the two points called die and case unless:

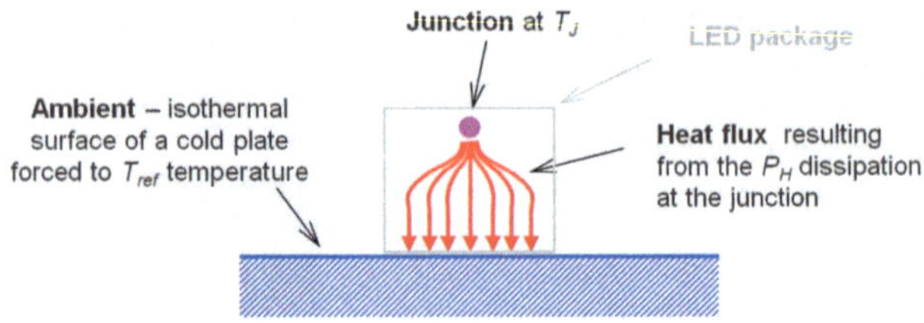

Fig. 6.5 Illustration of the junction-to-case thermal resistance (R_{thJC}) of LED packages according to the formal definition of the thermal resistance

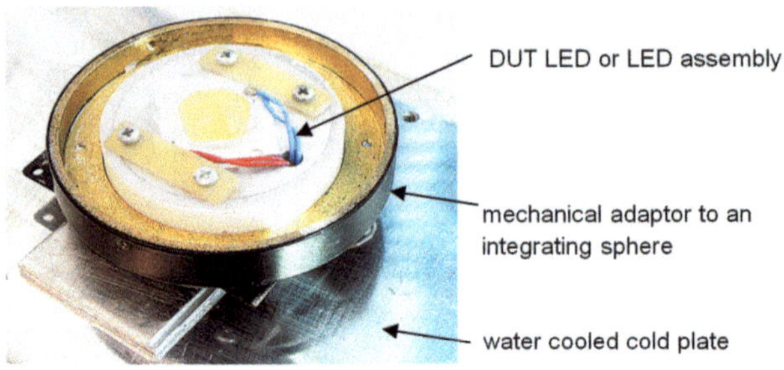

Fig. 6.6 Actual test setup for the thermal measurement of an LED on a liquid-cooled cold plate. (Source: [7], © JEDEC, reproduced with permission)

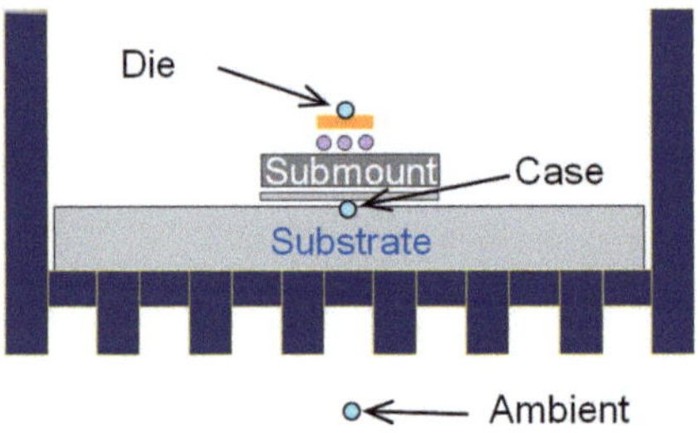

Fig. 6.7 Practical cooling solution for a typical LED-based product

- The die and case surfaces are at uniform temperature.
- The heat flux between die and case is known.

Regarding the first bullet point, except for high-power LEDs with large die size or multidie LED packages the assumption of a uniform die temperature is a correct approximation of the reality. It is the *case surface* that may cause problems. Although in a junction-to-case thermal resistance test setup (see Figs. 6.5 and 6.6) it is fair to assume the cold plate to be isothermal, under real application conditions where a heat sink substitutes the cold plate the heat spreader (or alternatively the board) cannot be considered to be at uniform temperature. The consequence is that the measured case temperature becomes dependent on the heat transfer coefficient h that describes the rate of heat transfer from the heat sink to the environment, usually including both radiation and convection. The second bullet point is usually also met, but should be checked in case some heat is leaking away through the optics, either by radiation directly from the source or by conduction and convection/radiation from the top surface. However, it should be stressed that in contrast to ICs the corrections involved are second-order.

In conclusion: *the main problem with the definition of the thermal resistance in the LED world is the often unchecked assumption of temperature uniformity in the case node region.* In testing, application of the method defined in the JESD51-14 standard [19] helps overcome this issue. For more details on this, please refer to Chap. 4 on thermal testing.

6.1.4.1 Extension of the Thermal Resistance Concept to Thermal Impedance

The way the EIA/JEDEC JESD51-1 standard [18] defines the thermal resistance is a rearrangement of Eq. (6.1)

$$R_{thJ-X} = \frac{T_J - T_X}{P_H} = \frac{\Delta T_{JX}}{P_H} \tag{6.2}$$

where T_X denotes the temperature of the reference point *X* and P_H denotes the power dissipated in the device. The ΔT_{JX} is the change of the junction temperature with respect to the reference temperature, in other words, a *difference of two temperatures measured at two different locations at one moment in time*.

Equation (6.2) conforms to creating a spatial temperature difference at thermal equilibrium: power the junction and measure simultaneously the junction temperature and the temperature at reference point *X*. This requires two "thermometers" that both need to be calibrated. When using the electrical test method discussed above, the junction is one of the thermometers through the calibration of the *TSP* of the semiconductor chip.[2] In case of LEDs—as for any other pn-junction—the TSP is the V_F, forward voltage of the device measured at forced, constant forward current. The other thermometer that measures the T_X reference temperature of point *X* is typically a thermocouple that also has to be properly calibrated.

If the reference temperature is the ambient temperature (that we can control if measurements are performed, e.g., on a cold plate), then in the unheated device, in its initial thermal state $T_{J0} = T_X$, i.e., the initial junction temperature and the reference point temperature are equal. After heating the device, it reaches its final thermal equilibrium. We can continuously record the T_J junction temperature transition during the heating, from a cold steady state to a hot steady state (or vice versa in a cooling process).

This continuous procedure—also known as the "static" test method—suggests another reformulation of Eq. (6.2). Suppose, in the initial steady state a known P_{H1}, heating power is applied, while in the final steady state another known heating power P_{H2} is applied (see Fig. 6.8).

For both cases, we can express the junction temperature based on the pattern of Eq. (6.2)

$$T_{J1} = R_{thJ} - X \cdot P_{H1} + T_{refX} \tag{6.3a}$$

[2] This is referred to as *K-factor calibration* in the JEDEC thermal testing standards, see [6, 18, 19]. See also Chap. 4 on thermal testing.

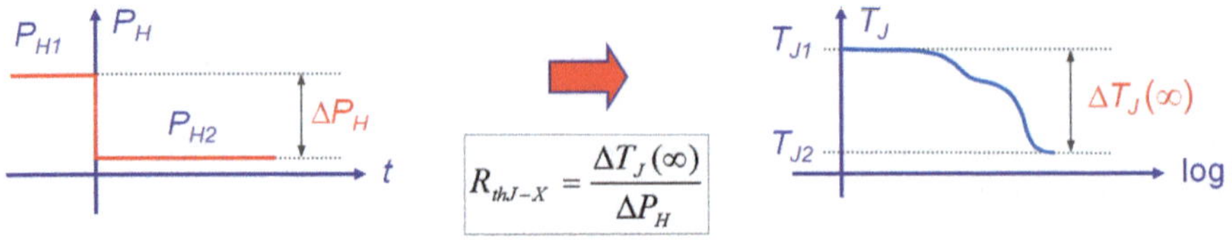

Fig. 6.8 Junction-to-X thermal resistance calculated from a temporal difference of the junction temperature and the power dissipated in the device. (Source: [6], © JEDEC, reproduced with permission)

$$T_{J2} = R_{thJ-X} \cdot P_{H2} + T_{refX} \tag{6.3b}$$

Subtracting Eq. (6.3a) from Eq. (6.3b), we obtain

$$T_{J2} - T_{J1} = R_{thJ-X} \cdot (P_{H2} - P_{H1}) \tag{6.4}$$

Equation (6.4) can be interpreted such that T_{J1} and T_{J2} are junction temperatures occurring at different time instances: $T_{J1} = T_J(t_1)$ and $T_{J2} = T_J(t_2)$. Substituting these and rearranging Eq. (6.4) yields

$$R_{thJ-X} = \frac{[T_J(t_2 = \infty) - T_J(t_1 = 0)]}{P_{H2} - P_{H1}} \tag{6.5}$$

or

$$R_{thJ-X} = \frac{\Delta T_J(t)}{\Delta P_H} \tag{6.6}$$

where $\Delta T_J(t) = T_J(t_2) - T_J(t_1) = T_{J2} - T_{J1}$ and $\Delta P_H = P_{H2} - P_{H1}$.

In other words, instead of using a spatial temperature difference along the junction-to-X heat-flow path, Eq. (6.6) suggests that the thermal resistance can also be defined as a metric, which can be calculated from the difference of the initial and the final steady-state value of the junction temperature caused by a known change in the heating power at the junction, provided the reference temperature is kept constant. This means that instead of taking the difference in temperature at two different locations at one moment in time, we take the difference in temperature at two different moments in time at one location, the junction only.[3] A major advantage of this approach is that calibration errors that are typical in temperature measurements are absent. Another advantage is that the reference temperature is no longer part of the equation, reducing the need of calibration of temperature measurement as well as eliminating possible ambiguities about the definition of the reference point of power LEDs. A third advantage is that the junction region is usually more uniform in temperature than the case region.

[3] The transient dual interface test method defined in the JEDEC JESD51-14 standard [19] for the measurement of the junction-to-case thermal resistance of power semiconductor device packages is also based on this approach.

In summary, a measurement at two points at the same time is mathematically equivalent to measuring one point at two different times. Equation (6.6) results in the classical expressions of the (steady-state) *thermal resistance* and *thermal impedance* as follows:

If $P_{H2} = 0$, then $T_{J2} = T_X$, thus, if $t_1 = 0$ and $t_2 = \infty$, we get back Eq. (6.2).

Equation (6.6) can be formulated such that a "thermal resistance" value can be obtained at any t time instance after switching the heating

$$R_{thJ-X}(t) = \frac{\Delta T_J(t)}{\Delta P_H} = Z_{thJ-X} \tag{6.7}$$

The quantity defined by Eq. (6.7) is time-dependent and is equivalent to the concept of the *thermal impedance* as described in MIL-STD 750D [20]. Thermal impedance describes the dynamics (time evolution) of the change of the junction temperature as a response to stepwise switching of the heating power. If $t_1 = 0$ and $t_2 = \propto$, then Eq. (6.7) defines $R_{thJ\text{-}X}(\propto)$, which is equal to the classical $R_{thJ\text{-}X}$, steady-state thermal resistance.

The thermal impedance as a time function is usually presented by a logarithmic time scale. It is equal to the time function of the junction temperature change $\Delta T_J(t)$ projected to an excitation of 1-W change in the heating power (assuming linearity, i.e., no temperature dependence). Thermal impedances can be described either in the time domain as shown in Eq. (6.7) or in the frequency domain (which is important, e.g., for AC LEDs); see, for example, [21]. Thermal impedances given in frequency domain are usually represented graphically by means of the so-called complex loci (also known as Nyquist diagrams). The so-called *structure function* can also be considered as a representation of a $Z_{thJ\text{-}X}$ junction-to-X thermal impedance. For further theoretical details on the structure function concept, please consult Appendix A of the JESD51-14 [19] or the very first original papers by Székely [22, 23] who more than two decades ago introduced the concept of structure functions for the characterization of the thermal properties of semiconductor device packages.

According to the so-called electrical test method [6, 18], the $\Delta T_J(t)$ function is measured indirectly through the temperature dependence of the pn-junctions' V_F, forward voltage under forced, constant forward current conditions. The relation between the forward voltage and the junction temperature is provided by the temperature sensitivity of the forward voltage $S_{VF} = \mathrm{d}V_F/\mathrm{d}T_J$. The reciprocal of this sensitivity is known as the K-factor: $K = 1/S_{VF}$ (see, for more details on the K-factor and the so-called K-factor calibration, Chap. 4 on thermal testing and Sect. 2.1.3.3 of Chap. 2 on the physical basics of LEDs).

6.1.4.2 Driving Point Thermal Impedance, Transfer Thermal Impedance, and Its Relation to Thermal Characterization Parameter

If there is an LED device with individually accessible LED chips that can be heated and measured individually (such as an RGB module)—its thermal properties can be

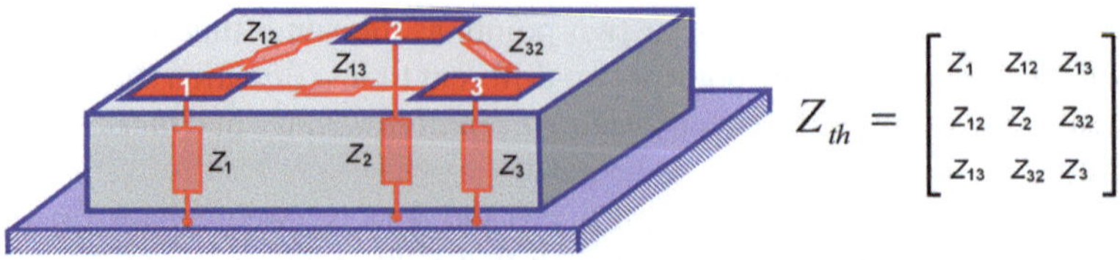

Fig. 6.9 Thermal impedance matrix of multiple LEDs on the same substrate

fully represented by its *thermal resistance matrix* [24] or *thermal impedance matrix* (see Fig. 6.9).

Such a matrix contains elements in the main diagonal that represent *driving point thermal impedances* and contains off-diagonal elements that represent *thermal transfer impedances*:

- **Driving point thermal impedance** is defined as the ratio of the change of junction temperature as a response to the change of power applied at the same junction, i.e., change in powering and change of temperature take place at the same physical location (same junction). If the applied change in power was 1 W, the value of driving point thermal impedance gives the value of the resulting junction temperature. *Structure function representation is defined for driving point thermal impedances only*. (Elements Z_1, Z_2, and Z_3 in Fig. 6.9 represent driving point thermal impedances.)
- **Thermal transfer impedance** between locations *i* and *j* related the temperature change at location *i* when power change was applied at location *j*. If the applied change in power at location *j* was 1 W, the value of the related thermal transfer impedance is equal to the value of the temperature change at location *i*. Thermal transfer impedances can be represented by time functions (time domain) or they can be represented in the frequency domain (complex loci or Nyquist diagrams) as well as time-constant spectra,[4] but they *cannot* be represented by structure functions. (Elements Z_{12}, Z_{23}, and Z_{13} in Fig. 6.9 represent transfer impedances.)
- If long time has elapsed after the abrupt change in powering, thermal impedances are equal to steady-state values:
  - Driving point values represent junction-to-reference *X* thermal resistances.
  - Transfer impedances reduce to *thermal transconductances*, which resemble the classical *junction-to-reference point thermal characterization parameters* [18].

Elements in the thermal impedance matrix are either time functions (or complex functions if impedances are represented in frequency domain). At $t = \infty$, impedance functions represented in time domain reduce to their counter parts in the thermal

[4] In case of thermal transfer impedances, both positive and negative values may be present in the $R(\tau)$ thermal time-constant spectra while driving point thermal impedances are represented by only positive values in the $R(\tau)$ functions. For details on thermal time-constant spectra refer to Appendix A of [19] or see papers [22, 23].

resistance matrix. In the following, when referring to matrix elements, we shall refer to elements of the thermal resistance matrix.

In theory, thermal impedance matrices or thermal resistance matrices must be symmetrical: $R_{thij} = R_{thji}$ but values measured in practice are asymmetrical [24]. The matrices may include locations different to LED junction, e.g., solder point or temperature monitoring points as suggested, e.g., by LED vendors for their users. This way, e.g., junction-to-solder point thermal characterization parameters may be measured and provided by LED vendors for their customers.

If passive locations (such as solder point or temperature monitoring point on the substrate) are included in the thermal impedance/resistance matrix representation, the driving point impedance for such a location is typically not measured and the corresponding off-diagonal elements of the matrix are also not identified.

Element values of the thermal resistance or thermal impedance matrices of multi–heat-source LED devices depend on the reference environment. Figure 6.9 illustrates the case when the common substrate is attached to an ideal heat sink (cold plate).

By assuming temperature-independent thermal properties of material parameters in the heat conduction paths, thermal impedance or resistance matrices can be used to calculate temperature responses to any combination of powers applied at the different heating locations (by using the superposition principle of linear systems). In a given physical arrangement/assembly, the validity of this assumption must always be checked by a few tests at different temperature values of the reference environment. Other nonlinear effects can have the same impact. Such can be air turbulence, heat radiation, or temperature dependence of the η_e, radiant efficiency (wall plug efficiency (WPE)) of the LEDs.

Further in this chapter, we focus on the driving point thermal impedances (dynamic case) or junction-to-reference *X* thermal resistances (steady-state case). Note, however, that procedures defined for junction temperature measurement and determination of heating power are also valid for the measurement of thermal transfer impedances (dynamic case) or thermal characterization parameters (steady-state case).

6.1.5 Additional Drivers Regarding LED Thermal Characterization

Unlike in the IC world, the T_J junction temperature of an LED is not just a performance indicator of the *thermal design* but also plays a major role in *lighting design* since all properties of the light output of an LED depend on the absolute junction temperature, as is clearly demonstrated in Fig. 6.10. One of the main reasons is that a noticeable color shift in RGB-based white light systems occurs (see Fig. 6.11) due to the different responses to temperature of different colored LEDs.

Chapter 2 on physical basics of LEDs' operation discusses in detail why the light output characteristics of LEDs depend on the operating conditions. The forward current applied to the LEDs is the primary variable—the higher the supplied current,

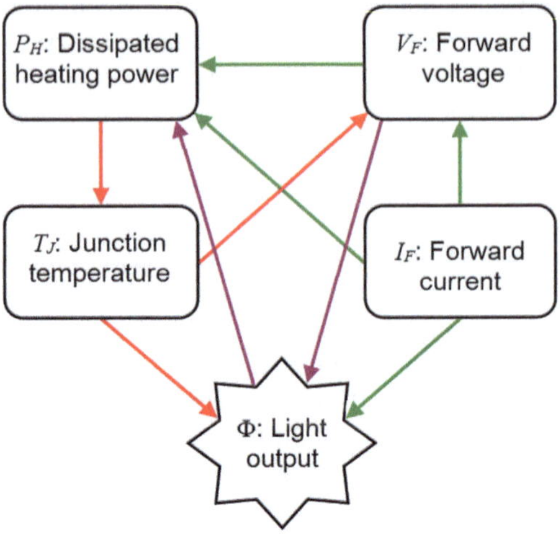

Fig. 6.10 Light output depends on "everything." (Source: [6], © JEDEC, reproduced with permission)

the more light is generated by the LED. Unfortunately, when an LED is driven by a constant current source, the light output drops with increasing temperature. This general feature of all LEDs is best illustrated by the dependence of their light output spectra as shown in Fig. 6.11.

In addition to the efficiency drop, the color of the LEDs' light also changes as it is shown by the shift of the peak wavelength. Consequently, since the junction temperature T_J of LEDs is more widely used in the design process of LED-based lighting solutions, well-established definitions of standardized thermal metrics and models are needed both by the LED manufacturers and the lighting system designers. As a response to this demand, in April/May 2012, JEDEC has published a series of standards about thermal testing of power LEDs [5–8]. In addition to this, the International Commission on Illumination/Commision Internationale de l'Eclirage (CIE) set up new technical committees (e.g., TC-2-63 and TC-2-64) to deal with the thermal aspects of LEDs. As of the date of writing this book, the work on final technical reports of these committees is still in progress.

The temperature and forward current dependence of LEDs' spectral properties means that Eq. (6.1) should be an integral part of the overall design of any LED-based lighting solution. The consequence is that the design of electrical and thermal conditions cannot be decoupled as is the case in classical electronic circuit design, leading to multidomain models [25, 26]. Despite suggestions in white papers [11–13], so far no real standardization activity has taken place in the field of LED compact modeling. Section 6.5.2 of this chapter is devoted to further details on this topic.

By keeping LEDs cool, high efficiency can be maintained. A thermal management solution that delivers better cooling also delivers more useful lumens in a given application. This means that the real junction-to-ambient thermal resistance of the LEDs in their application environment is a key factor in lighting design. Unfortunately, until recently different LED vendors reported their products' thermal resistance and

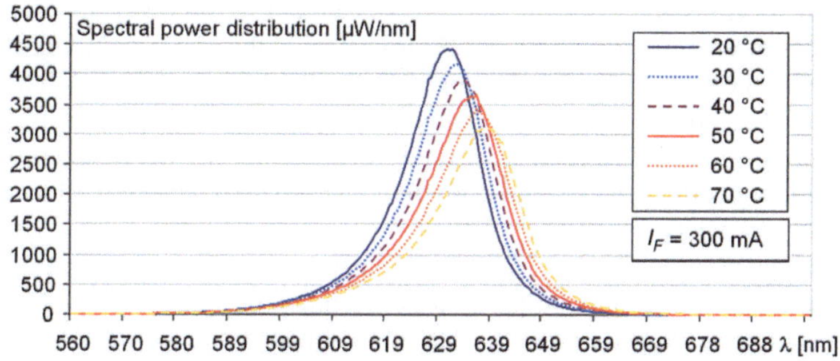

Fig. 6.11 Current and temperature dependence of the spectral distribution of the light output of a red LED. (Source: [62], © 2010 Society of Photo-Optical Instrumentation Engineers)

other temperature-related characteristics in different ways, prompting us to start thermal standardization activities within the framework of JEDEC in early 2008. These activities resulted in the "official" adoption of the measurement guidelines that will be discussed in this chapter in detail.

Among the different committees of the different standardization bodies, an agreement emerged that vendors need to consider the actual P_{opt}, emitted optical power (i.e., the radiant flux usually denoted as Φ_e) of the LEDs when calculating their real thermal resistance. A twofold interpretation can be defined, such as

$$R_{th-el} = \frac{\Delta T_J}{I_F \times V_F} \tag{6.8a}$$

$$R_{th-real} = \frac{\Delta T_J}{(I_F \times V_F) - P_{opt}} \tag{6.8b}$$

Here, the product of the LED's forward current and forward voltage ($I_F \times V_F$) represents the change of electrical power when the LED is switched on from zero current to a constant I_F forward current and ΔT_J denotes the change of the LED chip's junction temperature as a response to the change of heating power. Equation (6.8a) gives an estimate on the temperature reached when only the electric characteristic of the LED is known. Equation (6.8b) states that for describing the heat-removing capability of the physical structure correctly the *real thermal resistance* has to be calculated from the temperature change and the heating power. Obviously, the part of the electric power that leaves the device as light does not add to the heating, except in cases where a phosphor applied to the LEDs dome absorbs a non-negligible fraction.[5]

[5] As with practical measurement methods, there is no way to separate the cause of junction heating due to dissipating some of the supplied electrical power and junction heating due to conversion losses of the phosphor deposited on top of the blue LED chip. From a thermal characterization point of view, we shall not distinguish the "monochromatic" color LEDs from the phosphor-converted white LEDs.

Both definitions give back the same ΔT_J change if we multiply the proper R_{th} by the proper power value. However, as all parameters of the LED (including efficiency) change with the current level and actual temperature, $R_{th\text{-}real}$, the real thermal resistance of the LED package is a much more robust device descriptor because the $R_{th\text{-}el}$, *electrical-only thermal resistance* strongly depends on the actual operating point. Understanding both measures and using *real thermal data* are essential to successful LED design projects.

Nowadays new standards in the SSL industry refer to the real heating power of LEDs (see, e.g., [3, 4]). Recently developed standard drafts, e.g., from International Electrotechnical Commission (IEC) already include normative references to some of the LED thermal testing standards published by JEDEC [5, 6].

6.2 What the LED World May Learn from the Past

> Those who do not learn from history are doomed to repeat it.
> Santayana

6.2.1 Lessons Learned from the IC World

It is always instructive to look back and see if we can learn from the mistakes made in the past. In the field of thermal standardization, we see a remarkable parallel between what happened in the semiconductor world 20 years ago and what is happening right now in the LED world.

Around 1990, it became clear that thermal characterization of IC packages was a chaos. Manufacturers all over the world used different standards, and even within a single manufacturer's environment intolerable differences showed up, as Table 6.1 nicely demonstrates. Various dual in-line (DIL) packages were sent to three different sites of a large semiconductor manufacturer, with the question to return thermal data according to the standard that was in use at the thermal testing facility of the given site. The returned numbers clearly indicate the problem: any meaningful comparison was out of the question.

6.2.2 Why End-User Rebellion Started Around 1990

While the manufacturers started to complain about the lack of well-defined standards, end-users did rarely raise their voice before 1990. The main reason this situation changed was that a new tool arrived that made it possible for designers to predict for the first time the thermal behavior of their products in a specific application: computational fluid dynamics (CFD). The accuracy that can be expected from these

Table 6.1 Thermal resistance measurement round-robin data for a variety of dual in-line (DIL) packages. (Work done by Luiten at Philips in 1991)

R_{thJA} measurements DIL 40				
Leadframe	Heat spreader	Various Philips labs (data in K/W)		
		Lab 1	Lab 2	Lab 3
NiFe	Al	52	31	39
	Cu	55		
	None	115	67	83
CuFe	Al	39		27
	Cu	38		
	None	57		46

R_{thJA} junction-to-ambient thermal resistance

tools is not discussed here (see, e.g., [27]), but in principle the only data that are lacking for which the end-user has no responsibility are the component thermal data provided by the manufacturer. From this point of view, it is obvious that the accuracy that can be reached is ultimately limited by the information in these data sheets. End-users started to realize that it did not make sense to spend a lot of money in software and training to facilitate virtual prototyping, only to be seriously limited by the lack of reliable information provided by the manufacturers.

However, it is one thing to say: we cannot use the data sheets; it is quite another thing to formulate what *is* needed. Because the use of user-friendly CFD tools in the electronics industries started in Europe, end-users in Europe started also the initiative to combine efforts to try to answer the question what was required. These needs were clearly identified as early as 1991 resulting in a European project proposal championed by Philips, and finally led to the launch of the EU-funded DELPHI project in 1993. When this project was finished, the members (all end-users) had a good idea what they really wanted [28–30], and they invited the three largest semiconductor manufacturers in Europe to join the team (SEED 1996) to investigate if the manufacturers could indeed realize what the end-users wanted [31, 32]. SEED was followed by PROFIT (2000) [33–35] in which new members participated, focusing mainly on the extension to the transient domain and on continuation of the efforts to get the methods standardized. Indeed, from 1990 onward, European members played an active role in the JEDEC JC15 committee that mainly concentrated on the standardization of thermal characterization of IC packages.

A major breakthrough in thermal characterization emerging from the DELPHI project was the notion that we needed a boundary between the responsibilities of the manufacturers and the end-users:

- The *manufacturer* is responsible for the thermal model of the part and nothing else.
- The *end-user* is responsible for the specification of the thermal environment to which the part is exposed.

The consequence of this partition between responsibilities was that we needed a new way of thermal characterization, coined "Boundary Condition Independent Compact Thermal Modeling." For a recent overview of compact modeling issues, please consult [36] along with the latest JEDEC guidelines [37–39] for compact thermal modeling of semiconductor device packages. In short, the manufacturer should publish thermal data in such a way that the end-user could use these data for whatever application the component is intended. Obviously, this is the end of thermal metrics such as R_{thJA}, because by definition this thermal resistance includes the environment and hence is heavily dependent on it. Currently, various aspects of compact thermal modeling have been standardized (see [36, 38, 39]) or are near standardization (such as vendor-independent file format of compact model libraries).

6.2.3 Lessons to be Learned from the CFL World

The US Department of Energy (DoE) considers it critical to guide the market introduction of SSL products because of their energy saving potential. While many SSL products are reasonably good, a recent study revealed some concerns: lower light output than claimed is quite common, disappointing lifetimes contradict excessive claims, and color quality is inconsistent. Test results of MR16 replacement show a serious amount of misleading product literature. The LED market situation drove an important initiative launched by DoE in 2008 to build a community of Quality Advocates. Interestingly, the initiative is based on recognizing the fact that the introduction of compact fluorescent lighting (CFL) in the 1970s was seriously hampered by unclear, misleading, and exaggerated performance claims [40]. The main lesson learned is that selling underperforming products can discourage the early adaptation of a new technology. The team has defined five critical parameters that should be part of any quality analysis: luminous flux, efficacy, input power, correlated color temperature (CCT), and color rendering index (CRI). Most of them depend on temperature. Reliability among others will be added in the future. A key component of the initiative is a voluntary SSL Quality Pledge launched in 2009. SSL Quality Advocates pledge to use the Lighting Facts® label, similar to a nutrition label. The Lighting Facts label provides a quick and simple summary of product performance data as measured by the new industry standard for testing photometric performance, such as IES LM-79-2008 [41]. The Lighting Facts reports provide product performance results in five areas: lumens, efficacy, watts, CCT, and CRI.

6.3 On the Use of LED Thermal Characterization for Reliability and Performance Prediction

The most important reason to use thermal data for an LED-based product design is to estimate the junction temperature upon which to base any assessment of the reliability of the final product (apart from adhering to official regulations such as safety standard ANSI/UL 8750 [42]). It is useful to make a distinction between the device (the LED itself) and the package or system:

- Device reliability: intrinsic light output reduction under operating conditions
- Package reliability: failures caused by thermal stresses and ageing

6.3.1 Device and Package Reliability

Before starting the discussion, we would like to spend a few words on the definition of "reliability" because not everybody is aware of its defined meaning. The "official" definition is: *the probability that a product will perform a specified function under specified operational and environmental conditions for a specified interval.* Often, a distinction is made between "reversible" and "irreversible" failures. If reversible, the problem is often called a "performance" failure, otherwise a "reliability" failure. From a lighting perspective, it should be noted that there is a world of difference between LED and conventional lighting caused by the fact that the longevity of LEDs may exclude irreversible failures during its operational lifetime, requiring another definition of "lifetime."

The main temperature-related problem at the device level over time is reduction in light. Because of this phenomenon, the light output *at the system level* may decrease to an unacceptable level before a "real" irreversible failure occurs. Lumileds pioneered a way to present reliability data by two sets: lifetime and lumen maintenance. Lifetime (also called "mortality") is presented by the letter B followed by a number. For example, B50 is the time by which 50 % of the population is expected to fail before a specified condition is encountered. This condition is the lumen maintenance and is given by the letter L followed by a number, e.g., L70 means that any product with more than 30 % light output degradation is classified as a failure. The LM-80 standard [43] is a major step forward but there are still some thermal questions to be answered.

The standard defines case temperature as the temperature of the thermocouple attachment point on the LED light source package as defined by the manufacturer of the package. It is here where a major problem lies. Not only should the thermocouple be standardized, but also the exact location and the substrate, another lesson that can be learned from the IC world. Figure 6.12a shows an example of how leading manufacturers have been presenting their reliability data, showing the expected lifetime as a function of the junction temperature at the 90 % confidence interval.

As mentioned already, the LM-80 standard [43] defines the test procedure used to obtain input data to create Bxx-Lyy plots like the B50-L70 diagrams shown in Fig. 6.12a. But the raw measurement data are not yet sufficient; there must be a commonly agreed methodology that allows measured data (e.g. up to 6,000 h of operation) to be turned into actual lifetime prediction. The recently published IES TM-21 standard [44] provides such a definition. Application of this standard is supported by a free calculator spreadsheet [45, 46]. Figure 6.12b presents TM-21-based lifetime prediction diagrams from the same vendor.

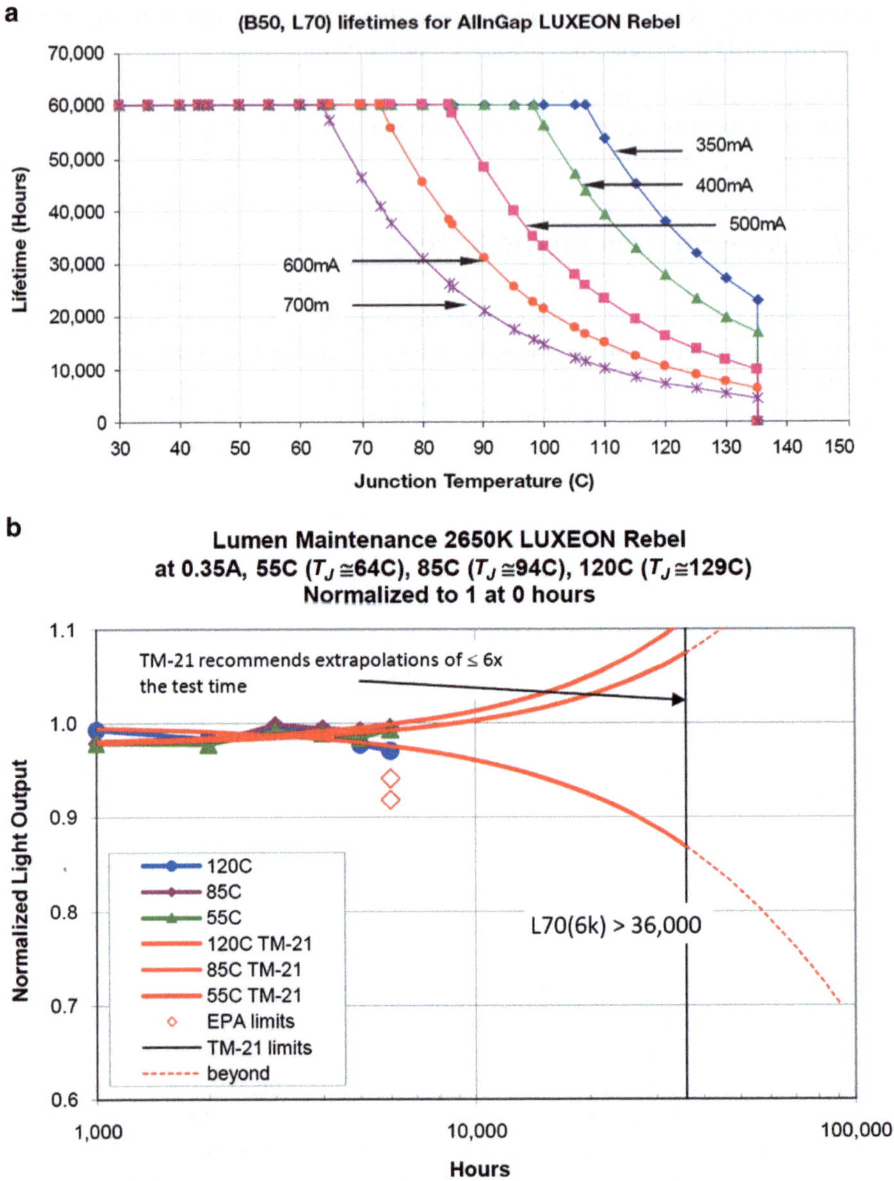

Fig. 6.12 Lifetimes for AlInGaP LED by means of **a** B50, L70 diagrams as used until recently and **b** TM-21-compliant lifetime prediction plots also indicating actual LM-80 test data points. (By courtesy of Philips Lighting. Philips started using TM21 plots, other vendors (like Osram) still publish Bxx-Lyy plots.)

Needless to say that data reporting on lifetime parameters are subject to standardization because the average customer will get lost if one manufacturer quotes reliability in terms of B50, L70 and another in terms of B10, L50.[6]

[6] As of late 2012, some LED vendors started publishing TM-21-compliant lifetime predictions instead of B50-L70 plots (see Fig. 6.12b), some others still publish B50-L70 plots.

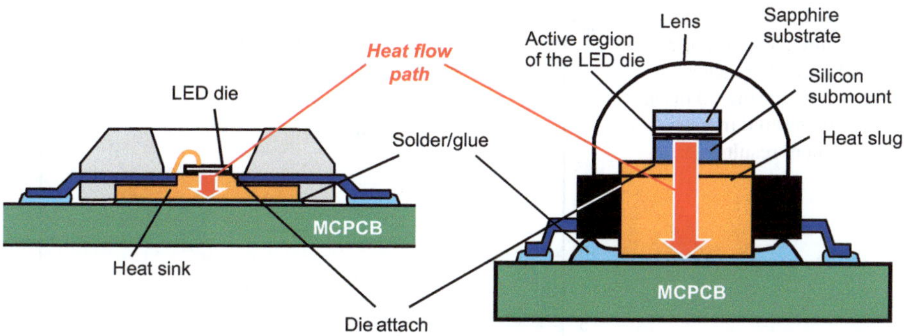

Fig. 6.13 Sketches of some mainstream power LED packages

Yet another aspect related to this topic is that in recent years, the light output stability (lumen maintenance) of power LEDs has been improved quite significantly. Despite this progress, or maybe even because of this, a customer may still want to know what happens when the LEDs are driven outside the recommended range, as happened in the IC world (uprating).

Let us now focus on package reliability with its many faces. Sketches of two packages containing LED dies are shown in Fig. 6.13.

While LED packages are not of the same complexity as state-of-the-art IC-packages, many locations exist where failures may occur, and having products consisting of tens or hundreds of LEDs, the need for accurate reliability prediction becomes obvious. Some degradation and failure mechanisms that are worth mentioning are:

- Yellowing of phosphor containing encapsulations
- Lens degradation
- Delamination of adhesive layers
- Solder joint failures

One should note that for high-quality LEDs, these degradations usually only occur when operating outside rated conditions. LEDs suffer also from current-dependent failure mechanisms, such as electromigration and Joule heating causing excessive local temperature rise in current-carrying tracks and wires. For further reading on LED reliability issues, please see Section 1.1 of Chapter 1 and consult references [47–53].

In summary, most of the degradation and failure mechanism that rule the lifetime of an LED-based product are temperature-assisted failure mechanisms. Consequently, in order to estimate lifetime, designers need reliable information about the expected temperature profile over time.

6.3.2 System Reliability

As is the case with the IC world, we have two separate reliability issues: one at the component level, one at the system level. While at the component level, we can live

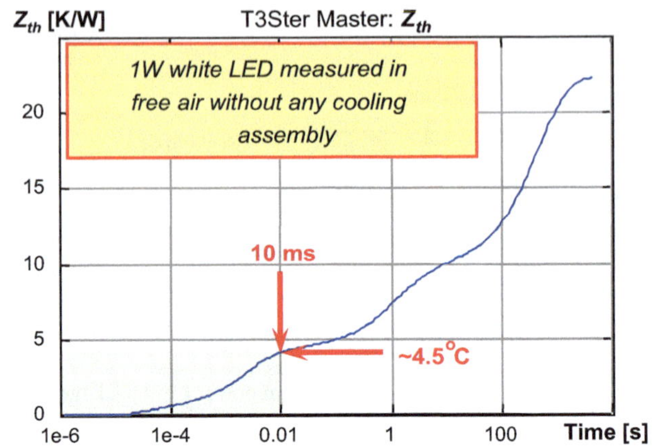

Fig. 6.14 Junction temperature elevation of a 1-W white LED reaches about 4.5 °C during 10 ms as shown by thermal transient measurement results

with quoting data under specified conditions, we need at the system-level data under operating conditions.

6.4 Drawbacks of Sloppy Data Sheets of LEDs

The most important things a designer of an LED-based product wants to know about are Φ_V, luminous flux (lm) and η_V, efficacy[7] (lm/W), not only at zero-hour but also over its expected lifetime. The problem is that both key parameters are not linearly related to driving current and temperature. Hence, it is not sufficient to report data only at some optimistic temperature. The question is: how bad are most of the current data sheets? The answer is: pretty bad, at least from the perspective of the experienced designer, for the following reasons (see also Grabner-Meyer [55]):

- Data sheets do not reflect real-life operation, especially regarding operating temperature. Often T_J is specified at 25 °C, which is misleading, because the efficacy at maximum rated power is often 50 % lower than at room temperature. Another problem with this claimed $T_J = 25$ °C statement is that—unlike the general misbelief—the duration of the measuring pulse applied in production testing is long enough to result in significant junction heating as illustrated by Fig. 6.14.
- They differ strongly in content (T_{ref}, I_{ref}).
- Often, a direct comparison between manufacturers is not possible.
- While nonthermal but temperature-related: Translation to useful lumens or non-nominal use is *not* “idiot”-proof and not an easy task even for an experienced

[7] As pointed out in other chapters of this book, there are multiple definitions of *efficacy*. *Efficacy of source of radiation* relates the emitted luminous flux of a light source to its radiant flux. The more often used meaning of efficacy is when the emitted luminous flux of a light source/luminaire is related to its input electric power. The second definition provides hint about how efficiently the given light source converts electrical energy into light. For official definitions of these efficacy terms, please consult, e.g., CIE’s ILV—the *International Lighting Vocabulary* [54].

designer. The reason is that the temperature dependence of the relative luminous flux is given only by means of graphs rather than by means of exact data and formulas suitable for calculations.

On the positive side: it should be noted that there is a tendency among the leading LED suppliers to improve upon this situation and some have started to quote values that are more realistic and already comply with the most recent LED thermal testing standards [5–7]. Additionally, one should mention here the tools of Future Lighting Solutions [56], which help designers in getting the data they need or the most recent MCAD-embedded CFD tool of Mentor Graphics with a demo library of compact models of LEDs, including a simple model aimed at hot lumens calculations of some popular LED types (FloEFD v12 with its LED module) [57].

6.4.1 Examples of Problems with Some Sloppy LED Data Sheets

6.4.1.1 The Definition of Power

Until recently, most LED vendors performed thermal testing according to the EIA/JEDEC JESD51-1 standard and they treated "power" in a sloppy way. The standard clearly mentions "power dissipated in the device" but does not give any definition how to calculate it. For silicon devices, there is no doubt that $P_H = I_F \cdot V_F$ (the electrical power calculated as the product of the total current and the voltage) while in case of LEDs an energy balance must be considered when calculating the power dissipated *in the device*. Hence, maybe the biggest problem today hampering a fair comparison is that LED-specific thermal testing standards are not yet widespread in all LED vendors' daily practice. Though the leading manufacturers take into account the efficiency of the conversion from current into light, there are still many other LED vendors who do not follow this practice.

To highlight the importance of this issue for modern power LEDs, let us compare two simple cases to address the consequences, and let us define $R_{th\text{-}el}$ as the thermal resistance based on the total electrical power supplied to the LED and $R_{th\text{-}real}$ as the thermal resistance based on the total dissipation minus the light output. Let us assume two LEDs with different radiant efficiencies:

Case 1 Radiant efficiency $\eta_e = 25\,\%$, supplied electrical power $P_{el} = 10\,\text{W}$, junction temperature rise $\Delta T_J = 50\,°\text{C}$

$$R_{th-el} = \frac{\Delta T_J}{P_{el}} = \frac{50}{10} = 5\,\text{K/W}$$

$$R_{th-real} = \frac{\Delta T_J}{P_{el} - P_{opt}} = \frac{\Delta T_J}{P_{el} \times (1 - \eta_e)} = \frac{50}{10 \times 0.75} = 6.7\,\text{K/W}$$

Case 2 Radiant efficiency $\eta_e = 50\,\%$, supplied electrical power $P_{el} = 10\,\text{W}$, junction temperature rise $\Delta T_J = 50\,°\text{C}$

$$R_{th-el} = \frac{\Delta T_J}{P_{el}} = \frac{50}{10} = 5\,\text{K/W}$$

$$R_{th-real} = \frac{\Delta T_J}{P_{el} - P_{opt}} = \frac{\Delta T_J}{P_{el} \times (1 - \eta_e)} = \frac{50}{10 \times 0.5} = 10\,\text{K/W}$$

From this simple numerical example, one can clearly see that in case of LEDs with higher efficiency, assuming the same junction temperature rise as for less efficient LEDs, the real thermal resistance of the package is higher. Therefore, as of today, some LED vendors report $R_{th\text{-}el}$ values on their data sheets (simply because $R_{th\text{-}el}$ is always lower than $R_{th\text{-}real}$ providing a better sales argument), misleading lighting system designers during the thermal evaluation of their designs.

In our view, any thermal resistance reported on data sheets should only be related to the physical properties and the dimensions and should be independent of the actual operating point (forward current, voltage, and junction temperature through the energy conversion efficiency), as is clear from the example above. The problem in practice is that the user should know the efficiency for the actual thermal design if the $R_{th\text{-}el}$ value is reported. Of course, the LED community could decide for a metric based on the total power input, resulting in the same LEDs differing only in efficiency to get different R_{th}'s. However, to prevent confusion, clear distinction in nomenclature is needed—as is the case with JEDEC's recent LED thermal testing guidelines [6] in which the *real thermal resistance/real thermal impedance* of DC-driven LEDs are clearly defined.

We are of the opinion that a higher efficiency is an important sales argument with the increasing interest in sustainability, and hence it makes sense to report this parameter on the data sheets too, as is also recommended in [7].

As shown in Fig. 6.15, the *radiant efficiency* (many times also referred to as WPE, which is P_{opt}/P_{el}) depends on temperature and current. This, of course, also results in the current and temperature dependence of the *efficacy*, which can be considered as the efficiency weighted with the $V(\lambda)$ visibility function defined by the CIE.

This is the reason why any metric of an LED reported in a data sheet should also be reported together with the current and temperature at which the given metric was identified. The value of the forward current to report is unambiguous but the definition of the temperature value to report needs careful discussion. In an ideal case, this must be the junction temperature provided there is an unambiguous way to identify the junction temperature of the LED in practice. The procedure defined in the JESD51-51 standard [6] now offers a clear procedure to identify LEDs' junction temperatures.

6.4.1.2 The Series Thermal Resistance Approach

As an example of the problems that are associated with a series resistance approach, let us have a look at the data sheets of one of the major LED manufacturers. In Fig. 6.16, we present a sketch taken from a typical application brief. (Calculation of the heating power in this example still does not consider the emitted optical power.) The assumptions underlying the series resistance approach are the following:

- The heat generated in the die follows the path sketched via the thermal pad, board/heat sink and finally the ambient.
- The resistances are defined locally, in other words, the thermal resistance from die to heat slug is only dependent on local parameters.

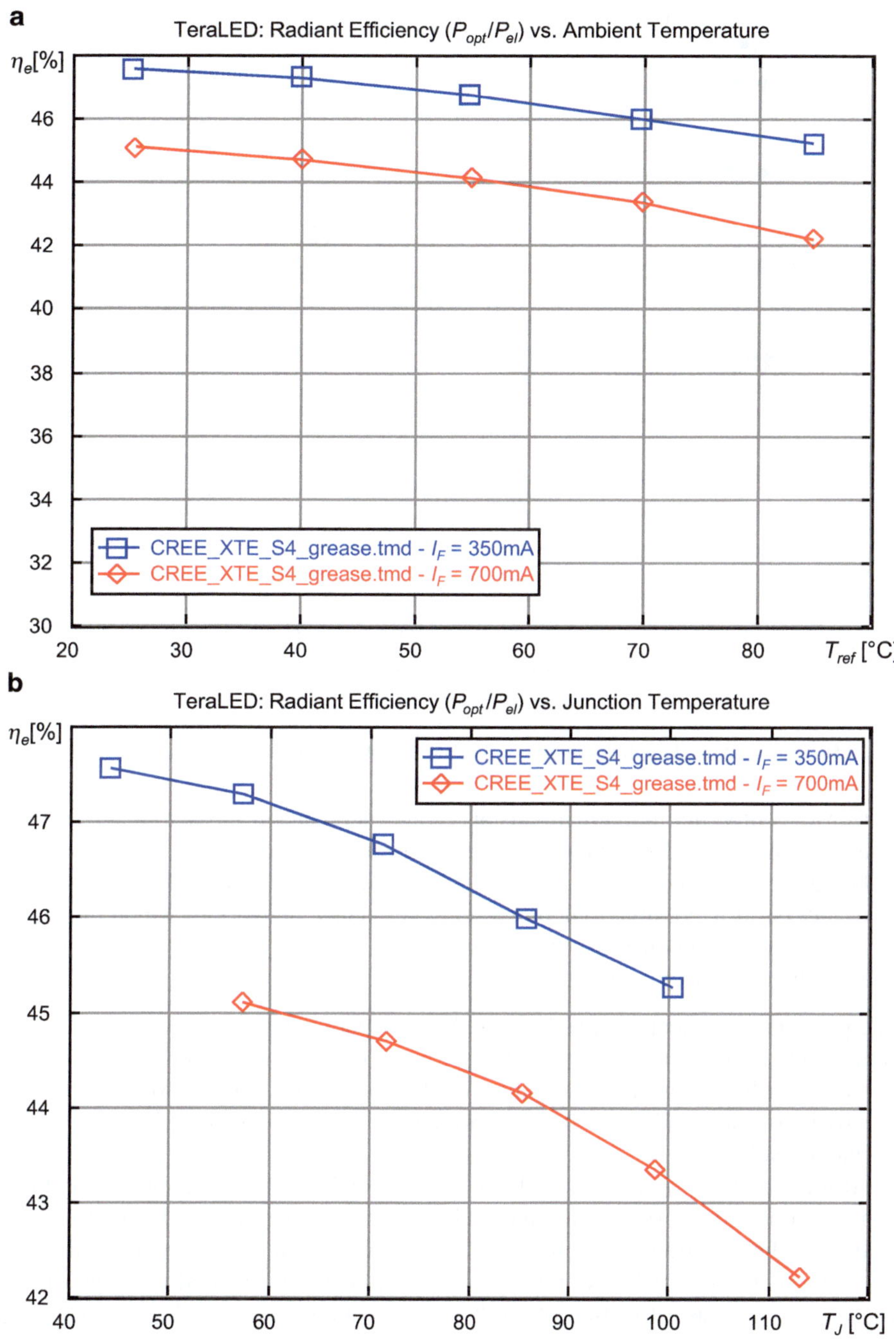

Fig. 6.15 Temperature and forward current dependence of the radiant efficiency of a sample of a Cree X-TE white LED shown as a function of **a** preset reference temperature values and **b** corresponding junction temperature values

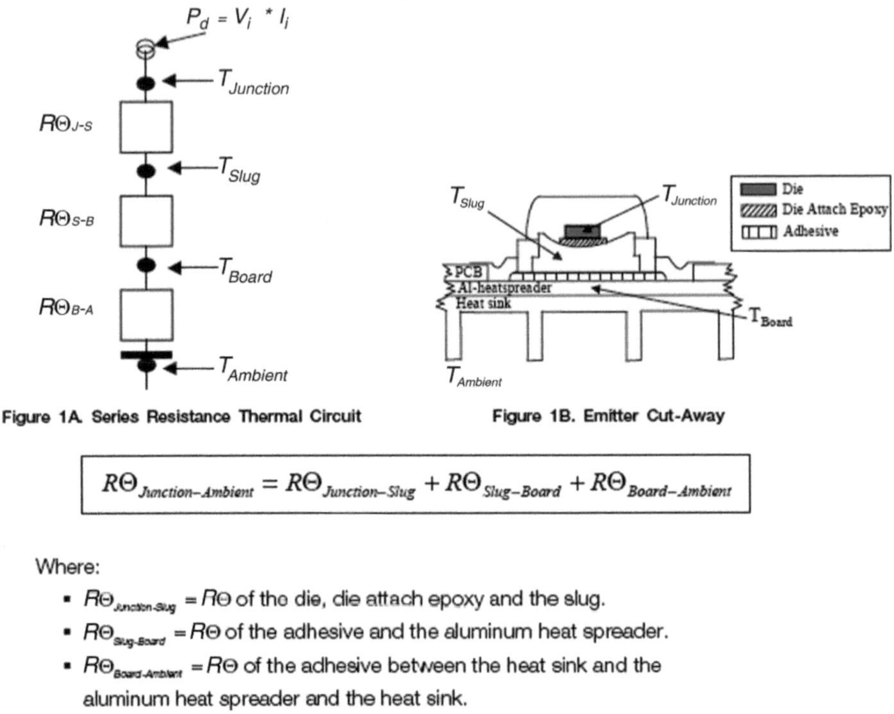

Fig. 6.16 From a typical application brief. (By courtesy of Philips Lighting)

- Consequently, the individual resistances are independent of each other. For example, the thermal resistance from die to heat slug is neither dependent on the board thermal conductivity, nor on the heat transfer coefficient. This may sound trivial, but it is possible to define a series resistance network of which the resistors are dependent on each other. This principle is underlying the well-known heat spreading approach proposed, e.g., in [58].

The procedure to check was as follows. Various parameters such as board thermal conductivity and heat transfer coefficient were varied over a wide range for the purpose of illustration, and the values of the thermal resistances were calculated according to the application brief mentioned. The graphs in Fig. 6.17 show the essential results.

At the top, $R_{th\ die\text{-}thermal\ pad}$ should be only dependent on the dimensions and the thermal conductivity of the thermal pad, *not* on the thermal conductivity of the board and the heat transfer coefficient. Over the whole range, we see a variation of 15 % that seems acceptable for most practical applications. The bottom graph shows $R_{th\ thermal\ pad\text{-}board}$ as a function of its thermal conductivity and h. In this case, this R_{th} should be proportional to k_{board} and independent of h. It is clear that we meet a problem when the effective board conductivity becomes smaller than about 5 K/W because the values of the published resistances become a function of the application

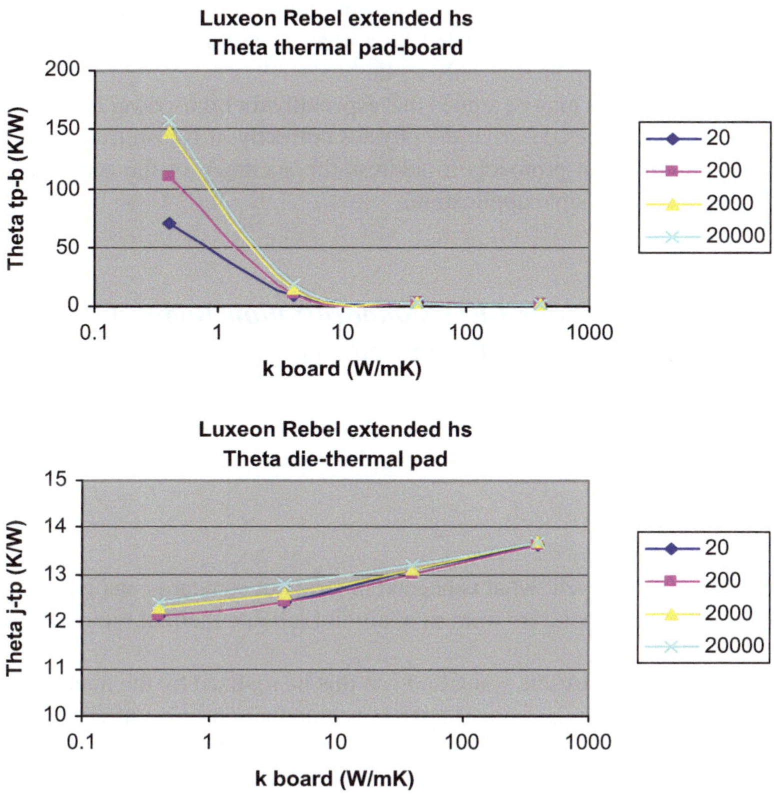

Fig. 6.17 Graphs demonstrating the drawbacks of the series resistance approach with h between 20 and 20,000 W/m^{2}K as parameter. *Top* Resistance thermal pad-board. *Bottom* Resistance die-thermal pad

and as such cannot be used with confidence, unless the application does resemble more or less the measurement conditions. This situation is the consequence of heat spreading, where it is impossible to separate the conduction and convection parts. In other words, a series resistance approach can never result in a boundary condition-independent thermal model. However, there are exceptions. For many high-power LEDs, the lateral temperature gradients can be neglected. In these cases, it can be shown that the series resistance approach is valid. For example, for state-of-the-art high-power LEDs the lower limit of 0.3 K/W in the graph is not realistic [59]. It should also be noted that a correct value of some resistances in the chain is not mandatory in all cases because these resistances may not be dominant in a real application. Often, it turns out that the resistance to ambient is dominating, hence, errors in the published data are not critical.

In order to check the validity of these assumptions, a Philips Luxeon Rebel® LED as depicted in Fig. 6.18 was modeled on a board using a commercial CFD-based thermal simulator in conduction-only mode.[8]

In summary, while it may be argued that especially for high-power LEDs the series resistance approach does describe the physics correctly, it is also true that we need thermal standardization protocols to address the accuracy of the series resistance approach for every possible application.

6.4.2 *Other Questions LED Standardization Bodies Considering Thermal Aspects Should Address*

Apart from the problems with sloppy LED sheets, many other thermal characterization issues also pose a challenge. A couple of years ago, the following questions were raised [11–13]:

Basic Questions

- For all parties involved: what is needed in terms of standards and guidelines?
- What do the manufacturers want as a fair and reliable thermal metric to compare products?
- What do trained end-users want, and can this be realized by the manufacturers?
- What about a purely numerical approach (conduction-only) after suitable calibration? Additionally, as an alternative to compact models, it may be feasible to provide the end-user with a fully detailed model because there are usually no proprietary data to be considered.
- What links with other international standardization bodies or national measurement laboratories would be required and how can proper cooperation between the various standardization bodies be warranted?

Thermal Measurement-Related Questions

- *What are the pros and cons of steady-state versus transient measurements in LED test conditions?*
- *How to measure the "thermal" power?* In theory, this may be based on spectroradiometric measurements or on the direct measurement of Φ_e the radiant flux of the LED (using a detector having a flat spectral response). One should be careful when selecting these methods: whatever type of total flux is measured, the measurement should be calibrated to a traceable standard. For example, for total

---

[8] The issues discussed, illustrated by the Luxeon Rebel® are especially severe at the component level of the LEDs themselves. Once assembled, e.g., to a standard star-shaped MCPCB, the physical dimensions of the Rebel® LEDs are by an order of magnitude smaller and the heat-flow distribution problem becomes negligible on board level, not to mention luminaire-level thermal analysis. In these cases, an effective thermal resistance (impedance) with appropriate footprint area at the board might meet the accuracy requirements. The detail and accuracy of modeling therefore always needs to be adjusted to the nature of the analysis problem: the issues discussed are important if the LEDs are assembled on an FR4-based PCB.

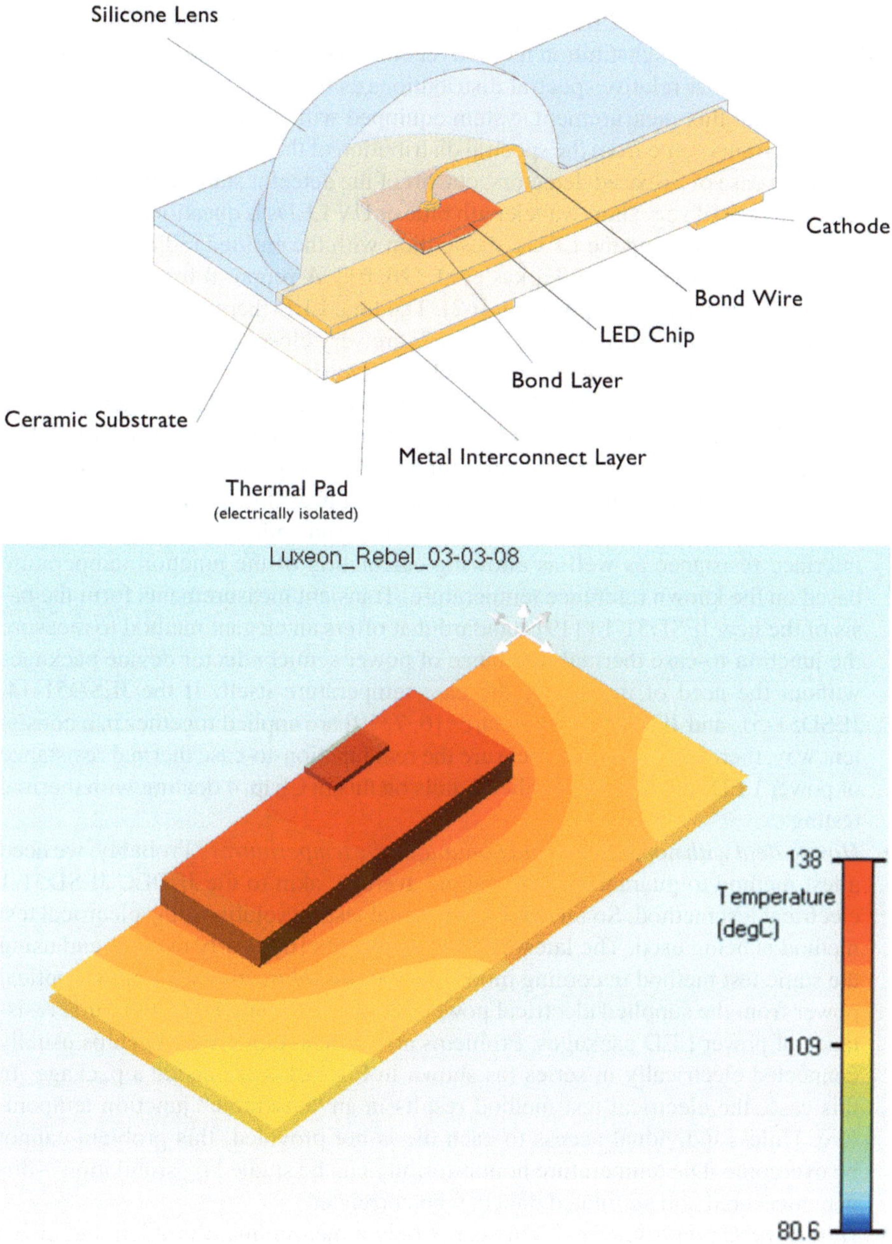

Fig. 6.18 *Top* Sketch of a Philips Luxeon Rebel® LED (by courtesy of Philips Lighting). *Bottom* Half-symmetry model on a PCB

flux measurements the CIE 127-2007 document about the measurement of LEDs [60] with the strict substitution method recommends using standard LEDs, which possess a similar relative spectral distributions as the LEDs under test. In case of using a total flux measurement system equipped with a radiometric detector, the physical limits come from the spectral distribution of the sensitivity of the detector in use. In case of Si-based detectors, cut-off of the detector starts at blue, therefore measurement of very short wavelength blue or UV LEDs is questionable. An early solution for correcting the LEDs' dissipation with the emitted radiometric flux is suggested in two papers by Farkas et al. [26, 61]. A proposal for a standard can be found in a paper by Poppe et al. [62]. The final LED thermal testing standards developed and published by JEDEC [6, 7] are very close to this proposal.

- *How to measure the case temperature?* Our opinion is that power LEDs can be best characterized on a cold plate. Then, though we do not have the case temperature itself, we have quite accurate information about the cold plate temperature. The interfacial layer between the cold plate and the heat slug should be addressed. Detailed information about the junction-to-ambient heat-flow path can be extracted from thermal transient measurements—providing an option to address the thermal interface resistance as well as allowing calculation of the junction temperature based on the known reference temperature. Transient measurements form the basis of the new JESD51-14 [19] standard that offers an elegant method to measure the junction-to-case thermal resistance of power semiconductor device packages without the need of measuring the case temperature itself. If the JESD51-14, JESD51-51, and JESD51-52 standards [6, 7, 19] are applied together in a consistent way, there is a method to measure the real junction-to-case thermal resistance of power LEDs as well. See further details on this in Chap. 4 dealing with thermal testing.
- *How to deal with nonuniform and nonunique die temperatures?* Probably, we need a test method to guarantee some unique average, akin to the JEDEC JESD51-1 electrical test method. So far, for LED thermal testing solutions, the electrical test method is being used. The latest JEDEC guidelines [6] clearly recommend using the static test method in cooling mode and require subtracting the emitted optical power from the supplied electrical power in order to obtain the real thermal resistance of power LED packages. Problems arise when multiple LED chips usually connected electrically in series (as shown in Fig. 6.1a) are inside a package. In this case, the electrical test method results in an "ensemble" junction temperature. Unless individual access to each die is not provided, this problem cannot be overcome. Die temperature nonuniformity can be studied by simulation—this way measured and simulated data can be correlated.
- *How to tackle multisources?* This is not only a measurement problem but also a data representation problem. Single-source compact modeling approaches cannot be used. A matrix representation is probably the preferred option as already mentioned in Sect. 6.1.4.2 and shown in Fig. 6.9. For LED applications, see, e.g., papers by Zhang and Treurniet [63] or Treurniet and Lammens [24]. One has to note that we can measure the elements of the matrices only if access to every individual LED chip is provided, i.e., if the LED-based system was also designed for *thermal testability*. If there is no individual (preferably 4-wire) access to all

the LED chips of the device under test (DUT), we can measure only an overall average junction temperature and we run into the problems of the "ensemble" junction temperature and "ensemble" thermal resistance as discussed already in relation to the junction temperature.

- *How to deal with phosphor-encapsulated dice?* These pose extra problems, caused by absorption of light resulting in an extra heat source. The location of this heat source is not clearly defined for end-users. The phosphor can be deposited on top of the blue chip or away from the junction (remote phosphor white LEDs). Information about the absorption and the spatial distribution of the heat generation would be rather difficult to obtain. On top of this, we would have a system with multiple heat sources where only the dissipation at the junction could be controlled, so one would not be able to measure the elements of the matrix of the junction-phosphor multi-heat-source system. Consequently, the extra heat originating from the phosphor would disturb somehow the measurement of the thermal properties seen from the junction and there is no method available to measure this distortion. Again, a possible work-around could be proper numerical simulation, but we believe this should remain the task of the LED manufacturers.
- *What about pulsed-type thermal measurements?* Such measurements (referred to as the dynamic test method in the JEDEC JESD 51-1 document) are to be avoided due to several reasons. One very important reason is that such measurements cannot be combined with CIE 127:2007-compliant total flux measurements. To assure consistency of thermal metrics and optical metrics, combination of JEDEC JESD51-1 static test method-compliant thermal measurement and CIE 127-2007-compliant total flux measurement seems to be the only feasible way—as recommended now by the latest JEDEC documents JESD51-51 [6] and JESD51-52 [7].
- *How to relate the short-pulse results to properties that can be measured under steady-state operating conditions?* In production testing, properties of LEDs are measured by short pulses at temperatures that usually do not represent the actual operating conditions. To tackle this problem, details of the complex LED behavior are required. The major issue is that the LED junction considerably heats up even within a short time (e.g., 10 ms) as it was already shown in Fig. 6.14.

Obviously, the answers to these questions are by no means trivial and are best approached by collecting relevant information from all parties involved. Fortunately, many issues could benefit from the work that has already been done in the IC world. There are two considerations that mitigate the problems compared to the situation we faced 20 years ago with IC packages: from a thermal point of view LEDs are a lot less complex than, for example, microprocessors, and we have all the experience from the past including many standardization templates. On the other hand, it might turn out that thermal-only models are not sufficient for LEDs and electro-thermal-optical multidomain models may be the preferred solution.

Some of the questions discussed in this section were already answered in [12], to some other questions progress in technology has provided additional answers since the first white papers dealing with thermal standardization issues of power LEDs were published. For example, the JEDEC JC15 committee developed new thermal testing guidelines for LEDs that are already referred to in new LED performance testing standards being developed nowadays. The new JEDEC LED thermal testing standards are natural adoptions of prior testing techniques. The combined thermal and radiometric/photometric measurement of LEDs recommended in the JESD 51-52 document [7] is a natural extension of CIE's earlier recommendations about the total flux measurement of LEDs [60].

6.5 Recent Developments

For a presentation about thermal characterization in general including the roadmap, see Guenin [64], showing that standardization of ICs covers not only the test methods but also related topics such as compact models and interface resistances. For the most recent status, see, e.g., [65].

Regarding thermal testing of general semiconductor devices, the JC15 committee of JEDEC developed a modular approach. The JESD51 document [17] is an *overview document*, providing a general description of different aspects of thermal testing of packaged semiconductor devices. These are: test methods and test environments including device mounting. The document includes the following statements that are still valid:

> Each group will have one or more applicable documents to reflect different thermal measurement requirements. Because environmental conditions, component mounting approaches and device construction techniques will change as technology changes, additional documents will be added to these groups as the needs arise and standards established.

This modular approach led to the development of new standards about compact thermal modeling—based on the results of the DELPHI and SEED projects as mentioned earlier. The new, transient measurement-based method of the recent JESD51-14 standard for the measurement of junction-to-case thermal resistance (R_{thJC}) of power semiconductor device packages with a single heat-flow path relies on techniques developed in the PROFIT project (structure functions—see Appendix A of the standard [19]).

6.5.1 *Overview of LED-Specific Thermal Testing Standards Published by JEDEC*

The modular approach of the JEDEC JESD51-* series of standards allowed accommodating new documents providing definitions specific to LEDs. These new documents added to the JESD51-* series are following:

JEDEC JESD51-50: *Overview of Methodologies for the Thermal Measurement of Single- and Multichip, Single- and Multi-PN-Junction Light-Emitting Diodes (LEDs)* [5]

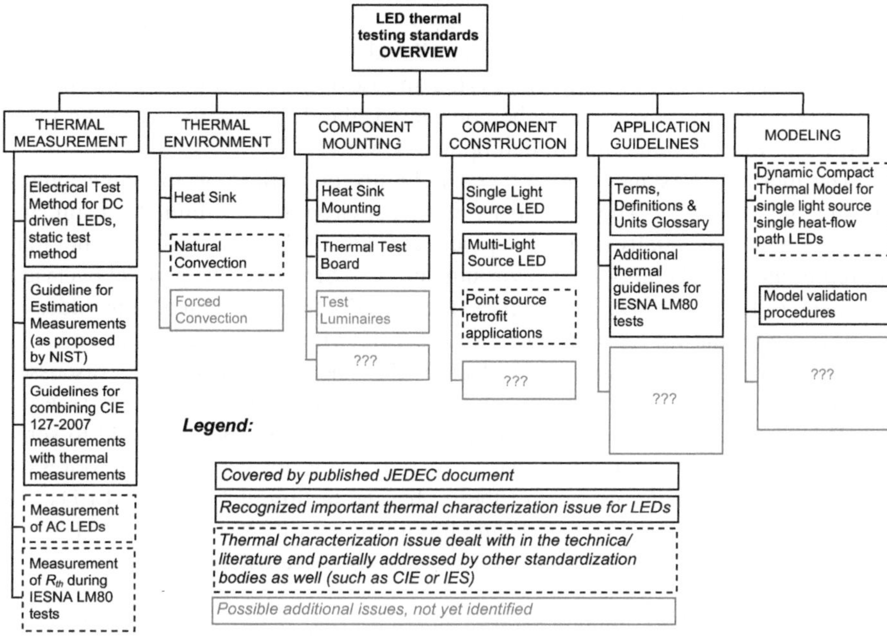

Fig. 6.19 JEDEC's overview chart of possible modules of LED thermal testing and characterization standards and guidelines. (Modified version of the original published in [5])

This is an overview document about the JESD51-5* series of LED-specific additions to the JESD51-* family of standards. The details of the relevant measurement methodologies are distributed among several documents so that the appropriate combination of documents can be selected to meet specific LED thermal measurement requirements. The JESD51-50 document provides the overview; the rest of the documents are grouped as shown in Fig. 6.19. It repeats the introductory sentence of the JESD51-1 document:

> Each group will have one or more applicable documents to reflect different thermal measurement requirements. Because environmental conditions, component mounting approaches and device construction techniques and processes will change as technology changes, additional documents will be added to these groups as the needs arise and standards established. As appropriate, each of these documents will contain terminology and symbolic definitions specific to the material covered by the individual document; this information will also be included in a single document to make for easy access. [5]

The JESD51-51 document provides the basic concept of LED thermal testing with the following equations:

$$T_J = P_H \cdot R_{thJ-X} + T_X \tag{6.9}$$

and

$$P_H = P_{el} - P_{opt} = I_F \times V_F - P_{opt} \tag{6.10}$$

where $R_{thJ\text{-}X}$ refers to the *junction-to-reference environment X thermal resistance*, T_X denotes the temperature of environment X and P_H denotes the power heating the LED's pn-junction. The basic message is that in case of LED thermal characterization the real P_H heating power has to be considered: the difference between the input electrical power (P_{el}) and the emitted optical power (P_{opt}).

"The thermal resistance term (R_{thJX}) is highly dependent on the environment surrounding the device. The two most common environments, infinite heat sink and natural convection usually define the practical limits of thermal resistance, but may not represent typical component environments" [5]. Similar to the general overview document of the entire JESD51-* series, the JESD51-50 document also includes an overview chart that we reproduce here in Fig. 6.19 with some minor additions.

As presented in this chart, some of the LEDs' thermal-related issues discussed also in this chapter are already covered by published JEDEC documents (see the JESD51-5* series). Some other fields are clearly identified and discussed to some extent in the technical literature, but as of today, are not yet formally addressed by JEDEC. Just to mention some of these: thermal testing of AC-driven LEDs, the role of thermal (transient) testing during lifetime/reliability tests such as LM-80, or compact modeling of LEDs. In subsequent sections, these will be addressed in more detail.

JEDEC JESD51-51: *Implementation of the Electrical Test Method for the Measurement of Real Thermal Resistance and Impedance of LEDs with Exposed Cooling Surface* [6]

> The purpose of this document is to specify how LEDs' thermal metrics and other thermally-related data are best identified by physical measurements using well established testing procedures defined for thermal testing of packaged semiconductor devices (published and maintained by JEDEC) and defined for characterization of light sources (published and maintained by CIE—the International Commission on Illumination).
> LED specific aspects of thermal testing procedures are defined in a generic way, which means, that whenever possible, no distinction is made between steady-state and dynamic (transient) thermal measurements. [6]

The JESD51-51 document specifies thermal testing procedures for power LEDs, which are typically used in the operating regime of the forward current of 100 mA and above and which emit visible light. When speaking about thermal testing, one can distinguish between laboratory testing and bulk testing. The scope of this document is laboratory testing of DC-driven power LEDs.[9] Recommendations given in JESD51-51 are valid for both steady-state and dynamic (transient) thermal measurements of LEDs, both relying on *JEDEC JESD51-1 static test method.* The major recommendations of the JESD51-51 document are the following:

- It is recommended to measure power LEDs on a cold plate as on component level the recommended thermal metric is the R_{thJC} junction-to-case thermal resistance. The cold plate test environment resembles best the typical application conditions of LED components that are attached to a heat sink (as discussed in Sect. 6.1.4).

[9] Although the problem of measuring directly AC (mains)-driven LEDs is recognized (e.g., in 2011 CIE has set up the TC2-76 technical committee to deal with this issue because many technical publications deal with different aspects of AC LEDs—see, e.g., [21, 62, 66, 67]), no testing standards specific to AC-driven LEDs have been published yet.

- The real thermal resistance—as defined by Eq. (6.8b)—must be measured. For the measurement of the R_{thJC}, the test method described in the JESD51-14 document [19] is recommended with the addition of the calculation of the real heating power as shown by Eq. (6.10).
- In consistence with the recommendations of the JESD51-14 document, the transient extension of the so-called JESD51-1 static test method [18] in cooling mode is recommended. Hence, the best practice in thermal testing of LEDs is to let them heat up by the normal operating forward current, and once they get stabilized at the operating junction temperature, switch them off to measure the cooling transient of the junction temperature. The thermal resistance or the thermal impedance is to be calculated as described by Eqs. (6.5) and (6.7).
- In order to assure consistency between the thermal measurement and the optical measurement providing the value of the P_{opt} emitted optical power (total radiant flux) value for the heating power calculations, it is recommended to use a single, combined thermal, and radiometric LED measurement station. The optical part of such a setup can be realized as shown in Fig. 5.11 of Chap. 5. See also the description of the JESD51-52 document further in this section.
- It is recommended to calculate the real junction temperature of the measured LED as follows:

$$T_{J-real} = P_H \cdot R_{thJ-ref} + T_{ref} \tag{6.11}$$

where P_H is the applied heating power as defined by Eq. (6.10), $R_{thJ\text{-}ref}$ is the measured real junction to cold plate thermal resistance and T_{ref} is the cold plate temperature. Paper [68] suggests an alternate method for determining LEDs' junction temperature. The accurate implementation of that method is not trivial, e.g. knowledge of the thermal time constants of the LED+cold plate assembly and information about the nature of the initial section of LEDs' junction temperature transient is needed.

The detailed test procedure defined in the JESD51-51 standard can be summarized by Figs. 6.20 and 6.21. First a constant, so-called I_H heating current (which is typically equal to the usual operating forward currents of LEDs, such as 350 or 700 mA, etc.) is applied: the LED under test is switched on.

When the LED gets thermally stabilized, its forward voltage remains constant (V_H in Figs. 6.20 and 6.21). This is the situation when the light output properties of the LED (including the emitted optical power) can be measured. It is the requirement of CIE's LED measurement procedures to measure LEDs in thermal steady state (see the CIE 127-2007 document [60] and the JEDEC JESD51-52 document [7]).

Switching off the LED during thermal testing means that its forced, constant forward current is switched from the operating current to a small, so-called I_M measurement current that is chosen such that no significant heating and no significant light output is provided while the current is still sufficient to keep the junction at a sufficient forward voltage. (Typical values are in the range of 1–10 mA.)

After a significant electrical transition between two operating points in the hot LED's I–V curve (transition from the (I_H, V_H) point to the (I_M, V_{Fi}) point—see Fig. 6.21), the junction temperature will drop resulting in a forward voltage increase.

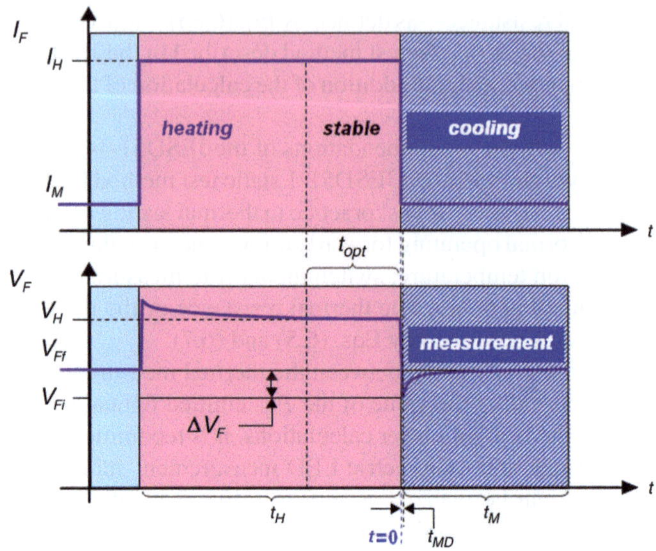

Fig. 6.20 Forward current and forward voltage waveforms of an LED during thermal testing. (Source: [6], © JEDEC, reproduced with permission)

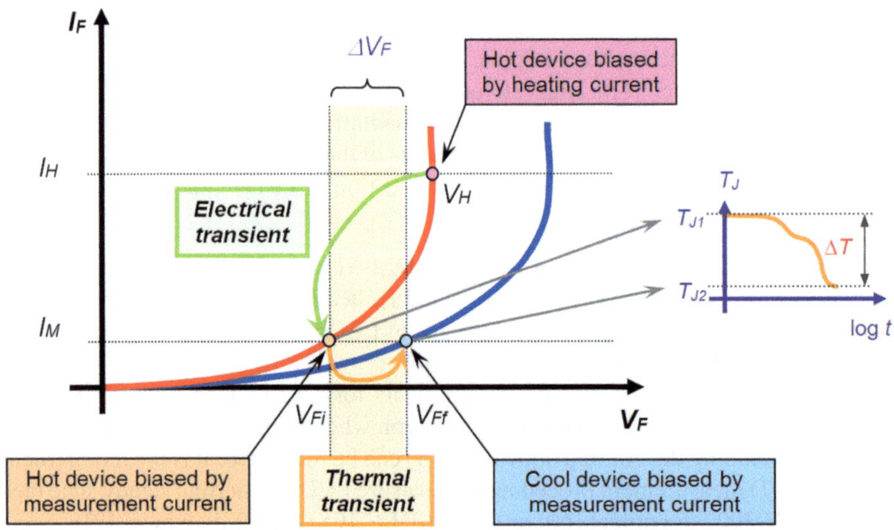

Fig. 6.21 The electrical and thermal transient processes taking place during the thermal testing of LEDs measured by the static test method in cooling mode. (Source: [6], © JEDEC, reproduced with permission)

The $V_{Fi} \rightarrow V_{Ff}$ transition is measured and is translated to the junction temperature transient: $\Delta T_J = V_F(t) \cdot K$. It is important that the K-factor is identified for the I_M measurement current (see further details in Chap. 4 about thermal testing). These transient processes are shown in Fig. 6.22 as the time function of the forward voltage.

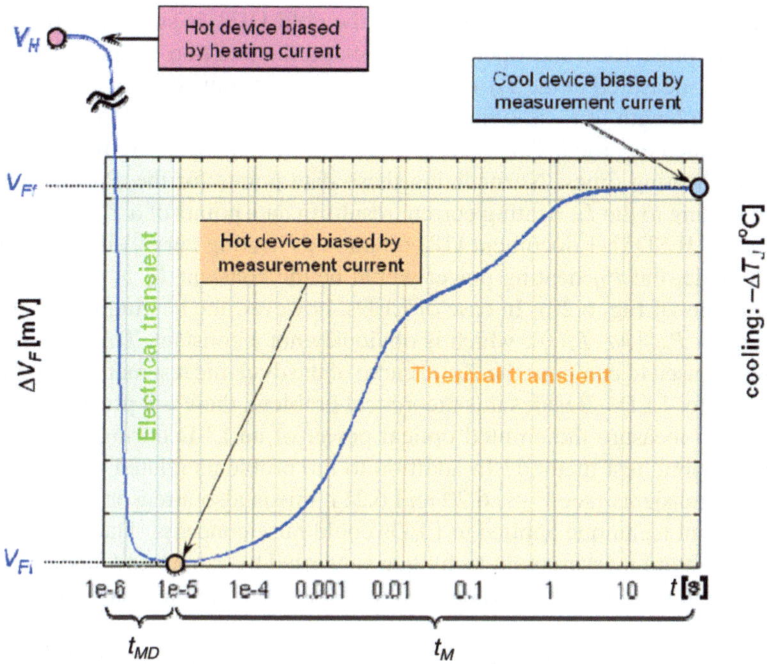

Fig. 6.22 Electrical and thermal state transitions of an LED shown as a time diagram after time instance $t = 0$ as indicated in the test waveform of Fig. 6.20, indicating timescales typical for LEDs measured on a cold plate. (Source: [6], © JEDEC, reproduced with permission)

The electrical transient inherently present at every current switching of pn-junctions is a parasitic effect from the point of view of thermal characterization based on the electrical test method. During this parasitic part of the forward voltage transient (which takes up to 10–50 μs) the junction temperature already changes. To account for this change, first the parasitic electrical transient must be cut off from the measured time function, and then using data points corresponding already to the real junction temperature transient the time function needs to be extrapolated back to the time instant of the switching. The JEDEC JESD51-14 document [19] provides a definition of this extrapolation procedure.

The requirement that for LED thermal testing the static test method of the JEDEC JESD51-1 document should be used needs some explanation. As shown in Fig. 6.15b, the energy conversion efficiency of LEDs, consequently the emitted optical power changes with junction temperature: these parameters drop at higher temperatures. As in the calculation of LEDs' real thermal resistance the emitted optical power appears, it is important that the emitted optical power is measured exactly at the same (absolute) junction temperature at which the JESD51-*-compliant thermal measurement was performed.

The JEDEC JESD51-1 document [18] defines two test methods. According to the so-called *dynamic test* method, a *heating curve* is composed during a lengthy measurement process in such a way that the cold DUT is subject to the I_H heating current pulse for a short period of time and the junction temperature elevation achieved is measured at the I_M measurement current immediately after the heating pulse with a length of t_H is switched off. This $\Delta T_J(t_H)$ value (after some data correction) is then recorded. This way, a series of $\Delta T_J(t_H)$ values are recorded for ever longer t_H heating times. Normalizing these data points for the P_H heating power corresponding to the I_H heating current results in data points of a $Z_{th}(t)$ curve. (That is what the JESD51-1 document [18] refers to as heating curve.) If the DUT was a silicon diode, the P_H heating power would be the constant $I_H \times V_H$ value (using the notations of Fig. 6.20). In case of LEDs, however, the heating power would be $I_H \times V_H - P_{opt}[I_H, T_J(t)]$, which is obviously not a constant. This means that the data points used to compose the $Z_{th}(t)$ curve of the dynamic test method would not be consistent for LEDs. Besides this theoretical problem, there is a practical one: when and how to measure the emitted optical power of an LED during thermal testing by the dynamic test method? In addition to the problems related to the switch-off electrical transients (see Figs. 6.20 and 6.21), this is also a reason why any pulsed measurement technique applied to LEDs could raise concerns. Therefore, the junction temperature measurement technique suggested by Zong and Ohno [66] is to be implemented with great care when used in connection with thermal characterization of LEDs.

In case of the transient extension of the static test method of JESD51-1 (as described in Sect. 6.1.4.1 and as defined by JEDEC standards JESD51-14 [19] and JESD51-51 [6]), there is single switching only and the junction temperature transient is recorded in real time, continuously.

Such a measurement of cooling transients in case of LEDs assures that the value of the heating power is well defined: $I_H \times V_H - P_{opt}[I_H, T_J(t = 0)]$ (again, using the notations of Fig. 6.20) and remains consistent throughout the entire $Z_{th}(t)$ measurement process. This fundamental difference in case of LED measurements between the dynamic and static test methods of JESD51-1 is illustrated in Fig. 6.23. For the measurement comparison, the same measurement hardware was used with two different measurement control software codes, using identical electrical settings and with additional measurement parameters providing the best possible match between the different kinds of test procedures. Besides the obvious mismatch for LEDs (due to the theoretical problem outlined here), a practical aspect has to be mentioned. The time interval of the real junction temperature transient of the measured LED was about 17 s. The minimal time interval needed to measure this with the transient extended static test method is 34 s (at least 17 s for heating up the junction and 17 s for measuring its cooling) with a data density of 200 points in a decade of time. The testing time needed for the measurement using the dynamic test method with a data density of 20 points per decade of time was about 30 min; about an order of magnitude difference in the required testing time. Further details of such a comparison can be found in a recent paper by Vass-Várnai et al. [69]. Such a comparison can be repeated by commercially available thermal testing hardware and related measurement software.

a

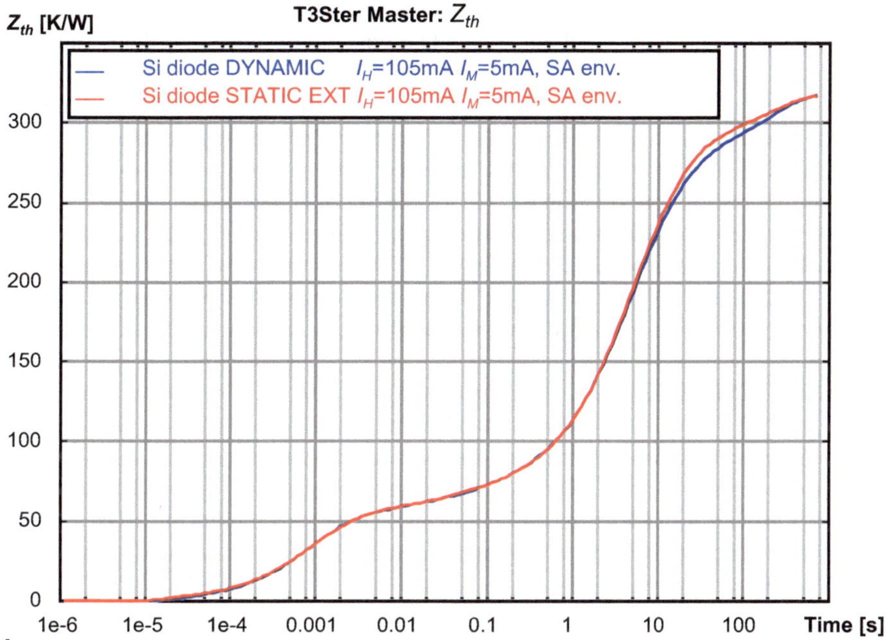

b

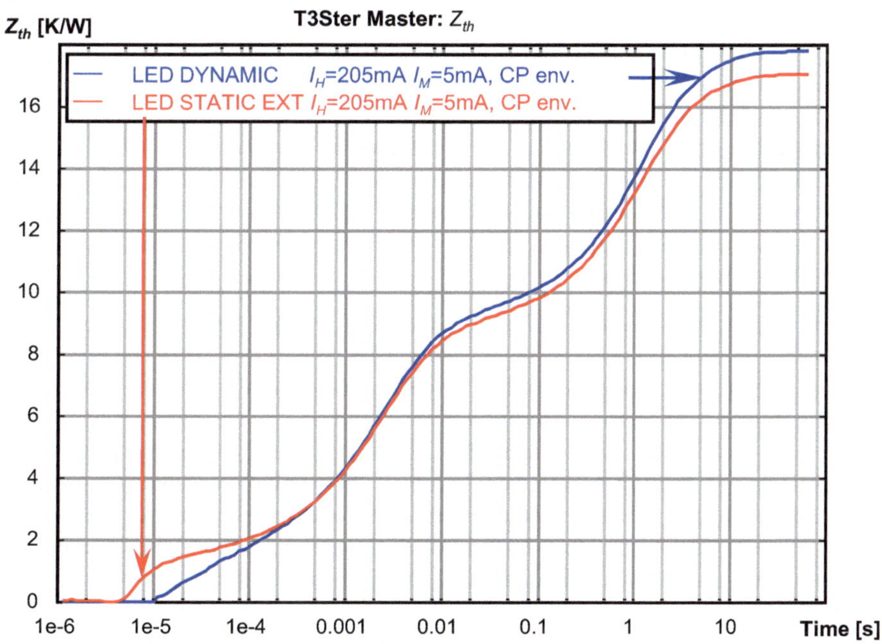

Fig. 6.23 Comparison of the JEDEC JESD 51-1-compliant "dynamic" test method with the JESD51-14-compliant cooling mode transient extension of the "static" test method **a** for a silicon diode measured in a JEDEC standard still-air chamber and **b** for an LED measured on a cold plate, using the same measurement hardware and identical electrical settings

There could be issues with LED devices that contain multiple pn-junctions (see Fig. 6.1). If the elementary pn-junctions are not equipped with electrical connections providing individual access to them, the LED is considered as a single device and the measured characteristics are so-called *ensemble characteristics*, such as an overall forward voltage of all pn-junctions in series, an overall ensemble junction temperature, an overall ensemble thermal resistance, etc. For related measurement details, see the published standard [6] and the glossary of terms and definitions related to LED thermal testing, provided in the JEDEC JESD51-53 document [8].

If the transfer thermal impedances among the different individual LED junctions or the self-impedances of these elementary LED devices need to be measured (see Sect. 6.1.4.2), then the LED devices (LED-based lighting system) need to be *designed for thermal testability*. Further discussion on this topic is beyond the scope of this chapter.

JEDEC JESD51-52: *Guidelines for Combining CIE 127-2007 Total Flux Measurements with Thermal Measurements of LEDs with Exposed Cooling Surface* [7]

This document is aimed to be used in conjunction with the JESD-LED series of standards especially with JESD-51-51 document. It focuses on the measurement of the *total radiant flux* of LEDs in combination with the measurement of LEDs' thermal characteristics and guidelines on the implementation of the recommendations of the CIE 127-2007 document [60].

These guidelines specify testing procedures and conditions for power LEDs that are typically used in the operating regime of the forward current of 100 mA and above and which emit visible light measured and considered as a *single light source*. This document is restricted to LEDs that have an exposed cooling surface. Guidelines provided refer to *laboratory measurements*. Issues of high-speed bulk measurements of LEDs (such as in-line testing aimed, e.g., for binning) are dealt with by other standardization bodies such as the relevant technical committees of CIE. Recommendations given in this document are valid for *LEDs powered by DC forward current* only. The light output measurements of LEDs in connection with thermal testing should be performed according to the recommendations of the CIE 127:2007 with the following additions:

- *Light output measurement techniques specified for laboratory testing shall be used.*
- Light output measurements need to be carried out with *DC current supply and under steady-state thermal conditions*.
- The *total radiant flux of the test LED has to be measured* for the purpose of considering in heating power calculation of the thermal resistance/impedance measurements, for the same forward current and junction temperature pairs, which take place during thermal test. In addition to the measurement of the radiant flux, photometric and colorimetric quantities can be also identified.
- *The thermal environment of the LED under test during light output measurements should be the same as the one used in thermal testing*. The recommended test environment is a liquid-cooled or a Peltier-cooled, temperature-controlled cold plate. Test fixtures like the one shown in Fig. 6.6 are suggested.

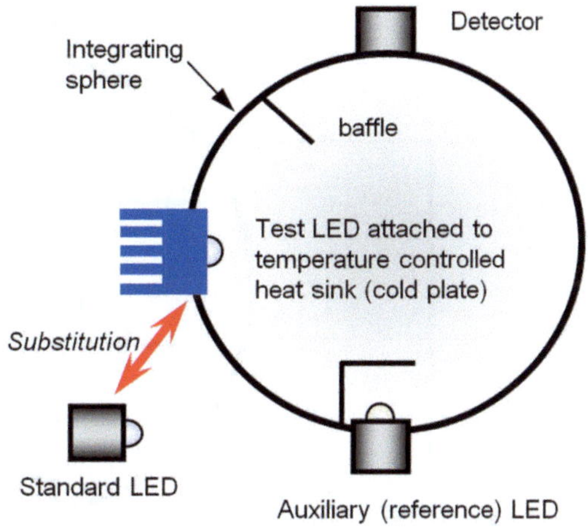

Fig. 6.24 Suggested integrating sphere setup for total radiant flux measurement of LEDs to be used in combination with thermal measurements. (Source: [7], © JEDEC, reproduced with permission)

- For light output measurements performed in connection with thermal test, *4-wire (Kelvin) electrical connections* to the test LED are needed.
- The light output measurement apparatus should be calibrated against a *current and temperature stabilized reference* light source also known as a standard LED. *Standard LEDs providing radiant flux in the same order of magnitude as the expected radiant flux of test LEDs shall be used.* Also, the r*elative spectral distribution of the reference standards should be as close to the relative spectral distribution of test LED* as possible. In other words, the *strict substitution method* described in CIE 127:2007 [60] shall be used. For more details on laboratory testing of LEDs' light output properties, see Chap. 5.
- *Measurement of the total radiant flux shall be preferably carried out using an integrating sphere* (aimed at measuring visible light). Since power LEDs do not have backward light emission, collecting light is sufficient from 2π steradian solid angle. Therefore, to allow proper cooling of the test LED through an actively cooled, temperature controlled cold plate, the so called 2π sphere geometry should be used. The arrangement suggested for use in combination with thermal measurements of LEDs is shown in Fig. 6.24.

The light output measurement system outlined above can be completed with any other type of detector, which provides photometry/calorimetry data of interest. In order to assure consistency between thermal and light output measurements, application of combined thermal and radiometric LED testing station is preferred; see Fig. 6.25. In such a setup, the DC current for the light output measurements is provided from the thermal testing apparatus (thermal test equipment) when it is switched to provide the I_H (heating) current (switch SW1 set to the I_H position). The electrical scheme of the combined setup needs another switch (SW2) with which the forward current of the test LED can be completely switched off—allowing the dark photocurrent offset compensation and the self-absorption compensation in the light measurement system.

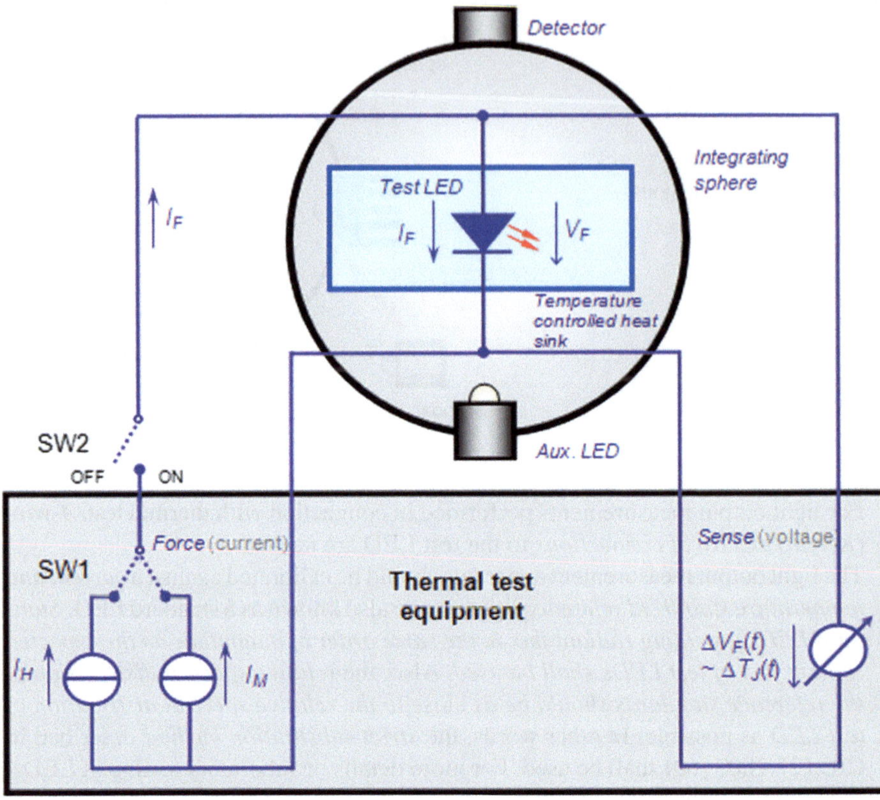

Fig. 6.25 Schematic diagram of a combined thermal and radiometric LED testing station. (Source: [7], © JEDEC, reproduced with permission)

The control of the heat sink temperature is also an essential part of such a combined LED testing station. The advantage of such a combined measurement setup is that the (junction) temperature dependence of the LEDs' light output properties (such as the optical power, total luminous flux, scotopic flux, [10] or color coordinates) can also be measured, along with derived quantities such as energy conversion efficiency or luminous efficacy.

Once the real $R_{thJ\text{-}ref}$ thermal resistance in the actual test setup, the real P_H heating power and the T_{ref} temperature of the cold plate are known, both the JESD51-51 and JESD51-52 documents recommend to calculate back the real T_J junction temperature with the formula given by Eq. (6.9). Both documents recommend including the T_{ref} and T_J junction temperatures in the test reports. It is also suggested to provide the

[10] Scotopic flux is obtained as the integral of the spectral power distribution of the light source weighted by the $V'(\lambda)$ function—the sensitivity of the human eye in the dark adapted state (scotopic conditions).

energy conversion efficiency (radiant efficiency) data in the form of plots—as we have already shown in Fig. 6.15. The JESD51-52 document explicitly recommends that "Light output metrics must be reported together with the applied forward current and the identified junction temperature" [7]. Furthermore: "It is preferred that LEDs' data sheets always include typical data of $\Phi_e(I_F, T_J)$, radiant flux values, $\eta_e(I_F, T_J)$, energy conversion efficiency values, and $\Phi_V(I_F, T_J)$, luminous flux values, $\eta_V(I_F, T_J)$, efficacy values, preferably in the form of plotted diagrams and numerical tables" [7].

Figure 6.26 shows such test results in the form of plots while in Fig. 6.27 the same results are presented in tabular format. Note in Fig. 6.26 that for the different forward currents the temperature sensitivity of the luminous flux is also provided ($S\Phi_V$ values in the legend)—useful information for hot lumen calculations.

JEDEC JESD51-53: *Terms, Definitions, and Units Glossary for LED Thermal Testing* [8]

"In order to facilitate the communication of thermal and radiometric/photo-metric measurement and data information of power LEDs a clear and well-defined language is necessary" [8]. The JESD51-53 document provides a collection of terms and definitions with the aim of helping both the thermal testing and the optical testing community to describe thermal and thermally related properties of LEDs more accurately. Though the JEDEC standard JESD77D provides definitions for light emitting devices, some of these definitions are not coherent with the some definitions common in the lighting industry. Since the JESD51-52 establishes a link toward CIE recommendations, it was logical to maintain the coherence through the JESD51-53 document: definitions, symbols, and notations regarding light output measurements used here are coherent with those defined by CIE, especially in the International Lighting Vocabulary [54] and in the CIE 127-2007 document [60] as well as in some other relevant standards of other standardization bodies from the SSL industry, such as IES RP-16-10 [1].

Since this document has been prepared by the *JEDEC JC15 Committee on Thermal Characterization*, it is focused on *thermal* characterization of LEDs—as special, packaged discrete semiconductor devices. It also provides reference to light measurement to the extent required for thermal characterization of LEDs. Thus, this document should be used in conjunction with the JESD51-5* series of standards. This document is also aimed at experts performing light output measurements of LEDs when dedicated attention is paid to thermal issues as well during such measurements. Therefore, terms commonly used in both thermal testing and optical testing are collected here with common definitions coherent with the ones usual in thermal and optical testing.

Besides the purpose of the JESD51-53 standard, there is no point in reproducing anything else from this document except for the summary of symbols used throughout the JESD51-5x series (provided in Table 6.2), especially because in this chapter most of these symbols are used.

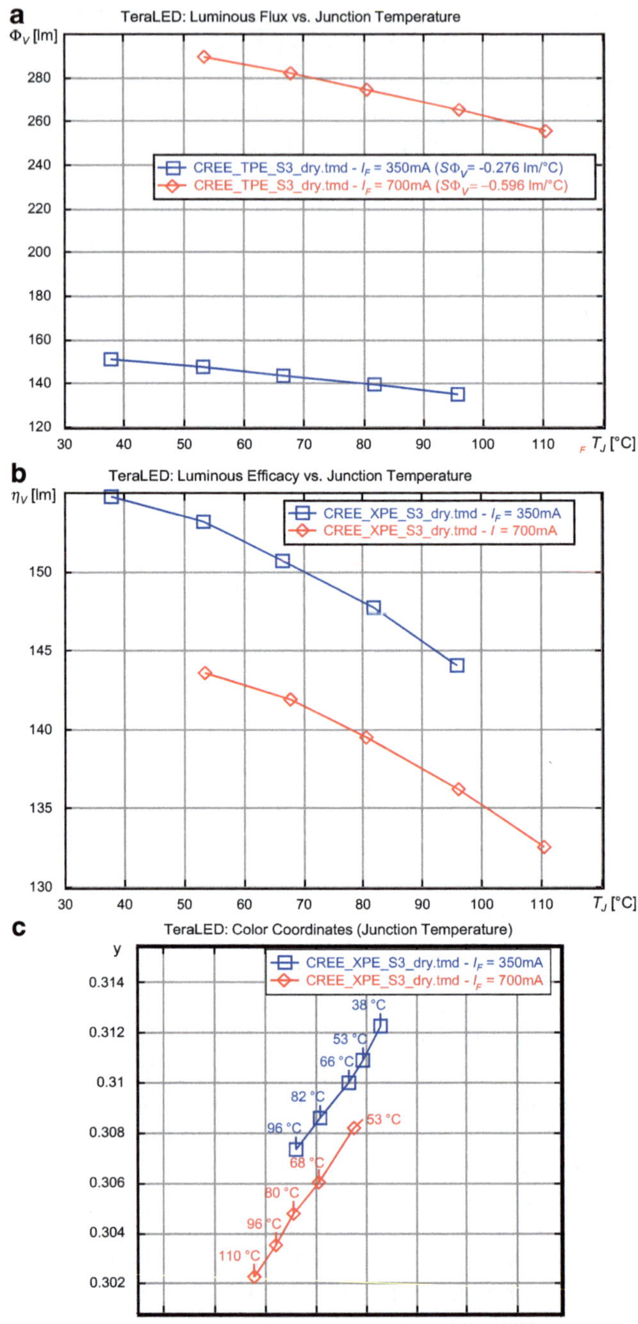

Fig. 6.26 Optical test results of a recent white LED provided by a combined thermal and optical LED testing station, meeting the data reporting requirements of JESD51-52. **a** Luminous flux versus junction temperature plots. **b** Luminous efficacy versus junction temperature plots. **c** Color coordinates (CIE 1931 2° observer) versus junction temperature plots

Optical Test Report - CREE_XTE_S3_dry.tmd

Optical Test Report	
Project:	CREE_XTE_S3_dry
Date:	Fri Jun 22 03:55:07 2012
Meas. Mode:	Combined(T3Ster + Booster)

Av.Temp.[°C]	Av. V_f /	I_f	Av. V_f /	I_f
	[V]	[A]	[V]	[A]
25.2	2.788	0.350	2.882	0.700
Popt [mW]	483.5		939.4	
η [%]	49.6		46.6	
Φ_v[lm]	151.0		289.7	
η_v[lm/W]	154.8		143.6	
Φ_v'[lm]	321.8		622.9	
CCT [K]	7445.2		7598.6	
x	0.3006		0.2995	
y	0.3123		0.3082	

Av.Temp.[°C]	Av. V_f /	I_f	Av. V_f /	I_f
	[V]	[A]	[V]	[A]
39.9	2.752	0.350	2.842	0.700
Popt [mW]	474.8		921.2	
η [%]	49.3		46.3	
Φ_v[lm]	147.5		282.4	
η_v[lm/W]	153.2		141.9	
Φ_v'[lm]	316.7		611.0	
CCT [K]	7521.3		7748.1	
x	0.2999		0.2981	
y	0.3109		0.3060	

Fig. 6.27 Optical test results of a recent white LED provided by a combined thermal and optical LED testing station, meeting the data reporting requirements of JESD51-52. All results are presented in tabular format

Table 6.2 Most frequently used symbols in the JESD51-50 through JESD51-52 series of documents

Symbol	Unit of measure	Name, description
T_J	°C	**Junction temperature** of the LED (see JESD51-1), denoted and referred to in CIE 127:2007 as T_C, the *chip temperature*. In the temperature range of interest, using °C is more common
ΔT_J	°C or K	**Change of junction temperature** (see JESD51-50, JESD51-1). For temperature differences, °C is commonly used
$R_{\theta JX}$, R_{thJX}	K/W	**Junction-to-specific environment thermal resistance** (see JESD51-50, JESD51-1) where X refers to the environment in question
Ψ_{JX}	K/W	**Junction-to-X thermal characterization parameter** (see JESD51-13)
V_F	V	**Junction forward voltage**
I_F	A	**Junction forward current**
P_H	W	**Heat dissipated at the junction of the LED** (see JESD51-50), also denoted as P_H and referred to as heating power in JESD51-1, JESD51-14 and JESD51-51
P_{opt}	W	**Emitted optical power of the LED** referred to as *total radiant flux* and denoted as Φ_e in CIE S 017/E:2011 ILV. It is also called *total radiant power*
P_{el}	W	**Electrical power supplied to the LED,** which is equal to the product of the forward voltage and the forward current: $P_{el} = V_F \cdot I_F$. This quantity is denoted as P in CIE 127:2007 and CIE S 017/E:2011 ILV
Φ_e	W	**Emitted optical power of the LED** as defined and referred to in CIE S 017/E:2011 ILV as *total radiant flux* or *radiant power*, alternate notation to P_{opt}
Φ_V	lm	**Total luminous flux**
λ	nm	**Wavelength of the emitted light**
$S(\lambda)$	W/nm	**Spectral power distribution** indicating the radiant power of the emitted light at a given wavelength
η_e, WPE	%	**Radiant efficiency** or *energy conversion efficiency* or *wall plug efficiency* of the LED: 100 × value of the P_{opt} emitted optical power divided by the P_{el} supplied electrical power
η_V	lm/W	**(Luminous) efficacy**, the value of the LED's emitted total luminous flux Φ_V divided by the P_{el} supplied electrical power
z	s	**Logarithmic time**, the absolute value of this quantity is defined as $z = \log(t)$
$Z_{\theta JX}$, Z_{thJX}	K/W	**Junction-to-specific environment thermal impedance**, the temporal change of junction temperature with respect to temperature of environment X, normalized to 1 W heating power and scaled in z logarithmic time
TSP	n.a.	**Temperature sensitive parameter**, in case of semiconductor diodes it is the V_F forward voltage
S_{VF}	mV/K	**Temperature sensitivity of the forward voltage**, measured at I_M measuring current (also called as sensor current)
K	K/mV	**K factor**, reciprocal of the S_{VF} temperature sensitivity of the forward voltage, identified at I_M measuring current (also called as sensor current)

Table 6.2 (continued)

Symbol	Unit of measure	Name, description
I_H	A	Value of the forward current of the LED applied as *heating current*
V_H	V	Value of the forward voltage of the LED when biased by the *heating current*
I_M	mA	Value of the forward current of the LED applied as *measuring current*
t_{MD}	s	**Measurement delay time**, time elapsed between the instance of switching the power applied to the LED under test and the instance of the first reading of the TSP not disturbed by electrical transients
V_{Fi}	V	Initial value of the forward voltage of the LED immediately after switching the power across the diode
V_{Ff}	V	Final value of the forward voltage of the LED when diode reached its final thermal steady state after switching the power

6.5.2 Compact Modeling of LEDs

As seen in Fig. 6.19, LED compact thermal modeling is within the scope of JEDEC's standardization activities. So far, LED compact modeling has not reached the same mature state as in case of semiconductor devices used in electronics. In the subsequent subsections, first we describe the status of compact thermal modeling of general semiconductor packages used in electronics, including the major concepts also described in compact thermal modeling guidelines published by JEDEC, followed by the short history of LED compact modeling. As this procedure is not restricted to the thermal modeling of the LED package only, it is rather called multidomain (compact) modeling of LEDs.

6.5.2.1 Introduction: Compact Thermal Modeling in Electronics

The motivation for compact modeling of LEDs is very similar to the motivation for compact thermal modeling of semiconductor device packages. About two decades ago in the IC world one aspect was speeding up CFD-based thermal simulations: instead of detailed models of IC packages with very fine mesh (e.g., at the chip area) a thermal network model with a few nodes and with a few network elements provided significant reduction in computational need (CPU power, memory).

Simulation models based on detailed device geometry are computationally expensive, and may reveal proprietary information about the part. Compact models provide correct thermal behavior within an acceptable level of error without internal part details, and are computationally efficient.

Nowadays, with advanced parallelized simulation codes running on computers with fast multi-core processors and multiple gigabytes of RAM, reduction of the need for computational resources became less relevant. The other aspects of compact models are still very important: by means of such models, semiconductor vendors can provide accurate thermal data to their end-users without the need of disclosing proprietary details about their packaging technologies/designs.

As mentioned in Sect. 6.2.2, the DELPHI and SEED projects [16, 28–32] led to a well-established modeling methodology and finally to end-user acceptance. DELPHI and SEED resulted in steady-state compact thermal models (CTM) while one of the targets of the PROFIT project [33–35] was to create dynamic CTMs. It took quite some time before use of package CTMs became widespread in the electronics industry [36] and the de facto standard compact models became de jure standards by JEDEC having published their compact thermal modeling guidelines [70–73].

There are two basic concepts in relation to CTMs of semiconductor device packages:

1. *Boundary condition independence* (*BCI*): If a CTM is aimed at CFD-based thermal simulation, then it is very important that the model used does not reflect any particular thermal boundary condition as these inherently would be calculated by the CFD program when the thermal properties of the system—containing the package described by its CTM—is simulated. A CTM, which is boundary condition-independent, is called a BCI model. The JEDEC guidelines cited earlier provide exact definition of BCI and a way to "measure" BCI of such simulation models.
2. *Test-based versus simulation-based models*: Test-based models as the JEDEC 2R models [72] many times reflect the particular test conditions of the physical measurement from which such models are derived and inherently lack any kind of error estimation and error analysis. The classical JEDEC 2R model is claimed to include, for example, the effects of the test board used during the measurements. In case of simulation-based models, such as the classical DELPHI-type models [73], the model generation algorithm inherently includes error analysis. Calculation of a metric of BCI index is also such a kind of error analysis.

The above cited standards are related to steady-state CTMs. As of today, there are no standards yet related to dynamic CTMs. Some vendors provide such models for their power semiconductor devices being used, e.g., under pulsed conditions. Note that the dynamic compact thermal modeling has been the subject of research for many years. While the theoretical aspects are well known, it is the lack of software that prevents the introduction of dynamic models in practice. However, there is some light at the end of the tunnel, see [74] for a recent discussion.

6.5.2.2 LED Multidomain Models

Compact modeling of LEDs, of course, includes the CTMs of LED packages. However, in case of LEDs the electrical, thermal, and optical properties are tightly coupled, therefore a single value estimate for the *dissipation* may not be sufficient for the thermal design of LED applications.

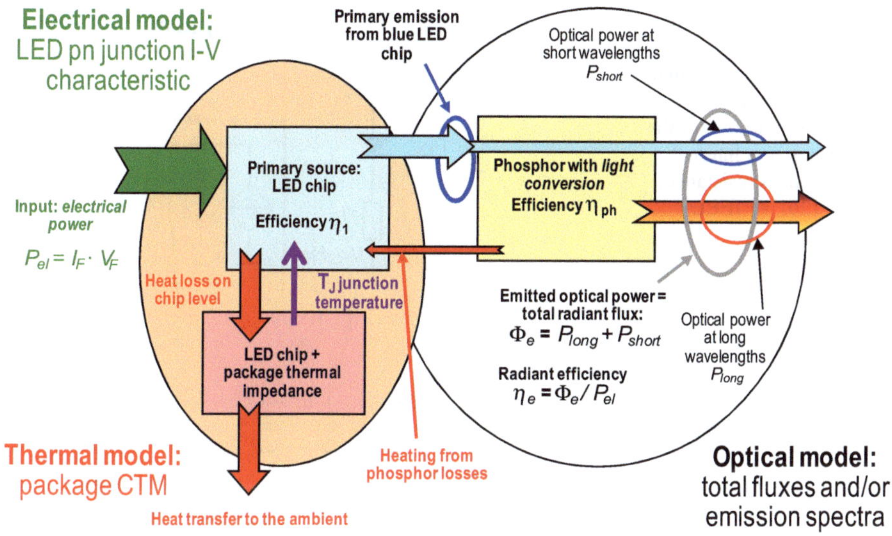

Fig. 6.28 Possible components of a multidomain model of white LEDs. (After Fig. 21.13 of Schubert's book on LEDs [75])

Many times, the electrical operation of the LEDs should also be modeled, requiring network models aimed for nonlinear electrical circuit simulation programs used in the electrical engineering practice (such as SPICE or its different commercial descendants).

Furthermore, the design objective is not only to provide a thermal management solution, which assures a *safe junction temperature* (as shown by Eq. (6.1) in Sect. 6.1.1), but also to assure the *prescribed level of luminous flux* at that junction temperature. This requires an accurate, yet simple model suitable for the estimation of luminous flux at operating junction temperature (popularly known as *hot lumens* calculation). Figure 6.28 provides a summary of LEDs' major characteristics that need to be captured by a multidomain model in order to allow proper self-consistent simulation of the electrical, thermal, and light output characteristics of LEDs.

LED Package Compact Thermal Modeling One of the first multidomain LED models was published by Farkas et al. in 2004 [76]. The reason why this model was developed and proposed was that the authors observed operating point (forward current and ambient temperature) dependence of locations of different package features of Lumileds Luxeon LEDs. Although they measured these LEDs in a combined thermal and radiometric measurement setup (recommended nowadays by the JEDEC LED thermal testing standards JESD51-51 and JESD51-52), the obtained structure functions scaled in the real thermal resistance of their measured LEDs still exhibited this behavior. In their suggested model (shown in Fig. 6.29), they built a dynamic CTM of the LED package in question by means of a Cauer-type thermal RC ladder the elements of which were identified by V. Székely's Network Identification by Deconvolution method [22], [23]. They also created an electrical model of the LED in which the internal electrical series resistance of the device is also considered. The

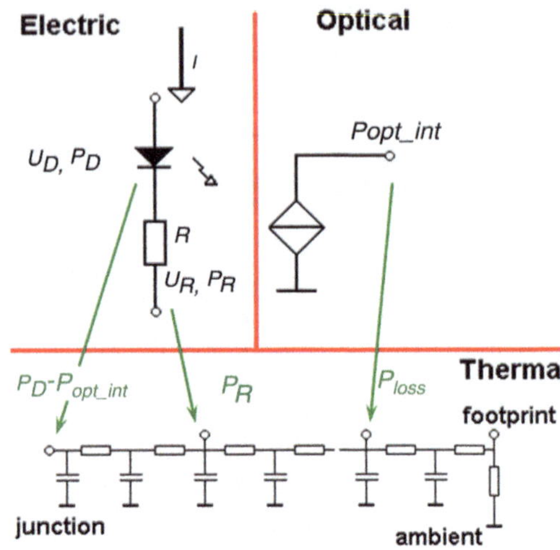

Fig. 6.29 Schematic of the topology of a multidomain LED model used to simulate the effect of the different thermal configuration of the LED pn-junction internal electrical series resistance. (Source: [26], © IEEE)

authors also suggested to consider possible heating effects of the optical losses of the LED. They realized this as a macromodel in their in-house electrothermal circuit simulator. They used the model to perform computer experiments to study the observed operating point dependence of the package structure functions obtained from their measured $Z_{th}(t)$ thermal impedance curves. With these simulation experiments, they found explanation for the observed device behavior:

- If the electrical series resistance of the LED is physically close to the active semiconductor pn-junction (this is what they called a *hot resistor model*), the heating effect of that resistor cannot be distinguished from the heating effect of the nonradiative recombination processes at the pn-junction; the total power loss is fed into the heat-flow path at the junction node.
- If the electrical series resistance of the LED is mostly physically detached from the semiconductor pn-junction, its dissipated heat is still conducted away by the LED package but that portion of the total dissipation is to be fed into the heat-flow path RC network model at a node different from the junction mode (this is what they called a *cold resistor model*). This is the situation when the structure functions obtained from simulated junction temperature transients also showed the experimentally observed operating point dependence.

In a subsequent paper [25], the emphasis still was on the compact thermal modeling of the LED package. This paper shows that the heat-flow from the LED junction toward the ambient strongly follows a single path; the same part was measured under two very different thermal boundary conditions: in a natural convection environment and on a cold plate. As shown in Fig. 6.30, up to the package case (found at 9.2 K/W) or even to some extent in the thermal interface between the package case and the metal core PCB (up to 13.7 K/W cumulative thermal resistance), the structure functions obtained for the two different thermal boundary conditions are identical. This means that the dynamic CTM obtained by a stepwise approximation of the structure functions

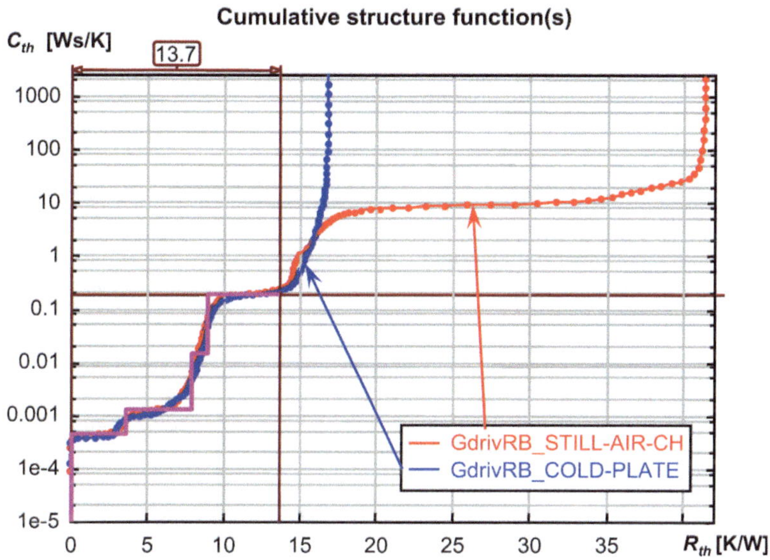

Fig. 6.30 Structure functions obtained for an MCPCB assembled 1-W LED measured on a cold plate and measured in a natural convection environment. The identified junction-to-case thermal resistance of the LED package along with the stepwise approximation of the structure function until this point is also shown in the diagram. (Based on the original published in [25])

(indicated by thick magenta lines in Fig. 6.30) can be considered as a boundary condition-independent model of the LED package (or LED assembly). For creating the *junction-to-case dynamic CTM* of the LED package, this work already used a method that later became known as the transient dual interface method—the basis of today's JEDEC standard [19] for the transient-based measurement of the R_{thJC} values of power semiconductor packages with a single heat-flow path and with an exposed cooling surface: JESD51-14.

In [25] such a test-based LED package model was used to study the behavior of an RGB LED module on board level. In a very recent paper [26], the steps of creating such LED package CTMs are shown.

A couple of comments to this method of LED package compact thermal modeling should be added:

a. The steady-state behavior of an LED package is fairly well represented by a JEDEC 2R model. Some LED vendors publish such models. The *junction-to-bottom* resistance of the 2R model can be replaced by the dynamic CTM obtained by the stepwise approximation of structure functions (see Fig. 6.31). The *junction-to-top resistance* of the JEDEC 2R model would correspond to the *junction-to-lens* thermal resistance of LED packages. Unfortunately, there is no way to measure this resistance; one needs to rely on numerical analysis to estimate this.
   It has to be mentioned that if one creates the Cauer-type RC ladder model based on structure functions generated from junction temperature transients of LED packages, one has to consider the junction-to-top thermal resistance of this extended 2R model in the created structure function. The measurement results reflect the

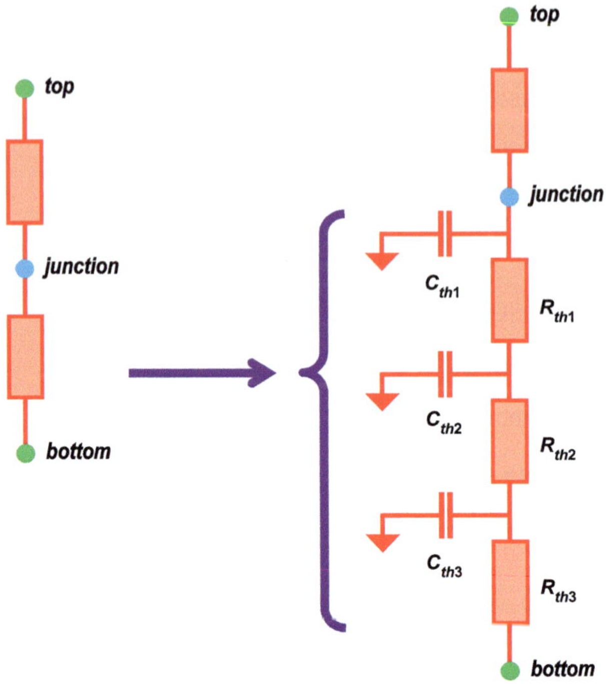

Fig. 6.31 Transient extension of the JEDEC 2R model for packages of power LEDs

net effect of the main heat-flow path (from the junction toward the bottom of the package) and the possible heat loss through the lens. This parasitic heat-flow path has to be accounted for by correcting the structure function used for creating the model of the junction-to-bottom heat-flow path. For this correction, the parallel heat flow toward the lens the $R_{th\_junction\text{-}to\text{-}top}$ value needs to be "subtracted" using algorithms published by Rencz et al. [77, 78].

b. LED chips are made of compound semiconductor materials, the thermal properties of which show a temperature dependence. Therefore, it is justified to suspect that the elements of the RC ladder model (or the structure functions from which such models are created) also show a temperature dependence. Therefore, it is advised to check this in daily practice. As seen in Fig. 6.32, structure functions used for modeling a white power LED do not show any significant change in a temperature range of 60 °C. The only change seen in the diagram is the change in the quality of the thermal interface material (TIM) between the measured LED assembly and the cold plate used during the measurements. In compliance with the JESD51-51 and JESD51-52 standards, the emitted optical power of the LED under test was considered in these measurements. Work by Poppe et al. [79] specifically deals with such temperature dependences and in this work it was also the TIM quality that showed temperature dependence.[11]

[11] See Fig. 11 of the cited paper. The ambient temperature range covered in this paper was 70 °C.

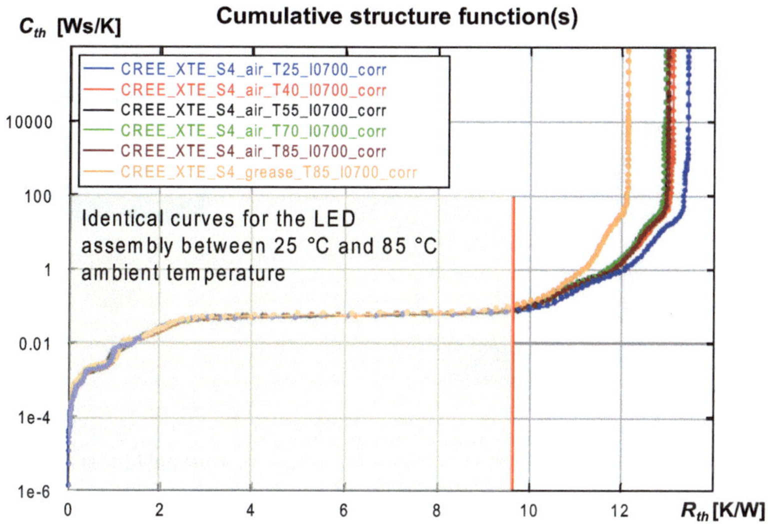

Fig. 6.32 Cumulative structure functions obtained for the same white power LED in a temperature range of 60 °C

c. A modeling methodology is useful only if it is supported by proper software tools available to LED vendors and end-users who are interested in LED compact modeling. The concept of this test-based compact model generation of power semiconductor device packages with a single heat-flow path and with an exposed cooling surface is implemented in the data processing software of a commercially available thermal transient tester [80] and using such models is supported by the CFD simulation tools [81, 82] of the same vendor. Figure 6.33 illustrates this implementation. In the figure, CFD simulation results obtained for an LED-based street-lighting luminaire are shown in which LED CTMs have been used [83].

d. LED package compact models can be used for different reasons. One application is performing simulation experiments (as it was the case in the paper of Farkas et al. [76]). Such experiments can be performed with circuit simulators like SPICE. The previously mentioned data processing software supports this option too. For steady-state problems, dynamic models are not needed but there are LED applications where there is periodic heat load on the LEDs. Examples are pulse width modulation-based dimming or directly AC mains-driven LEDs. For both situations, dynamic CTMs of LED packages can be used. A recent paper on the characterization of directly AC mains-driven LEDs [21] provides an example. Last but not the least, CTMs of LED packages are very useful when no reliable data on LED package geometry are available for system (luminaire)-level CFD analysis. In such a situation, the test-based compact modeling approach (see Fig. 6.33) may help to provide the missing information for a successful CFD analysis.

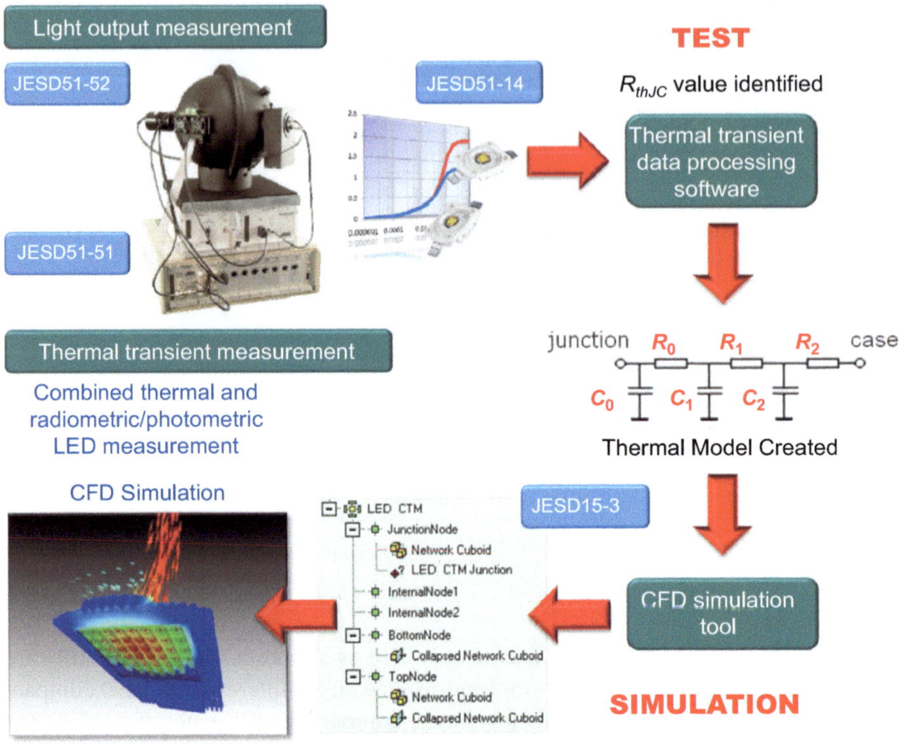

Fig. 6.33 Implemented workflow of the test-based LED CTM generation and model application in CFD-based thermal analysis of LED-based lighting solutions. The relevant JEDEC standards are indicated in the diagram

LED Electrical and Electrothermal Modeling As mentioned in the introduction, in many cases electrical or electrothermal simulation of LEDs is needed in order to provide the right value of the heating power for thermal simulation.

LED vendors publish SPICE diode model parameters for their LEDs but unfortunately the standard SPICE diode model is not an electrothermal one and does not provide any means of predicting any light output property of LEDs. LED vendors' SPICE model sets thus are valid at a given junction temperature and are mostly limited to a given current range.

To extend a standard SPICE electrical-only diode model to an electrothermal one needs adding extra features to the model. These include an extra thermal branch and a thermal node to describe the heating power and represent the junction temperature, proper heating power calculation and local junction temperature dependence of the parameters of the electrical model equation, and calculation of the electrothermal and thermoelectrical transconductances between the two parts of the electrothermal model—as illustrated in Fig. 6.34. Implementation of such models requires lot of experience in semiconductor device modeling and programming practice, not to mention the need of having access to the source code of the circuit simulator in question.

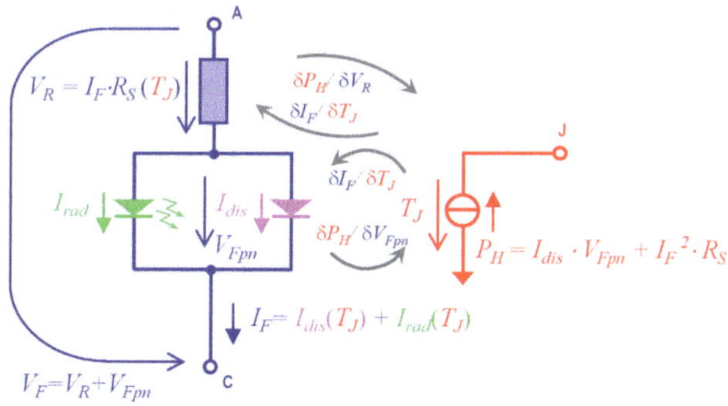

Fig. 6.34 Network topology of a possible LED pn-junction electrothermal circuit model

This is the reason why the rather limited technical literature on LED multidomain modeling includes papers describing SPICE macromodels. For example, Górecki [84] suggests a model in which the emitted luminous flux is represented by a controlled voltage source. The equation representing the luminous flux in Górecki's model is a fully empirical exponential model, describing the forward current and temperature dependence. The odd nature of this model is that a "current"-like quantity is represented by a "voltage" source. In the electrical part of his model, Górecki uses two current generators: one representing the diffusive current of the diode (based on the Shockley model), the second representing the generation-recombination current originating from the space charge region of the junction. As this second current component has significant effect on the diode characteristics at low total forward currents only, from a practical perspective of power LEDs operated in the many hundred milliampere range its modeling is not too important. Górecki also models the reverse characteristics of LEDs, which is a feature nice to have but not relevant from a practical point of view either.

Negrea et al. [85] also suggest a multidomain LED model for SPICE. Similarly to Górecki, they also describe the luminous flux by a voltage-controlled voltage source but their model is a product of a junction temperature-dependent and a forward current-dependent polynomial.

Figure 6.34 suggests yet another multidomain LED model suitable both for equation-level and macrolevel realization.

For this model, we take a practical approach. First, we focus on the relevant sections of the LEDs' I-V characteristics: large current region of the forward current, where the Shockley model solely does not properly represent the electrical behavior, hence considering the internal series resistance of the LEDs is a must. The second goal is to provide a comprehensive LED model the parameters of which can be identified from JEDEC JESD51-51- and JESD51-52-compliant combined thermal and radiometric/photometric measurement results, completed with basic electrical measurements regarding the LED's I-V characteristics.

As Fig. 6.34 shows, an LED is considered as two diodes with the same forward voltage connected in parallel. One diode is like a silicon diode with zero light emission; this diode dissipates heat only. The other diode resembles a 100 % efficient LED with no heat loss at all but emits only light and its emitted total radiant flux (emitted optical power) is equal to its current multiplied by its voltage. The electrical series resistance is also included in the model—its V_R voltage drop due to the total I_F forward current also contributes to the heating power of the LED.

The forward current component associated with radiative recombination processes is denoted by I_{rad}. On the one hand, this current component can be identified from the LEDs measured P_{opt} optical power (also known as Φ_e radiant flux)

$$I_{rad}(V_{Fpn}) = \frac{P_{opt}}{V_F} = \frac{\Phi_e}{V_{Fpn}} \tag{6.12}$$

On the other hand, it can be modeled as a normal diode, for the sake of simplicity neglecting the effect of the series resistance, using the Shockley model of diodes only

$$I_{rad}(V_F) = I_{0\_rad}\left[\exp\left(V_F/(n_{rad}V_T)\right) - 1\right] \tag{6.13}$$

where $V_T = kT/q$ is the so-called thermal voltage (roughly 26 mV at room temperature, k is Boltzmann's constant, T denotes the absolute temperature of the junction and q is the elementary charge), n_{rad} is the so-called ideality factor (for normal diodes typically between 1 and 2), and I_{0_rad} is the saturation current; the last two are model parameters depending on the T_J junction temperature. In this model, all losses in connection with light generation and emission (described by efficiency factors such as quantum efficiency, extraction efficiency, phosphor efficiency[12]; see Fig. 6.28) are inherently included in the other current component denoted by I_{dis} in Fig. 6.34. This other diode in the model of Fig. 6.34 is primarily associated with *nonradiative recombination processes*, thus, causing the self-heating of the LED. It can be identified from physical measurements as follows

$$I_{dis}(V_F) = I_F(V_F) - I_{rad}(V_F) = I_F(V_F) - \frac{P_{opt}}{V_F} \tag{6.14}$$

It can also be described by the well-known diode equation as

$$I_{dis}(V_F) = I_{0\_dis}\left[\exp(V_F/(n_{dis}V_T)) - 1\right] \tag{6.15}$$

where parameters I_{0_dis} and n_{dis} are similar as described in Eq. (6.13).

The total forward current of the LED is described by the following equation:

$$I_F(V_F) = I_{0\_dis}\left[\exp(V_F/(n_{dis}V_T)) - 1\right] + I_{0\_rad}\left[\exp(V_F/(n_{rad}V_T)) - 1\right] \tag{6.16}$$

Thus, the current component corresponding to the internal losses can be expressed from measured quantities I_F, V_F, and Φ_e as follows:

$$I_{dis}(V_F) = I_F - \frac{\Phi_e}{V_F} \tag{6.17}$$

[12] For details on physics of LED operation, refer to Chap. 2.

Figure 6.35 presents the current components of an LED for $T_J = 55\,°\mathrm{C}$ where I_F and V_F were measured directly, I_{rec} and I_{dis} were identified for every operating point according to Eqs. (6.12) and (6.17), respectively.

In the high current regime, the diode internal series resistance cannot be neglected, losses on this resistance contribute to the overall energy conversion efficiency of the LED. The effect of the series resistance is well visible in the lin-log plot of the I-V characteristic in Fig. 6.35b.

The LED electrical model shown in Fig. 6.34 is completed with the internal series resistance. At high currents, significant voltage drop occurs on the series resistance—it adds to the forward voltage of the internal pn-junction providing the external V_F forward voltage measured between the anode and cathode electrode.

Thus, the Shockley model—at high currents neglecting the $+1$ next to the exponential term reads as

$$I_F = I_0 \cdot \exp\left(\frac{V_F - I_F R_S}{n V_T}\right) \tag{6.18}$$

The Shockley model expresses the forward current from the voltage drop between the anode and cathode electrodes of the diode—in the same way as circuit simulators based on the nodal solution method expect the constitutive equations of circuit branches to be formulated. Unfortunately, if we include the effect of the series resistance, our diode equation is no longer an explicit function of the forward voltage. However, Eq. (6.18) can be formulated such that the forward voltage is an explicit function of the forward current and the model parameters

$$V_F = n V_T \cdot \ln\left(\frac{I_F}{I_0}\right) + I_F R_S \tag{6.19}$$

This relationship can be used to check the correctness of the model parameters. This is especially useful in our case since the LED measurement systems are typically current-controlled: they measure the forward voltage at preprogrammed, forced, constant forward currents.

Multidomain LED Compact Models In a comprehensive LED compact model, the LED package thermal model is driven by the junction node: the heat generated by the LED is fed into the package CTM (see Fig. 6.36). This modeling approach is suitable for both SPICE-like circuit-level simulation (either with macromodels of the LED pn-junction or with equation-level implementation in the circuit simulation code) and detailed 3D CFD analysis.

If SPICE-like circuit simulation is performed, then the case node of that package model is connected to the compact RC model of the environment and the whole LED model—connected to its electrical environment through the anode and cathode nodes (denoted by A and C in Fig. 6.36) of the LED pn-junction model—can be simulated by a circuit simulator.

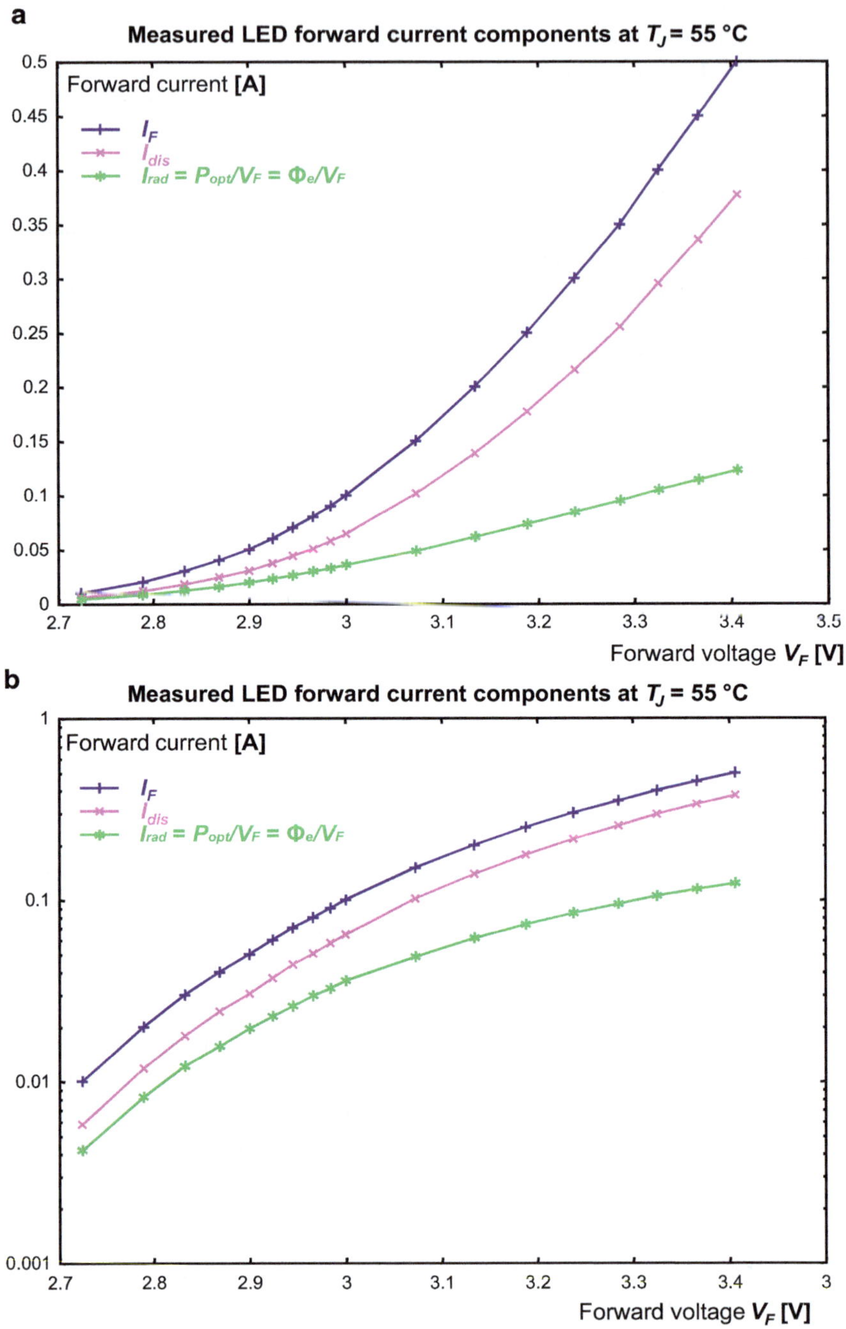

Fig. 6.35 Forward current components of a 1 W LED **a** in a lin-lin coordinate system and **b** in a lin-log coordinate system, measured at 55 °C junction temperature. (Measured with the help of a JESD51-51/JESD51-52-compliant test setup)

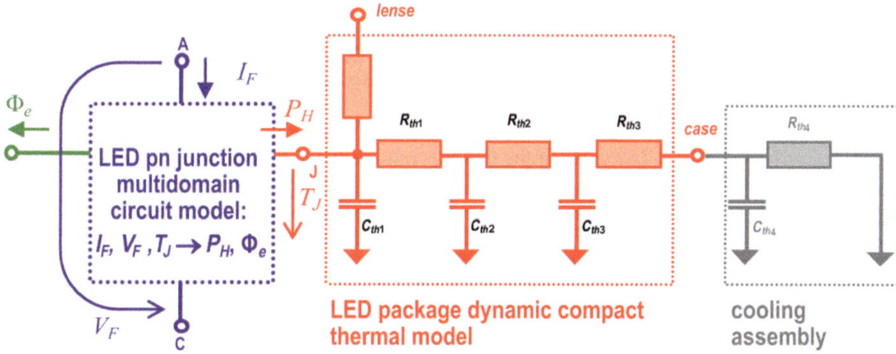

Fig. 6.36 A possible multidomain LED model with some details of a package dynamic compact thermal model (CTM)

In an "ideal" simulation setup, both the pn-junction is described by an electrothermal circuit model and the thermal environment is considered by a 3D thermal simulator, ideally a CFD tool and the system is solved self-consistently. This principle and its realization by a simple and fast conduction-mode-only thermal simulator were presented in [25].

6.5.2.3 Simplified Multidomain LED Compact Model and an Application Example

If studying the thermal aspects is more relevant, the electrical model is simplified and the LED package CTM is connected to a detailed 3D geometrical model of the LED's environment and a CFD-based thermal simulator is used for the analysis.

An advantage of the LED package compact model (shown in Figs. 6.31 and 6.36) is that such compact models are practically boundary condition-independent, thus can be completed with a CTM of the LED's environment (as shown in Fig. 6.36) and can be treated by a SPICE-like network solver. Alternatively, such a model can be embedded into a 3D CFD tool by completing it with geometric information regarding package external dimensions and area(s) of external node(s) through which heat transfer toward the environment takes place.

A good practical trade-off between the oversimplified constant heating power assumption and the full electrothermal solution is assuming linear junction temperature dependence at a constant $I_F = I_{F0}$ forward current for the LED. This follows the typical electrical operation of DC-driven LEDs: both the LED drivers are basically constant current sources and in thermal testing the LED under test is driven by current sources. The electrothermal model in this case reduces to the quasi-linear dependence of the V_F forward voltage on the T_J junction temperature. Thus, any forward voltage value can be easily calculated if the $V_{F0} = V_F(T_0)$ forward voltage value measured at a fixed T_0 reference value of the junction temperature is known and the $S_{VF0} = \mathrm{d}V_F/\mathrm{d}T_J$ temperature sensitivity of the forward voltage at T_0

Table 6.3 Contents of a practical multidomain LED model

Symbol	Meaning
I_{F0}	Nominal value of the forced, fixed forward current, e.g., 350 mA, 700 mA, etc.
T_0	Reference junction temperature value, e.g., 80 °C
V_{F0}	The value of the forward voltage at I_{F0} and T_0
S_{VF0}	The temperature sensitivity of the forward voltage at I_{F0} and T_0
Φ_{e0}	The total radiant flux (emitted optical power) at I_{F0} and T_0
$S_{\Phi e0}$	The temperature sensitivity of the total radiant flux at I_{F0} and T_0
Φ_{V0}	The total luminous flux at I_{F0} and T_0
$S_{\Phi V0}$	The temperature sensitivity of the total luminous flux at I_{F0} and T_0

temperature and I_F nominal forward current is also known. Consequently, the approximate value of the forward voltage can be expressed as follows:

$$V_F(I_{F0}, T_J) = V_{F0} + S_{VF0} \cdot (T_J - T_0) \tag{6.20}$$

The S_{VF0} sensitivity is easily provided from K-factor calibration-type measurements.

In practical cases (assuming typical forward currents and operational temperature ranges), both the radiant flux and the luminous flux exhibit a fairly linear dependence on junction temperature (see Fig. 6.26a). Both can be expressed with formulas similar to Eq. (6.20)

$$\Phi_e(I_{F0}, T_J) = \Phi_{e0} + S_{\Phi e0} \cdot (T_J - T_0) \tag{6.21}$$

$$\Phi_V(I_{F0}, T_J) = \Phi_{V0} + S_{\Phi V0} \cdot (T_J - T_0) \tag{6.22}$$

Thus, a practical simplified multidomain LED model includes the LED package CTM completed with a set of scalars as listed in Table 6.3. The advantage of this practical multidomain LED model is that it includes parameters that can be easily obtained from a series of JEDEC JESD51-51- and JESD51-52-compliant combined thermal and radiometric/photometric measurements. In other words, the workflow shown in Fig. 6.33 needs to be completed with postprocessing the measured light output characteristics (linear regression analysis of the measured total radiant and luminous fluxes).

The above multidomain LED model is implemented in the latest version of an MCAD-embedded CFD tool [82]. Figure 6.37 shows simulation results of a street-lighting LED luminaire prototype developed within the Hungarian government-sponsored KözLED project.

6.5.2.4 Calibration of LED Compact Models

Calibration of compact models needs some attention. In case of electronics (IC) packages, the JEDEC compact modeling documents provide some guidance, especially for the DELPHI models [73]. For the 2R models [72], this seems to be quite simple: are the models correctly predicting the junction temperature or not.

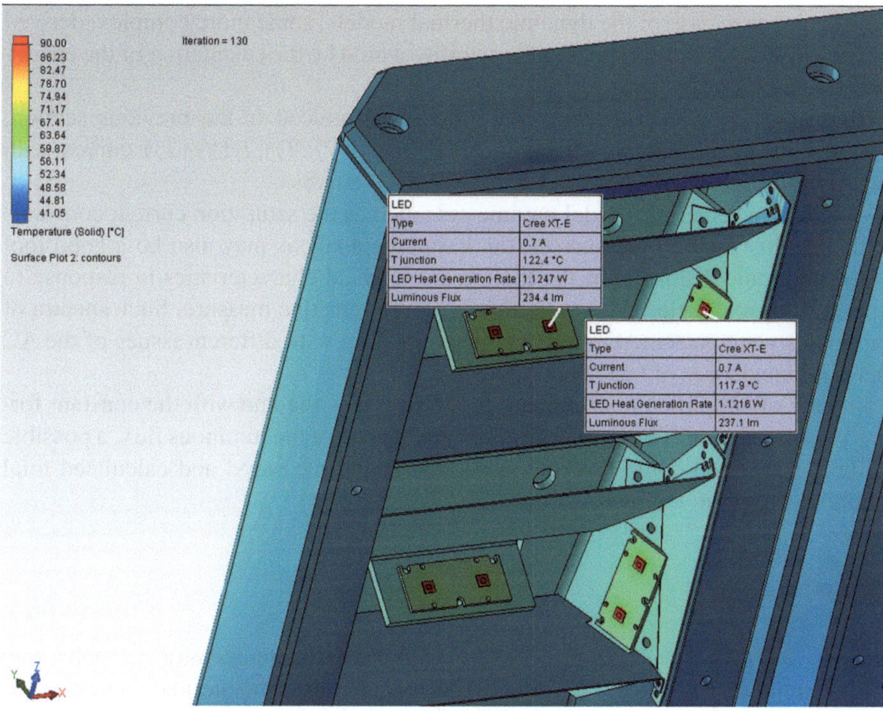

Fig. 6.37 Close-up view of MCAD-embedded CFD simulation results of a street-lighting luminaire with LED hot lumens calculations using a constant forward current-based multidomain LED compact model. (By courtesy of Optimal Optik Ltd. and Hofeka Ltd., Budapest, Hungary)

As LED package CTMs are derived from the generic 2R models, their calibration should also be similar. For steady-state situations again, the steady-state junction temperature could be the metric useful for model calibration. If the LED package CTM is a dynamic model, we expect the model to provide the same unit-step response as the real LED package, i.e., the $Z_{th}(t)$ provided by the model and the $Z_{th}(t)$ curve measured should be as close as possible. Comparing the raw $Z_{th}(t)$ curves might be misleading as behind small differences of the curves there could be significant discrepancies of the model. A better metric would be to compare the time-constant spectrum derived from the measured thermal impedance curve with the discrete time-constant pattern of the model. A discrete network model would only have discrete time constants—as many as the RC stages the model contains. The model is considered OK if the discrete time-constant pattern matches the maxima of the continuous time-constant spectrum of the measured thermal impedance.

In case of multidomain models, the question is at least twofold. On the one hand, the applied thermal model should be correct—this can be checked decoupled from the electrical model as described above. For the correctness of the I–V characteristic, simple comparison of the measured and calculated curves might not be sufficient.

Here again, as in case of the dynamic thermal models, some more complex, derived metrics would be appropriate. One possibility would be the calculation of the energy conversion efficiency.

In case of the electrothermal LED model introduced in the previous section, it would mean to compare $\eta_{e-sim}(V_F, T_J) = I_{rad}(V_F, T_J)/I_F(V_F, T_J)$ curves with measured $\eta_{e-meas}(V_F, T_J) = \Phi_e(V_F, T_J)/(I_F \cdot V_F)$ curves.

To check whether the model parameters (such as the saturation current constants or the ideality factors) are correct, the harmonic analysis may also be a good tool as the harmonic distortion of the nonlinear electrical characteristics in response to a purely sinusoidal electrical excitation is a very sensitive measure. Such spectra of harmonics were presented in a recent paper dealing with different issues of the AC thermal impedances of LEDs [21].

In case of simplified multidomain models such as the one with the constant forward assumption and linear temperature dependence of the luminous flux, a possible calibration method could be the comparison of the measured and calculated total luminous fluxes of well-defined test cases.

6.5.2.5 Summary

Because the forward V_F voltage decreases with increasing temperature in such a way that it cannot be neglected, the best approach is to couple the thermal and electrical calculations. In case of pn-junctions, it is relatively simple to construct an electrothermal model, though its implementation is not that straightforward. One may start with the classical diode equation encompassing a few temperature-dependent parameters (I_0, nV_T). It has to be completed with the thermal model of the package and the model of the light output. The light output can be approximated by means of fitting a polynomial approximate model to describe the $\Phi_e(I_F, T_J)$ or $\Phi_v(I_F, T_J)$ functions. In most practical cases, assuming a linear junction temperature dependence of total luminous flux (total radiant flux) might be sufficient.

The ultimate goal would be to define a standardized, general LED model that includes:

- An electrical model of the (internal) pn-junction including a few temperature-dependent parameters and the electrical series resistance
- A model of the light emission depending on the forward current (taken from the electrical model) and the junction temperature $P_{opt} = \Phi_e(I_F, T_J)$
- A thermal model of the package (complexity depending on the package type)—ideally a BCI model

The model outlined above can be implemented in SPICE-like circuit simulators that are in daily use by electrical engineers. For thermal designers, however, the complexity of such a unified LED model should not exceed the capacity of a spreadsheet application. Simplified models for constant forward current seem to be appropriate.

In this section, LED compact models with different complexity have been shown. The most complex ones are meant for use in the academic research only, while other models with a more practical approach are good candidates for widespread industrial use. Here, the question is how quickly LED vendors and end-users as well as simulation tool providers come to an agreement about what is really needed. Consensus not only on the right model complexity but also on the corresponding model calibration technique is also needed.

One obvious approach given the successes in the past regarding IC thermal characterization could be an EU-funded project, in which European LED manufacturers, test equipment and simulation software vendors, lighting system/luminaire manufacturers, and academic research teams would be involved. Such a consortium should tackle the technological issues and prepare proposals for subsequent discussion within JEDEC and other international standardization bodies involved.

6.6 Thermal Aspects During Light Measurements of LEDs

As can be concluded from the previous sections, thermal aspects of LEDs and SSL products are very important; therefore these aspects need to be seriously considered, also in the case of light output measurements.

This has been realized by different standardization bodies such as IEC and CIE.

For example, CIE Division 2 has created a number of technical committees to revisit different metrology aspects related to LEDs including the aim of providing thermal specifications for the test conditions. The scope of TC 2-63 and TC 2-64 is the closest to the topics addressed in this book. Testing of AC-driven LEDs is covered in Chap. 4—the corresponding CIE technical committee is TC 2-76. The scopes of these committees are the following:

- TC 2-63 Optical Measurement of High-Power LEDs: To develop a CIE recommendation on methods for the operation of high-power LEDs in DC and in pulse mode, at specified junction temperatures, for optical measurements. (Related topics in this book are discussed in Chap. 4 on thermal testing and Chap. 5 on laboratory methods of optical testing of LEDs.)
- TC 2-64 High-Speed Testing Methods for LEDs: To prepare a technical report on high-speed testing methods for electrical, thermal, and optical quantities during the production of LEDs and the conversion of the values to DC operational conditions including the related time-dependent functions. (See Chap. 11 of on the related topic: testing methods of LEDs in manufacturing.)
- TC 2-76 Characterization of AC-driven LED Products for SSL Applications: To prepare a technical report on the measurement of characteristic quantities of AC-driven LED products, including operational conditions, and photometric quantities focused on those influenced by the effect of flicker. (Chap. 4 on thermal testing of LEDs includes a section on the AC thermal impedance of directly AC mains-driven LEDs.)

Each of these committees focuses on its own technical reports. As of today, the reports are in draft state, not yet available to the general public.

6.7 Conclusions

The chapter provides a comprehensive overview of problems associated with thermal characterization of power LEDs. Problems related to lack of standards and possibly sloppy data sheets have been highlighted: the current data sheets should be improved from the perspective of both the manufacturer and the system designer. The similarities from a historical point of view between LED and IC thermal characterization and standardization are demonstrated. A detailed overview of the recently published JEDEC LED thermal testing standards is provided. Another section is devoted to an overview of the state of the art of the LED package compact modeling and multidomain (compact) modeling of LEDs. Compact modeling issues need a closer cooperation of LED vendors, test equipment and simulation tool vendors, as well as SSL designers in order to define what is really needed and how the LED modeling tasks could be best distributed among the major stakeholders. To conclude, we can state the following:

- Without standards, manufacturers who cannot resist swindling have a competitive edge, creating burned soil for the manufacturers who are at least willing to provide their customers with useful thermal data.
- When manufacturers do not take the initiative, end-users are going to demand reliable thermal data for their applications at some moment in time. The manufacturer who can provide these data in time has a competitive edge.
- LED thermal metrics are not only required for reliability and lifetime prediction but also for proper lighting design because all aspects of light output are temperature-dependent.
- Accurate thermal modeling and testing requires dedicated expertise that is probably not available within the existing standardization bodies that focus mainly on their own specific fields (thermal, electrical, optical). Standards for LEDs require a complex approach: good understanding of semiconductor operation and modeling, knowledge of heat transfer, thermal testing, modeling and simulation, and last but not the least, expertise in optical testing of LEDs.
- To define proper standards, manufacturers of components, systems, software, and test equipment together with system designers should cooperate right from the start to gain precious time.
- Based on the successes in the past, it is worthwhile to check how a research consortium could be initiated to address the scientific and technological challenges of LED thermal characterization and to prepare proposals for the yet missing thermal standards.

References

1. ANSI/IESNA IES Nomenclature Committee, IES RP-16-10, Nomenclature and Definitions of for Illuminating Engineering, ISBN 978-0-87995-208–2
2. http://www.zhagastandard.org/specifications/lightengine.html
3. Zhaga Interface Specification Book 1: Overview and common information. http://www.zhagastandard.org/specifications/book-1.html

4. Zhaga Interface Specification Book 3: Round light emitting surface 9 mm–23 mm. http://www.zhagastandard.org/specifications/book-3.html. (October 2012)
5. JEDEC Standard JESD51-50, "Overview of methodologies for the thermal measurement of single- and multi-chip, single- and multi-PN junction light-emitting diodes (LEDs)". http://www.jedec.org/sites/default/files/docs/jesd51-50.pdf. (April 2012)
6. JEDEC Standard JESD51-51, "Implementation of the electrical test method for the measurement of the real thermal resistance and impedance of light-emitting diodes with exposed cooling surface", www.jedec.org/sites/default/files/docs/JESD51-51.pdf. (April 2012)
7. JEDEC Standard JESD51-52, "Guidelines for combining CIE 127-2007 total flux measurements with thermal measurements of LEDs with exposed cooling surface". www.jedec.org/sites/default/files/docs/JESD51-52.pdf. (April 2012)
8. JEDEC Standard JESD51-53 "Terms, definitions and units glossary for LED thermal testing". http://www.jedec.org/sites/default/files/docs/jesd51-53.pdf. (May 2012)
9. Lasance C, Poppe A (2008) On the standardization of thermal characterization of LEDs compared to IC packages: differences, similarities and proposal for action. White paper version September 2008, submitted to JEDEC JC15 committee
10. Lasance C (2008) On the standardization of thermal characterization of LEDs Part I: Comparison with IC packages and proposal for action. Proceedings of the 14th International Workshop on THERMal INvestigation of ICs and Systems (THERMINIC'08), 24–26 September 2008, Rome, Italy, pp 208–212
11. Lasance C, Poppe A (2008) On the standardization of thermal characterization of LEDs Part II: Problem definition and potential solutions. Proceedings of the 14th International Workshop on THERMal INvestigation of ICs and Systems (THERMINIC'08), 24–26 September 2008, Rome, Italy, pp 213–219
12. Lasance C, Poppe A (2009) On the standardization of thermal characterization of LEDs. Proceedings of the 25th IEEE semiconductor thermal measurement and management symposium (SEMI-THERM'09), 15–19 March 2009, San Jose, USA, pp 151–158
13. Lasance C, Poppe A (2009) Challenges in LED thermal characterisation. Proceedings of the 10th International Conference on Thermal, Mechanical and Multiphysics Simulation and Experiments in Microelectronics and Microsystems (EuroSimE'09), 27–29 April 2009, Delft, The Netherlands, pp 1–11
14. Lasance C (2008) Heat spreading: not a trivial problem. ElectronicsCooling 14(2)
15. Lasance C (2008) The practical usefulness of various approaches to estimate heat spreading effects. Proceedings of The Second International Conference onThermal Issues in Emerging Technologies (ThETA'08), 17–20 December 2008, Cairo, Egypt, pp 149–158
16. Rosten H, Lasance C (1995) DELPHI: the development of libraries of physical models of electronic components for an integrated design environment. In Berge J-M, Levia O, Rouillard J (Eds) Model Generation in Electronic Design. Kluwer Academic Press, pp 63–90
17. JEDEC Standard JESD51, Methodology for the thermal measurement of component packages (single semiconductor devices). This is the overview document for the JESD51- series of specifications. www.jedec.org/sites/default/files/docs/Jesd51.pdf. (December 1995)
18. JEDEC Standard JESD51-1, Integrated circuits thermal measurement method—electrical test method (single semiconductor device). www.jedec.org/sites/default/files/docs/jesd51-1.pdf. (December 1995)
19. JEDEC Standard JESD51-14, Transient dual interface test method for the measurement of thermal resistance junction-to-case of semiconductor devices with heat flow through a single path. www.jedec.org/download/search/jesd51-14.pdf. (November 2010)
20. MIL-STD-750D Method 3101.3, Thermal impedance (response) testing of diodes
21. Poppe A, Siegal B, Farkas G Issues of thermal testing of AC LEDs. Proceedings of the 27th IEEE Semiconductor Thermal Measurement and Management Symposium (SEMI-THERM'11), 20–24 March 2011, San Jose, USA, pp 297–304
22. Székely V, Bien TV (1988) Fine structure of heat flow path in semiconductor devices: a measurement and identification method. Solid-State Electron 31(9):1363–1368

23. Székely V (1998) Identification of RC networks by deconvolution: chances and limits. IEEE Transactions on Circuits and Systems I—Fundamental Theory and Applications 45(3):244–258
24. Treurniet T, Lammens V (2006) Thermal management in color variable multi-chip LED modules. Proceedings of the 22nd IEEE Semiconductor Thermal Measurement and Management Symposium (SEMI-THERM'06), Dallas, USA, 14–16 March 2006, pp 186–190
25. Poppe A, Farkas G, Székely V, Horváth Gy, Rencz M (2006) Multi-domain simulation and measurement of power LED-s and power LED assemblies. Proceedings of the 22nd IEEE Semiconductor Thermal Measurement and Management Symposium (SEMI-THERM'06), Dallas, USA, 14–16 March 2006, pp 191–198
26. Poppe A (2012) A step forward in multi-domain modeling of power LEDs. Proceedings of the 28th IEEE Semiconductor Thermal Measurement and Management Symposium (SEMI-THERM'12), San Jose, USA, 18–22 March 2012, pp 325–330. (ISBN: 978-1-4673-1109–0)
27. Lasance C (2005) CFD simulations in electronic systems: a lot of pitfalls and a few remedies. ElectronicsCooling 11(2)
28. Rosten HI et al (1997) Final report to SEMI-THERM XIII on the European-funded project DELPHI-the development of libraries and physical models for an integrated design environment. Proceedings of the 13th IEEE Semiconductor Thermal Measurement and Management Symposium (SEMI-THERM'97), Austin, USA, 28–30 January 1997, pp 73–91
29. Rosten HI, Lasance CJM, Parry JD (2005) The world of thermal characterization according to DELPHI-Part I: background to DELPHI. IEEE Tr. on Components, Packaging, and Manufacturing Technology Part A 20(4):384–391
30. Lasance CJM, Rosten HI, Parry JD (1997) The world of thermal characterization according to DELPHI-Part II: experimental and numerical methods. IEEE Tr. on Components, Packaging, and Manufacturing Technology Part A 20(4):392–398
31. Pape H, Noebauer G (1999) Generation and verification of boundary independent compact thermal models for active components according to the DELPHI/SEED methods. In: Proceedings of the 15th IEEE Semiconductor Thermal Measurement and Management Symposium (SEMI-THERM'99), San Diego, USA, 9–11 March 1999, pp 201–211
32. Lasance CJM, den Hertog D, Stehouwer P (1999) Creation and evaluation of compact models for thermal characterisation using dedicated optimisation software. In: Proceedings of the 15th IEEE Semiconductor Thermal Measurement and Management Symposium (SEMI-THERM'99), San Diego, USA, 9–11 March 1999, pp 189–200
33. Lasance CJM (2003) Final report on the EC-funded thermal project PROFIT. In: Proceedings of the 9th International Workshop on THERMal INvestigations of ICs and Systems (THERMINIC'03). Aix-en-Provence, France, 24–26 September 2003, pp 283–289
34. Lasance CJM (2004) Highlights from the European thermal project PROFIT. J Electron Packag 126(4):565–570
35. Lasance CJM (2001) The European project PROFIT: prediction of temperature gradients influencing the quality of electronic products. In: Proceedings of the 17th IEEE Semiconductor Thermal Measurement and Management Symposium (SEMI-THERM'01), San Jose, USA, 20–22 March 2001, pp 120–125
36. Lasance C (2008) Ten years of boundary condition independent compact thermal modeling of electronic parts: a review. Heat Transf Eng 29:149–168
37. JEDEC Standard JESD15-1 (2008) Compact Thermal Model Overview. www.jedec..org /download/search/jesd15-1.pdf. (October 2008)
38. JEDEC Standard JESD15-3 (2008) Two-Resistor Compact Thermal Model Guideline. www.jedec.org/download/search/jesd15-3.pdf. (July 2008)
39. JEDEC Standard JESD15-4 (2008) DELPHI Compact Thermal Model Guideline. www.jedec.org/download/search/jesd15-4.pdf. (October 2008)
40. Brodrick J, CFLs in America: lessons learned on the way to market. http://apps1.eere.energy.gov/buildings/publications/pdfs/ssl/cfls_july_lessons.pdf
41. IES LM-79-08 Standard, "Approved method: electrical and photometric measurements of solid-state lighting products". http://www.ies.org/store/product/approved-method-electrical-and-photometric-measurements-of-solidstate-lighting-products-1095.cfm

42. ANSI/UL 8750 Standard, "Safety standard for light emitting diode (LED) equipment for use in lighting products"
43. ANSI/IESNA IES-LM-80-08 Standard, "Approved method for measuring lumen maintenance of LED light sources". ISBN 978-0-87995-227-3 (2009)
44. ANSI/IESNA IES-TM-21-11 Standard, "Projecting long term lumen maintenance of LED light sources". ISBN: 978-0-87995-259-4
45. Peters L (2012) "EPA introduces TM-21 calculator". LEDs Magazine, http://ledsmagazine.com /news/9/2/8. Accessed Dec 2012
46. US Environmental Protection Agency's TM-21 calculator. http://www.energystar.gov/TM-21calculator. Accessed Dec 2012
47. Ueda O (1999) Reliability issues in III-V compound semiconductor devices: optical devices and GaAs-based HBTs. Microelectron Reliab 39:1839–1855
48. Kim H, Yang H, Huh C, Kim S-W, Park S-J, Hwang H (2000) Electromigration-induced Failure of GaN Multi-quantum Well Light Emitting Diode. Electron Lett 36(10):908–910
49. Narendran N, Gu Y, Freyssinier J, Yu H, Deng L (2004) Solid-state lighting: failure analysis of white LEDs. J Crystal Growth 268:449–456
50. Uddin A, Wei C, Anderson T (2005) Study of degradation mechanism of blue light emitting diodes. Thin Solid Films 483:378–381
51. Krames MR, Shchekin OB, Mueller-Mach R, Mueller GO, Zhou L, Harbers G, Craford MG (2007) Status and future of high power light-emitting diodes for solid-state lighting. J Display Technol 3(2)
52. Hu J, Yang L, Shin MW (2007) Mechanism and thermal effect of delamination in light-emitting diode packages. Microelectron J 38:157–163
53. Hsu Y-C, Lin Y-K, Chen M-H, Tsai C-C, Kuang J-H, Huang S-B, Hu H-L, Su Y-I, Cheng W-H (2008) Failure mechanisms associated with lens shape of high-power LED modulus in aging test. IEEE Trans Electron Devices 55(2):2008
54. CIE S 017/E:2011 ILV: International Lighting Vocabulary
55. Grabner-Meyer (2007) LED data sheet comparison. LED-Professional Review, September 2007
56. http://www.futurelightingsolutions.com/en/development/Pages/index.aspx
57. http://www.mentor.com/products/mechanical/products/floefd/
58. Lee S (1998) Calculating spreading resistance in heat sinks. ElectronicsCooling 4(1):30–33
59. Yu J, Oepts W, Konijn H, PCB board thermal management of high-power LEDs. In: Proceedings of the 24th IEEE Semiconductor Thermal Measurement and Management Symposium (SEMI-THERM'08), 16–20 March 2008, San Jose, USA, pp 63–67
60. CIE 127-2007 Technical Report, "Measurement of LEDs"
61. Farkas G, Poppe A, Schanda J, Muray K (2004) Complex characterization of power LED-s: simultaneous measurement of photometric/radiometric and thermal properties. Proc. CIE LED Conference, Tokyo, CIE x026:2004, pp 92–95
62. Poppe A, Farkas G, Molnár G, Katona B, Temesvölgyi T, He JW (2010) Emerging standard for thermal testing of power LEDs and its possible implementation. In: SPIE Proceedings 7784, Tenth International Conference on Solid State Lighting (Eds: Ian Ferguson; Matthew H. Kane; Nadarajah Narendran; Tsunemasa Taguchi), 1 August 2010, San Diego, CA, USA, p 778414
63. Zhang L, Treurniet T (2008) On the challenges of thermal characterization of high-power, high-brightness LED packages. ElectronicsCooling 14(2)
64. Guenin B, Update on Thermal Standards Work by JC15.1. http://www.ewh.ieee.org/soc/cpmt/presentations/cpmt0303a.pdf
65. Guenin B (2012) Update on JEDEC thermal standards. ElectronicsCooling Vol.18, September 17th, 2012, http://www.electronics-cooling.com/2012/09/update-on-jedec-thermal-standards/
66. Zong Y, Chou P-T, Lin M-T, Ohno Y (2009) Practical method for measurement of AC-driven LEDs at a given junction temperature by using active heat sinks. In: SPIE Proceedings 7422, Ninth International Conference on Solid State Lighting (Eds: Tsunemasa Taguchi, Ian T. Ferguson, Christoph Hoelen, Jianzhong Jiao), 2 August 2009, San Diego, CA, USA, p 742208

67. Liu Y-w, Jayawardena A, Klein TR, Narendran N (2010) Estimating the junction temperature of AC LEDs. In: SPIE Proceedings 7784, Tenth International Conference on Solid State Lighting (Eds: Ian Ferguson; Matthew H. Kane; Nadarajah Narendran; Tsunemasa Taguchi), 1 August 2010, San Diego, CA, USA, p 778409
68. Zong Y, Ohno Y (2008) New practical method for measurement of high-power LEDs. In: Proceedings of the CIE Expert Symposium 2008 on Advances in Photometry and Colorimetry, CIE x033:2008, pp 102–106
69. Vass Varnai A, Parry J, Toth G, Ress S, Farkas G, Poppe A, Rencz M (2012) Comparison of JEDEC dynamic and static test methods for thermal characterization of power LEDs. In: Proceedings of the 14th Electronics Packaging Technology Conference (EPTC' 12). 5–7 December 2012, Singapore, Paper 229
70. JEDEC Standard JESD15, "Thermal Modeling Overview" (October 2008). www.jedec.org/sites/default/files/docs/JESD15.pdf
71. JEDEC Standard JESD15-1, "Compact Thermal Modeling Overview" (October 2008). www.jedec.org/sites/default/files/docs/JESD15-1.pdf
72. JEDEC Standard JESD15-3, "Two-Resistor Compact Thermal Model Guideline" (October 2008), http://www.jedec.org/standards-documents/docs/jesd-15–3
73. JEDEC Standard JESD15-4, "DELPHI Compact Thermal Model Guideline" (October 2008)
74. Sabry M, Dessouky M (2012) Framework theory for dynamic compact thermal models. In: Proceedings of the 28th IEEE Semiconductor Thermal Measurement and Management Symposium (SEMI-THERM), San Jose, USA, 18–22 March 2012, pp 189–194
75. Schubert EF (2006) Light-emitting diodes, 2nd ed. Cambridge University Press, Cambridge
76. Farkas G, Haque S, Wall F, Martin PS, Poppe A, Voorst Vader Q van, Bognár Gy (2004) Electric and thermal transient effects in high power optical devices. In: Proceedings of the 20th IEEE Semiconductor Thermal Measurement and Management Symposium (SEMI-THERM'04). San Jose, USA, 9–11 March 2004, pp 168–176
77. Rencz M, Poppe A, Kollár E, Ress S, Székely V, Courtois B (2004) A procedure to correct the error in the structure function based thermal measuring methods. In: Proceedings of the 20th IEEE Semiconductor Thermal Measurement and Management Symposium (SEMI-THERM'04). 9–11 March 2004, San Jose, USA, pp 92–98
78. Rencz M, Poppe A, Kollár E, Ress S, Székely V (2005) Increasing the accuracy of structure function based thermal material parameter measurements. IEEE Tr on Comp and Pack Techn 28(1):51–57
79. Poppe A, Molnár G, Temesvölgyi T (2010) Temperature dependent thermal resistance in power LED assemblies and a way to cope with it. In: Proceedings of the 26th IEEE Semiconductor Thermal Measurement and Management Symposium (SEMI-THERM'10), 21–25 February 2010, Santa Clara, USA, pp 283–288
80. Mentor Graphics T3Ster® Master software v2.1. http://www.mentor.com/products/mechanical/products/t3ster/
81. Mentor Graphics FloTHERM® V9.1 CFD based thermal analysis software. http://www.mentor.com/products/mechanical/products/flotherm
82. Mentor Graphics FloEFD/ v12 MCAD embedded CFD based thermal analysis software LED module. http://www.mentor.com/products/mechanical/products/floefd/
83. Poppe A, Szalai A, Parry J (2012) Advanced thermal characterization improves LED street-light design. LEDs Magazine No. 53:51–56. (July 2012), http://ledsmagazine.com/features/9/7/16
84. Górecki K (2012) Electrothermal model of a power LED for SPICE. Int J Numer Model 25:39–45
85. Negrea C, Svasta P, Rangu M (2012) Electro–thermal modeling of power LED using SPICE circuit solver. In Proceedings of the 35th International Spring Seminar on Electronics Technology, pp 329–334

Part III
Advances in Cooling Technologies

Chapter 7
Air Cooling for LED Lighting

Raghav Mahalingam

Abstract A Light Emitting Diode (LED) is a semiconductor that converts electrical energy into light and heat. Typically, energy conversion efficiencies, i.e., the percentage of input energy converted to light, are in the 20–40 % range, resulting in a significant amount of heat being generated in the pn-junction of the LED. In most applications, this heat has to be conducted away from the junction and then convected and/or radiated to the ambient air. The convective part of the heat transfer usually requires extended surfaces or heat sinks and in many cases, a method of creating airflow over the heat sink to transport the heat away into the ambient, and is generally referred to as air-side heat transfer. This chapter will discuss the various methods that can be employed for air-side heat transfer. The chapter is divided into five major sections. Section 2 deals with the system-level thermal management of LEDs and discusses the importance of cooling LEDs as well as the various thermal paths and resistances involved in a typical LED system. Section 3 describes the fundamentals of both natural and forced convection heat transfer, including some basic relations and equations used in convective heat transfer. Section 4 describes the different technologies that exist today for air-side heat transfer. Section 5 compares the different technologies with respect to the system-level metrics such as acoustics, power consumption, reliability, etc., required for designing a cooling solution. Finally, Sect. 6 summarizes the chapter.

7.1 System-Level Thermal Management of LEDs

An LED converts electrical energy into light and heat. As of today, about 60–80 % of the input energy is converted to heat (Fig. 7.1). For example, given typical efficacies of about 90–100 lm/W for current LEDs, a 1,600 lm LED replacement for a 100 W incandescent bulb needs about 18 W of input power, which translates to about 15 W of heat. As the required level of emitted luminous flux increases, heat dissipation

---

R. Mahalingam (✉)
Nuventix, 4635 Boston Lane, Austin, TX 78735 USA
e-mail: raghav@nuventix.com

C. J. M. Lasance, A. Poppe (eds.), *Thermal Management for LED Applications*, Solid State Lighting Technology and Application Series 2,
DOI 10.1007/978-1-4614-5091-7_7, © Springer Science+Business Media New York 2014

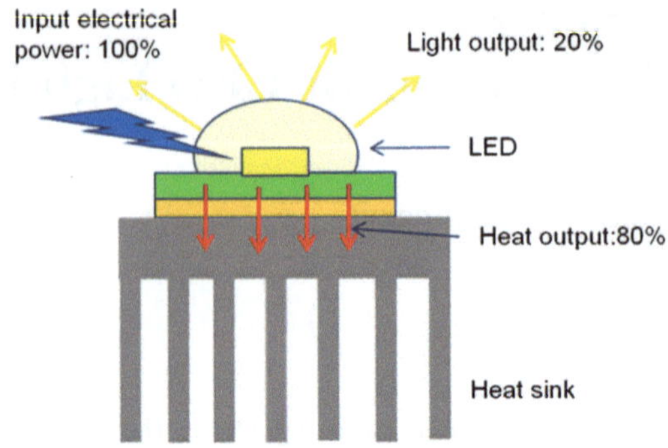

Fig. 7.1 Typical energy distribution in an LED light

levels of 40–60 W are not uncommon (and continuing to increase), and this heat has to be cleverly managed to prevent the LED light from deteriorating over time or failing altogether. Additionally, in designing a thermal management solution for High-Brightness LED (HB-LED) lighting systems, a cooling solution must have high reliability (at least as high as the LEDs themselves), low acoustic output (since people want to see lights, not hear them), low power consumption (in order to maintain high system efficacy), be as small as possible and allow for lighting architects to have maximum design and installation flexibility in implementing cooled LED lights.

One of the main attractions of LEDs over incandescent or fluorescent lamps is their long lifetime. The Alliance for Solid-State Illumination Systems and Technologies (ASSIST) defines the useful life of a luminaire as the amount of time for the luminaire to have a 30 % reduction from the initial light output or a 50 % reduction from the initial light output. This reduction in emitted luminous flux is known as lumen maintenance and they are denoted by $L_{70\,\%}$ and $L_{50\,\%}$ [1]. HB-LED products are being designed to have a useful life that ranges from 25,000 up to 50,000 h. As of 2011, the DoE (US Department of Energy) ENERGY STAR Program Requirements for SSL Luminaires [2] states that for residential lighting, the minimum useful life should be 25,000 h and 35,000 h for indoor and outdoor luminaires, respectively and 35,000 h for all commercial lighting. In order for the luminaire to achieve such lifetime hours, the LED junction temperatures (T_J) must be maintained according to the product specification. In general, a 10–15 °C increase in T_J can reduce the lifetime by 50 % [3]. A so-called L70-B50 diagram for a Philips Lumileds Luxeon K2 LED, which illustrates this idea is shown in Fig. 5.16 of Chap. 5. Retrofit bulbs and down light products are especially difficult from a thermal management perspective due to their smaller form factor and the fact that many such devices are installed in flow-restricted fixtures—as discussed in Chap. 11.

Keeping the temperature of LEDs within reasonable limits requires a combination of thermal management processes. Figure 7.2 shows the heat transfer path for the heat produced within an LED.

Heat that is produced in the junction of the LED gets spread in the LED chip (and its possible submount), conducted to an integrated back-plate (which could be Aluminum or Copper). An electrical analogy is used to describe the flow of heat across a temperature gradient (Fig. 7.2).

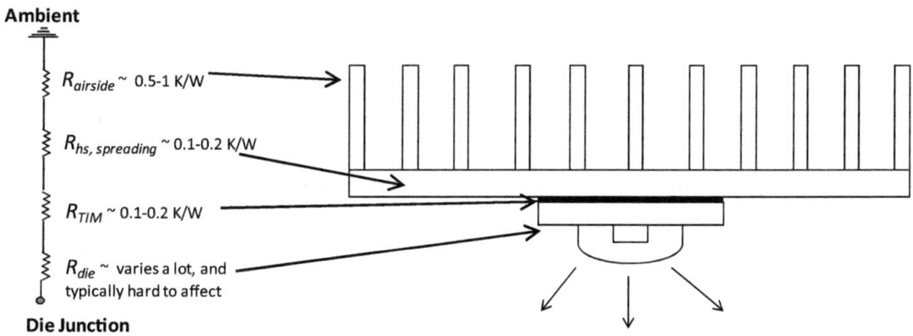

Fig. 7.2 Heat transfer path in a typical air cooling scenario

The resistance of heat flow up to the back-plate is referred to as junction-to-case thermal resistance or R_{die} or R_{jn}, and varies a lot from one type of LED to another due to the internal packaging in the semiconductor package. LED vendors make every effort to reduce this figure. In most cases, this is a quantity that the cooling solution provider cannot affect but it is critical to know this value, because the junction temperature depends on this. (A recent testing standard, JEDEC JESD51-14 provides a new methodology with which $R_{th\,JC}$ can be easily identified. More details on this are provided in Chap. 3.) However, since the focus in this chapter is on air-side cooling, it will be assumed that the cooling solution only addresses the $R_{th-airside}$, and consequently, the temperature at the back-plate of the LED.

Once the heat has been conducted and spread to the back-plate of the LED, it must be transferred across an interface to the heat sink. The resistance associated with this heat flow is called contact resistance and is due to microscale air gaps between the two contacting surfaces. With proper selection and application of Thermal Interface Materials (TIMs), this contact resistance can be minimized to about 0.1 °K/W. (Chapter 8 deals with advanced TIMs used to reduce this resistance while Chap. 3 provides some information on the most recent TIM testing possibilities).

The next step in the heat transfer path is the spreading of the heat from a surface that is the size of the back-plate of the LED to a surface that has an area many times more, i.e., a heat sink. The term *heat sink* is a misnomer, since the metal structure that is typically called the heat sink only serves the function of taking the heat from a small surface to a large extended surface exposed to air (or to another cooling fluid). The heat sink is not actually a sink of heat, i.e., it does not absorb the heat,[1] it only transfers it. Historically, however, the extended surfaces are called heat sinks and this chapter will adhere to this nomenclature. As it shall be explored later in the chapter, the amount of heat transferred away from the LED strongly depends on the extended

[1] The huge thermal capacitance, however, plays an important role in dynamic mode, e.g., when an LED is AC-driven or when it is PWM dimmed—resulting in reduced absolute value of the AC thermal impedance at higher operating frequencies or reduced pulsed thermal resistance values when the frequency of the PWM dimming is higher. See details on this in Chap. 3.

surface area of the heat sink, regardless of the cooling technology used. However, when heat is spread from a small area to a large one, a resistance to the heat spreading is introduced called R_{hs} or $R_{spreading}$. This can be carefully minimized by heat sink design and can be kept to within 0.1–0.2 K/W. Several methods have been described in the literature for calculating the spreading resistance for heat sinks [4, 5]. Please see Chap. 9 for a detailed methodology as to how to calculate the spreading resistance. Several modern finite element analysis (FEA)/computational fluid dynamic (CFD) softwares can directly compute heat spreading in complicated geometries and it is recommended to use such tools for spreading resistance calculations since they are readily available and easy to use.

Finally, the heat must be removed from the extended surfaces of the heat sink to the ambient, called air-side heat transfer. Air-side heat transfer could be a combination of convection and radiation. Discussion of radiation is not in the scope of this chapter. It is important to note that while in most cases radiation is a small fraction of forced convection, it is often not negligible compared to natural convection heat transfer and should be considered by heat sink designers. As far as convection is concerned, there are two processes involved here. The first one is the transfer of heat from the surface of the heat sink to the local fluid surrounding the heat sink. The second mechanism is the transfer of heat from the local fluid to the ambient air (which is the ultimate sink for the heat). These two processes are by far the most critical for cooling HB-LEDs. The total thermal resistance budget (defined at the thermal resistance from the back-plate of the LED to ambient) is dominated by the air-side heat transfer. The thermal resistance at the heat sink to air interface in many cases is responsible for over 80 % of the overall thermal resistance budget. This chapter focuses on this important aspect of cooling LEDs, i.e., air-side heat transfer or convective heat transfer (called air-cooling for brevity). The next section describes the fundamentals of convective heat transfer and the main parameters that influence the effectiveness of the convective heat transfer.

7.2 Convective Heat Transfer

Whenever any system, natural or man-made, gets an input of any form of energy, it converts the energy into one or more forms, kinetic, potential, chemical, electrical, thermal, optical, acoustic, etc. Usually, only one of the above forms is the desired output and the rest of the forms are wasted energy. In many systems, thermal energy may be the desired output, such as heating coils. However, in most electronic systems such as microprocessors or LEDs, thermal energy is a wasted by-product and it has to be either reused or dissipated away to a large sink of thermal energy to prevent the system from failing. The ultimate sink of thermal energy may be air, water, or the ground, but in a large fraction of applications it ends up being ambient air.

The process of convection from a surface can either be buoyancy-driven, as designed by nature (hence, called natural convection) or can be aided by "forcing" air to move past the surface (hence called forced convection). Forced convection can

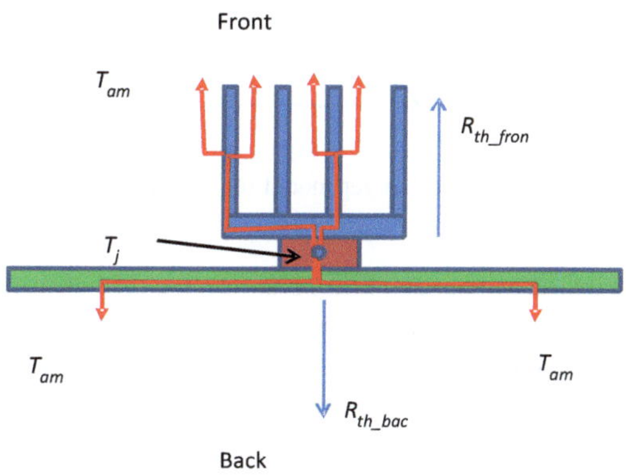

Fig. 7.3 Heat transfer path in a typical electronic system

also be split into a couple of different methods. In many electronic systems such as servers, there are fans that blow air through the whole system and heat sinks are used locally to enhance heat transfer. Quite often, these heat sinks are just cooled by the system flow. However, it is also possible to add local air movers to heat sinks such as synthetic jets or fans. When heat sinks have dedicated localized air movers, they are called active heat sinks. In most practical applications, LED lights will be cooled either using natural convection or active heat sinks. This chapter will therefore focus on these two methods, i.e., natural convection and active heat sinks. Also, there are countless books and articles written on the fundamentals of heat transfer as well as methods to calculate heat transfer on simple geometries such as flat plates and cylinders. Since this book is designed to help engineers in the LED industry design thermal management solutions, this chapter will discuss the critical parameters in the design of heat sinks for convection cooling. The next section will discuss variables and equations that are common to both natural and forced convection.

7.2.1 Commonly Used Relations in the Calculation of Convective Heat Transfer

Since the overall thermal performance of any cooling solution is typically expressed as a thermal resistance (with lower thermal resistances being better cooling solutions), thermal engineers need a method of calculating thermal resistance for different kinds of configurations. The overall thermal resistance of any system can be modeled as a series/parallel combination of thermal resistances.

For example, as seen in Fig. 7.3, heat produced in the junction of a semiconductor has several paths it can take, and the amount of heat traveling through different paths is dependent on the thermal resistance of that path.

The total thermal resistance of a system like this can be written as

$$1/R_{th\_tot} = 1/R_{th\_front} + 1/R_{th\_back} \tag{7.1}$$

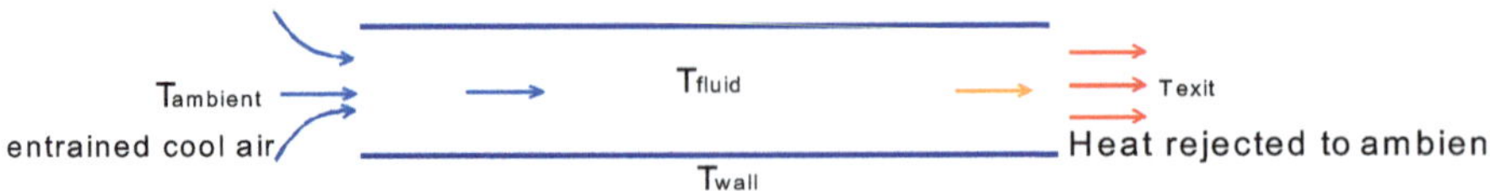

Fig. 7.4 Heat exchanger model for calculating thermal resistance

For the purposes of this chapter, we will assume that the backside thermal resistance is large compared to the frontside thermal resistance (i.e., not much heat flows to the backside) and focus only on the heat sink side of the system.

The heat transfer processes on the heat sink side of the system can be broken into three major parts, heat transfer through conduction within the heat sink and convection and thermal radiation from the heat sink. Chapter 9 deals with the heat spreading and thermal resistance networks in detail, so in this chapter, we will focus on the convective heat transfer only.

Heat transfer through convection is carried out by two processes. A simplified form of looking at this heat transfer process is using a channel flow (as is the case in many extruded heat sinks), as seen in Fig. 7.4. The same rules can be applied to first-order accuracy regardless of the shape of the fins.

Once heat is transferred to the fins of a heat sink by conduction, it is convected to ambient air through two processes:

1. Heat transfer from the wall to the local fluid. The rate of heat transfer from the surface of a body to a fluid is calculated using the equation,

$$Q = h\,A\,(T_w - T_f) \tag{7.2}$$

where

h Heat transfer coefficient
A Surface area exposed to the heat transfer
T_w Average temperature of the surface exposed to the heat transfer
T_f Local temperature of the fluid

T_f is the temperature of the fluid that is near the fins. It is not the ambient temperature, which is the temperature of the air before it has picked up any heat from the system. Many a times, designers make the mistake of assuming the local fluid temperature equal to the ambient temperature, resulting in a calculated heat dissipation that is higher than reality.

T_w is typically the average temperature of the fins of the heat sink. Many a times, designers make the mistake of using the temperature of the base of the heat sink or even the temperature of the surface of the die, which both lead to higher than actual values of the heat dissipated and therefore an under-designed heat sink.

A is the surface area of all the fins of the heat sinks that are exposed to the ambient air. Many a times, depending on the density of the fins, the base of the heat sink should also be included in the heat transfer calculations.

h is the heat transfer coefficient and the main parameter which determines which cooling technology can be used for a specific application, once the allowable size of the cooling solution is fixed. Most LED lights where the size and/or weight of the heat sink are not constrained can be cooled with natural convection. However, as power density of LEDs increase, the size of the heat sinks becomes larger and larger, to a point where the heat sink either does not fit into the available space (such as in ceiling cans) or the heat sink is too heavy for the structure it is mounted on. Under these circumstances, a localized air mover is employed to increase the heat transfer coefficient and therefore the ability of the heat sink to take out more heat from the system.

2. A second equation that is also important in discussing heat transfer from a heat sink is the amount of mass flow through the heat sink. A certain volume of any fluid has only a fixed amount of heat it can absorb as it flows through a heat sink. This is determined by the specific heat capacity of the fluid. The heat transported by a fluid as it moves along the fins of a heat sink is governed by,

$$Q = \dot{m}\, c_p(T_e - T_a) \tag{7.3}$$

where

$\dot{m}$ Mass flow rate through the system (or the fins)
c_p Specific heat capacity of the fluid
T_e Temperature of the fluid at the exit of the fins
T_a Ambient temperature of the fluid as it enters the fins

Using the equations for heat transport, heat transfer, and thermal resistance, the overall air-side thermal resistance can be written as

$$R_{th} = 1/(\eta \dot{m} c_p) \tag{7.4}$$

This model of heat transfer is called the heat exchanger model [6] and includes an effectiveness parameter

$$\eta = [1 - \exp(-hA/\dot{m}\, c_p)] \tag{7.5}$$

Therefore, the main parameters needed for calculating the air-side thermal resistance of a system are the mass flow rate through the systems, the heat transfer coefficient, the area exposed to heat transfer, and the specific heat of the fluid.

It is useful to describe another parameter that is used by heat transfer engineers, called the Nusselt number, *Nu*. The Nusselt number is a nondimensional parameter, which represents the ratio of convective to conductive heat transfer, and is defined as

$$Nu = h\, D/k \tag{7.6}$$

where

h Heat transfer coefficient

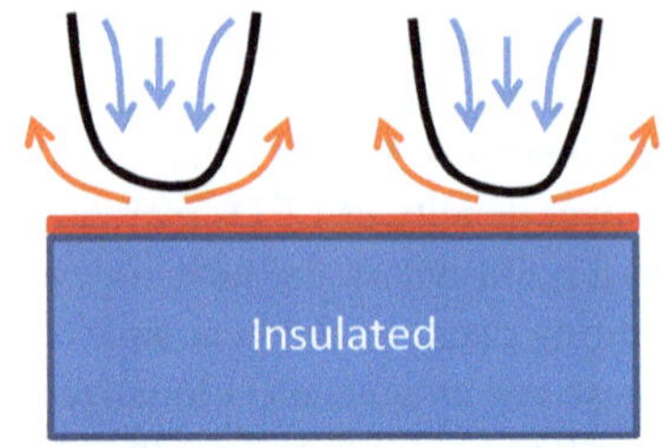

Fig. 7.5 Natural convection from a horizontal surface. Since cold air is heavier than hot air, it descends, pushing away the hot air from near the heated surface. Pockets of circulation are formed resulting in a buoyancy-driven flow that carries heat away from the surface

D Characteristic dimension of the flow (such as diameter of a pipe)
k Conductivity of the fluid

Although direct computation of heat transfer coefficients is complicated and time consuming in mixed flows, empirical correlations for Nusselt number have been developed for many configurations, which can be used by engineers in field applications. In most practical applications, the Nusselt number is known from correlations and the heat transfer coefficient is backed out from the Nusselt number equation, as

$$h = Nu\, k/D \tag{7.7}$$

7.2.2 Natural Convection

When any fluid (for the purposes of this chapter, henceforth the fluid shall be assumed to be air) is in contact with a hot surface, thermal energy or heat is transferred from the hot surface to the layers of the fluid next to the hot surface. This transfer begins through the process of conduction, where heat is conducted from the molecules at the surface to the molecules of air through collisions, thus increasing the internal energy of the layer of air next to the solid surface. This increase in internal energy is reflected as a rise in temperature of the bulk air. Figure 7.5 provides a good example for illustrative purposes.

As the layer of air near the solid surface heats up, its density decreases, resulting in a region of varying temperature and density called the thermal boundary layer. The cooler layers of air further away from the hot surface, being denser, are heavier and thus, buoyancy pulls the cooler air towards the surface and as the cool air moves closer to the surface, it displaces the warm air. Over time, this process repeats continuously, with the hot air getting displaced away from the surface and being replaced with cooler air, creating a pattern of circulating flow that constantly provides cool air to the hot surface. As shall be seen later in the chapter, the pattern of circulation is a strong function of the inclination of the surface, but the fundamental concept of cool air replacing the hot air by the action of buoyancy is what sets up natural convection flows.

Natural convection flows have been studied extensively since the beginning of thermal science. As buoyancy is the driving factor in natural convection, it is customary to design natural convection heat sinks with the fin surface parallel to the direction of gravity. However, it is important to keep in mind that many LEDs might be located in systems where there are mean flows present. For example, an LED in the presence of other heat-producing components may see the buoyancy-driven convection from the other components. LEDs may be situated in locations where there are other air currents (such as high voltage alternating current (HVAC) systems). It is also critical to note that in many cases the local air the LED is exposed to might be preheated by the presence of other heat-generating components. As power levels increase, many designers intentionally move to forced convection.

7.2.3 Forced Convection

Forced convection is a concept that we use quite often in our daily lives, sometimes without even being aware of it. The act of blowing on a forkful of hot food is forced convection. When air is forced over a hot surface, the process of heat transfer is actually not fundamentally different from natural convection, in that heat is transferred by conduction from the surface to the first layer of air next to it. The heat transfer from the layer of air next to the surface to layers further away occurs by mixing of the hot and cold air. In the case of natural convection, the mixing is created by buoyancy. In the case of forced convection, the mixing is created by the imposed airflow.

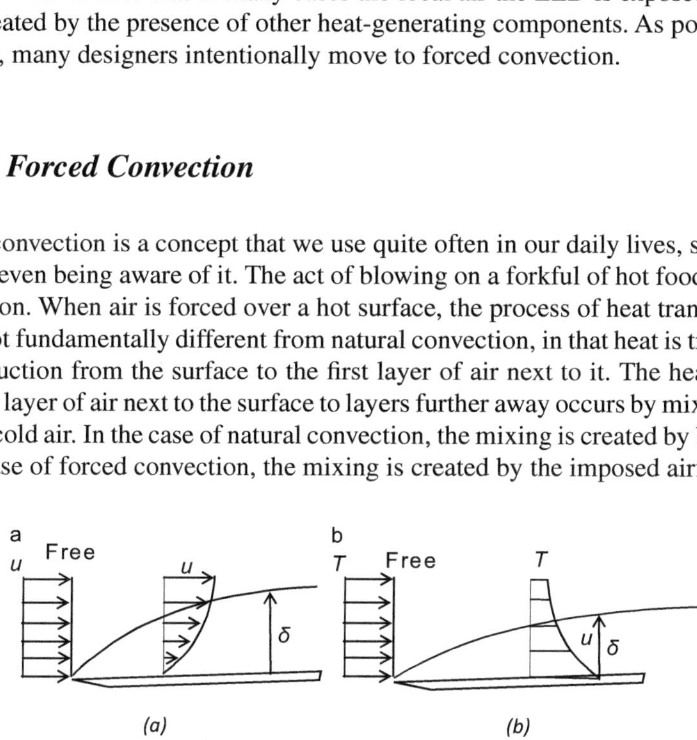

Fig. 7.6 Steady laminar forced convection from flat plate [7]

As seen in Fig. 7.6a, as flow goes over a heated surface [7], the no-slip condition on the surface creates a velocity boundary layer, with the flow velocity zero at the surface increasing to freestream velocity far away from the surface. A corresponding thermal boundary layer also exists (Fig. 7.6b), where the temperature of the fluid at the surface equals that of the surface and slowly drops off (in the case of the surface being hotter than the fluid) to freestream temperature. The boundary layer acts as a layer of insulation (or a thermal resistance to the heat transfer from the surface to the freestream). This thermal resistance is a function of the thickness of the boundary layer, and in most macroscale flows, the thermal resistance is inversely proportional to the thickness of the boundary layer. The goal of forced convection is to minimize this thermal resistance. One item of note is that the boundary layer shapes in the figure are illustrative and not the actual slopes. For example, the slope of the curve for the thermal boundary layer is normal to the plate at $x = 0$. Also, the relative thicknesses of the boundary layers depend on the Prandtl number of the fluid, with $Pr < 1$ having thicker thermal boundary layers than velocity boundary layers and $Pr > 1$ having thinner thermal boundary layers than velocity boundary layers. Air, which is the most common fluid used, has a Prandtl number of 0.7, and so the thermal boundary layer for air is thicker than the velocity boundary layer.

Forced convection cooling can be of several types: steady laminar convection, steady turbulent convection, and unsteady turbulent convection, in increasing order of effectiveness. Steady laminar convection is typically seen in electronics with system fans such as servers, where the fans are not close to the heat sink and provide a stream of mostly laminar flow through the heat sinks. A newer technology that also provides a steady laminar flow over fins is called Ionic Wind or Corona Drive Air Propulsion, where a flow is created by charged ions dragging along air as they move across an electric field. Steady turbulent convection is typically seen in active fan sinks where a fan is placed close to the fins of a heat sink, with the turbulent flow from its blades impinging on the fins. The most effective method of heat transfer is unsteady turbulent convection, where an unsteady, turbulent flow is created near the fins of the heat sink. Synthetic jets and piezofans fall in this category. The next section will describe the above technologies in more detail.

7.3 Key Air Cooling Technologies for LEDs

As seen in the previous section, there are two main types of convective heat transfer, free convection or natural convection and forced convection. The following subsections will illuminate using examples the various existing technologies in the market place that can be used for air cooling.

7.3.1 Heat Sink Technology for Natural Convection

Natural convection (also called free convection) cooling solutions have often been the go-to solution for LED markets until now because of low power density and moderate cooling requirements. Products did not require the higher heat transfer capability of forced convection cooling. However, now the market is demanding luminous flux output and quality of light in the same form factor as that of traditional incandescent or halogen-based lamps. This demand requires higher power LED arrays and higher-power density coupled with low junction temperature (T_J) requirements for long life and is driving the need for forced convection cooling solutions.

Natural convection heat sinks typically consist of a heat spreader plate on which a fin structure is mounted. Depending on the heat dissipation level (and therefore, the size of the heat sink), the spreader plate used could be a solid block of aluminum

Fig. 7.7 Example of heat sink technology for natural convection cooling of LEDs

or copper, or it could have embedded heat pipes that use a two-phase heat transfer to conduct heat from the spreader plate to the fins. It is also possible to use vapor chambers that act as the spreader plate for even larger heat sinks. The fin structure for natural convection heat sinks could be of many types, and some examples are shown below. The goal of designing fins for natural convection is typically to maximize the fin area for a given volume, without choking the buoyancy-driven flow between the fins. Advances in natural convection heat sink technology have primarily been in the optimal packing of fins to minimize choking of the buoyant flow. An excellent description of modeling natural convection heat sinks is provided in [8]. There are countless number of heat sink vendors that provide all shapes and sizes of natural convection heat sinks. Some examples are: Wakefield, Cool Innovations, Coolermaster, and Aavid Thermalloy. The Buyers Guide of Electronics Cooling Magazine has a more comprehensive list of suppliers [9]. Figure 7.7 shows a natural convection heat sink for an A19 lamp.

7.3.2 Synthetic Jets

Synthetic jets are formed by periodic suction and ejection of fluid out of an orifice bounding a cavity by the time periodic motion of a diaphragm that is built into one of the walls of the cavity (Fig. 7.8a, [10]). During the ejection phase (Fig. 7.8b, c, d), a coherent vortex, accompanied by a jet, is created and convected downstream from the jet exit. Once the vortex flow has propagated well downstream, ambient fluid from the vicinity of the orifice is entrained (Fig. 7.8e, f). The bulk of the high-speed air

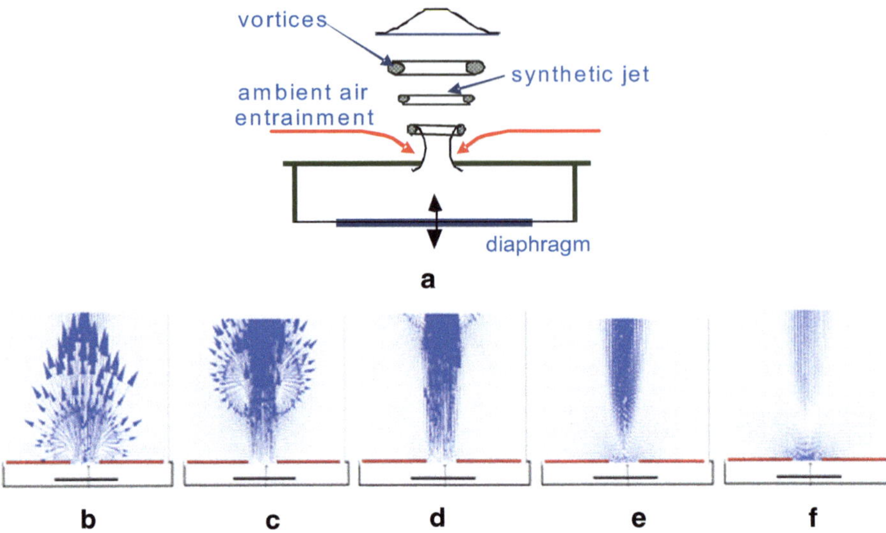

Fig. 7.8 Particle Image Velocimetry data of formation of a synthetic jet

has moved away from the orifice, avoiding re-entrainment, while quiescent air from around the orifice is sucked into the orifice. Thus, a synthetic jet is a "zero-mass-flux" jet comprises entirely of the ambient fluid and can be conveniently integrated with the surfaces that require cooling without the need for complex plumbing. The evolution of a two-dimensional synthetic jet has been studied in detail by Smith and Glezer [11]. The far-field characteristics (e.g., rate of lateral spreading and streamwise decay of centerline velocity) are similar to conventional turbulent jets.

7.3.2.1 Synthetic Jet Ejectors

The principle of jet ejectors or jet pumps [12] has been known for several decades now. A jet ejector consists of a primary high-momentum jet driving a secondary airflow through a channel as shown in Fig. 7.11. The low pressure created by a primary jet discharging into the channel results in entrainment of quiescent ambient flow, thus creating an increase in overall flow rate at the channel exit. This is also shown in Fig. 7.9 where the computed induced flow is plotted as a function of channel width in a channel flow driven by a high-momentum jet. The overall flow rate can be an order of magnitude higher than the jet flow itself, depending on the operating conditions. In conventional jet ejectors, the primary jet is created using a pressure source ducted into the entry of a channel. The use of synthetic jets as the primary jet is an attractive option since the only input to the primary jet is electrical, requiring no plumbing and pressure supplies. During the blowing stroke of the synthetic jet, the jet ejector phenomenon is similar to steady jet ejectors, wherein a primary high-momentum jet creates a low pressure in a channel resulting in the entrainment of fluid from the quiescent medium. During the suction stroke, the low pressure in the jet

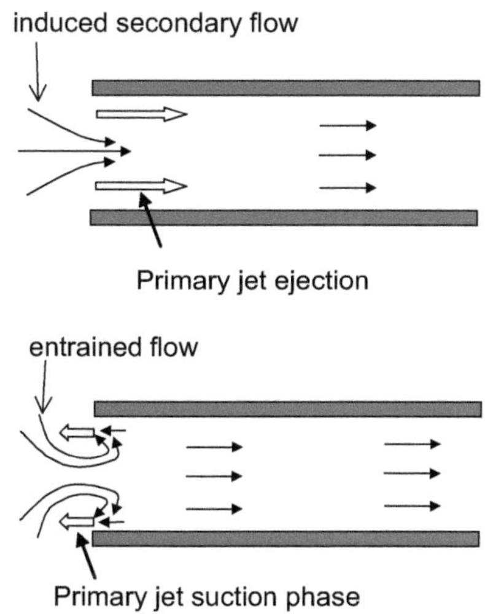

Fig. 7.9 Principle of operation of a jet ejector and calculations of ratio of induced secondary channel flow to jet flow in a jet ejector

cavity results in additional flow entrainment, which is forced out during subsequent blowing stroke.

7.3.2.2 Synthetic Jet Thermal Performance Data

Research performed over the last several years at Georgia Tech and Nuventix has shown significant improvements in air-side heat transfer compared to steady flows or fan-type flows. In channel cooling experiments, synthetic jet ejectors have been shown to have much higher heat transfer coefficients than steady flows of same Reynolds numbers based on the mean channel flow [13–15]. Figure 7.10 shows that

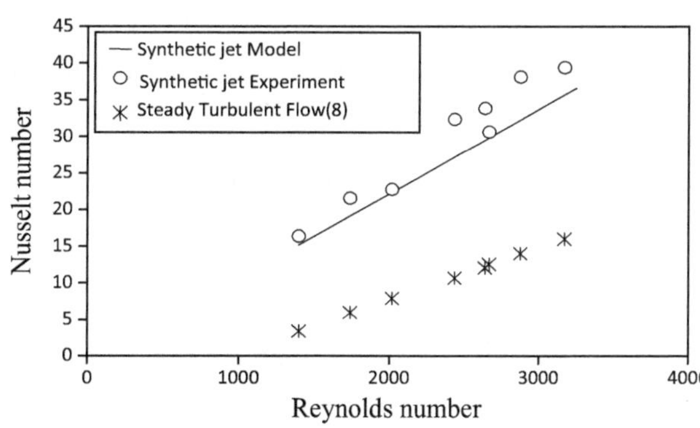

Fig. 7.10 Comparison of predicted and measured Nusselt numbers for the synthetic jet-driven channel flow with a correlation by Gnielinski [16]

synthetic jet-driven channel flows have higher Nusselt numbers than conventional turbulent channel flow [16].

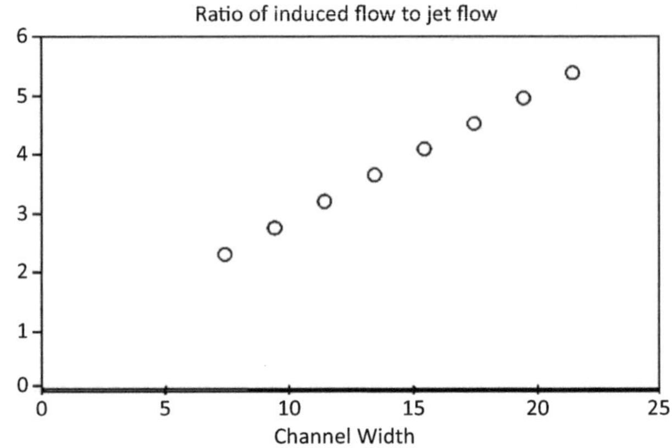

Fig. 7.11 Ratio of induced flow to the primary synthetic jet flow in a channel

Prediction of the overall thermal performance of the jet flow within the channels requires calculation of the induced flow rate due to jet ejector action, calculation of the heat transfer coefficient (from Nusselt number correlations shown above) followed by addition of the component resistances to obtain the overall thermal resistance as described in previous sections. The simplest way to calculate the flow created by a synthetic jet in a channel is by using jet ejector physics. A jet ejector is created when a primary high-momentum jet is injected into a channel. The primary jet, in turn, induces a secondary flow of ambient air into the channel. This secondary flow can be several times the primary flow, and is a function of the primary jet velocity and the channel parameters. The simplest jet ejector model predicts the flow rate by solving one-dimensional, linearized, mass and momentum conservation equations for the flow within the channel. These equations are fairly elementary to setup, but need to be solved iteratively, and are beyond the scope of this chapter. The pressure drop and heat transfer coefficients are modeled using modified correlations found in heat transfer literature for steady flows and applying the relative correlations described in the Nusselt number graph above. Figure 7.11 shows an example of how the induced flow rate of a synthetic jet ejector is a multiple of the primary synthetic jet flow.

Currently (2012), Nuventix is the only company manufacturing and selling synthetic jet cooling modules for LED lighting. Figure 7.12 shows several LED cooling solutions that have been developed using the principles described earlier, for different types of light form factors and lumen levels.

7.3.3 *Piezoelectric Fans for LED Cooling*

Another active cooling technology that has been of interest is the piezoelectric fan. This device achieves its cooling potential by continually waving a flexible fan back and forth near a heated surface. This helps to disrupt the thermal boundary layer as well as produce some amount of bulk flow to aid in ultimately transporting the

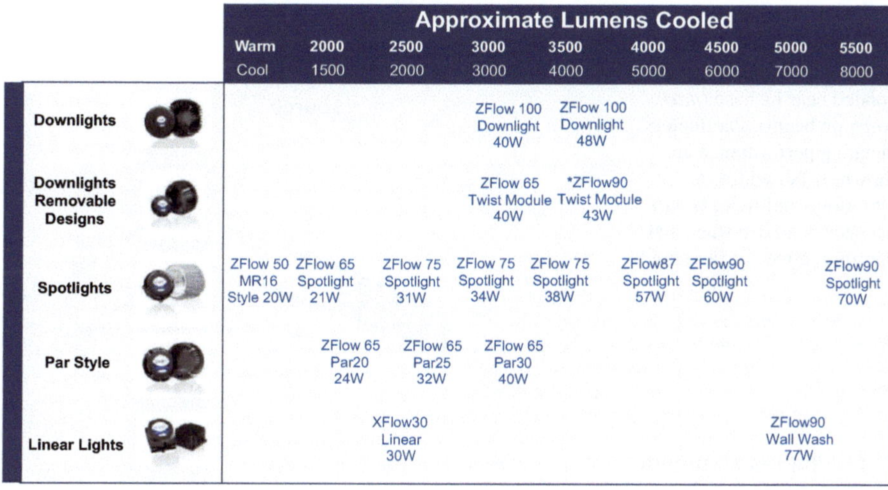

Fig. 7.12 Examples of LED lighting solutions using synthetic jets

heat to the ambient environment. These devices typically consist of thin flexible cantilever blades upon which is bonded a piezoelectric material. An alternating input signal causes the piezoelectric material to contract and expand, producing oscillations at the free end of the cantilever blade. As these fans operate at resonance, power requirements are minimal (typical published values range from 1 to 40 mW), and they can also be built to meet rigorous geometric constraints all the while remaining relatively noiseless (assuming the operational frequencies are below $\sim$ 100 Hz). Multiple investigators have shown the merit and viability of implementing such devices in various application-specific geometries [17–24]. However, in order to place this thermal management technique firmly in the toolbox of available heat dissipation methods, the performance characteristics must be quantified and expressed in terms of operating conditions and other geometric parameters. Correlations will be presented here which are based on extensive experimental testing and can be used to predict the thermal performance of a vibrating cantilever itself.

7.3.3.1 Thermal Performance Analysis

Gaining insight into how the thermal response (measured as the heat transfer coefficient, h) is influenced by a number of input variables is of primary importance in creating a usable correlation. The goal is to determine the functional relationship for the following statement:

$$h = f(\omega,\ A,\ D,\ L,\ G, \nu, k) \tag{7.8}$$

where

ω Oscillating frequency [rad/s]

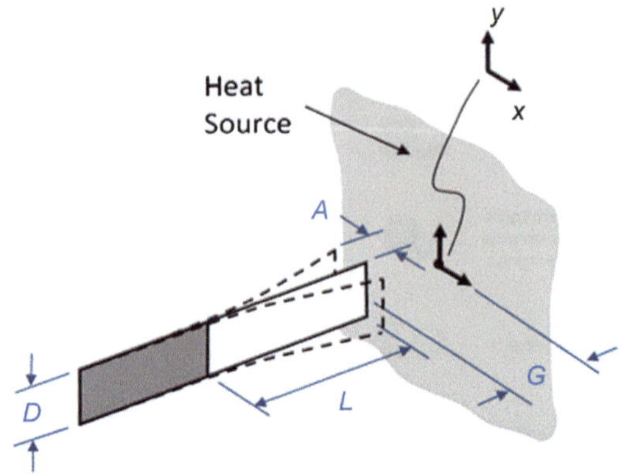

Fig. 7.13 Illustration of vibrating cantilever actuated from piezoelectric material bonded near its base (*dark patch* on beam). Parameters dictating performance are shown. *D* fan width, *A* vibration amplitude, *G* gap between heated surface and fan tip, *L* effective fan length

A Oscillation amplitude
D Width of the fan
L Effective length of the fan (portion of the blade not covered by the piezoelectric material)
G Gap (i.e., separation distance) between the heated target and the tip of the cantilever
ν Kinematic viscosity of the fluid (in this case, air)
k Thermal conductivity of the fluid (in this case, air)

Considering a total of eight variables (one response and seven inputs), one would expect there to be a total of five dimensionless parameters. These are the Nusselt number ($Nu = hA/k$), a dimensionless gap ($G' = G/A$), Keulegan–Carpenter number ($KC = 2\pi A/D$), frequency parameter ($\beta = \omega D^2/2\pi\nu$), and the aspect ratio of the beam ($\lambda = D/L$). Four of these five either take the form of well-accepted expressions (Nu, KC, β) or are fairly straightforward and unique to the problem under investigation (G'). The fifth parameter (λ) provides an estimate on how appropriate a two-dimensional analysis might be. A very large λ denotes a fan whose width is much greater than its length, thereby suggesting a two-dimensional analysis would be sufficient.

Six different cantilevers were tested at multiple vibration amplitudes and across a range of gaps. The only fluid tested was air and therefore k and ν were constants. The apparatus designed for these experiments is described in detail in [18], and consists of a constant heat flux surface and a piezoelectric fan mounted in a normal orientation as illustrated in Fig. 7.13. The surface thermal contours are captured using an infrared camera and subsequently used to extract the full field forced convection coefficient due to the fluid motion generated from the fan. From 160 different experiments, data can be collapsed into the following expression:

$$\overline{Nu} = C_1\left(\frac{KC^2 \cdot \beta}{2\pi}\right)^r \cdot \exp\left(C_2 \cdot G' \frac{\lambda}{2\pi \cdot \beta}\right) \tag{7.9}$$

From a least squares analysis of the experimental data, it is found that $C_1 = 0.140$, $r = 0.71$, and $C_2 = -2605$. The overbar included with Nu in Eq. 7.9 signifies that the result is an area-averaged heat transfer coefficient (as opposed to the stagnation coefficient). The area over which this data is averaged is the envelope of vibration of the fan blade as seen by the heated surface ($2A \times D$).

After substituting the expressions for KC, β, and λ into Eq. 7.9 and redefining two new dimensionless numbers, the result can be expressed as:

$$\overline{Nu} = C_1 (Re_A)^r \exp\left(C_2 \frac{G'}{B}\right) \tag{7.10}$$

where the oscillating Reynolds number (Re_A) and the Beam number (B) are defined as

$$Re_A = \frac{\omega A^2}{\nu} = \frac{KC^2 \cdot \beta}{2\pi} \text{ and } B = \frac{\omega DL}{\nu} = \frac{2\pi\beta}{\lambda} \tag{7.11}$$

Equation 7.11 can be considered valid for $350 < Re_A < 2{,}500$ and $G'/B < 3 \times 10^{-4}$. It is interesting to note that since the beams will typically operate at resonance, ω is more or less fixed, and therefore the parameter B will remain constant for a given beam. Hence, the terminology Beam number is adopted here, and suggests that holding all other variables constant, whether one doubles the width or length, the same response would be observed as either modification will double the Beam number (assuming the frequency remains unchanged). Experimental performance curves are shown for a particular fan ($B = 11{,}310$) in Fig. 7.14a where each data set represents a different vibration amplitude, and the thermal performance is plotted against the gap, which increases from left to right. Taking similar curves for all six fans and collapsing into a single curve reveals the result shown in Fig. 7.14b. Also included in the plot is the expression from Eq. 7.11. The exponential decay of the data signifies that the heat transfer performance suffers as the gap (distance from the cantilever tip to the heated target) is increased, a behavior that agrees well with one's intuition. The dimensionless numbers offered in Eq. 7.11 can be used to predict the thermal performance of a vibrating cantilever, regardless of properties of the beam under investigation.

It is also important to note from Fig. 7.14a the existence of an optimal gap between the fan tip and heated surface where performance is maximized. In other words, the fan should not be positioned too close or too far from the heated surface for the best possible results. This optimal gap $G'opt$ is found to be between 0.1×10^{-4} and 0.5×10^{-4} times the Beam number (B). If the analysis is restricted to characterizing the data from this optimum gap only (i.e., picking off the maximum points from the curves shown in Fig. 7.14a as well as additional curves from other fans), the result follows a power law relationship as a function of Re_A only according to:

$$\overline{Nu}_{\max} = C_3 (Re_A)^q \tag{7.12}$$

where the constants C3 and q are found from a least squares analysis to be 0.128 and 0.71, respectively. The experimental data is shown in Fig. 7.15 as well as the curve

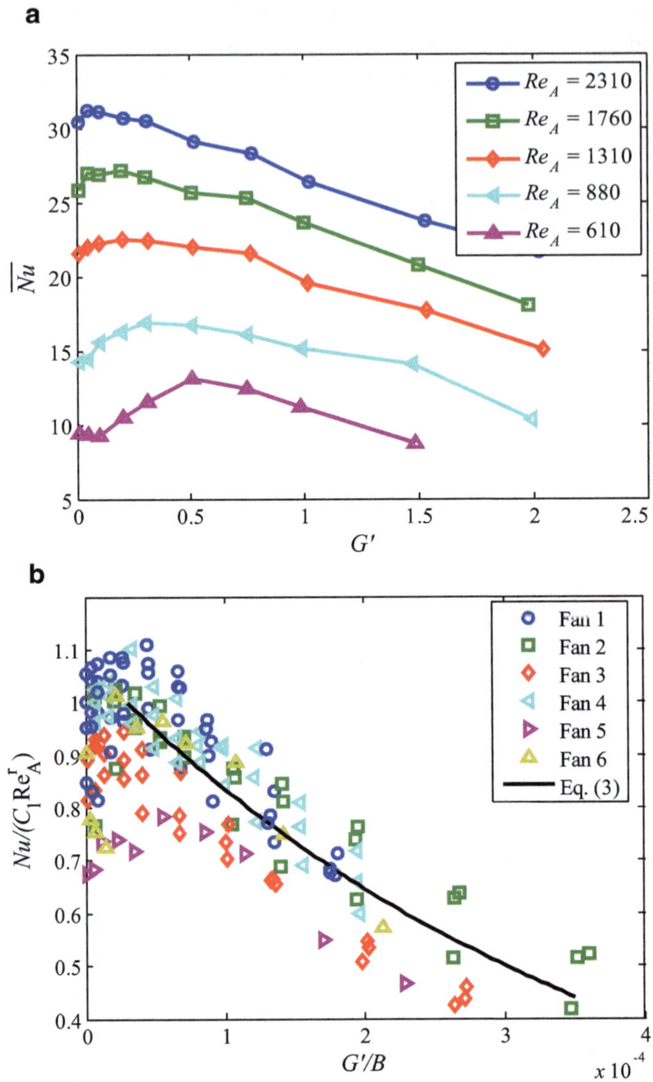

Fig. 7.14 Curves for thermal performance for **a** $B = 11{,}310$ and **b** all six fans tested [19]. Note that the x-axis in (**b**) has been further normalized for all fans by their respective B

from Eq. 7.12. Regardless of the fan under investigation, the results can be described with relatively simple expressions. This is true for data shown in Fig. 7.14b as well as Fig. 7.15.

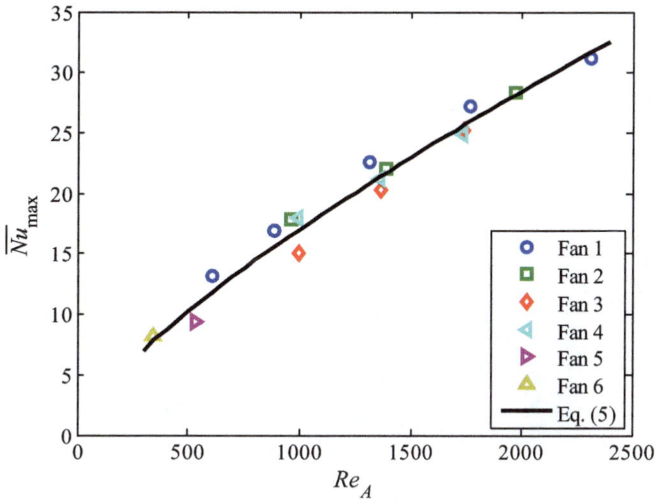

Fig. 7.15 Maximum Nusselt number as a function of oscillating Reynolds number for data from [20]

After substituting expressions for $\overline{Nu}_{max}$ and Re_A, Eq. 7.12 can be presented in terms of the dimensional quantities of interest according to

$$\overline{h}_{max} = 0.128k\left(\frac{\omega}{\nu}\right)^{0.71} A^{0.56} \tag{7.13}$$

It is important to note that this analysis of peak performance suggests that the heat transfer rate is dependent only on the frequency and amplitude of oscillation. In other words, factors such as length and width will dictate the actual position where one should place the fan for optimal performance, but once in that position, only the frequency and amplitude are important. A contour map of Eq. 7.13 is shown in Fig. 7.16 using air as the working fluid where each contour step depicts an order of magnitude change in the convection coefficient measured in W/m^{2}K. The data from Fig. 7.15 are shown for reference, and reveal that most of the results are clustered near 100 W/m^{2}K. Even conservative estimates would suggest an order of magnitude increase over natural convection conditions at the expense of tens of milliwatts.

The thermal performance of a piezoelectric fan can be predicted based on its geometry and operating parameters. If the fan is oriented normal to the heated surface and fixed at the optimal spacing, then the magnitude of the convection coefficient is only a function of the vibration amplitude and frequency. Other factors to consider when designing for piezoelectric fan cooling include the presence of enclosure walls and their pressure and flow rate capabilities (fan curve specifications). Currently (2012), the authors are unaware of any production cooling solutions available for this technology.

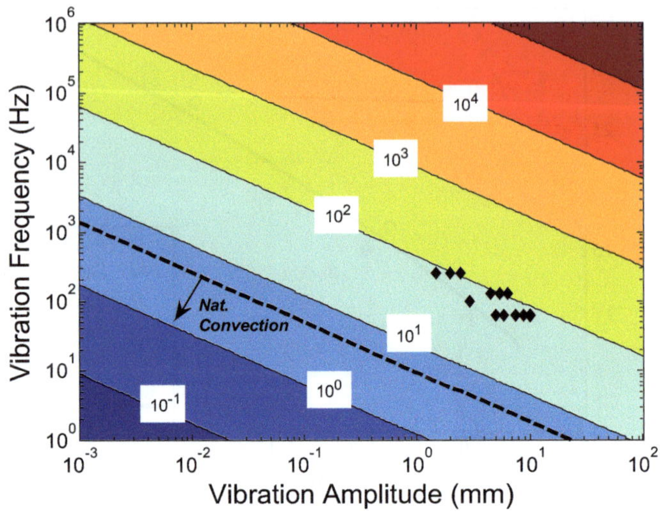

Fig. 7.16 Thermal performance of vibrating cantilevers based on oscillation frequency and amplitude. The contour values shown are the convection coefficient given in W/m^2K. Generated from Eq. 7.13

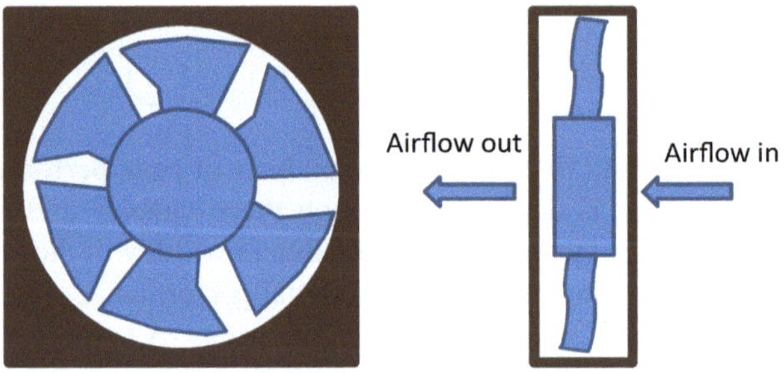

Fig. 7.17 The flow in an axial fan enters and exits along the axis of the rotor

7.3.4 *Fans*

Rotary fan technology has a ubiquitous presence in thermal management. Every system, which needs cooling through it usually has one or more fans providing the airflow. Most software codes in the industry have fan curves that can be used to design fans into any system. There are different types of fans (axial fans, centrifugal blowers, cross flow fans, etc.) but the main type of fans used in LED cooling is an axial fan. Figure 7.17 shows the structure of such a fan; the flow enters and exits along the axis of the rotor.

The performance of an axial fan, i.e., the flow created by it, is dependent on the pressure drop in the system. The operating point of an axial fan can be established by comparing the fan pressure curve and the system pressure curve. This is called a p–q

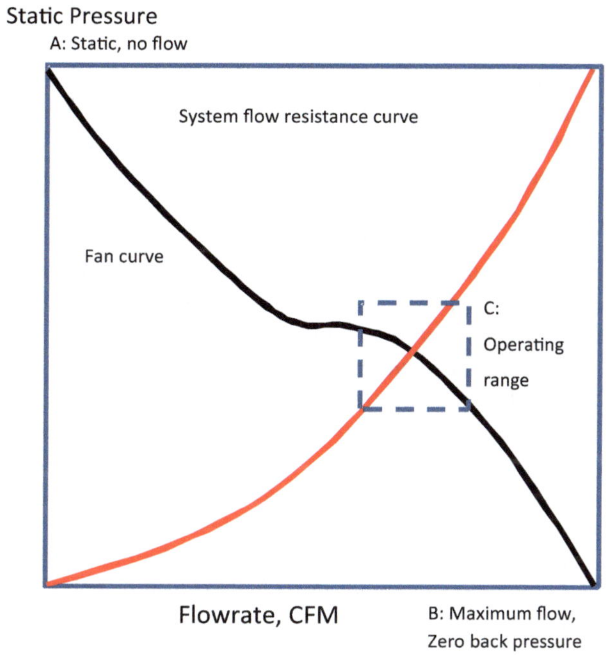

Fig. 7.18 The flow in an axial fan enters and exits along the axis of the rotor

Fig. 7.19 Example of fans available for LED cooling in the market

curve and can be obtained experimentally for fans. Figure 7.18 shows an example of a p–q curve. In the figure, region C is where the fan operates, due to the back pressure imposed by system flow resistance.

There are many fan cooling solutions available in the market for LED cooling from vendors such as EBM-Papst, Delta, Sunon, Comair Rotron, etc. An example is shown in Fig. 7.19. A very clear description of cooling fans and their implementation is shown in [25]

7.3.5 Ionic Wind or Corona-Driven Air Propulsion

Another air moving technology that has seen considerable research in the recent years is called Corona-Driven Air Propulsion. The principle of ionic air propulsion with corona-generated charged particles has been in literature for almost as long as electricity. Gas molecules surrounding the corona electrode are ionized by the high-intensity electric field at the tip, forming an ion stream between the corona and collector electrodes, thus creating airflow. The mechanism of DC-positive corona-induced ionic wind with pin-rod geometry is illustrated in Fig. 7.20. Application of voltage between a high tip-curvature corona electrode and a low tip-curvature collecting electrode creates a high electric field gradient at the corona electrode, ionizing its surrounding air molecules. The ionized air molecules are then propelled by the electric field, transferring momentum to neutral air molecules via collisions, thus resulting in bulk air movement toward the collecting electrode. The operating voltage range of an electrohydrodynamic (EHD) device lies between the corona discharge onset and the complete air gap breakdown voltages. Modeling of convective heat transfer of an EHD device is described in [27]

Some of the potential advantages of a corona-driven air pump are that it is a solid-state device with no moving parts and can be potentially microfabricated into silicon and can be very low-noise. However, there are also several barriers the technology must overcome to achieve mainstream adoption: these include, low lifetime due to electrode corrosion, difficulty to create flow against pressure drops, ozone creation, and dust accumulation. Details on the physical principles and development of Corona-Driven Propulsion Coolers are provided in [26–31]. Although the technology has been demonstrated as a proof of concept [26], the product is not yet commercially available. Ventiva has developed demos of ionic wind cooling for light bulbs. Another company, Tessera, has demonstrated coolers for laptops using this technology, but none for the LED market yet.

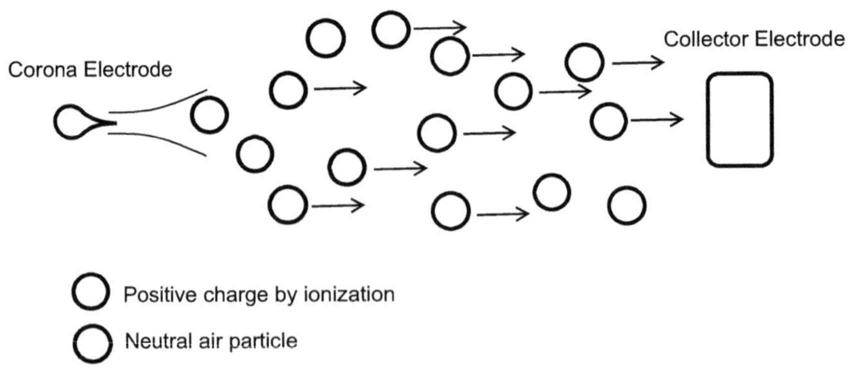

Fig. 7.20 Ion stream of a DC electrostatic air pump, where the corona and collector electrodes are powered by a high-voltage DC source [26]

7.4 Comparisons of Different Technologies Across System Metrics

The design of any thermal management system for commercial use cannot focus only on thermal performance. Most commercial products have a variety of requirements, with varying degrees of importance. Some of the system metrics or trade-offs in LED cooling are listed below:

- Thermal resistance: total heat dissipation, flux density
- Cooling volume and weight: smaller is usually better
- Reliability: mean time to failure (MTTF), harsh environments
- Acoustics: loudness, psychoacoustics
- Power consumption: coefficient of performance (COP), controllability
- Form factor flexibility
- Manufacturability
- Cost: mainstream adoption will happen only at low cost
- Environmental effects

The next section compares some of the existing technologies with respect to these metrics.

7.4.1 Cooling System Thermal Performance and Size

As discussed before, thermal designers have preferred to use natural convection wherever possible, since natural convection heat sinks have infinite life, consume zero power, and make no noise. However, as the lumen outputs increase, the constraint of size and results in the switch to an active cooling solution. The next section describes experiments that were conducted to compare the thermal performance, size, mass, and lumen output of typical natural convection versus synthetic jet-based forced convection cooling solutions. Note that the same reasoning will also apply when comparing natural convection against any other forced convection system.

Various luminaire form factors were tested such as a PAR38, PAR30, and Down Light, which are typical of traditional lighting and LED products currently on the market or under development [32]. Figure 7.21 shows pictures and dimensions of the devices that were tested. Items of interest for comparison were the thermal resistance ($R_{th\,s\text{-}a}$), volume, mass, and increased lumen output.

7.4.2 Mass and Volume

A Philips DLM module was used for comparing the mass and volume of a synthetic jet-cooled heat sink with that of a natural convection- cooled heat sink available in the market. The natural convection heat sink was an extruded aluminum heat sink with a straight parallel fin design. The synthetic jet-cooled heat sink was die-cast

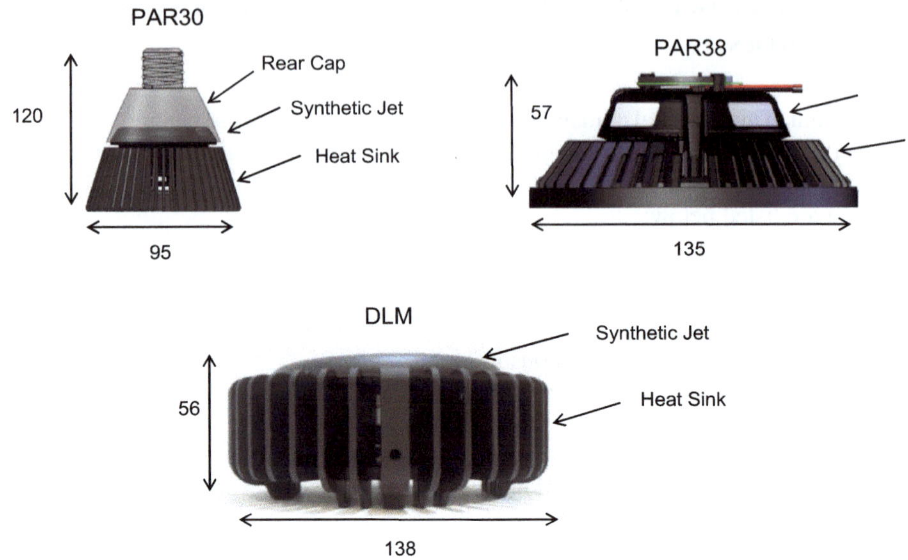

Fig. 7.21 Devices used for testing with dimension shown in millimeters

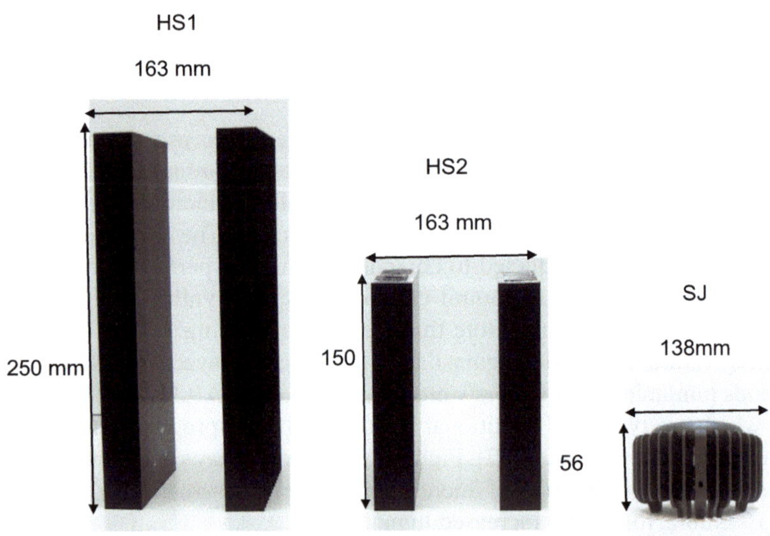

Fig. 7.22 Natural convection and synthetic jet thermal solutions for LED DLM

aluminum with radial fins and an embedded synthetic jet device. Figure 7.22 shows the natural convection heat sinks along with the synthetic jet device.

Table 7.1 compares the mass and volume of the natural convection solutions with the synthetic jet-based solution of equal $R_{th\,s\text{-}a}$ (0.63 K/W and 0.75 K/W). For

Table 7.1 Comparison of mass and volume of natural convection and synthetic jet DLM Solutions

	HS1	HS2	SJ
$R_{th\,s\text{-}a}$ (K/W)	0.63	0.75	0.63
Mass, g	2,392	1,435	606
Volume, cm^3	5,012	3,007	1,714

HS heat sink, *SJ* synthetic jet

example, the synthetic jet solution compared with HS1 natural convection solution was 76 % lower in mass and 66 % lower in volume. The larger mass and volume of the natural convection heat sink would limit the applications for which the luminaire could be installed. The compact synthetic jet solution will have fewer issues for mechanical fit. The $R_{th\,s\text{-}a}$ for the synthetic jet solution can be modified by adjusting the airflow rate. Therefore, the form factor will remain the same but the device can achieve both $R_{th\,s\text{-}a}$ values.

Table 7.2 shows the percentage improvement of a PAR30 synthetic jet-cooled heat sink in open air.

The improvement for the synthetic jet-cooled heat sink ranges from 42 to 52 % over natural convection for T_s and 55–73 % for $R_{th\,s\text{-}a}$.

In addition to improvements in both open-air and recessed fixtures, synthetic jet solutions are not orientation-dependent as are natural convection solutions. Figure 7.23 shows that the natural convection solution can vary as much as 15 % due to not being in the optimal orientation for natural convection. Orientation dependence is an important design constraint, especially in retrofit bulb form factors

Table 7.2 PAR30 open-air testing results

Open Air Testing

Q, W	T_s at 25 °C T_a SynJet	T_s at 25 °C T_a Natural Convection	% Improvement in T_s over natural convection
10	37.6	72.0	47.8 %
15	43.9	89.4	50.9 %
20	50.2	105.5	52.4 %

Q, W	ψ_{sa} at 25 °C T_a SynJet	θ_{sa} at 25 °C T_a Natural Convection	% Improvement in ψ_{sa} over natural convection
10	1.3	4.7	73.2 %
15	1.3	4.3	70.6 %
20	1.3	4.0	68.7 %

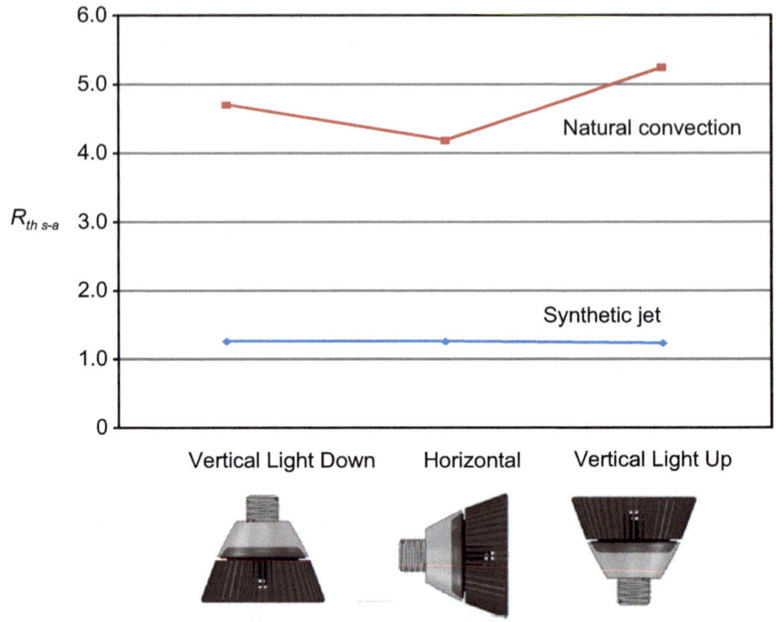

Fig. 7.23 Dependence of the thermal resistance on the orientation

Table 7.3 Total luminous flux values obtained for synthetic jet and natural convection-cooled PAR30 LED lamps

Parameter	Synthetic jet	Natural convection
Sink to ambient thermal resistance, $R_{th\ s\text{-}a}$, K/W	1.3	4.7
Total electrical input power, W	26	15
Thermal power (Q), W	21	12
Efficacy, lm/W	58	58
Luminous flux, lm	1,508	870

where the orientation is not always fixed in the final installation and, in fact, may need to be adjustable to direct the lumens on the proper areas.

7.4.3 Emitted Luminous Flux

Table 7.3 compares the emitted total luminous flux of the LED lamps equipped with synthetic jet thermal solution against the natural convection solution for the PAR30 replacement. As it can be seen, because $R_{th\ s\text{-}a}$ for the synthetic jet is a factor of 3.6 less than that of the natural convection solution, more thermal power can be dissipated even while maintaining a 27 °C lower T_S, which allows 73 % more luminous flux emitted. In fact, if the T_s for the synthetic jet solution were allowed to be 79 °C, Q could have been increased to 42 W, resulting in an increase in emitted luminous flux of 3,000 lm or 244 % more than the natural convection solution.

7.5 Reliability and Acoustics

Once the heat dissipation levels have warranted the use of active cooling solutions, the next critical metrics become reliability and acoustics, particularly for LEDs. This section will compare some of the active cooling solutions with respect to reliability and acoustics. Although several driving mechanisms can be used to generate synthetic jets, the most common are electromagnetic actuators. Electromagnetic actuators have an inherently reliable design advantage over traditional air movers such as centrifugal blowers and axial fans due to the fact that they have no rotating parts, bearings, or parts in frictional contact [33]. Additionally, the actuators are driven at low frequency, which has additional psychoacoustic benefits compared to air movers that operate at higher frequencies. Acoustical data has been collected for fans and synthetic jet devices and the differences have been compared below. The Sound Pressure Level (Lp) was measured in an anechoic chamber with the microphone placed at 1 m from the source.

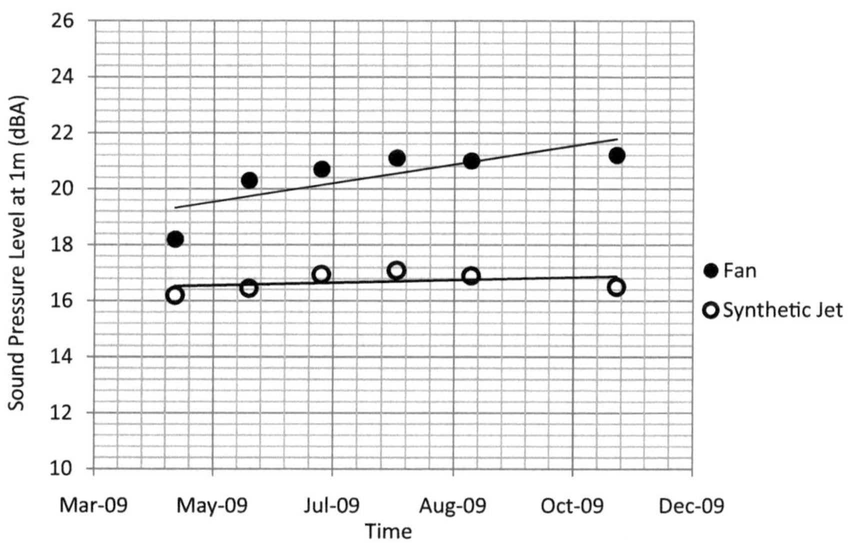

Fig. 7.24 Lp values for devices in ALT for 4,000 h

Figure 7.24 shows a graph that compares fan and synthetic jet air movers that have undergone Accelerated Life Testing (ALT) to illustrate the increase in acoustics over time. The devices were operated continuously in a 75 °C chamber for 4,000 h. Samples were removed periodically for acoustic measurements. As you can see from the linear fit of the plotted data, the fan Lp is increasing at a higher rate and has increased by greater than 3 dBA over the course of testing, while the synthetic jet devices remained relatively flat and stayed within 1 dBA of the initial values

Table 7.4 shows how orientation can affect the Lp. The images below the table show the orientations as viewed from the microphone; a fan image is shown, but the

Table 7.4 Lp as a function of orientation

Orientation	Synthetic jet			Fan A			Fan B		
	SJ 1	SJ 2	SJ 3	Fan A-1	Fan A-2	Fan A-3	Fan B-1	Fan B-2	Fan B-3
Standard	16.7	17.0	16.6	21.6	24.9	20.9	17.5	18.0	17.5
0 degree	16.2	16.8	16.9	21.7	22.1	21.8	17.2	18.2	17.7
45 degree	16.2	16.9	17.0	21.9	34.4	21.0	16.9	17.8	18.3
90 degree	16.2	16.6	16.5	23.0	27.9	21.2	17.7	17.6	17.3
135 degree	16.5	17.1	16.7	21.9	27.0	21.3	17.6	18.6	17.9
180 degree	16.2	16.9	17.0	22.1	29.8	23.3	17.5	18.2	18.3
Min	16.2	16.6	16.5	21.6	22.1	20.9	16.9	17.6	17.3
Max	16.7	17.1	17.0	23.0	34.4	23.3	17.7	18.6	18.3
%Maximum over	3 %	3 %	3 %	6 %	56 %	11 %	5 %	6 %	6 %

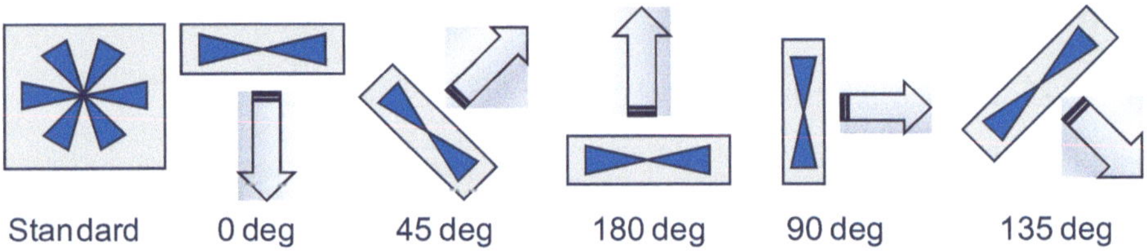

synthetic jet was also tested in the same orientations. You can see that the synthetic jet maintained a low percentage difference between minimum and maximum values and stayed within 0.5 dBA between orientations. Fan B also showed relatively stable behavior, although it exhibited a higher percentage difference between minimum and maximum and stayed within 1 dBA between orientations. Fan A showed erratic behavior and had as much as a 56 % change in Lp between orientations.

L10 life (not to be confused with L70 % or lumen maintenance) is a commonly used reliability metric for expressing the lifetime of traditional air movers. L10 is the time at which 90 % of the population survives within the specification at the maximum operating temperature for the device. In other words, it is the time at which 10 % of the population will be outside of the specification at the maximum operating temperature. Figure 7.25 shows the L10 of a synthetic jet device compared to traditional air movers with various bearing technologies over a wide temperature range. The synthetic jet devices show significantly higher L10 values over the various bearing options with traditional air movers available on the market [34].

It has been shown that the prominent mechanical failure mode of fans and blowers is bearing failure due to the loss or deterioration of the bearing lubricant and it is this failure mechanism most used to define the useful life. The lubricant loss and deterioration increases over time and with temperature and while the mechanical wear of the bearing will eventually lead to a catastrophic failure (loss of cooling capability), it usually is first heard by end-users because of the significant increase in acoustic signature. In addition to deterioration, lubricants can settle due to nonuse

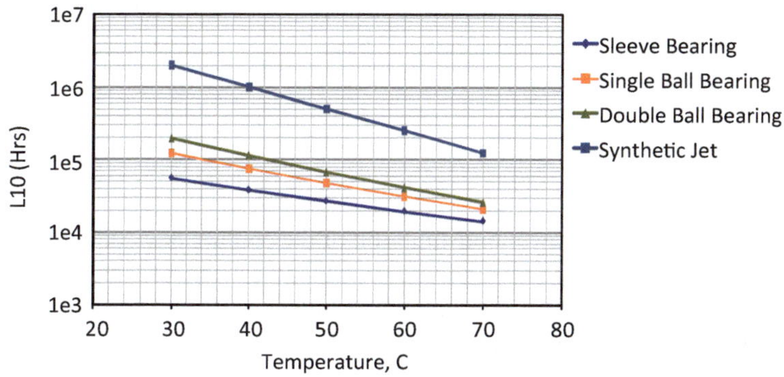

Fig. 7.25 L10 for synthetic jet devices and fans with various bearing technologies [30]

resulting in acoustic artifacts upon initial use. Bearings can also wear unevenly depending on the orientation of the air mover, so acoustics can be dependent on orientation.

Since there have not been any piezoelectric-based coolers or ionic wind-based coolers in production, experimental reliability and acoustic data do not exist for them. However, from a qualitative standpoint, some comments can be made regarding their reliability and acoustics. From an acoustics standpoint, piezoelectric and ionic wind coolers can both be potentially made very quiet. Piezoelectric fans are typically run at 100 Hz or lower, resulting in very little acoustic emission in the audible frequency range for humans. Ionic wind coolers are solid-state in nature and do not produce high acoustic emissions. However, aside from some proof of concept devices, high-performance cooling solutions have not been forthcoming from these basic concepts. Reliability is a much more critical issue with both these technologies. Piezoelectric fans are constructed by bonding thin layers of piezoceramic to metal or plastic substrates and are prone to cracking, delamination, and resulting breakdown of the ceramic, particularly at the high displacements required for airflow. Ionic wind coolers also have reliability concerns due to corrosion of electrodes as well as dust accumulation due to charged particles.

7.6 Power Consumption and Drive Electronics

One of the primary advantages of LED lights is their energy efficiency. A typical incandescent bulb provides an efficacy of about 15–17 lm/W, whereas typical efficacies of LEDs are about five times better, producing at least 75–85 lm/W. It is very important that any cooling solution for LED lighting also be energy efficient, i.e., consume a small fraction of the LED power to dissipate the heat produced by the LED. In addition, the electronics required to drive the cooling system should not be too big or too expensive. A common metric used for assessing the energy efficiency of air moving systems is *Coefficient of Performance* or COP, which is the

Table 7.5 Metrics for drive voltage and power for different active cooling technologies

Active cooling technology	Synthetic jets	Fans	Piezofans	Ionic wind
Power consumption (W)	0.5–1	0.5–1	0.1	1–5
COP	30–70	30–70	150	10–20
Drive voltage (V)	5 or 12	5 or 12	20–40	> 1000

ratio of heat dissipated to power consumption of the cooling device. Table 7.5 shows typical COPs for some of the technologies discussed earlier. Also shown is the drive voltages required for these technologies. The higher voltages required for some of the technologies may delay adoption into mainstream.

7.7 Cost and Manufacturability

Synthetic jets and fans are both mainstream technologies with established production facilities and volume manufacturing in place. They are comparable in cost. There is not much information about the production availability of either piezofans or ionic wind technology, so not much can be said about costs or manufacturability.

7.8 Summary

Since a large fraction of the energy input into LEDs is converted to heat, thermal management of LED lights has become a critical component in the design of light fixtures. This chapter discusses air cooling of LED lights. Some of the critical aspects of LED cooling are touched upon, such as thermal paths and cooling methodologies encountered in these thermal paths. The main focus of the chapter is air-side cooling, with a discussion on both natural convection and forced convection fundamentals. Some of the main relations used in convection cooling are described. Following that, the five commonly known methods of air cooling are discussed, these being natural convection heat sinks, synthetic jets, fans, piezofans, and ionic wind. The next sections compare these technologies using system metrics relevant to LED lighting.

In summary, as the LED lumen outputs increase, the need for heat dissipation causes a migration from natural convection to forced convection, driven by the size and weight constraints and in some cases even the cost of the material. Among the active cooling solutions, while fans have existed as a dependable active cooling technology, synthetic jets have established themselves as a front runner for cooling LED lights, due to their high reliability and low acoustic performance, while remaining on par with fans on power consumption and cost. Two newer cooling technologies, namely piezofans and ionic wind have also shown some promise, but are still not in volume production.

Acknowledgments Section on Piezoelectric fans contributed by Mark Kimber, University of Pittsburgh.

References

1. Bullough J, Gu Y, Narendran N, Taylor J (2007) LED life for general lighting: definition of life. Lighting Research Center ASSIST publication. Vol 1, Issue 1, pp 4–5. http://www.lrc.rpi.edu/programs/solidstate/assist/pdf/ASSIST-LEDLife-revised2007.pdf. Accessed 5 Mar 2012
2. United States Department of Energy (2009) Program requirements for SSL luminaires. Version 1.1, pp 3–4. http://www.energystar.gov/ia/partners/product_specs/program_reqs/SSL_prog_req_V1.1.pdf. Accessed 5 Mar 2012
3. Seoul Semiconductor (2008) Life time graph of Z-Power LED. pp 7. http://www.symmetron.ru/suppliers/seoul/files/pdf/seoul/Z-power/Reliability.pdf. Accessed 5 Mar 2012
4. Lasance C (2008) Heat spreading: not a trivial problem. Electronics Cooling 14:24–30
5. Lasance C (2010) How to estimate heat spreading effects in practice. J Electron Packag 132:031004
6. Moffat R (2008) Modeling air-cooled heat sinks as heat exchangers. Electronics cooling. Vol 14. February Issue http://www.electronics-cooling.com/2008/02/modeling-air-cooled-heat-sinks-as-heat-exchangers/. Accessed 5 Mar 2012
7. Incropera FP, Dewitt DP (1990) Fundamentals of heat and mass transfer. 3rd Edition, Wiley, pp 496–502
8. Teertstra P, Yovanovich MM, Culham JR (2000) Natural convection modeling of heat sinks using web based tools. Electronics Cooling. Vol 6. Sept Issue http://www.electronics-cooling.com/2000/09/natural-convection-modeling-of-heat-sinks-using-web-based-tools/. Accessed 5 Mar 2012
9. Electronics Cooling Buyers Guide (2012) http://www.electronics-cooling.com/buyers-guide-d-h/#Anchor29. Accessed 5 Mar 2012
10. Glezer A, Amitay M (2002) Synthetic jets. Ann Rev Fluid Mech 34:503
11. Smith BL, Glezer A (1998) The formation and evolution of synthetic jets. Phys Fluids 10(9):2281–2297
12. Gosline JE, O'Brien MP (1934) The water jet pump. U of California Publ Engrg 3(3):167–190
13. Mahalingam R, Rumigny N, Glezer A (2004) Thermal management using synthetic jet ejectors. IEEE Trans Compon Packag Technol 27(3):439–444
14. Mahalingam R (2007) Modeling of synthetic jet ejectors for electronics cooling. 23rd Annual IEEE SEMI-THERM Symposium, pp 196–199
15. Mahalingam R, Glezer A (2005) Design and thermal characteristics of a synthetic jet ejector heat sink. J Electron Packag 127(1):172–177
16. Gnielinski V (1976) New equations for heat transfer in turbulent pipe and channel flow. Int Chem Eng 16:359–368
17. Yoo JH, Hong JI, Cao W (2000) Piezoelectric ceramic bimorph coupled to thin metal plate as cooling fan for electronic devices. Sens Actuators 79:8–12
18. Kimber M, Garimella SV, Raman A (2007) Local heat transfer coefficients induced by piezoelectrically actuated vibrating cantilevers. ASME J Heat Transfer 129(9):1168–1176
19. Açıkalın T, Wait SM, Garimella SV, Raman A (2004) Experimental investigation of the thermal performance of piezoelectric fans. Heat Transf Eng 25(1):4–14
20. Açıkalın T, Garimella SV, Raman A, Petroski J (2007) Characterization and optimization of the thermal performance of miniature piezoelectric fans. Int J Heat Fluid Flow 28(4):806–820
21. Kimber M, Garimella SV (2009) Measurement and prediction of the cooling characteristics of a generalized vibrating piezoelectric fan. Int J Heat Mass Transf 52:4470–4478
22. Kimber M, Suzuki K, Kitsunai N, Seki K, Garimella SV (2009) Flow rate and pressure characteristics of piezoelectric fans. IEEE Compon Packag Technol 32(4):766–775

23. Sauciuc I (2007) Piezo actuators for electronics cooling. Electronics Cooling, February Issue, Vol 13. http://www.electronics-cooling.com/2007/02/piezo-actuators-for-electronics-cooling/. Accessed 5 Mar 2012
24. Sarpkaya T (1977) In-line and transverse forces on cylinders in oscillatory flow at high Reynold's numbers. J Ship Res 21:200–216
25. Cooling Fans Overview (2012) http://www.orientalmotor.com/technology/articles/cooling-fans-overview.html. Accessed 5 Mar 2012
26. Jewell-Larsen NE, Ran H, Zhang Y, Schwiebert M, Honer KA, Mamishev AV (2009) Electrohydrodynamically cooled laptop. 25th IEEE SEMI-THERM Symposium, pp 261–266
27. Jewell-Larsen NE, Karpov SV, Krichtafovitch IA, Jayanty V, Hsu CP, Mamishev AV (2008) Modeling of corona-induced electrohydrodynamic flow with COMSOL multiphysics. Proc. ESA Annual Meeting on Electrostatics, Paper E1
28. Jewell-Larsen NE, Tran E, Krichtafovitch IA, Mamishev AV (2006) Design and optimization of electrostatic air pumps. IEEE Trans Dielectr Electr Insul 13(1):191–2003
29. Schlitz DJ, Garimella SV, Fisher TS (2004) Microscale ion-driven air flow over a flat plate. Proceedings of ASME heat transfer/fluids engineering summer conference, pp 463–468
30. Jewell-Larsen NE, Hsu CP, Krichtafovitch IA, Montgomery SW, Dibene JT, Mamishev AV (2008) CFD analysis of electrostatic fluid accelerators for forced convection cooling. IEEE Trans Dielectr Electr Insul 15(6):1745–1753
31. Go D, Tirumala R (2012) Ionic winds—a new frontier for air cooling. To appear in Electronics Cooling, Spring Electronics Cooling, March 2012. http://www.electronics-cooling.com/2012/03/ionic-winds-a-new-frontier-for-air-cooling/. Accessed May 29, 2013
32. Noska B, Jin H, Mahalingam R (2010) Enabling new LED designs through advanced cooling technology. 26rd Annual IEEE SEMI-THERM Symposium, pp 305–310
33. Tian Xijin (2006) Cooling fan reliability: failure criteria, accelerated life testing, modeling and qualification. Annual reliability and maintainability symposium, pp 380–384
34. Schwickert M (2012) Synthetic jet thermal management technology increases led lighting system reliability. IEEE Reliability Society Annual Activities Report. http://rs.ieee.org/images/files/Publications/2009/2009-12.pdf. Accessed Mar 5 2012

Chapter 8
Advances in Thermal Interface Materials for Power LED Applications

David L. Saums

Abstract Thermal interface materials (TIMs) serve a critical function in light-emitting diode (LED) assemblies and electronic systems, although the cost of some is very modest. The basic function is to provide an effective thermal path between two dissimilar surfaces, often the base of the LED array and a heat sink or a metal heat-dissipating surface. The TIM material in the simplest definition is intended to reduce air gaps between mating metal surfaces, given the very poor thermal conductivity of air. Proper selection and application of a TIM will compensate for relative surface roughness and surface imperfections on one or both mating surfaces that create loss of surface-to-surface contact for the transmission of heat.

Addition of a thermal interface material adds a modest cost and process complexity in a product design. Increasing machining requirements for the surface of a heat sink or heat pipe assembly, by contrast, will generally add more cost than the cost of the thermal interface material (TIM) and the application process. Substantial improvements in heat transfer between two mating surfaces in the assembly will improve overall thermal performance efficiency of the completed product.

Reducing thermal resistance in light-emitting diode (LED) assemblies requires evaluation and selection of TIMs, which are available in a very wide variety of forms and types. As improvements have been made in other elements in the overall LED assembly from a thermal perspective, it is clear that an increasingly important contribution to reducing overall thermal resistance can be made by reducing losses across the various interfaces between different components in a final product assembly.

The large number of differing forms of such materials that can be considered for a new product design can become more complex as we begin to investigate potential material selection and become aware of the importance of different application methods, fastening requirements, and the various advantages and disadvantages of different types of materials.

D. L. Saums (✉)
DS&A LLC, 100 High Street Amesbury MA 01913, USA
e-mail: dsaums@dsa-thermal.com

C. J. M. Lasance, A. Poppe (eds.), *Thermal Management for LED Applications*,
Solid State Lighting Technology and Application Series 2,
DOI 10.1007/978-1-4614-5091-7_8,

8.1 Purpose of Thermal Interface Materials

TIM materials in an electronics assembly enable the use of packaged semiconductor devices and other electronic components in standard formats, assembled into electronic systems of many kinds, without requiring highly customized packaging. Standardization of components assists in lowering overall assembly and final product cost, a practice that has helped to propel the dramatic reductions in electronics systems costs for decades across the globe.

Development over the last 70 years of many different types of standardized package types for various components has provided the ability to mix and match individual components to develop electronic systems of all types with increasing functionality and lower cost. The TIM material is an inexpensive thermal transfer mechanism at each packaging level, to assist in minimizing thermal losses.

Any properly selected and applied TIM will reduce the thermal resistance between two mating surfaces by eliminating air gaps and providing efficient heat transfer from a semiconductor or other heat-dissipating device to a heat sink. Since heat is a major contributor to device failure, the heat transfer mechanism is intended to minimize build-up of heat within the electronic component and improve overall reliability of the complete system. The goal for efficient product design is to select the TIM material that meets thermal performance, product design, cost, manufacturing process, and reliability requirements most efficiently.

TIMs are ubiquitous in electronic systems. Two principal categories of applications exist for TIM materials in a very broad market definition:

A. Materials used in manufacturing and assembling an LED or semiconductor device into the form in which the device is supplied to an original equipment manufacturer (OEM); this is the "packaged" device which is a heat source in a final product assembly. In this application, the joining material used may in fact be a reflowed solder, which in addition to providing the joining solution also serves an important thermal function by transferring heat from the die to the package body; or it may be a traditional TIM material which either also acts as an adhesive or which requires mechanical fastening hardware.
B. Materials used by the OEM to transfer heat from the LED or other heat source, to a heat dissipation component such as an aluminum finned heat sink or to a luminaire or lamp bezel. This is the primary packaging level where a TIM material is selected for a lighting assembly.

Many TIM materials exist today in the global electronics industry in what may seem at times to be an overwhelming number, from many vendors. Continuous development of new materials is occurring both in industry and research institutions to meet existing problems and as more specialized needs arise, and as new chemistries, carriers, and concepts are explored and developed. Although a TIM may seem an insignificant cost factor in a complex electronic assembly, the focus at every packaging level to reduce thermal losses and enhance device reliability results in ever-increasing scrutiny for each TIM and packaging material.

Myriad applications and design requirements exist for TIMs for many different types of semiconductor components to improve heat dissipation. Applications exist for every packaging level in electronic systems and LEDs are only one such application market. TIM materials have been developed and manufactured globally for more than 60 years for electronic systems.

The primary objective in selecting an appropriate thermal interface material is to evaluate the principal application characteristics, at each packaging level within a lighting fixture where efficient heat transfer must occur; determine the manufacturing process requirements that are necessary; determine acceptable cost and thermal performance levels, as these frequently require a compromise in all but the highest performance systems; and determine what level of reliability is necessary for the product design. The use of thermal conductivity as the sole determinant of an appropriate TIM material is rarely a successful design approach for a manufacturable and successful product.

8.2 Common Nomenclature

Θ	Thermal resistance per unit area, also referred to as unit thermal resistance (°C-cm^2/W or °C-in.^{2}/W)
T	Temperature (°C)
A	Area (cm^2)
Q	Heat flux (W/cm^2)
t	Thickness (cm)
k_{bulk}	Bulk Conductivity (W/m-°C)
TIM	Thermal Interface Material
IR	Infrared, as applied to thermal imaging for thermal analysis
PCTIM	Phase-change TIM
TTV	Thermal Test Vehicle
TIM1	Thermal Interface Material, die to lid or heat spreader (generally, internal to a semiconductor package)
TIM2	Thermal Interface Material, case to heat sink base (generally, external to a semiconductor package)

8.3 Classification of Thermal Interface Material Types

More than 10 broad categories of TIMs are available from hundreds of suppliers globally, with thousands of individual supplier material designations. Why is there such an abundance of these seemingly inconsequential materials? The answer to this question includes determining what the purpose of a TIM material is intended to be in a specific application. If the material must act as an adhesive, as well as a transfer mechanism for removing heat from a component such as a power semiconductor, then the design of the particular TIM selected will be very different from a TIM that is to be clamped in place mechanically. If the problem requiring a TIM material is that there is quite a large air gap between the heat source and a metal component that can act as a heat sink (such as an enclosure wall or other metal surface or component), then a very compliant and relatively very thick material must be selected to bridge that large gap. For applications where high-speed, high-volume assembly requires minimized placement time for assembly, then a dispensable thermal compound and a dispensing system or a preform, stamped to a very specific shape and supplied on large rolls, may be used. The many different application requirements for these materials have driven development of an ever-increasing number of different categories, with different performance levels, application methods, costs, thicknesses, adhesive qualities, electrical conductivity values, dispensability, reworkability, and other related characteristics.

Establishing a classification system for TIM materials will be helpful by identifying basic differences in physical properties (such as thickness and attachment and assembly process required), as well as thermal and electrical properties. Once an overview of material types is analyzed and understood, a second step would then be to identify which material types are applicable in a specific type of LED assembly. The third step is determination of which specific TIMs within this material type, attachment method, and assembly process may offer the best combination of thermal, mechanical, and electrical performance; cost, suitable application process, and availability.

The simplest description of the most common TIM materials is a carrier material that is typically flexible and compliant, such as a fluid or a polymeric material, filled with one or more thermally conductive particles. In the case of a fluid carrier such as silicone oil, sufficient filler constituents are added and mixed to create a paste. Examples of thermally conductive fillers used include aluminum oxide, boron nitride, copper, aluminum, and silver. Different shapes of particles, sizes of particles, combinations of fillers, and the interface between the various filler particles and the carrier have been the subject of much exploration in order to develop TIM materials that meet quite a surprising variety of requirements in electronic systems.

What appears to be a rather simple material intended to perform a simple function within an electronic assembly has in fact developed into a worldwide industry made up of many manufacturers, suppliers, and research organizations because of the very wide range of requirements found for TIM materials (Table 8.1 and 8.2).

8.3.1 Thermal Compounds (Thermal Greases)

Several terms refer to this category of TIM, including thermal grease, thermal paste, and heat sink paste. These compounds, available for more than 60 years, and certain light oils are generally considered to be the first materials intended to function as what we consider today to be a TIM.

Table 8.1 Thermal interface material classification outline—traditional material classes. (Source: DS&A LLC)

Thermal interface material classification outline			
Traditional TIM material classes			
TIM material class	Typical thickness ranges	Important subcategories	Available reinforcing carriers
Elastomeric pads, graphite pads, other die-cut preforms	0.076–0.76 mm (0.003–0.030″)	Non-dielectric, dielectric	Fiberglass Aluminum foil Kapton
Pressure-sensitive adhesives (*PSA*), Thermally-conductive	0.127–0.381 mm (0.005–0.015″)	Non-dielectric, dielectric	Free film Fiberglass Aluminum foil Expanded aluminum Kapton
Phase-change materials (*PCTIM, PCM*)	0.012–0.254 mm (0.0005–0.010″)	Non-dielectric, dielectric	Free film Fiberglass Aluminum foil Expanded aluminum Kapton
Gap-fillers	0.254–7.62 mm (0.010–0.300″) (*Pre-assembly thickness*)	Silicone rubber preforms, molded components	Internal fiberglass External fiberglass Polyester Aluminum foil Molded components
Cure-in-place compounds, thermally-conductive	N/A		N/A
Thermal compounds (*Thermal pastes, Thermal greases*)	0.0127–0.127 mm (0.0005–0.005″) (*Post-assembly thickness*)	Silicone-based vs. silicone-free, paste vs. dry preform	N/A
Thermally-conductive adhesives (dispensed)	0.0762–0.635 mm (0.003–0.025″)		N/A

N/A not applicable, *PCM* phase-change materials, *PCTIM* phase-change thermal interface material, *PSA* pressure-sensitive adhesives, *TIM* thermal interface material

8.3.1.1 Silicone Oil-Based Thermal Compounds

Thermal compounds today frequently consist of silicone oil as a carrier with one or more solid filler materials, often spherical forms of aluminum oxide or boron nitride, which provide enhancement of the thermal conductivity of the compound. The rheology employed for development of TIM materials is critical to overall thermal performance and suitability for use in electronic systems. Silicone oil-based thermal compounds (often termed as "thermal grease") are generally considered to be the single most commonly used TIM material globally and, for the majority of typical applications, are viewed as very low in cost. Silicone oils are very commonly available from a large number of sources globally and are inexpensive, leading to

Table 8.2 Thermal interface material classification outline. (Source: DS&A LLC)

Thermal interface material classification outline			
Recent material development classes and non-traditional material classes			
TIM material class	Typical thickness ranges	Important subcategories	Available reinforcing carriers
Die-attach adhesives, highly filled	0.051–1.02 mm (0.002–0.04″)	–	None
Liquid-metal alloys; low melting point metal preforms	0.05–0.20 mm (0.002–0.008″)	Preforms, shims, liquid metal; reactive clad metal foils	N/A
Polymer-solder hybrids (*PSH*)	0.05–0.20 mm (0.002–0.008″)	–	N/A
Solder TIMs (reflowed, typically as TIM1)	0.05–0.31 mm (0.002–0.012″)	Lead-free, lead-containing	N/A

N/A not applicable, *PSH* polymer-solder hybrids, *TIMs* thermal interface material

continued usage as a carrier for TIM compounds. Principal advantages of silicone-based thermal greases are low cost, excellent wetting and conformability to imperfect and out-of flat surface conditions, lack of a need for curing or activator agents, the ability to load fillers to obtain high bulk thermal conductivity, ease of rework, and wide availability.

Available from dozens of suppliers in many different formulations (and also available from other vendors under private label supply agreements) silicone-based thermal greases suffer from very common application issues and have well-known failure mechanisms that impact final product reliability. It is these issues, with a dominant type of TIM that have driven the development of many newer and different types of TIM materials that are now in use globally. Extensive testing by vendors and by OEMs of silicone-based thermal greases has documented specific failure mechanisms. Also, note that there are continuing programs to develop silicone thermal compounds that do not exhibit such behavior and will allow continued use of such widely used compounds.

Silicone oils have traditionally been used as the most common carriers for thermal greases, the primary useful property of silicone being a high propensity to wet mating surfaces and migrate into microscopic surface voids. However, this same wetting and migration characteristic may also be found to appear in a container that has been in inventory and is opened: silicone oil, a clear fluid, often will be noted as having separated out of mixture. If separation is observed, the oil must be remixed with the conductive solid particles prior to application.

After application, separation between the silicone oil and the solid constituents of the TIM may also occur. This can be observed as a visible clear liquid that appears at the periphery of a joint, such as when a semiconductor component has been mounted to a heat sink with a silicone-based thermal grease as the TIM. The degree of separation and migration of the oil from the TIM will vary with different constituent materials, different silicone oils, and differing mounting attitudes and other specifics of a system design. These are reasons why many silicone-based thermal greases typically suffer significant performance degradation with time and should be avoided.

The separation and migration of silicone oil can affect other components within an electronic system. Migration of silicone oils from silicone-based thermal greases caused significant problems for optoelectronic lenses and components as this new technology was deployed within telecommunications equipment developed in the early 1980s. Migration and/or outgassing of silicone oils can have important negative impact in optical systems, such as in LED arrays and in telecom systems, where oils can coat lens materials and cause reduction of light out-coupling efficiency and/or degradation of system performance. Outgassing of silicone oils can also interfere with joining material processes (such as solder vapor reflow) and deposit on and interfere with electrical contacts.

Discovery of these problems led to the publication of internal engineering specifications at certain telecommunications equipment companies that banned the use of silicone oils in any electronic system designed by or purchased by those companies, to be installed in company network equipment. These specifications (including those issued by AT&T Technologies, now Alcatel Lucent) were instrumental in pushing forward the development of nonsilicone-based thermal greases and phase-change TIMs (PCTIMs), to solve this problem.

A second documented reliability concern with silicone-based thermal greases is referred to as the pump-out phenomenon. The primary reference for this term historically has been mechanical pumping action at the interface, frequently with highly viscous silicone thermal grease, as the assembly is exposed to repeated thermal cycling. Thermal cycling occurs in operation with a high heat flux device, such as a microprocessor or power semiconductor, which is constantly being switched or accessed during operation. If all semiconductors are thought of as primarily being an electrical switch, regardless of the particular function, the repeated electrical switching action has a corollary, as a result: repeated heating and cooling of the device. As this occurs, thousands of times per unit of time, the semiconductor package materials are heating and cooling and subjected to differential thermal expansion rates for each material. Therefore, the baseplate of a power semiconductor is heating and cooling and expanding and contracting at one rate, determined by the physical characteristics of the material, while the aluminum or copper cold plate or heat sink to which it is attached is expanding and contracting at a second rate. Especially if the assembly is mounted vertically, the net effect is for an exacerbation of the silicone oil separation problem experienced with silicone-based greases, where the mechanical pumping action forces the oil from the interface.

In addition to the problems created by the presence of silicone oils in an electronic system, silicone oil separation and migration results in a phenomenon known as dry-out for TIM materials. The loss of the fluid carrier is evident after disassembly and visual inspection of the joining surfaces; in the worst case, only the dry particulate fillers remain at the interface. Surface wetting characteristics are lost and the TIM loses the ability to function at the level seen at time zero, when the material was first applied. The loss of the fluid carrier is also evident in testing for thermal performance over time, where thermal resistance at the interface rises with time and operating cycles. In the worst-case examples, of course, failure to thoroughly remix the TIM paste in the original container, prior to application and assembly, will mean that an insufficient amount of the fluid carrier is present even at time zero.

A useful discussion of both the silicon migration and pump-out phenomena can be found in [1], with a concise description of reliability testing and resulting failure mechanisms and empirical data depicting the impact of environmental stressing on thermal resistance of these materials.

A further discussion of reliability testing (high-temperature bake, humidity, thermal cycling, power cycling, and high-temperature cycling) can be found in [2], with corresponding, before and after data for several different TIM material types. More information on aging can be found in a recent paper [3].

Well-formulated and relatively high-performance thermal greases are available from several manufacturers who have attempted to rectify common reliability problems [4].

More recent developments in silicone oil-based thermal compounds include those that also contain aluminum, copper, silver, or diamond fillers. Development programs have focused on the use of filler particulates of differing diameters, following compaction theories for improvement of the overall thermal conductivity of the final product. These developments have been driven by the desire to improve the overall thermal conductivity; several of the most successful silicone oil-based thermal compounds in industry today contain aluminum as a primary filler. Extensive testing of these thermal compounds has been demonstrated in industry literature, which is especially useful to provide a comparative basis for evaluation of newly developed TIMs.

Thermal performance of silicone-based thermal compounds is exceptionally good relative to many other types of TIMs and relative to material cost. This factor is a strong reason for considering this class of TIM. Review of current silicone-based thermal greases shows thermal resistance values for well-performing thermal greases to be in the range of 0.07–0.10 °C-cm^2/W in applications with clamping force applied typical of power semiconductor applications.

An important advantage in many types of electronics applications for a TIM such as silicone thermal compounds is that this type of TIM allows for rework. Rework requirements for design include two types of rework during the product expected life cycle: *factory* rework, where there may be need for disassembly during the manufacturing process of a product, typically after a quality control check, and *field* rework, where a technician or the end-user is expected to be able to disassemble and replace or repair a component. The use of silicone thermal compounds and similar organic materials, such as phase-change compounds, has traditionally met requirements for both factory and field rework and component replacement. The reworkability question becomes increasingly important with higher value components or at a stage in the manufacturing process where a high level of overall product assembly cost has been reached, before the specified replacement or repair must occur.

A type of thermal grease is the cross-linked silicone polymer gel, which requires curing. Gels demonstrate a low modulus after curing but are also difficult to rework after curing. This form of grease typically offers high bulk thermal conductivity and is frequently used as a TIM1 material. Fillers may be either metals or ceramics such as zinc oxide.

8.3.1.2 Nonsilicone Thermal Compounds

Thermal compounds that are designed to avoid the use of silicone oils are available in two basic formats:

- Dispensed compounds (commercially available in bulk quantities, to be applied using common dispensing equipment used with adhesives)
- Dry-to-the-touch preforms and sheets (commercially available in finished form, not requiring any dispensing equipment)

The development of thermal greases in this classification that do not contain silicone oils has included development of dry-to-the-touch preforms which can be handled, applied, and removed and repositioned readily. This simplicity is very important for both low-volume and high-volume applications where cost of assembly is viewed as a critical element of the overall TIM material assessment for design-in. The ability to handle these materials manually for placement or, with simple vacuum pick-up tools or automated placement systems, in high-volume manufacturing operations offers important opportunities to reduce total cost for an LED assembly. These materials have quickly become preferred "drop-in-place" alternatives to the screening, dispensing, and stencil printing operations that are used with thermal compounds in bulk forms.

The use of a die-cut preform (in other words, cut to the shape specified by a design engineer for the outline of, for example, an LED array to be mounted on an aluminum extruded or die cast heat sink base), not only greatly simplifies the manufacturing assembly operation but also has an important performance attribute. The use of a die-cut preform of a known, predetermined thickness eliminates the potential for overapplication of a paste form of thermal grease or other bulk compounds.

Since a primary driver of TIM thermal performance is the use of the thinnest possible application of material that fills the void between two surfaces, applying a specified thickness of a TIM in a preform sheet, reliably and repeatedly without variation due to operator error, is a necessary critical control step on the manufacturing floor.

Also, note that these nonsilicone thermal compounds that are supplied in sheet form and in die-cut preforms are available in three basic versions:

- Nonsilicone thermal compound without a carrier material, in sheet or die-cut shapes,
- Sheet or die-cut preform as a compound applied to both surfaces of a thin aluminum foil carrier,
- Sheet or die-cut preform as a compound applied to both surfaces of a dielectric carrier (such as Dupont Kapton® MT polyimide).

The proper selection of a dielectric carrier with sufficient so-called cut-through resistance (the ability to resist cutting or tearing from a sharp particle or tool), thickness, and nonflammability rating is critical to the ability of the dielectric versions to meet regulatory agency certifications. This is also true for other types of dielectric TIM material types. A note regarding dielectric materials appears in a later section.

Interestingly, these nonsilicone thermal compounds have relatively excellent bulk thermal conductivity values, as compared to that of several of the best-performing silicone-based thermal greases. Comparative thermal performance testing has been shown to be essentially equal to the best silicone greases when applied over the normal range of pressures that are typically seen for power semiconductors, LED assemblies, and radio frequency (RF) devices [5].

8.3.2 Elastomeric Pads

Elastomeric pads are a very common TIM format, developed and sold for electronics system assembly for decades in a very wide range of formulations, prices, thicknesses, thermal conductivity, and thermal performance. An elastomeric pad is typically a polymeric material that may be either a thermoset or thermoplastic, may or may not require some type of cure cycle, have a low Young's Modulus, and is relatively flexible and soft at room ambient temperature. These are generally viewed as relatively simple, easy to use, and low-cost TIM materials. Elastomeric pads are also robust, an excellent characteristic for larger preforms which must be applied manually.

The elastomer is used as a carrier to which higher conductivity fillers are added, as is true for other classes of TIM materials. Many elastomeric pad TIMs are electrically conductive; due to the presence of high thermal conductivity, fillers are also electrically conductive. Fillers include metals, graphite, and ceramics. Filled elastomeric pads can be found with thermal conductivity values that are relatively modest (at best for most commonly available commercial products, not more than 2.0 W/mK).

Typical characteristics of most elastomeric pad TIMs (wide availability, very moderate cost, simplicity, robustness for handling, relative compliancy, and moderate thermal performance) have enabled this class of product to become a very widely used, high-volume thermal solution for a very broad range of electronic system applications.

8.3.3 Pressure-Sensitive Adhesives

Pressure-sensitive adhesives (PSAs) are considered for those designs where mechanical fasteners are not to be used to attach the LED assembly to the heat sink, which may be due to the need for minimum total cost or other limitations. The use of an adhesive, typically applied in a preform, is frequently found in low-power LED assemblies where only a relatively modest thermal resistance value for the TIM is required. Consequently, PSAs are not generally a direct alternative to silicone-based thermal greases and similar higher-performance TIMs. Where high thermal performance is not required, the use of a thermally conductive PSA may simplify product design and reduce assembly time, with reduced component count and minimized assembly cost. An added advantage of most PSAs (often referred to as adhesive tapes, also) is that there is generally no need for postcuring after assembly.

There are many types of PSAs available from many vendors. The number of alternative carriers and types of adhesives used is too great to be described in detail here. In general terms, these TIMs are typically supplied in a die-cut form, with an adhesive coating applied to both surfaces of a carrier material. The presence of an adhesive coating can significantly reduce the overall thermal performance of the TIM material; the choice of a low-quality or low-conductivity carrier material may have the same result.

There are a selection of carrier materials, thicknesses, fillers, and overall thermal conductivity values for PSAs. A typical example would be a filled elastomer, again using a thermally conductive filler such as aluminum oxide ("alumina"); the carrier may be simply silicone rubber. The pressure-sensitive coating is applied to both surfaces and the final TIM product is supplied either in a continuous roll form or in die-cut preforms on a release liner. A second release liner is added to provide protection and to make handling, shipping, and storage practical. PSAs in preforms demonstrate test performance generally in the range of 0.30–0.75 °C-cm^2/W in applications with clamping force applied typical of power semiconductor applications.

Common failure mechanisms can include adhesive failure due to lack of or improper cleaning of contaminants on surfaces prior to application; creep of certain adhesives under stress; and cohesive failure due to coefficient of thermal expansion (CTE) mismatch and resulting stresses during cycling. PSAs applied to plastic integrated circuit (IC) packages have a history of failure where mold release agents are present on the package surface and/or continue to outgas from the plastic mold compound.

For low-power assemblies where a moderate thermal resistance value is adequate, PSAs can be a low-cost, easy-to-apply, and robust solution.

8.3.4 Phase-Change Materials

The term phase-change applied to TIMs refers to developments of organic compounds that change phase from a solid to a liquid at a predefined temperature. Introduced by the initial developer in 1984 for power semiconductor applications, as an alternative to traditional thermal greases, this material type was initially found to be very successful for power semiconductors and, by 1999, became well-established for microprocessor applications due to a combination of thermal performance, reliability, and simplicity of storage and application processes.

PCTIMs have moderate bulk thermal conductivity values, conform very well to surface disparities, do not require any curing or activator other than the need to change phase at a designated temperature, can be reworked, and are dry to the touch for handling and placement. Typical phase-change temperatures for PCTIMs (also termed PCMs) are 45, 51, and 60 °C (selection of a desirable phase-change temperature for a TIM may be influenced by expected storage temperatures or by consideration of expected semiconductor operating temperature).

Electrically nonconductive and electrically conductive versions of PCTIMs are available from many manufacturers. Most major TIM manufacturers now offer PCTIM preforms, given the global acceptance of this material type.

Relatively few types of nonelectrically conductive PCTIM preforms are available, as these are commonly used only with certain types of nonelectrically isolated power semiconductor package types. Electrically nonconductive (dielectric) PCTIMs are not typically used for LED applications but are found in power supplies. Dielectric PCTIM preforms that are most commonly found are manufactured with a phase-change compound coated on both sides of a material such as Dupont Kapton® MT. Kapton and Kapton MT are polyimide films that act as extremely durable carriers that resist punch-through from metal particles and sharp edges, resist damage from handling and improper assembly, highly temperature-resistant, and meet electrical nonconductivity testing requirements. These properties explain why this family of films has proven to be so useful for these specialized types of TIM requirements where thermal conductivity is required but durable electrical isolation must be provided.

The proliferation of phase-change materials (PCMs) has expanded the range of these very practical TIM solutions for electronic assemblies. Most vendors supply only die-cut preform formats. A small number of dispensable phase-change thermal interface material (PCTIM) compounds have been developed for high-volume, automated dispensing and screening application processes. Phase-change compounds and preforms are available with performance typically in the range of 0.07–0.30 °C-cm^2/W in applications with clamping force applied typical of power semiconductor applications, depending on the compound selected and type of PCTIM [6].

8.3.5 Gap-Fillers

This category of TIM materials may be thought of as very thick versions of the elastomeric pads previously described. Gap-fillers are frequently termed as viscoelastic, indicating that such materials have relatively high viscosity under either sheer or tensile stress, while exhibiting an elastic tendency. These materials are rather tacky and thick yet conform under pressure. These are sold in sheets and pads and may commonly be found placed in an LED assembly between a metal component in the heat transfer path and a metal sidewall of a fixture.

The term "gap-filler," unfortunately, is also frequently applied to all forms of TIM materials, which can be quite misleading. While the basic function of all TIM materials is to eliminate voiding between two surfaces, filling the void by replacing air with a material of known (and presumably higher) thermal conductivity, a primary objective is to use the thinnest material possible.

Very thick elastomeric pads are typically defined as those materials with thickness greater than 0.254 mm (0.010″) as a general approximation; some vendors may use a moderately higher or lower minimum thickness value to designate their particular gap-filler product families. Materials of greater than 7.62 mm (0.300″) may be heavy and rather difficult to work and handle and may in fact be of dubious value in an assembly.

8.3.6 Thermally Conductive Adhesives (Dispensed)

Similar to thermal greases, there are many types of adhesives available with varying degrees of thermal conductivity enhancements and many of the filler materials are quite similar to fillers for thermal greases, gap-fillers, and other classes of TIM materials. Both one-part and two-part adhesives are available. It is important to work with clean surfaces, to remove any contaminants and to promote proper adhesion and curing.

The most common adhesive binders are two-part (requiring an initiator material to be introduced to the base material) epoxies and heat-activated epoxies, acrylic adhesives, and room temperature vulcanizing (RTV) silicones. The RTV silicones require an activator material to initiate a chemical reaction and many are highly temperature-resistant and robust in environmental stress testing, have low viscosity, and flow well. Heavy concentrations of fillers applied to a base RTV silicone will increase thermal conductivity but restrain viscous and flow behaviors.

Fillers that are added to the base material to create a thermally conductive adhesive are aluminum, ceramics such as aluminum nitride, carbon black, and silver. Examples of thermally conductive adhesives (frequently, referred to as TCA materials) are an aluminum-filled two-part epoxy, a second is a silver-filled thermoplastic.

An excellent discussion of thermal conductivity testing for TCAs can be found in [7].

8.3.7 Die-Attach Adhesives, Highly Filled

Die-attach adhesives are developed and manufactured for, as the term suggests, the attachment of silicon die to substrates and printed circuit carriers. Typically heavily filled with metal particles with high thermal conductivities, such as silver, these adhesives are advertised with thermal conductivity values stated to be as high as 60 W/m-K. The metal filler for some commercial thermoplastic silver-filled die-attach adhesives may exceed 85 % by weight [8].

The high filler ratio of die-attach adhesives is also perceived as offering prospects for use as a TIM in some types of IC, power RF, and high-performance LED module designs. Electrically conductive filled die-attach adhesives for LED array manufacturing may be dispensed by stencil printing, screen printing, or by needle dispensing.

These are relatively expensive adhesives due to the high percentage of silver alloy contained in these adhesives. The high filler ratio also requires careful design of dispensing systems. Filled die-attach adhesives offer electrical conductivity as well as thermal conductivity and can act as a mechanical interface between two dissimilar materials with very different CTE. These attributes are important for some types of package design.

Further development with filled die-attach adhesives that offer stable thermal resistance values, high thermal conductivity, excellent adhesion, and carefully matched CTE values may offer a path to added improvement in power LED packaging.

An innovative concept is a bond-line control mechanism, designed to yield precision dispensing and bond-line definition. This concept may also be used to prevent uneven component placement or tilted die placement [9].

8.3.8 Cure-in-Place Thermal Compounds

Cure-in-place compounds are typically applied with automated needle dispensing systems to form a gasket or seal at the periphery of an assembly or module. Cure-in-place compounds are often adhesives and contain ceramic fillers such as boron nitride, with a rubber binder. These materials are easily dispensed, cure at room temperature, and excess material at the joint is easy to clean up after curing is complete. Certain of these materials may be considered to be a cure-in-place thermal grease, containing silicone and having similar thermal properties to thermal greases with the addition of an adhesive.

Sometimes referred to as gasket compounds, these may require either ultraviolet (UV) curing or heat curing. Thermal conductivity values are similar to those of many common thermal greases [10].

8.3.9 Nontraditional Thermal Interface Material Categories

8.3.9.1 Low-Melting Point Alloys

Low-melting point alloys (LMAs) are a relatively new concept for potentially new TIM material developments and include materials which may be in liquid or near-liquid state at room temperature. indium metal and Indium alloys have been used as TIM materials in certain applications, including for computing systems in rare instances in high-volume production.

These LMA materials are not generally considered to offer combinations of cost and handling and retention characteristics that are appropriate for LED applications. As is true in each of these types of statements for TIM material categories, exceptions may be found in very specialized applications. An excellent reference for liquid indium TIM materials development and application may be found below, for further reading [11].

8.3.9.2 Polymer-Solder Hybrids

Development of a newer form of thermally conductive gel, containing metallic particles to improve overall thermal conductivity, is indeed a hybrid form of TIM.

These gels utilize metallic particles that are typically a low-melting solder alloy, such as indium-tin or indium-tin-bismuth. The use of dissimilar-sized filler particles (both metallic and ceramic or metallic only) provides an extension of the concept of greater particle compaction to reach higher thermal conductivities. Higher bulk thermal conductivity values have been demonstrated with higher loading of a low-melting point metal such as indium and no metallization of surfaces to be joined is specifically required to achieve good wetting of surfaces. A curing temperature is required, postulated on the specific requirement of the polymer utilized as the base carrier, and the reflow temperature of the metal alloy chosen for the filler spheres or particles must fit within the curing process temperatures specified.

This newer category of TIM material has not been identified in widespread use and may be considered to require processing requirements and cost tolerance that is greater than can be accepted for significant implementation in LED application markets. Additional development work may be required to achieve necessary improvements to allow for reworkability requirements, given the base polymer requirement for a cure cycle for proper processing, and additional investigation is needed for reliability analysis in high relative humidity environments.

8.3.9.3 Metallic Thermal Interface Materials—Solders

The most common and widely used form of a metallic TIM may be the broad use of reflow solders for semiconductor device attachment. The use of a reflowed metallic joint yields a thermal path of relatively very high thermal and electrical conductivity; typical bulk thermal conductivity values range between 30 and 80 W/m-K, for many solder alloys.

The implementation of so-called "TIM1" reflowed solder TIM materials, replacing what had been thermal greases, thermally conductive gels, and phase-change TIMs, began in microprocessor production in November 2003, by Intel Corporation. Widespread understanding of the value of reflowed solders as a TIM material, in effect, has been increased by this development in the high-volume microprocessor market.

Low thermal resistance values observed for reflowed indium alloy solders have been demonstrated in the range 0.02–0.05 °C-cm^2/W [12].

The excellent bulk thermal conductivities obtainable with reflowed solder alloys meets market requirements for very high conductivity TIM solutions, but also requires specific processing steps and costs, including the application of metallization to both surfaces to be joined in order to achieve the desired surface wetting. Metallization is not an insignificant process cost for high-volume electronic components and is a critical factor in LED manufacturing, if a metalized joining surface is not already available for other reasons [13].

The use of reflowed solder alloys as a TIM also introduces potential reliability concerns induced by higher temperature operation, where poor compliance between expansion rates at temperature of different materials joined with a solder alloy may cause long-term reliability factors due to mismatched CTE. LED manufacturers and luminaire and lighting device manufacturers are typically deeply familiar with solder processes and solder materials, which is advantageous for implementation of a solder TIM. However, the higher temperature-induced CTE stress within a packaged LED array or within a completed light source may be identified as a potential failure mechanism.

These materials are also not considered to be easily reworkable without availability of trained personnel and capability for assembly heating to facilitate removal. Intermetallics may form between two dissimilar materials during the initial reflow heating process and these intermetallics should also be removed in order to properly clean the surface prior to replacement.

8.3.9.4 Graphitic Pads and Sheets

Another category of high-performance TIMs developed recently are flexible graphite die-cut pads that, when laminated with an adhesive, have proven to be very useful additions to the range of materials available for LED assembly design. Excellent in-plane thermal conductivity values have traditionally been measured with graphitic forms of materials; these recent developments now demonstrate through-plane bulk thermal conductivity values for the graphite material of up to 16 W/m-K. Also manufactured in laminated plastic, die-cut pads are available which do not exhibit delamination or flaking of electrically conductive flakes; this improvement has been well received for applications requiring interface materials with high conductivity values as thin as 0.140 mm (0.0055″) [14].

8.3.9.5 Metallic Thermal Interface Materials—Indium Alloy Foils

Indium alloys have been utilized as TIMs for several specific applications, although typically not for LED applications. Indium alloys have been used traditionally as what are termed shims, as an example for radio frequency components in telecommunications systems.

The value of indium alloys in these applications is that indium metal offers significantly higher thermal conductivity as compared to most common TIM materials and compounds; a primary value of indium metal also is compliancy under higher clamping forces.

Indium metallic TIMs have also recently been introduced with patterning of the foils. The purpose is to provide additional compliancy under pressure. Applications for TIMs using this concept include very high-performance enterprise server processors, power semiconductors, wireless telecommunications RF power semiconductor devices, and similar components.

These high-performance materials are often higher in cost, given the inherent cost of indium metal, and therefore are not frequently found in high-volume LED market applications when only the base cost of the TIM is considered in a cost analysis. What is frequently overlooked is the simplicity of assembly using this class of product and the robust and durable nature of metallic TIMs, important for long-term reliability. These alloys do not contain a polymer, silicone oil, or other less robust constituents

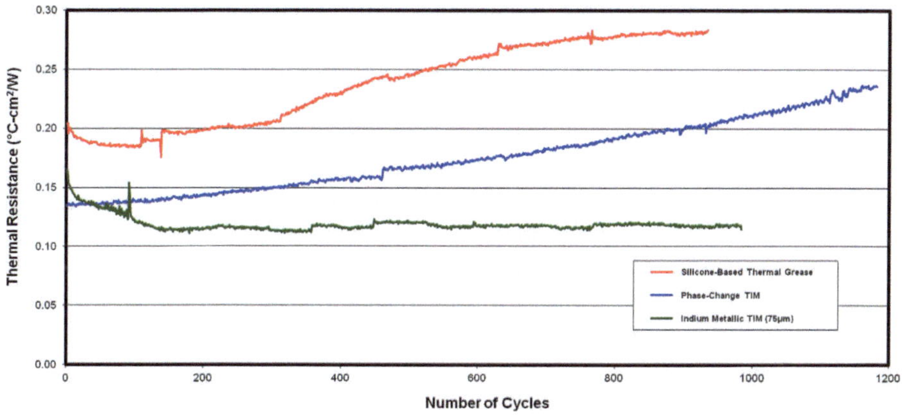

Fig. 8.1 Thermal resistance performance of three classes of thermal interface materials (TIMs) under power cycling. (Source: Indium Corporation)

that may be subject to some of the typical failure mechanisms (due to temperature, time, or silicone outgassing) that are found in traditional forms of TIMs.

An example of comparative power cycling reliability test data for three classes of TIM materials (a common aluminum-filled silicone-based thermal grease, a phase-change TIM compound, and an indium alloy foil), is shown in Fig. 8.1. The results indicate increasing thermal resistance for the two polymeric materials, compared to relatively stable resistance of the metallic TIM, as the number of power cycles increases.

An interesting and useful phenomenon with patterned indium alloy foils is that there is an inherent characteristic that may be referred to as a "set" performance enhancement: when the TIM is applied and the heat source and heat sink are fastened together, a performance improvement occurs as the clamping force causes the material to set up in the microscopic surface voids. This phenomenon occurs without need for a high-temperature cure or other process. Power cycling tests of semiconductor devices have also demonstrated that, over several thousand cycles, the thermal resistance of the material will be reduced sharply from the initial (so-called "time zero") value.

Product development for TIMs includes continued evaluation of alternative alloys and patterns that may result in improved thermal performance, improved performance in relatively large voids created by lack of flatness over large surface areas, and assessment of different alloys to realize lower-cost solutions.

8.3.9.6 Reactive Clad Foils for Low-Voltage Ignition Joining Processes

A development in metallic TIMs in the last decade is a multilayered concept which is similar to a solder interface, but that does not require a complete solder reflow application process as the material reacts to a low-voltage electrical pulse that initiates joining. This is a flux-free process; an instantaneous low-voltage ignition is applied to a TIM manufactured as aluminum and nickel nanoscale layers sputtered on a thin indium foil, creating a room-temperature metallic bonding. These are currently sold as a commercial TIM product for exceptionally large-area applications where dissimilar metals are to be bonded uniformly, such as joining backing plates to sputtering targets for process equipment. This promising technology offers the potential for practical future application potential in high-power LED applications where reliability over time is critical. To make practical the use of this material and process for high-volume applications (such as power LED assembly), the next development areas being pursued include automated application processes that provide rapid, high-volume preform placement and voltage initiation. These developments will make available for the LED assembly designer a robust and practical metallic TIM material [15, 16].

8.3.9.7 Carbon Nanotube Arrays

Adaptation of nanotechnology and the extremely high thermal conductivities for such potential concepts as carbon nanotube (CNT) arrays, although of intense interest at the university level, have not as of this writing yielded a practical and successful new CNT TIM material. Many industry experts view successful implementation of nanotechnology to developing new TIM materials as the single most important and a promising research and development (R&D) area.

CNTs and nanoparticles have been evaluated for use in many different electronic materials. Examples of research investigations in materials areas similar to TIM materials includes the potential for use of CNTs and nanoparticle sizes in other electronics packaging materials, such as underfill materials. A large amount of R&D with CNTs and related technologies is underway globally [17].

The bulk thermal conductivity of CNTs is extremely high. This is the reason for interest in these structures to develop new TIM materials. A primary challenge is the high thermal interfacial resistance between the CNT and the contacting surface of the carrier material needed as a support; this high contact resistance may prove to impact the total thermal resistance of the structure negatively.

Previous development and commercialization programs utilizing a nanotechnology or nanosized filler particles have been announced. Typical research programs at TIM manufacturers and research institutions have included the addition of chopped nanotubes or nanosized particles as fillers, within an existing carrier fluid or thermal compound. The challenges of these developments have meant relatively few commercial materials have been announced by vendors. One announced material was a nanofiber-filled thermal grease developed and commercialized but subsequently withdrawn by the manufacturer [18].

More recent nanotechnology developments include more holistic concepts utilizing grown CNTs in a uniform array, as compared to random dispersion of nanotubes or nanoparticles within a carrier polymer. These material concepts typically are described as a metal foil which acts as a carrier (or a combination of a metal foil with a low-melting temperature indium alloy filler) upon which vertical arrays of CNTs have been grown. One example of such a development program in the United States is a Defense Advanced Research Projects Agency (DARPA) TIM development program with several competing teams. These development teams include defense and aerospace systems manufacturers and university laboratories and have produced very low-rate prototype quantities of wafers of arrayed CNTs which have potential for development of processes which yield extremely high thermal conductivity TIM materials. Initial thermal performance testing results indicate that such materials offer very low thermal interface resistance values, similar to solder alloys and liquid metal TIM materials, although significant hurdles remain for development of practical TIM materials.

The future direction for these development programs is not yet evident. Such materials development programs are not clearly identifiable as offering commercially viable TIM materials in the near future that can be considered for LED applications in commercial product development.

Similar to the US DARPA TIM development project, the European Union's 7th research framework also sponsored the NANOPACK project [19] where a consortium of computer, aerospace, and automotive electronics companies, TIM developers and manufacturers, universities, independent research laboratories, and test equipment manufacturers worked on the development of new nanoparticle-based TIMs and new test methodologies and test apparatus. This resulted in new, commercialized test equipment, the principles of which are discussed in a later section of this chapter. The work of the NANOPACK consortium is continued by the NANOTHERM project [3], aiming at an investigation of nanotechnology-based potential new TIMs. This effort is starting from studies on nanoscale heat-transfer and ending with TIM application demonstrators, including an LED application and a reliability testing demonstrator. As the NANOTHERM project was launched in 2012, the project results are not expected before 2015.

A very significant development issue within the DARPA program mentioned above is the need for development of thermal resistance testing capabilities with the required resolution and repeatability necessary for very low thermal resistance values [19].

8.4 Defining Applications and Selecting Thermal Interface Material Types

Choosing the appropriate material type for an application with excellent surface wetting and low contact resistance properties, the thinnest possible application with complete coverage (or the thinnest preform), and determining how to apply an optimized clamping force, are all important factors affecting thermal performance of polymeric TIMs *in the assembly*. Maximum operating temperature, avoidance of compound runout or silicone oil separation and pump-out, assembly process requirements, and similar practical application requirements are also important to the proper selection of a TIM.

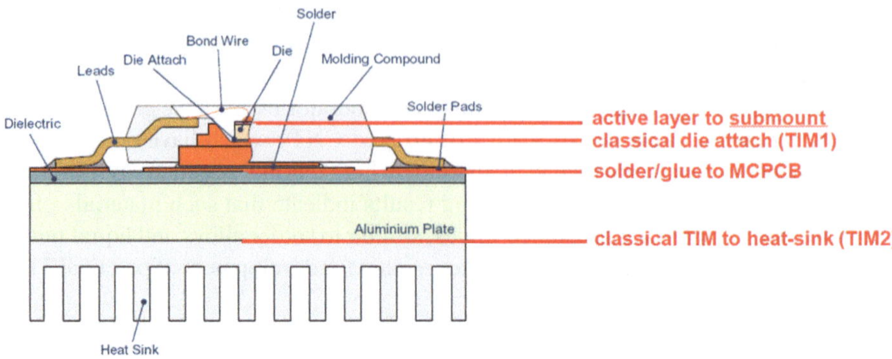

Fig. 8.2 Illustration of recent common design of light emitting diode (LED) package and thermal path

It is important as to describe where we find applications most commonly for TIMs in LED light sources and lighting assemblies. For the purpose of illustration in Fig. 8.2, the application areas may be generalized to portray a very common configuration of the light source, firstly, so as to identify why we find different types of applications and application requirements, and then identify what types of TIM materials may fit those generalized application requirements.

We see that the semiconductor die is attached typically with a reflowed solder or with die-attach adhesive (Fig. 8.2), to provide both electrical and thermal connection to the package material. In the case where a solder is used, the solder selected may be either leaded or lead-free and may also include solder-like materials that are applied with novel manufacturing processes. Increasingly, in order to meet environmental requirements, leaded solders are replaced with either lead-free solders or with die-attach adhesives and similar solutions that do not contain lead. If a metal heat slug forms the base of the LED, the metal is most typically copper, copper-tungsten, copper-molybdenum, or a similar metal component having certain thermal conductivity and, ideally, a CTE which is similar to that of the semiconductor die in order to avoid delamination or voiding due to the mechanical stress caused by different thermal expansion[1].

The implementation of a metal slug or base structure provides a thermal path by conduction from the power LED semiconductor, often via a solder or die-attach adhesive to a metal-core printed circuit board (MCPCB) substrate.

Given this depiction of a common high brightness LED light source and mechanical and optical lens structure, we now proceed to a simplified table of TIM materials appropriate for each level of packaging or type of application. Table 8.3, illustrates how general classes of TIM materials may be applied in the structure of a luminaire

[1] Kovar, copper–tungsten (more properly, tungsten–copper as the larger constituent volume is typically tungsten), copper-molybdenum, and AlSiC (aluminum-silicon-carbide metal matrix composite) are examples of materials having CTE values which are more closely matched to silicon than is true for copper.

Table 8.3 General outline of thermal interface material applications for light-emitting diode (LED) arrays within a luminaire. (Source: DS&A LLC)

Thermal interface material general outline for LED array applications		
General application type	*Typical* TIMs used	*Typical* thickness ranges
Attachment of LED array substrate to "heat sink slug	Solders	0.05–0.10 mm (0.002–0.004″)
	Solder and solder-like materials	
	Die-attach adhesives	
Attachment of "heat sink slug" to finned heat sink	Thermal pastes	0.0125–0.125 mm (0.0005–0.005″)
	Non-silicone thermal pastes	
	Non-silicone dry thermal paste preforms	
Heat sink to luminaire housing interior surface	Gap-fillers	0.25–3.80 mm (0.010–0.150″)
Heat sink to heat pipe, other applications	TCAs	0.075–0.63 mm (0.003–0.025″)
	Thermally-conductive cure-in-place compounds	

LED light emitting diode, *TCAs* Thermally-conductive adhesives, *TIMs* thermal interface materials

or other electronic system which utilizes high-brightness LED light sources, such as an HDTV display. The term "heat sink slug" in column 1, Table 8.3 is taken from the general term used for the thick metal base used in some types of LED packaging. The TIM material classes listed as typical for each of these application areas and the typical thickness ranges shown are drawn from a classification system for TIM material types that is described in tables 8.1 and 8.2 of this chapter the significant differences in the material classes are the type of material (and therefore manufacturing process in volume assembly operations) and the typical thickness ranges of these material classes. Note that while thermal performance of TIM materials is important when selecting a material for a specific design, the type of application considered is a primary determinant of which materials may be used and the type of application strongly influences the thickness of material employed. This is a basic reason why so many types of TIM materials have been developed with widely varying compositions and thermal conductivity values, as well as other properties.

8.4.1 Primary Determinants for Thermal Interface Material Selection

The determinants of improved thermal performance for a TIM material in any single application include three primary performance drivers:

- Relative TIM material thickness in an interface, after clamping force is applied in the assembly by the fastening method used;

- Surface wetting and contact resistances under pressure;
- Bulk thermal conductivity of the TIM.

It is the combination of these major factors that results in the final interfacial resistance in a given application. Measurement of the combined resistance can be made with state-of-the-art in-situ thermal transient testing.

The above statement of the primary determinants of performance presumes that the gap between two surfaces is known and that a reasonable initial thickness of TIM material has been applied, so that all air is forced from the interface as clamping force is applied. The above list constitutes the traditional definition of the significant performance mechanisms for a well-performing TIM. These are the initial values that should be considered that affect initial thermal performance (so-called "time zero" thermal performance upon completion of factory assembly of the product with the incorporated TIM material).

Thickness in this context is highly relative to the type of application and material employed. Selecting an appropriate TIM material for what is termed a gap-filling application, for example, in a luminaire where a large gap of between 100 to 1,000 μm requires a substantially different TIM than would be chosen for a significantly smaller gap. While this statement may appear to be intuitively obvious, it is an important distinction to be made in explaining the large number of different categories of TIM materials which exist, understanding how classification of materials is important in making initial selection of a limited number of material types to be evaluated, and why there are differing performance factors which apply to different TIM categories.

It is important to remember that thermal conductivity alone for a specific TIM material is *not* the primary determinant for thermal performance and published values for individual materials should not be used as the solitary selection criteria. The three performance drivers listed above include the bulk thermal conductivity of the material, but relative material thickness, surface wetting, and contact resistances are very important factors for finished end-product thermal performance.

Each material has a thermal conductivity value (k) which is an inherent value of the material, assuming a uniform structure. Note that thermal resistance (R or Θ_{TIM} or R_{th}, depending on country and regional preferences) may be thought of as an inherent property for each thickness of a material that has a given bulk thermal conductivity value. A thermal resistance value for the same material will vary with thickness, area, interfacial contact resistances, and the test methodology that is used to determine the value (some thermal conductivity and thermal resistance test methodologies are not appropriate for all material types, especially those that are not monolithic materials, and may in some instances be misapplied; this is a subject beyond the scope of this discussion).

8.4.2 Secondary Determinants for Thermal Interface Materials Selection

Additional determinants for selection of a TIM material other than factors directly related to the specific primary attributes listed above include mechanical, electrical, chemical, and application format requirements. Such determinants include:

- Long-term life and reliability, including thermal performance over time; referred to as "end-of-life" performance, as a measure of thermal resistance achieved at end of product life. This is an important factor that is determined in part by factors which include:
  - Presence of silicone oils or other constituents which may tend to dissipate with time, temperature, or mounting attitude;
  - Tendency of a compound composition to become brittle with time or temperature effects;
  - Tendency of a composition to harden if improperly mixed prior to dispensing or application;
  - Tendency to flake or decay or to shed particles, especially if electrically conductive, as may occur with certain types of graphitic materials.
- Appropriate temperature range for intended application
- Dielectric performance, for those TIM materials where electrical isolation must be created and maintained;
- Cut-through resistance, for dielectric materials;
- Suitable regulatory approvals from standards organizations, such as CSA (Canadian Standards Association, VDE (Germany), and UL (Underwriters' Laboratories, USA), as appropriate;
- Reworkability, if required, defined either as factory rework or for field rework and replacement of a failed component or TIM material;
- Suitable carrier materials for die-cut preform TIM materials where handling, automated dispensing, and/or vacuum pickup is necessary—and where the carrier material selected must meet assembly requirements;
- Proper dispensability for compounds, driven by relative unit volume production and process requirements for manual assembly versus automated assembly;
- Viscosity for compounds which must be dispensed with dispensing nozzle systems;
- Viscosity and stencil or screening performance for compounds which must be dispensed and screened or stencil-printed to achieve a final placement on a component surface;
- Appropriate thixotropicity for dispensed materials, phase-change compounds, and other TIM materials which may be required to form a fillet or to act as a sealant or which must not suffer from mechanical pumping action or inappropriate flow due to mounting attitude;
- Vulnerability to high- or low-temperature causal factors, such as an inappropriately low phase-change transition temperature for a PCTIM compound, or freeze

damage for a compound when subjected to air shipment or application in extreme temperature environments;
- Adhesive properties, as required or to be avoided;
- Pot life (length of time the working material can be left open during use on the manufacturing floor, without requiring extensive clean-up of equipment or the open container) and clean-up requirements and cost after use;
- Lack of toxic or environmentally harmful chemicals which may outgas with time or temperature;
- Lack of constituent chemicals which may outgas and cause discoloration or damage to optical devices and components, such as lens materials for certain types of LEDs and lighting devices;
- Environmentally-suitable constituent materials to enable proper recycling or disposal at end of product life without environmental damage;
- Necessity for and cost of low-temperature storage and/or shipment (such as occurs with certain die-attach materials which must be shipped and stored at specified temperatures which are below normal office ambient conditions);
- Necessity for and cost of two-part material mixing time and mixing equipment (for those materials which are designed with a base material and a separate activator, to be mixed immediately prior to use);
- Compliancy, typically for gap-filling TIM materials;
- Cost and availability of dispensing, screening, stenciling, or pick-and-place equipment;
- Cost and complexity of application operator training, if required;
- Material purchase cost and total application process cost.

8.4.3 Selecting the "Best" Thermal Interface Material

A common question when thermal and mechanical design of an LED assembly is first taken up is the question of what defines the single "best" TIM. The preceding lists illustrate the wide range of factors that influence development and design of TIM materials by manufacturers and the range of topics to be considered when selecting a single material for use in a product design. A simple answer is that there is no simple single definition of the best material until the specific parameters for the assembly and the manufacturing requirements are determined; the answer is then a function of the overall product design, preferred manufacturing process, life and reliability desired, and cost requirements.

Maximum thermal performance achievable for the "best" material is not typically the criterion for selecting such a TIM in industry. This is an important distinction, as industry must consider more than a laboratory or theoretical performance maximum for a single material in terms of bulk thermal conductivity or some similar, single metric. Every OEM design application for a TIM must typically be considered within the confines of existing manufacturing practices, assembly processes, and cost targets available in the production facility. Cost limitations, for example, preclude machining and polishing of contacting metal surfaces to eliminate surface defects for virtually any assembly for commercial use. While a laboratory test condition may seek to use ideal surfaces in order to eliminate extraneous variables in performance measurement, such surface finishes are not cost-effective in a commercial product. Surface finish and flatness of a heat sink surface, for example, may be specified in the procurement specification but typically within a range or to a given condition. The function of the TIM selected is to provide maximum heat transfer across the resulting joint between two imperfect surfaces. Using flatness as an example, the range of requirements encountered across the electronics industry for the variety of metal component manufacturing processes employed, such as die casting, extruding, rolling, and stamping of components explains in part why such variety exists in TIM material types.

Only when a problem has been identified with an existing TIM material, or the application and assembly process, is a new form of material that requires a different process typically considered. This reflects the need in industry to minimize disruption of existing processes and logistics, manufacturing experience, and overall product quality measured for a given set of manufacturing facilities.

8.5 Defining and Examining Parameters for the Application

The first task of the designer is to address the relative position of the TIM material(s) in the total heat transfer scheme for a given assembly. For example, in natural convection applications it could well be the case that the TIM is not the dominant restrictive factor. Considering the various trade-offs that are available between components, material selection, machining characteristics, surface treatments, fastener schemes, locations of fasteners, and cost elements is important in developing a product design that meets the desired product goals.

8.5.1 Mating Surfaces

TIMs improve heat transfer between two components, replacing air in any gaps that exist due to dissimilarities in surface condition and surface roughness. Air has very low thermal conductivity value and is therefore not desirable in any interface where a heat source, such as the die, must be provided with a suitable heat transfer path to some heat sink or enclosure wall, to maintain device reliability. Surface roughness of two metal components that must be joined can be improved by increasingly fine levels of machining; however, machining operations even when fully automated add cost in a manufacturing process.

TIMs provide a compromise in all applications between the costs of surface machining, applied clamping force, and attaining maximum possible heat transfer with ideal surfaces. The cost of machining ideal matching surfaces to eliminate any possibility of surface imperfections or irregularities is typically not warranted in all but the most critical electronic systems. Application of high clamping forces to attempt to achieve a perfect interface is also impractical, introducing new stresses and other detrimental characteristics. Therefore, the assumption always must be made that manufacturing of two components that must be joined together will result in minute surface irregularities with the potential to entrap air when the two components are joined, and the proper TIM selection is the compromise.

The solution normally employed in design of electronic assemblies is to specify the following where a TIM is required:

a. Surface roughness and flatness requirement for each metal surface that must be joined to another;
b. Clamping or fastening method to be used, with components specified to meet various design requirements for clamping force applied, cost of components and assembly, and preferred fastening process;
c. Maximum expected gap between surfaces;
d. Known heat load to be transferred across the interface to a heat sink of some kind.

In the design of LED assemblies, possible shadowing or light absorption by clamping components must also be considered (Note: In this discussion, the use of the term "heat sink" is intended to include any natural convection, forced-air cooled, or liquid-cooled device. This may be a complex liquid cold plate for an extremely high-performance system such as high-bay industrial lighting fixtures or be as simple as the sheet metal "can" of a high-volume LED luminaire for residential use. The particular type of heat dissipating device or surface is of less importance in this discussion, although mechanical requirements such as fasteners and clamping force and surface roughness are critical to successful TIM selection and assembly).

TIMs are often viewed as a very inexpensive component in any electronic system, yet these materials provide a critical function that is directly related to adequate heat removal and system reliability. The low cost of many types of TIM materials often can also result in insufficient attention being given to proper selection, testing, and application in series manufacturing. It is suggested that the inexpensive nature of these materials should not be interpreted as meaning that there is no real importance attached to selection; the same is true for planning and implementing well-designed manufacturing processes.

TIMs provide a critical solution. Applications within an LED-based electronic system include more than the LED devices themselves. In a complete system-level view, there may also be other components which generate heat and for which a TIM will be needed to provide a thermal path to a heat dissipation surface or device.

The best selection of a TIM material for a particular application is made by assessing and prioritizing the most important requirements for the component and assembly specification. Most electronics design and manufacturing companies do not devote resources to testing and evaluating TIM materials as they are announced by TIM vendors; frankly, the cost of equipment and the engineering time necessary for competent evaluation can be significant and many companies devote engineering resources to larger components within an assembly.

Therefore, a practical approach to TIM material selection is important to many and must include consideration of these factors.

8.5.2 Need for Assembly Rework and Repair

A factor that is frequently found in many electronics applications but which may not be immediately obvious is that rework of assemblies may be required. The electronics industry consists largely of many, many building blocks of various components, materials, and devices that are selected individually and built into a finished system. The need to repair, replace, or upgrade certain components in such complex assemblies establishes basic design criteria for how this is to be handled. In some simple and low-cost systems, the assembly is assumed to be discarded if a failure occurs. In more complex and costly systems, and especially in mission-critical system design, design for rework becomes an essential and in many cases mandatory system design goal.

Rework generally consists of two primary categories: factory repair and replacement of a module or device, following testing of a module after assembly; and repair or replacement after the system is shipped to the customer. Requirements for reworkable system design vary by industry segment and application, but this is a factor that may have considerable influence on TIM selection.

In general, a requirement for rework will drive the selection of fastener types or adhesives, rather than a permanent joint such as soldering or brazing. Designing where reworkability is a requirement may therefore mandate selection of a thermally conductive adhesive or a set of mechanical fasteners and a thermal grease or phase-change TIM, in place of what may otherwise be a permanent solder joint. This is a simple example, but the need for rework in system design and assembly process design may drive selection of TIM materials that are not obvious and not the theoretical maximum thermal performance achievable in a more perfect environment.

Rework for a heat sink, TIM material, and semiconductor or LED array should always, as a general statement, include complete removal of the TIM material first applied during initial product manufacture and cleaning of the mating surfaces. Replacement of the TIM material with a TIM that is of equivalent or of higher performance would always normally be recommended by equipment manufacturers.

In practice, identification of the original TIM material may prove to be exceedingly difficult for many types of products; in this case, replacement with a TIM material known to be a tested and proven high-performance TIM, of an identical type, would be desirable.

8.5.3 Comparison of Thermal Interface Materials with Shims and Solders

The use of reflowed solder and other permanent metallic joining materials and processes in effect also is solving similar problems as with traditional definitions of a TIM: a joining between two unequal metal surfaces is made and heat is transferred across the joint. In certain segments of the electronics industry, the use of metallic shims also provides, in effect, the equivalent of a TIM. A primary example is the use of indium alloy shims between an RF power amplifier and a copper or aluminum heat sink in a telecommunications system design or in a microwave transmitter. Mention should be made in any discussion of TIM material types of the availability of such metallic shim materials, but the use in LED applications is not common.

Solders of many alloys and forms are used in large volumes throughout the electronics assembly industry as a primary means of joining. Solders also are acting as TIMs, carrying a known heat load across the two surfaces to be joined. However, the field of metallic joining materials and the necessary processes and equipment is large and generally should be treated separately from this discussion of TIMs for proper thermal design of LED assemblies.

Wherever it may be possible to select a solder material and process to join two components, especially for high-volume applications, the use of a solder joint will typically be expected to provide improved heat transfer given the ability to use a metal alloy in a permanent joint. Reflowed solders cannot be used, of course, where the potential need for reworkability must be considered.

8.5.4 Dielectric and Nondielectric Materials

Certain applications for TIM materials require selection of an interface material which has the ability to transfer a heat load yet not conduct an electrical current. A basic assumption for TIM materials, as a general statement, is that any material that is thermally conductive is also electrically conductive. This may not be obvious on first examination of, for example, a selected thermal grease or other form of TIM. While no value of electrical conductivity may be found in a data sheet for the material under evaluation, an awareness of constituents will be necessary to determine the degree of electrical conductivity. As an example, traditional thermal grease which contains alumina and silicone oil may also contain aluminum particles in the dispersion. Therefore, a degree of electrical conductivity must be assumed. Another example is a phase-change compound applied to both sides of an aluminum foil; the foil acts as the carrier and allows the material to be easily handled for processing, assembly, and rework, but also provides electrical conductivity.

Certain categories of TIM materials also have important subcategories, identified as dielectric versions, to meet the need for a nonelectrically conducting material. Elastomeric pads, phase-change, and gap-filler TIMs are excellent examples of where both electrically conducting and nonconducting versions may be found in

Table 8.4 Example properties and test methodologies, dielectric materials. (Source: DS&A LLC)

Typical dielectric TIM material properties		
Property	Value	Test methodology
Dielectric strength	V/mm (V/0.001″) VAC	ASTM D-149 IEC 80243
Dielectric constant	@ 1 KHz	ASTM D-150 IEC 60250
Volume resistivity	Ohm-cm	ASTM D-257 IEC 60093

TIM thermal interface material

industry. In certain categories of materials, the dielectric materials are dominant by usage.

Continuing the example of one TIM category where both dielectric and nondielectric versions exist, there are forms of PCTIMs which utilize an electrically nonconductive carrier or substrate, such as Dupont® Kapton™ MT; this material is available in a variety of forms and thicknesses with ratings assigned by testing laboratories such as CSA, VDE, and UL. A PCTIM can be manufactured with a phase-change compound coated on an aluminum foil as the carrier, for applications where electrical conductivity is useful or desirable. The same compound may be found coated on a Kapton MT carrier, in place of the aluminum foil, to create a dielectric PCTIM with a known dielectric rating.

These materials have specific thicknesses, cut-through ratings, and dielectric test values that are published by the manufacturer. A typical dielectric TIM material vendor data sheet will provide, in addition to thermal performance values across a range of pressures, a set of data that is specific to the electrical insulating properties of the material, with references to standardized test methods. Examples of these values and the relevant test methods are shown in Table 8.4.

As with all TIM materials, it is important to review the specific product data sheet to determine the version that meets desired requirements.

8.5.5 Working with Adhesives

Adhesives with specific levels of thermal conductivity are included in this category, with several subcategories to indicate the primary basic forms that these materials are available in. This overview is not intended to include all types available for electronics applications.

As a general statement, most adhesives designed for joining with some level of thermal conductivity offer overall performance that is typically lower than that of many other TIMs, a result of the dual role which TCAs are intended to fill.

A category of highly TCAs is also included in this discussion, die-attach adhesives. This adhesives family, typically silver-filled, is in certain instances used as a TIM for processors (between the backside of the processor silicon die and the underside of

Table 8.5 Selected typical bulk thermal conductivity of materials. (Source: DS&A LLC)

Material	Typical range (W-mK)
Silicone oil	0.1
Silicone-based thermal greases	0.2–10.0
Phase-change compounds	0.3–10.0
Indium	86
Aluminum (@25 °C)	205
Pyrolytic graphite (in-plane)	300
Pyrolytic graphite (through-plane)	3.5
Copper (@25 °C)	401

the lid, most commonly, for those processor packages which have a lid). Die-attach adhesives are used to attach LED die to metal heat spreaders and substrates.

These silver-filled die-attach adhesives offer relatively high levels of thermal conductivity but have certain other characteristics (including relative cost) which limit their broader use. High filler loading ratios may indicate a die-attach adhesive which will be difficult to dispense properly, for example, due to the tendency of a silver or other metallic filler to agglomerate and potentially block dispensing equipment. Die-attach adhesives provide useful properties for selected applications and are an adhesive variant which may be considered for certain types of LED packaging applications.

8.5.6 Relative Bulk Thermal Conductivity of Materials

Relative thickness of a TIM material is a significant design factor for overall performance. A general rule may be considered: the thinner the TIM material in the application, after clamping force is applied, the better the overall heat transfer achieved (assuming reasonable surface wetting and reasonable thermal conductivity). The relatively low thermal conductivity values for virtually all TIM materials as compared to a soldered or metallic joining material is the primary reason for this general guideline.

Relative bulk thermal conductivity is important. Note that some materials are anisotropic, with a different value for through-plane versus in-plane thermal conductivity; the difference can be substantial. A brief example of relative typical thermal conductivity values is shown in Table 8.5.

The values shown in the adjacent chart should give a clear indication why sharp improvement in bulk thermal conductivity of TIM materials has been a primary focus of many research and development projects recently, as the role of TIM materials has gained greater attention in industry.

Bulk thermal conductivity of material cannot stand as the sole determinant of overall TIM performances, as has been stated, but must be considered in combination with other factors that can strongly influence final performance.

8.5.7 Rheology, Particle Shape, and Thixotropicity

It should be noted that the rheology of thermal interface material compounds and the shape of particles used as fillers are critical to thermal performance and mechanical behavior of these compounds. The use of diamond particles has been examined by many vendors and research institutions as a logical choice for a filler (given the high thermal conductivity value range for all forms of manmade diamond powders) to create a high thermal conductivity thermal compound; this has also been examined as a constituent for PCTIMs. The shape of common diamond particles, which are typically platelets, has proven to be an obstacle to development of such improved TIMs; the platelets tend to have high propensity for agglomeration, which presents significant difficulties in maintaining proper mixing and suspension for use in dispensing systems.

Similarly, silver particles as a filler for higher conductivity thermal compounds have proven challenging as the silver metal is heavy and the maintenance of a properly dispersed suspension for preparation, handling, and dispensing during manufacturing assembly of finished electronic systems is a difficulty. Silver and diamond particles also are aggressive and have a propensity to abrade interior surfaces of dispensing equipment, depending on the shape of the particle chosen for manufacturing of a compound.

Relative thixotropicity of compounds refers to the propensity to hold shape as dispensed, as opposed to a relative propensity to flow once dispensed. A highly thixotropic compound that is dispensed from a needle dispensing system will hold the diameter and form of the dispensing needle, until a component is pressed into place on the compound. This high degree of thixotropicity is highly valued for cure-in-place compounds, thermal greases, phase-change compounds, PSA TIMs, and other TIM materials where this characteristic allows a dispensed dot or line of such a compound to be placed, prior to assembly of the LED or heat sink into the final product assembly.

Thixotropicity is also important and useful for a PCTIM or a thermal grease, for example, when a dispensed TIM is placed on a surface and the heat sink or other mating component is then placed and the assembly is fastened. The highly thixotropic compounds will form a fillet of any surplus TIM compound at the periphery, if properly dispensed, and the highly thixotropic nature of a properly selected TIM will remain in place at the periphery of the joined components, without exhibiting run-out or migration even if the assembly is installed in a final product in a vertical or other mounting orientation. This is a critical property of properly designed thermal compounds and phase-change thermal compounds.

Mention should be made of the range of possible filler materials, used to introduce a constituent with high thermal conductivity to a carrier. Common fillers in current TIMs can be carbon-based fillers and graphitic materials, zinc oxide, boron nitride, other naturally occurring materials, metals in a variety of particulate shapes, particulates, and powders of differing shapes and diameters, and low-melting point metallic spheres.

8.5.8 Relative Costs

Thickness of material is a distinguishing factor due to the large differences between certain types of application requirements. As a general rule, very thick TIM materials are often required to "bridge the gap" in a contemplated design, between a heat-dissipating device (typically of relatively low power, such as 3 W) and a metal component such as an enclosure sidewall that is in close proximity. In these cases, which are very common in many types of equipment, the enclosure sidewall is seen as a very attractive and cost-effective heat dissipating surface; if the TIM can be used to fill the gap between the device and the wall surface; the heat load is conducted to the enclosure case. This eliminates the potential added cost of a separate heat sink and the assembly process labor. TIM materials that are designed for these general types of applications are frequently classified as gap-fillers, with a range of designed values for compliance, bulk thermal conductivity, and often a tackiness or other adhesive coating to provide the attachment and retention mechanism at minimum cost.

Gap-fillers are often relatively poor overall in thermal performance values but fulfill a very useful function and there is a large market for such materials. Often, a theoretical view would suggest that such materials offer little value as a TIM; in practice, the low relative thermal performance value is of considerably less importance as compared to the cost and simplicity of eliminating an additional heat sink.

The reason for elaborating on these distinctions is simple: a general rule of thumb for TIM material effectiveness and selection is to choose the thinnest possible material for a given application. In gap-filling applications, this general rule tends to be overridden completely in order to serve the intended function of providing a relatively thick material of some known bulk thermal conductivity to fill the gap, often with an adhesive layer applied on both contacting surfaces (which may further reduce apparent thermal performance in the theoretical model). As the British might say, "Mind the Gap!" In these gap-filling roles, a material that may be generalized as greater than 250 μm (0.010″) in thickness is considered to be classified in the gap-filler category. Again, practical considerations are the rule of the day.

Gap-fillers are available from many suppliers and have a rather different set of design and development criteria, reflecting their unusual role in relation to more traditional, very thin forms of high-performance TIMs. The importance, however, of gap-filler TIMs in many industry segments (such as automotive, consumer audio, LED luminaire, and memory components should not be underestimated.

8.6 Development Areas for TIM Materials

There are three major areas of focus for TIMs, all of which are important:

1. *Thermal performance and assembly performance improvements*: Continued and substantial effort is made by both industry and research universities to develop

TIM materials with improved performance characteristics. As may be evident to the reader already, continual development of electronic systems introduces new market needs and requirements for thermal and packaging materials of many types. Miniaturization, a hallmark of electronic industry progress, is one factor; increases in heat flux, component density, and many other changes across the industry impact TIM material requirements. Improvements in thermal material performance can have a significant impact on electronic component reliability. Product areas receiving intensive focus for research and development in industry today include:

a. Investigation of new material concepts to meet requirements for large increases in bulk thermal conductivity, to drive thermal resistance of a high-performance TIM to significantly lower levels (assuming interfacial resistances can be resolved). This theoretical approach does risk ignoring other contributing factors to overall thermal performance, in initial stages of research and academic studies, but has brought more understanding in research institutions of the importance of requirements for surface wetting, compound stability at temperature, practical manufacturing application methodologies, reworkability, and long-term stability under operating conditions. Performance-focused development programs globally have focused on the implementation of nanotechnologies within thermal interface and other thermal materials, given the extremely high bulk thermal conductivity of various forms of carbon nanomaterials. Laboratory and theoretical investigation of carbon nanotechnologies has become a primary focus area for research organizations globally, both within industry and academia. Research in this area includes a central concept of note, the development of a "bed-of-nails" array of vertically-aligned CNTs with a practical method of providing an interconnect path at the heat source contact surface and the exit surface of the array. This research area continues to receive significant funding and attention globally and may yet contribute a breakthrough material concept that can be demonstrated to be practical for industry use.
b. Metallic and other higher-conductivity stand-alone materials, including reflowable metals, and higher-conductivity constituent fillers that can be utilized to develop higher thermal performance TIM materials. New chemistries for compounds that exclude the use of silicone oils, improve TIM stability over time and temperature, resist compound dry-out and embrittlement, or which support new fillers and other constituents are similar areas of research. The use of diamond particles and platelets, long a subject of investigation by R&D groups within industry vendors and universities, includes examination of particle shapes to reduce tendency to agglomeration, for example. Liquid indium metal and low melting point metals as fillers have been investigated as a potential TIM material for certain computing system applications, and in some cases commercialized as new interface products, but may not offer properties and cost levels supportable for series production of LED assemblies. Requirement for a retention mechanism for an electrically conductive LMA, for example,

adds an additional mechanical component to the assembly and to the cost structure.

c. Practical solutions that meet requirements for ease of use and practicality in manufacturing assembly, including high-temperature-capable materials. Increasingly, market requirements in several electronics market segments require new thermal materials developments for high-temperature operation. Continuous high temperature operation is important, for example, for wide band gap power semiconductors such as silicon carbide, for aerospace applications and for subterranean well-drilling instrumentation, at temperatures approaching 300–450 °C. The power LED market needs for high-temperature operation already indicate a need for materials operating continuously in the 150 °C range, for example. These values generally are higher than operating junction temperatures for ICs and many power semiconductor die.
d. Material chemistries that offer improvements in long-term reliability and improved chemical and temperature stability, especially under conditions of environmental stress.

2. *TIM material testing procedures* are also receiving continued attention. As improvements are made in reducing thermal resistance of other components and materials within a semiconductor module, the thermal resistance of the TIM materials becomes a larger percentage of the total. More focus is placed on reducing the thermal resistance of each of the materials used in an assembly. This includes the TIM; as the improvement increasingly is focused on driving performance values to smaller and smaller numbers, the need for TIM testing systems that can adequately distinguish between increasingly thin materials with exceptionally small thermal resistance values is increasing.

Overall semiconductor system efficiency is an increasingly important focus in the global effort to improve electrical energy usage and efficiency, across all industries from hybrid vehicles to solar power generation and transmission to industrial motor drives and power LED assemblies. Even the efficiency of a material as minute as the TIM is a part of the overall goal of continued improvement.

An industry-standard procedure exists as a testing methodology for TIM materials: American Society for Testing and Materials (ASTM) D 5470-12. This standard is available from ASTM International and is discussed in a later section of this chapter.

Other test methodologies include transient or dynamic testing technique with specialized test equipment and procedures developed by highly specialized TIM material test equipment vendors; the use of thermal test vehicles (TTVs; some of which are made available from certain semiconductor manufacturers on a limited basis); infrared (IR) imaging cameras to determine temperature differences for an assembly or heat source; and others. Testing methodologies and description of the differences between each is described elsewhere in this volume.

3. *Education*, to improve understanding in the electronics design, manufacturing, and service sectors, of the importance of appropriate selection of TIM materials has grown in importance as the focus on improving system efficiency has grown.

The purpose of this primer is to aid in developing a better understanding of how and why TIM materials of different type function, in order to improve the selection of materials that are appropriate for a given electronic system design. The distinctions between the many different types and forms of TIMs is of particular importance to us, in assisting to improve understanding about why some materials are very useful from both strictly thermal performance value but also from a manufacturability and usability perspective, which is critically important to successful electronic system design.

8.7 Technical Challenges for Industry Development

New challenges for TIMs are driven by several important and widespread industry trends:

- Heat flux: Increased heat fluxes for semiconductor devices drive requirements for reduced thermal resistance by achieving reduced resistance or thinner material application, or both. In LED applications, the semiconductor devices include LEDs, drivers, and certain power components within the complete system design.
- Package materials improvements: Coincident with TIM material performance improvements which address rising heat flux is the need for development of overall improvement in LED packaging and heat dissipation components and fastening mechanisms, to lower manufacturing costs and streamline assembly costs. Fastener design is important because continued improvement in fastening component designs that increase clamping force applied and reduce assembly time will yield both reduced thermal resistance values and reduced overall assembly cost.
- Miniaturization: Across the semiconductor industry, continuing trends toward further miniaturization of heat-dissipating components of all types, especially for hand-held devices, result in demands for improvements in thermal materials.
- Large surface area applications: Conversely, there are needs for improved TIM materials which are applicable to assemblies with large contacting surfaces (such as for the attachment of a multiple-LED array on a single substrate to a heat sink or other metal surface), with adequate thermal performance without requiring precision machining of the two joining surfaces. The precision machining to achieve improved flatness, to lessen the need for a high-performance TIM, is an expensive solution that is not practical in a cost-sensitive market such as LEDs.
- Life testing and reliability: Market awareness continues to grow regarding the need for improved TIM performance *at end of life*, as compared to time zero, especially for paste forms of TIMs containing silicone oil and similar carriers that may be subject to outgassing, mechanical pump-out, or fluid migration and resulting dry-out at the interface. Typically, improvements in materials can be found by moving to different materials; in highly cost-sensitive applications, however, this solution may not be acceptable if different materials are higher cost solutions. A further discussion of aging of LED assemblies and materials used, including the impact of thermal grease degradation with temperature and time can

be found in [3]. Temperature range: Operation under increased temperature levels, for power semiconductor devices, including GaN and SiC devices capable of die operating temperatures in the range of 125–200 °C, for military, aerospace, and vehicle electrification. There are also requests for TIM materials that will operate reliably at much higher temperatures (to 290 °C and higher); applications for downhole well drilling are a good example of where exceptionally high operating temperatures for certain components may be found.

- Assembly: Simplified and automated handling and assembly requirements for high volume manufacturing.

8.8 Thermal Interface Material Testing Methodologies

Characterization of TIMs is a subject that has received significant attention in the last two decades, driven by the growing importance of TIM materials in electronic systems; the need for greater precision in test instruments; continuing industry needs to test an increasingly diverse range of materials types including two-phase materials; and the desire to accurately replicate test results under similar conditions. An early overview can be found in [20].

The demand for greater precision in test instruments arises from the very small thermal resistance values that must be measured with the highest performing TIM materials. The development of newer materials includes those that ideally offer very high thermal performance and which may be required to have extremely small in-situ thickness values; test methods are therefore needed to measure very small values quite precisely.

An important point for this subject is the increased range of characterization goals needed beyond simple thermal performance: material behavior over time (and, for example, in high temperature and/or high moisture or other harsh environments); accelerated life testing; potential for outgassing, toxicity, embrittlement, and for environmentally harmful constituents; and similar advanced testing capabilities.

Qualification testing by an equipment manufacturer of a selected thermal interface material for an intended product assembly design must include in the most basic terms:

1. Determine that the selected TIM material will provide the needed heat transfer performance when the complete product assembly is switched on for the first time.
2. Determine that the selected TIM will provide essentially the same tested heat transfer performance, without degradation, through all expected operating conditions and without evidence of loss of material due to temperature change, mechanical stress placed on the assembly, or due to inherent material properties for the expected use and operating temperatures, transportation and storage temperatures, relative moisture, mounting attitude, and similar conditions suitable for the complete assembly.

3. Specified production application process requirements (available application equipment and appropriate dispensing process technique, suitable storage and open-container life times, lack of toxic or environmentally damaging constituents, and similar).

There are many different types of electronic products that must incorporate TIM materials, designed for many types of manufacturing, assembly, and operational environments. Certain types of electronic products operate in more rigorous environments. An LED assembly used in an industrial lighting fixture designed to deliver very high luminous flux may be installed in a non-air conditioned warehouse near a metal roof exposed to full solar conditions, for example, in a remote location; operating temperatures of the high-power device in a high ambient temperature location may expose certain types of TIM materials to extremely high operating temperatures. Conversely, an LED warning beacon designed for mountaintop antenna masts or for marine use on ocean buoys may be exposed to extremely low ambient temperature conditions, below the ability of an organic TIM to function as expected.

8.9 Purposes for Thermal Performance Testing

The overall purpose of thermal performance testing is to determine performance over a defined range of specified parameters largely intended to replicate expected usage conditions and product requirements. There should be an awareness of two basic purposes for which thermal performance testing is used, and the reasons for them and the differences in the data generated.

Thermal interface material testing methods for thermal performance serve these two basic purposes:

a. Standardized controlled conditions testing in the laboratory: Precise measurement of individual materials under standardized, carefully controlled laboratory conditions is used for the purpose of TIM material development, by vendors, to determine if a potential new material under test meets rigorous market requirements. In these tests, as many test parameters as possible are delineated with precise control to ensure that possible sources of test error are eliminated. As examples: uniform test head materials are used, mounting surfaces are precision machined, precision temperature sensors are applied properly in known locations, clamping force applied is measured precisely, pressure is applied with parallel mating surfaces, a known one-dimensional heat flow is applied to the system, and the test system is calibrated prior to beginning testing. Careful control of all parameters is made to ensure that every potential source of variation from test to test and test error is eliminated to produce directly comparable empirical results under known conditions. The goal of this type of test is to produce standardized test results that can be compared to a known material, to determine if a development material meets thermal performance requirements and therefore is worthy of continuing development work to move toward release as a new TIM material product offering from that vendor. This same standardized test data with known parameters is then added to TIM material technical data sheets when the material is commercialized.

b. Product design "in-situ" testing: Practicing thermal engineers test TIM materials to determine performance of a selected material in an LED product design under development. The first step is to choose a reduced set of candidate TIM materials, which requires using the data sheet values provided by TIM vendors using the standardized controlled-condition test methods (identified as Purpose "A," above). This first selection step with data sheet values is made to reduce the many available materials to a reasonable group (perhaps 3–6 candidate materials) that can be cost-effectively and efficiently measured during the design and material selection process. Now, this is the typical test requirement which OEM mechanical engineers face, where the test conditions are not necessarily as carefully controlled and where compromise must be recognized on mating surface flatnesses, surface roughness, precise control of clamping pressures, manufacturing process controls, and other factors. For example, the cost of machining metal surfaces to provide a polished and flat surface typically is higher than can be considered for the manufacturing requirements for a commercial LED fixture. Therefore, reasonable specifications must be written for surface finish and flatness, knowing that both will impact TIM material performance in production assemblies. When a product design engineer responsible for selecting a TIM for a given LED system tests a selected material under these conditions, a common statement is that the thermal performance test results do not compare well with the published performance value available from the vendor data sheet for that material. This is a challenge in understanding TIM thermal performance test results and is not often well understood. Nonetheless, the OEM product design engineer must make a selection and typically time and resources are limited, which in turn limits the number of materials that can be tested under these "in-situ" conditions.

A simple way to state the use of test data from these two very different testing environments is that the vendor produces test data in standardized and idealized circumstances with as much random variability removed as possible. This idealized Purpose "A" data are typically inserted into a TIM product data sheet by the vendor, for use by the product design engineer who is considering use of a TIM material for module design, to compare thermal performance of that material to another from a different vendor, in order to reduce the number of materials to be moved to the next stage of consideration. Once a small set of materials to be tested is selected on this basis and, knowing what other specific conditions exist in the LED module design in process, the product design engineer then proceeds to "in-situ" testing of this much-reduced number of possible candidates. Completion of in-situ testing will result in "Purpose B" test values that more closely reflect actual performance in that specific LED module design in series production, an important distinction from the initial data sheet test value. The design engineer then selects from the in-situ results the one material that appears to provide the best results under the actual product usage conditions expected. Note that maximum thermal conductivity and/or maximum thermal performance values alone are not always the only guideline for selection of a TIM. Note also that both test purposes have significant value and understanding the difference between these purposes is important.

8.10 Thermal Performance Test Variables

Let us compare the Purpose "A" directly comparative all-variables-removed laboratory testing methods with what is experienced in a defined set of application conditions. What is frequently termed as "in-situ" testing is the simple expedient of applying a TIM material in a physical product design under consideration and determining the performance of that material in that application. The values that result from this type of "in-situ" testing can frequently be significantly different from the published test values that appear in the vendor data sheet for that same material.

This is a point that is very commonly misunderstood. The "in-situ" situation allows for introduction of what may be a wide range of variables that impact overall performance of the TIM material under test; this approach is also considered to be highly application-specific. For example, testing of an LED array that is to be fastened to a heat sink for an industrial lighting fixture will include variables in the application that are less well controlled or, in fact, which in use by the final customer may not be controlled in any way. Examples of such variables in an application, occurring either during product assembly or during operation in a final installation, include:

- Heat sink surface flatness which is unknown or poorly controlled may lead to TIM voiding
- Failure to properly control application of TIMs which require dispensing, screen printing, or similar processes may result in inappropriate amounts of TIM material present (either too little or too much, or unevenly applied)
- Failure to torque fasteners sufficiently will lead to improper clamping force applied to the TIM
- Improper assembly of mating surfaces leading to lack of coplanarity and uneven TIM thickness
- High-bay mounting in an industrial facility may include exposure to high ambient temperatures
- Non-air conditioned environments may introduce exceptionally high moisture content in the air
- Warping of surfaces may occur due to temperature changes over time or uneven heating
- Chemicals, gases, or oils may be present in the ambient atmosphere in a manufacturing operation during product assembly or may be present in a final product installation in an industrial facility
- Disassembly or replacement of a component may mean that the TIM material is dislodged or mishandled by a service person.

It is therefore desirable to understand the differences between test results under actual product usage conditions and the controlled laboratory conditions of a standardized test method.

Using the vendor-supplied thermal performance values generated with a standardized test method in the laboratory under controlled conditions should therefore be viewed as the idealized values possible with the material under consideration. Operational influences for actual product applications may therefore lead to reduced performance. Testing under these "in-situ" conditions in a product assembly, such as the LED lighting fixture design for industrial plants (the example used earlier), is the step which a product design engineer considering use of one or more TIM materials will wish to undertake.

8.11 Steady-State Thermal Performance Test Methods

Measurement techniques assume laboratory controlled conditions in a steady-state condition, wherein a temperature equilibrium must be reached with a defined single-dimensional heat flow passing between two calorimeter bars equipped with resistance temperature detectors (RTDs), one bar being the hot zone and the second bar being cooled. Each calorimeter bar mounts two to five RTDs spaced equally through the bar; the RTD incorporates an electrical element and resistance of the element is correlated with temperature. Heat is passed in one-dimensional flow from the hot zone through the material under test to the cold zone.

American Society for Testing and Materials D 5470-12 Standardized Steady-State Testing The best-known of the controlled condition laboratory test methodologies is ASTM D 5470-12, a standard published and revised with improvements since 1993. This is the primary steady-state method. Most competent TIM material vendors publish reference data sheet thermal performance values that are developed using this methodology. This may be said to be a technique that attempts to provide relatively idealized test results, or best achievable with tight control over all facets of the test set-up. This method can be used for rigid, fluid, and composite TIM material types.

The technique applies the calorimeter principle by utilizing two isothermal meter bars with temperature sensors embedded in each, separated by a known thickness of the material under test. A known heat load is applied and flows across the two contacting meter bar surfaces and through the test material while a measured clamping pressure is applied. The meter bars may be thought of as heat flux meters and are typically machined from aluminum or copper; brass can also be used; an alignment and coplanarity vision system may also be used to ensure the greatest control over precise measurement of uniform material thickness.

Once the system has reached a temperature equilibrium, voltage, current, and temperature readings are taken. The essence of this methodology is to establish a steady-state condition of heat flux across the two surfaces of the hot meter bar, the test material, and the cold meter bar. The thermal resistance is defined per unit area[2] (what may be termed *R*-value, K-cm^2/W) and includes both the thermal resistance of the material and the contact thermal resistances at the two interfaces (the interfacial contact resistance) at the meter bar surfaces.

Thermal resistance per unit area is calculated for a single layer of material of a known thickness, given the known area (A) of the contacting surfaces. The method can be used to determine a thermal conductivity for the material, as the material thickness is known, by measuring the thermal resistance values for different thicknesses and plotting these values versus thickness. As the plot of resistance versus thickness is linear, the slope of the line is proportional to the inverse of material bulk thermal conductivity. Once plotted, the intercept value where thickness is zero is the sum of the contact resistances.

For preparation of thermal performance data that are to appear in a TIM manufacturer's technical data sheet, it is most useful to include the test method used (in this case, ASTM D 5470-12), the meter bar or test head area, surface finish and flatness conditions, the amount of power (W) applied, as well as the calculated thermal resistance versus thickness graphical data for each material tested. In this manner, a potential user equipped with a similar test stand can replicate data under similar conditions.

The latest revision of ASTM D 5470 was published in 2012. Note that this document describes the concept and the general form of the technique but does not attempt to describe the test equipment in complete detail; while many companies have built test equipment to follow this technique, only recently have several test equipment vendors entered the market with off-the-shelf equipment designs that are available for purchase. The current version of this test method is published by ASTM International [21].

The original ASTM standard was intended for use in testing solid TIM materials that were prevalent at that time, such as filled rubber pads with only modest compliancy and so-called gap-fillers, both types of which were rather thick and required high clamping forces. Several categories of materials were identified (e.g., "Type I," "Type II") as broad identification of characteristics. Subsequent revisions made to the standard now include provisions for testing with a wide range of clamping force applied. Typical TIM manufacturers' data presentations show thermal resistance versus clamping force; a common range is 1–13 bar (15-200 psi).

An important development in the most recent decade has been the development of commercial test equipment that follows ASTM D 5470-12, available from several vendors, and which incorporates increasingly sophisticated software and control mechanisms, increasing throughput and reproducibility of results.

A distinction in thermal characterization that is made at the beginning of this characterization section is the difference between testing under highly controlled conditions and that done under what is termed "in-situ" testing for final product

---

2 Often improperly termed "thermal impedance". See, e.g. Lasance et al. [2], "Challenges in Thermal Interface Material Testing", Proceedings, 22nd IEEE SEMI-THERM Conference, Dallas TX USA, March 2006, and the Editorial of the September 2010 issue of Electronics Cooling by Lasance.

design. A primary difficulty that TIM vendors must face in preparing thermal performance test data for a data sheet or similar purposes is that the vendor does not know how a material once commercialized may be used by every potential customer design engineer and does not control the many variables that apply to differing end product designs. Note that in microprocessor applications, as an example, the clamping force applied in attaching a heat sink assembly to a processor may be as low as 0.6 bar (8.7 psi), to avoid damaging fragile leads for an organic package. In a second example, in an end product design where a thermal solution must be designed for power semiconductor modules, the clamping force specified for heat sink attachment may exceed 9.0 bar (130.5 psi); these modules are designed with heavy copper baseplates that must adequately dissipate much higher loads and are intended to be attached with high forces applied to maximize TIM performance.

Given the very wide range of application clamping forces which TIM materials may be exposed to, in different electronic product markets, the TIM vendors therefore must publish thermal performance over a wide range of values.

Steady-State Testing with TTVs Development of application-specific test vehicles has become important in the IC market, including for microprocessors, graphics processors, and certain types of more specialized ICs. The concept for TTVs is to develop a semiconductor package which replicates the package type, materials, and functional appearance of what would otherwise be a production IC package or module, but which incorporate a specialized die with 4–32 sensors incorporated into the semiconductor die itself. These sensors can typically be independently controlled, to generate heating at different location and different levels across the surface of the IC; this is intended largely to replicate the expected thermal map (representing varied heat fluxes across the different functional areas of the IC die) and enables temperature readings to be made directly, without introducing any intrusive external measurement devices (such as a thin wire or thermocouple or RTS) in the interface.

TTVs are developed by some microprocessor and other semiconductor manufacturers for this very specialized thermal resistance measurement technique. These TTVs modules replicate the curvature of the surface of the silicon semiconductor die, which is the contact surface for the TIM1 material, and other features (roughness, clamping force for the designed heat sink assembly, etc.) and provides the ability to measure very small changes in heat flux across multiple zones on the die.

This type of thermal performance testing has not been made generally available for RF, power, and LED semiconductor devices as the relative heat flux across the die surfaces for power semiconductor devices is more uniform.

For LED design, many design engineers may find that the specified LED itself may be used for effective testing of a TIM, using the intended hardware design with selected fasteners and heat sinks; alternatively, a power diode or power transistor with a heat slug or metal package body similar in size to the LED base may be sufficient and may also be a rather inexpensive approach. This is the case as due to reduced concern regarding nonuniform die heat flux, as compared to ICs.

Thermal test die have been developed and made available commercially from at least one specialized thermal performance testing equipment manufacturer, along with thermal test wafers and associated thermal test equipment, and the LED design engineer should be aware of this testing option [22, 23].

8.12 Transient Test Methods

The development of time-dependent (so-called "transient" or "dynamic") test methods for TIM materials has been driven in large part by the desire to increase accuracy, reduce uncertainty, and increase reproducibility of data from test to test and lab to lab. Increasingly, transient techniques allow for testing with a specific semiconductor package and therefore more closely approach what has been termed in-situ test techniques. A significant value of the use of transient techniques is the ability to produce test results rapidly, without a requirement to reach temperature equilibrium. Transient testing techniques are considered to also yield results with improved reproducibility, from lab to lab, with greater accuracy. An additional advantage is that only temperatures at one location need to be measured; the heat flux does not enter the equation. Importantly, as dimensions shrink it becomes increasingly more difficult to apply steady-state techniques and transient testing becomes the only remaining practical technique. A disadvantage is the increased use of mathematic functions in deriving final data, but this problem can be solved as increasingly user-friendly interfaces for these systems are introduced.

Transient testing techniques included such methods as:

- Laser flash diffusivity
- Structure function mathematics applied with specialized software, from electrical data
- Model fitting
- Modified hot-wire (MHW; following ASTM C 1113 and ISO 8894)

One important transient testing technique also utilizes two isothermal meter bars with temperature sensors to measure temperatures at several points. The use of specialized software results in the development of so-called differential structure functions; measuring with and without the material under test provides a shift in position of the resulting curve, reflecting the thermal resistance of the TIM material under test [24]. In the original publication, the shift of a peak appearing in the differential structure function was suggested for use as an indicator of TIM quality. This principle later became the basis of any structure function-based TIM testing methodology. As the peak position may be uncertain due to numerical issues of the data processing software, the cumulative structure functions are used today to assess TIM quality: the "length" of the structure function section corresponding to the TIM layer in the heat-flow path is a more exact indicator of the thermal interfacial resistance. Transient measurements turned into cumulative structure functions combined with the principles of the ASTM D5470-12 test standard resulted in a very robust TIM testing technique[3].

A strong advantage of this transient technique is the ability to measure an assembly, such as a semiconductor package, TIM, and heat sink—with whatever fasteners and clamping pressures may be specified—and the functional semiconductor device is powered. With controlled ambient conditions, this technique can identify thermal resistance per unit area (*R*-value, or R_{th}); this technique and the associated equipment is highly specialized but exceptionally useful for thermal and electrical analysis of a complete LED package and, perhaps even more significantly for system designers, for a full LED assembly including certain optical parameters. In other words, this equipment and technique can be utilized for significantly more functions than simply measuring thermal performance of TIM materials alone.

A further strong advantage of this transient testing technique and the associated equipment is the ability to make measurements with variations of test material thickness, in-situ [25].

Thermal transient testing in combination with a well-defined measurement environment may also provide an accurate and highly repeatable method for bulk thermal conductivity testing. Applying the principle of the ASTM D5470-12 test standard, this can be accomplished by replacing the hot meter bar with a functional packaged semiconductor and the cold meter bar with a temperature-controlled pedestal that matches the heater/sensor semiconductor device package's geometry. Placing the TIM to be tested between the grips and setting up different thickness levels with high accuracy results in R_{th} versus thickness plots as defined in the ASTM standard. Although the derived data are based on steady-state results, the transient part of the measurement ensures that the data are consistent by indicating immediately any change in the test setup. Observing the transient curves, the user can also assure that the steady-state temperature data are reached and can optimize total measurement time. Although the tests are in principle thickness-controlled, three different measurement modes are proposed for the three different classes of materials referenced in the ASTM D-5470-12 standard:

For Type I materials (such as thermal compounds), strict thickness control is recommended, without maintaining any pressure on the sample, assuming that due to its low viscosity, the excess material leaves the void between the grips as the thickness decreases during the test.

For Type II materials (viscoelastic solids, such as gap-fillers) thickness control with continuous pressure is recommended, making sure that the material is kept at the target thickness. Setting up a pressure limit is advised, so the system can identify the minimum achievable thickness. As this value influences the in-situ R_{th} of the material, it should be identified in the results.

For Type III materials (noncompressible solids), pressure control is advised. This way the measurement of different samples at different thicknesses is possible, assuming that the contact resistances remain the same due to the comparable pressure

[3] This technique and the resulting commercial test apparatus [28] is one of the results of the NANOPACK project [18, 29].

at each tested thickness. The use of a high thermal conductivity compound is recommended between the surfaces of the sample and the grips in order to reduce the contact resistances. A disadvantage of this technique is the relative cost of the highly specialized equipment necessary. For more information on transient testing please consult references [26–28]. On the other hand, with the transient test equipment necessary for the test setup not only TIM testing capability is provided but any common semiconductor thermal testing according to the JEDEC JESD51 series of standards is also feasible.

Laser flash diffusivity is another transient testing technique that is used for measuring thermal conductivity of solid materials. This technique requires the generation of a short laser pulse that heats a thin layer of the sample material surface. Temperature on the backside of the sample material is monitored versus time and a curve of these variables is fit to a one-dimensional heat flow equation. The purpose is to measure the transit time of the heat pulse generated, through the sample material. For certain types of materials, a coating may be necessary applied on the surface for measurement purposes. A thermal resistance value may be calculated for thicknesses of a known sample size, although the primary use is typically for thermal conductivity testing of solid materials alone.

Note that laser flash methods are not well suited for attempting application-specific (in-situ) testing, as the samples are isolated. Laser flash diffusivity test systems are generally relatively expensive and in-situ testing is not usually possible.

A modified hot wire test methodology for measuring thermal conductivity of dielectric TIM materials having moderate thermal conductivity follows the general concept of the ASTM C 1113-09 standard, is a transient system, and utilizes a very fine platinum wire embedded in the sample material. While there are several drawbacks to this technique for use in measuring what are typically thin thermally conductive TIM materials, this method is most useful for relatively thick gap-filler materials that are electrical insulators and for other materials that may be useful in certain types of LED fixture designs.

This methodology has two important drawbacks: only isotropic materials can be measured using this technique and only electrically nonconductive materials may be measured.

The MHW technique employed for one test equipment manufacturer's test systems measures temperature versus time and does not require use of a known conductivity reference material, can be used with materials of considerable thickness, and is used to measure thermal conductivity in a technique which can be automated. (Note that there are certain exclusions in the ASTM and ISO 8894 standards that must be considered, including the specific exclusion of graphite-containing materials; graphite is used in a number of TIM materials of various types.) See also [29, 30].

Although the use of die-attach adhesives and solders as TIMs has not been discussed in detail in this chapter, a set of comparative test data is available which describes results from differing systems for placement of eutectic solder die-attach materials in preform and paste formats as what could be considered to be TIM1 applications for LEDs in a die-attach function, assessed in comparison to a eutectic AuSn metallization with differing deposition techniques [31].

Other performance testing for TIM materials can include thermal cycling and power cycling for reliability analysis. This type of testing can demonstrate important information regarding how a material behaves over hundreds and thousands of cycles, which are designed to stress the material, and how the constituent fluids and solid fillers behave with temperature extremes.

8.13 Conclusions

TIMs are frequently an insignificant cost component within a semiconductor package or a completed electronic product, such as an automotive LED forward lighting assembly. The relative insignificant position on the accountant's tally of bill of materials cost is often misleading as to the importance of this odd and often overlooked material, for a TIM material failure can appreciably affect the life and reliability of the product.

Starting with what today appear to be rather simplistic thermal greases and silicone rubber pads, the development of a striking array of applications across every segment of the electronics industry has led to an explosion of TIM materials concepts in the most recent decade. The many different applications in which these modest-appearing materials are found has also meant a continuing stream of requirements for increased thermal performance, improved reliability under increasingly challenging operating conditions, and eternal demands for rapid handling and placement techniques, improved reworkability, and lower total costs.

Graphitic materials, new forms of nonsilicone compounds, metallic TIMs, and phase-change compounds have brought new performance levels and significant improvements recently. Introduction of various forms of nanoparticle sizes and nanofibers appears to be imminent, with global innovation efforts to find practical structures and carriers for these concepts with such promise for large improvements in thermal conductivity.

Requirements for new material solutions with continual reductions in thermal resistance values, coupled with needs for greater precision and repeatability and the ability to test in-situ within a specific electronic product has generated renewed development in testing methods and commercially available test equipment. Understanding has grown rapidly of the need to consider the entire requirements envelope, not simply an increase in bulk thermal conductivity.

Improved tools for testing and analysis are now leading to improvements in industry understanding of the function and value of these rather unobtrusive materials that allow continued increases in electronic system operating performance and reductions in physical volume.

References

1. Viswanath R, Wakharkar V, Watwe A, Lebonheur V (2000) Thermal performance challenges from silicon to systems. Intel Technol J, Quarter 3 http://noggin.intel.com/technology-journal/2000/43/microprocessor-packaging
2. Chen S, Lee N-C (2012) High performance thermal interface materials with enhanced reliability. Proceedings, IEEE SEMI-THERM 28 Conference, San Jose CA USA, 22–24 March 2012
3. Poppe A, Molnár G, Csuti P, Szabó F, Schanda J (2011) Ageing of LEDs: a comprehensive study based on the LM90 standard and thermal transient measurements. Proceedings of the 27th session of the CIE (International Committee on Illumination), Sun City, South Africa, 9–16 July 2011, Vol I, Parts 1 & 2, Paper OP57, pp 467–477 (ISBN: 978 3 901906 99 2)
4. Hough P et al (2010) New developments for a no pump-out high performance thermal grease. Proceedings, IMAPS France 5th European advanced technology workshop on micropackaging and thermal management, La Rochelle, France, 3–4 February 2010
5. Papanu V Shaw R (2011) Comparative test data for TIM materials for silicon and silicon carbide substrates IMAPS France 7th European micropackaging and thermal management workshop, La Rochelle, France, 2–3 February 2011 http://france.imapseurope.org/index
6. Rauch R, Understanding the performance characteristics of phase-change thermal interface materials. Proceedings of 2000 Inter Society Conference on Thermal Phenomena
7. Teertstra P (2007) Thermal conductivity and contact resistance measurements for adhesives. Proceedings of ASME InterPACK 2007, IPACK2007-33026, Vancouver BC, Canada, 8–12 July 2007 ISBN-13: 978-0791842775, http://www.mhtlab.uwaterloo.ca/pdf_papers/mhtl07-6.pdf; http://www.sensorprod.com/news/white-papers/2007-07_tcc/wp_tcc-2007-07.pdf
8. DM6030 Series high thermal conductivity epoxy adhesive pastes, data sheet, DieMat Inc. 2005. Websites: www.diemat.com; www.namics.co.jp/e/diemat/index.html; http://www.icproto.com/typeroom/assets/uploads/pdf/DIEMAT%20DM6030.pdf
9. Die-attach Adhesives: Taking the lead with know-how. data sheet, 2010, Panacol-Elosol GmbH. Website: www.panacol.de; http://www.panacol.de/fileadmin/panacol/pdf_en/pdf_LED_en/bro_led_10_v1_gb.pdf
10. Cho-Therm® 1641 & 1642 Thermally Conductive Silicone Compounds, technical bulletin 62, Chomerics Division of Parker Hannifin Company, 1999. Website: http://vendor.parker.com/Groups/Seal/Divisions/Chomerics/
11. Martin Y, van Kessel T (2007) High performance liquid thermal interface for large volume production. Proceedings of IMAPS Symposium 2007, San Jose CA USA, 11–15 November 2007
12. Furman BK, Lauro PA, Shih DY, van Kessel T, Martin Y, Colgan E, Zou W, DeHaven P, Iruvanti S, Wakil J (2006) Indium based thermal interface materials. Proceedings, IMAPS Advanced Technology Workshop on Thermal Management 2006, Los Altos CA USA, 10–13 September 2006
13. Dani A, Matayabas C, Koning P (2005) Thermal interface material technology advancements and challenges—an overview. Proceedings, ASME InterPACK 2005 Conference, IPACK2005-73384, San Francisco CA USA, 17–22 July 2005 ISBN 0-7918-3762-9
14. HiTherm™ Flexible Graphite, data sheet, GrafTech International (USA). Website: www.graftech.com
15. He J, Van Heerden D, Subrumanian J (2006) Direct die attach using a room temperature soldering process. Tech Brief, Electronics Cooling Magazine, May 2006
16. Mattheau J (2010) Nanofoil localized heat source for precision bonding. Proceedings, IMAPS advanced technology workshop on thermal management 2010, Palo Alto CA USA, 26–28 November 2010
17. Schattenmann F (2004) Functionalized nano-particles for electronics packaging applications. GE global research/Binghamton University electronics packaging symposium, Niskayuna NY USA, 20 October 2004
18. Electrovac Wärmeleitpaste/Thermal Grease elNano® S27Z-2, Data sheet, Electrovac AG, Version 07 August 2006

19. Cola B (2011) Nanoscale polymer coatings for enhanced bonding and thermal conductance in carbon nanotube array interfaces. Proceedings, IMAPS advanced technology workshop on thermal management, Palo Alto CA USA, 7–9 November 2011
20. Lasance C (2003) Problems with thermal interface material measurements: suggestions for improvement, electronics cooling, September 2003
21. ASTM Standard D 5470, 2012 (2012) Standard test method for thermal transmission properties of thin thermally conductive solid electrical insulation materials. ASTM International, West Conshohocken PA USA, 2012, doi:10.1520/D5470-12; www.astm.org
22. Thermal Engineering Associates (2009) TEA TIM Test System. data sheet. Website: www.thermengr.com
23. Thermal Engineering Associates (2012) TEA TTB-5101, direct attach thermal test board. data sheet. Website: www.thermengr.com
24. Rencz M, Szekely V, Morelli A, Villa C (2002) Determining partial thermal resistances with transient measurements, and using the method to detect die attach discontinuities. Proceedings, 18th IEEE SEMI-THERM Conference, San Jose CA USA, pp 15–20, 12–14 March 2002
25. Vass-Várnai A, Sárkany Z, Rencz M (2012) Characterization method for thermal interface materials imitating an in-situ environment original. Microelectron J 43(9):661–668.
26. Lasance CJM, Murray CT, Saums DL, Rencz M (2006) Challenges in thermal interface material testing. Proceedings, 22nd Annual IEEE SEMI-THERM Symposium, Dallas TX USA, pp 42–49, 14–16 March 2006
27. Website: www.mentor.com/products/mechanical/products/t3ster/
28. Website: http://www.mentor.com/products/mechanical/products/dyntim
29. ASTM C 1113-09 (2012) Standard test method for thermal conductivity of refractories by hot wire (platinum resistance thermometer technique). ASTM international, West Conshohocken PA USA, 2012, doi:10.1520/D5470-12; www.astm.org
30. Chandler C et al (2001) Validation of the algorithm for direct thermal conductivity measurement with modified hot wire on film, insulation, and liquid. Proceedings, international thermal conductivity conference, Boston MA, 6–8 August 2001. Also, Anter Corporation, Pittsburgh PA USA, Website: www.anter.com; specifically: Anter Corporation Model 3141 thermal conductivity measuring system
31. Indium Corporation, Hartnett A Evans D Jr., Beck D, Homer S (2011, October 2011), Evaluation of test protocol for eutectic die attach using high-power LEDs. Technical Disclosure. Website: www.indium.com; http://www.indium.com/technical-documents/whitepaper/evaluation-of-test-protocol-for-eutectic-dieattach-using-high-power-leds#ixzz2VODogzDJ

Chapter 9
Heat Sink Basics from an Industrial Point of View

Clemens J. M. Lasance

Abstract Several other chapters (especially, Chap. 1 and 2) focus on the complexity of the physics that determines the performance of light-emitting diode (LED)-based systems. It is evident that temperature plays a significant role in optimizing light output and reliability. In many cases, the area available for cooling for a medium to high power LED is insufficient and hence area extension is required, usually by means of a local heat sink (using heat pipes to transfer the heat to where sufficient area is available is another option). Hence, an important way of controlling the temperature is by a proper choice of a heat sink. While in many applications the design freedom is limited, such as in retrofit LED lamps, the designer is still confronted with many questions: shape, number of fins, fin thickness, gap between fins, base dimensions, material, etc. Since literally thousands of heat sinks are available, many designers are confronted with the question: which one? Very often, the designer's choice is based on cost and manufacturer's data. Unfortunately, these data cannot be used with confidence because they are almost exclusively based on measurements in a closed duct, thereby disregarding bypass effects and inflow conditions.

The designer is faced with the following problem areas:

- Many new developments, in manufacturing as well as in layout (extruded, corrugated, pin fin, metal foam, narrow channel, enhanced plastic, etc.)
- Many criteria to base an optimal choice on (performance, weight, volume, cost)
- Manufacturer's data are often rather limited in practice because they often are derived for closed ducts
- Data need to include pressure drop quotations
- Standard correlations fail for many practical cases (usually only valid for confined heat sinks)
- Detailed CFD modeling with the purpose of arriving at the best heat sink for the application, especially at the system level, is in many designer's environments no option unless expert knowledge, time, a supercomputer and a calibration lab are available.

This chapter makes the designer familiar with the basics of heat sink heat transfer, shows first-order estimations, discusses the pros and cons of CFD modeling and treats a series of approaches in various stages of a design.

C. J. M. Lasance (✉)
Philips Research Emeritus, SomelikeitCool, Nuenen, the Netherlands
e-mail: lasance@onsnet.nu

C. J. M. Lasance, A. Poppe (eds.), *Thermal Management for LED Applications*, Solid State Lighting Technology and Application Series 2,
DOI 10.1007/978-1-4614-5091-7_9, © Springer Science+Business Media New York 2014

9.1 Heat Sink Basics

9.1.1 *Introduction*

This section starts with a short discussion of rules-of-thumb, to continue with showing that heat spreading is not a trivial issue but should be treated correctly. Many familiar design equations are based on often hidden simplifications, the so-called Murray–Gardner assumptions (as explained in latter sections), and a designer should at least be familiar with this approach that often fails in practice. One well-known notion in heat sink theory is the concept of fin efficiency, but several definitions exist. Finally, the role of radiation will be outlined.

9.1.2 *Rules-of-Thumb*

It is common practice for designers to try to increase heat dissipation to the limits of what is physically possible. In natural convection this limit is evident. The final limit is the area that is exposed to the environment and the maximum allowable ambient temperature. In other cases, the touch temperature of most metal enclosures is limited to about 60 °C as a consequence of safety reasons. With an ambient temperature of 30 °C, and a total heat transfer coefficient of 10 W/m^{2}K, the dissipation of a completely sealed enclosure can never be higher than 300 W/m^2. In forced convection-driven systems, the choices are greater but the noise generated by fans or jets imposes a severe limit in practice, not to mention the additional reliability issues introduced by moving mechanical parts and the decrease of the overall system efficacy of the actively cooled LED-based lighting systems. In liquid cooling the sky is the limit but this option is almost exclusively the domain of power electronics and space agencies, and here is a preference for cold plates, not heat sinks.

Let us first define what is meant by "rules-of-thumb." Rules-of-thumb are rules that provide only first-order estimates, and are especially useful in early design phases. Examples of useful rules-of-thumb are:

- Heat transfer coefficient for natural convection: 10 W/m^2 K
- Heat transfer coefficient for radiation ($\varepsilon = 1$): 5 W/m^2 K
- Thermal conductivity of multilayer board: 10 W/m K
- Formulae for flat heat sinks
- Integral methods based on conservation laws.

Rules-of thumb that are not recommended are:

- For every 10 °C increase in temperature, reliability decreases by a factor of 2
- Various formulae for extruded heat sinks
- Thermal resistance data, as published in vendor data sheets
- Natural and forced convection heat sink data, as published in vendor data sheets.

9.1.3 Heat Spreading

The basics of conduction, convection, and radiation heat transfer are covered in Chap. 3. This section focuses on the basics of heat spreading. Heat spreading is essentially area enlarging: the larger the area that is in contact with the heat exchanger (often the ambient air), the more heat can be removed at the same temperature difference. Unfortunately, even the simple case of heat spreading of a single source on a plane with uniform boundary conditions does not have an explicit mathematical solution. Hence, we have to rely on clever approximations or suitable computer codes.

A recent review paper [1] discusses many issues that are related to heat spreading. The paper compares the usefulness of a number of approaches that are generally in use to analyze heat spreading effects. It is shown that the popular series resistance approach has severe limitations in many cases when heat spreading plays a significant role. When we focus on heat spreading in heat sinks, the effects are less than when considering heat spreading in PCBs, but are not absent. Everywhere where we have a heat sink area that is larger than the source area, we face potential heat spreading problems. As a rule-of-thumb, we may state that when we have nonnegligible temperature gradients in the base of the heat sink we should take into account heat spreading. Fortunately, in many cases of practical interest the temperature gradients can be neglected. *Note*: we are not talking here about the gradients in the fins! Exceptions are: large-area heat sinks serving many light emitting diodes (LEDs) at the same time, heat sinks with a very thin base due to space restrictions, heat sinks that are much larger than the LED plus (metal core (MC)) printed circuit boards (PCB). However, designers tend to forget another important role of heat spreading: the one at the other side of the heat sink base. One should realize that a heat sink performs two very different functions:

1. To enlarge the surface area.
2. To spread heat.

It is the second function that causes a problem when we face a temperature gradient over the surface of an LED module that requires a heat sink. Suppose that we split the surface of our LED module into two areas: the central area, and the periphery. The heat spreading reduces the maximum temperature of the central area and increases the minimum temperature at the edges. Effectively, in terms of heat transfer coefficients, this means that the heat transfer coefficient of the central area is significantly increased. This effect is different from the effect caused by the first function of the heat sink, namely enlarging, effectively inducing a higher heat transfer coefficient over the whole area. The only way to solve this problem is to divide the top surface in at least two nested areas and to assess different average heat transfer coefficient values to both of them. In doing so, one needs to know which area to allocate to the central part. This is further discussed in [2]. The paper shows the enormous increase in the average central heat transfer coefficient compared to the total average value: 100 vs. 8 $W/m^2\,K$ for natural convection. Another important observation is that while the total heat flux through the top surface is not affected by the choice, the local heat flux is. Splitting the surface solves this problem also. In summary, it is recommended for analyses in early design phases to split the area connected to the heat sink to allow significantly different heat transfer coefficients accounting for the heat spreading effect of the heat sink base.

When heat spreading should be taken into account is mainly dependent on the ratio between the source and baseplate areas. The reader who is interested in calculating this effect is recommended to consult papers [3] and [4]. The final conclusion of these papers is that only in very special cases it is recommended to use analytical approaches, by far the safest way for a designer is to use specialized software such as a CFD code in conduction-only mode.

9.1.4 The Murray–Gardner Assumptions

About 70 years ago, Murray and Gardner formulated limiting assumptions to allow relatively simple fin analysis, which are now known as the Murray–Gardner assumptions. Designers need to be aware of these simplifications and should check regularly if their heat sinks still stay within these limits, especially for heat sinks that deviate from standard parallel plate extruded heat sinks. Notably, for LED applications often heat sinks are chosen with short and widely spaced fins for which the base area plays a relatively important role. These basic assumptions are the following:

1. The heat flow into the fin and the temperature at any point on the fin remains constant with time.
2. The fin material is homogeneous; its thermal conductivity is the same in all directions and remains constant.
3. The heat transfer coefficient between the fin and the surrounding medium is uniform and constant over the entire surface of the fin.
4. The temperature of the medium surrounding the fin is uniform.
5. The fin width is so small compared with its height that temperature gradients across the fin width may be neglected.
6. The temperature at the base of the fin is uniform.
7. There are no heat sources within the fin itself.
8. Heat transfer to or from the fin is proportional to the temperature difference between the fin and the surrounding medium.
9. There is no contact resistance between fins in the configuration or between the fin at the base of the configuration and the prime surface.
10. The heat transferred through the outermost edge of the fin (the fin tip) is negligible compared to that through the lateral surfaces (faces) of the fin.

9.2 Fin Efficiency

The effectiveness of enlarging the cooling surface area depends on the balance between conduction *in* the fin and heat removal *from* the fin. This section discusses the basics of fin heat transfer. Consider a longitudinal fin with a constant temperature T_w at the wall. Figure 9.1 is a sketch of the configuration.

Fig. 9.1 Sketch of a longitudinal fin

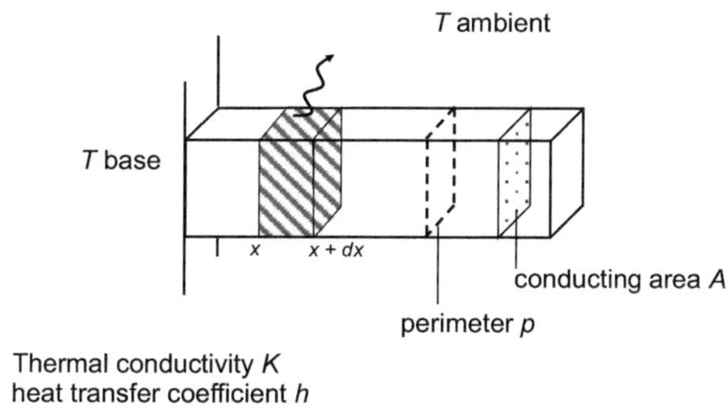

We consider the Murray–Gardner assumptions to be valid. After formulating the differential equation and solving, it is found that the general solution to this equation has the form:

$$T - T_{ambient} = c_1 \cosh(x/L_c) + c_2 \sinh(x/L_c) \tag{9.1}$$

with,

$$L_c = \sqrt{\frac{Ak}{hp}}$$

L_c, the characteristic length, has the dimension [m]. Let us consider the case of a fin with length L with an adiabatic tip end. We get the solution:

$$T(x) - T_{ambient} = (T_{base} - T_{ambient})\frac{\cosh(L/L_c - x/L_c)}{\cosh(L/L_c)} \tag{9.2}$$

The total amount of heat that is removed by the fin is equal to:

$$q_{total} = hpL_c(T_{base} - T_{ambient})\tanh(L/L_c) \tag{9.3}$$

From this formula, we can find the physical meaning of L_c. The amount of heat removed by the fin increases as the fin becomes longer. However, above a certain length adding more length does not result in more heat removal. For three times the characteristic length L_c, adding extra length results in only 1 % improvement, as is obvious from the table. In other words: for the 1D fin "infinity" starts around $3L_c$ (Table 9.1).

The maximum amount of heat that can be removed by the fin is given by:

$$q_{total} = hpL_c(T_{base} - T_{ambient}) \tag{9.4}$$

This is equal to the amount of heat that can be removed from a fin of length L_c at a uniform temperature T_{base}.

In literature, we often find the fin performance parameter defined as:

$$m^2 = \frac{hp}{kA} = \frac{1}{L_c^2} \tag{9.5}$$

Table 9.1 Dependence of tanh(L/L_c) on the value of L/L_c

L/L_c	tanh(L/L_c)
1	0.76
2	0.96
3	0.99

Note that m is not a dimensionless number. Two definitions are commonly used to describe fin performance:

Fin Efficiency (η_a) The relation between the heat removed by the fin and the heat removed by an ideal fin ($k = \infty$).

$$\eta_a = \frac{hpL_c \tanh(L/L_c)\ \Delta T}{hpL\ \Delta T} = \frac{\tanh\ (L/L_c)}{L/L_c} \tag{9.6a}$$

Or in a more common notation using *m:*

$$\eta_a = \frac{\tanh(mL)}{mL} \tag{9.6b}$$

Fin effectiveness (η_b) The relation between the heat removed by the fin and the heat that would be removed without a fin.

$$\eta_b = \frac{hpL_c\ \tanh(L/L_c)\ \Delta T}{h\ \mathrm{A}\ \Delta T} = \frac{L_c p\ \tanh(L/L_c)}{A} = \frac{k}{h} m\ \tanh(mL) \tag{9.7}$$

Obviously, η_b should be larger than one to render a fin effective. For very large h, for example as happens with boiling liquid, adding fins can result in a negative effect.

Kraus [5] challenges these two definitions and proposes a third one that would better suit the physics of the heat transfer. His basic problem with the standard definition is in his own words:

> The general idea of an efficiency as a performance parameter is sound; it is a dimensionless ratio comparing performance with a certain standard. However, the particular way that efficiency has been defined for fins compares every fin with a different standard; the actual fin performance as compared with what a geometrically identical fin could do if conditions were perfect. Two fins of different dimensions in the same environment may have the same efficiency but they may transmit different quantities of heat.

Kraus goes on by showing an example where a certain fin has a higher efficiency but transfers less heat than the other one. He points out that the fault is probably not in the efficiency concept but in the definition. The efficiency should have been called the *surface utilization factor*, because this is exactly what it is. He proposed two new parameterizations, one of them pertains to fin arrays, the other to singular fins and is coined the *thermal transmission ratio*, the ratio of the base heat flow to the base temperature excess:

$$\mu = \frac{q_b}{\theta_b} \tag{9.8}$$

In fact, this parameter could also be called the *fin thermal conductance*.

The validity of the assumptions of Murray and Gardner are challenged by a number of authors. One example is a paper by Maheswaran and Sekhar [6], who found by applying Monte Carlo simulation methods that the conventional formulations to evaluate the fin efficiency either underestimate or overestimate the efficiency of typical fins. However, most of these papers discuss the problem of a constant heat transfer coefficient for scales that are not common in our LED world.

9.2.1 Radiation

Usually the designer should not bother too much about the emissivity of the heat sink material because its contribution is generally less than for flat plates, unless the distance between fins is relatively large while the fin height is relatively small, such as often occurs in natural convection-driven LED systems. It is difficult to give some rules-of-thumb. We recommend a simple test when a global choice has been made: measure the temperature drop after painting the heat sink. Alternatively, when CFD codes are available, use the radiation option. However, in most forced convection applications radiation can be neglected. One should realize that even for heat sink materials with a low emissivity the effective emissivity would approach 1 because the radiation entering the space between the fins cannot escape.Also, please realize that the effective emissivity for short-wide channels could well be larger than one because the area available for radiation is larger than the projected area.

9.3 Heat Sink Modeling Approaches from a Designer's Perspective

9.3.1 Introduction

In this chapter we focus on heat sinks at the system level, not heat sinks per se. This introduces a number of complications for which the vendors cannot be held responsible, such as the fact that in many cases of practical interest we cannot speak about natural convection to characterize a heat sink because of the presence of many other sources, resulting in buoyancy-driven forced convection. Take an LED display. It is immediately obvious that, when addressing the reliability of a single LED, one cannot use its own dissipation to calculate the natural convection-driven flow. In addition, there is a big difference in ambient temperature between LEDs at the bottom and at the top. Also, most academic studies leading to the often-used (semi)empirical correlations focus on one heat sink only, with a well-defined approach flow, and either fully ducted or fully nonducted flow. Testing is only a serious alternative if a prototype is available, usually too late in the design process.

How about using CFD? Let us make an estimation of the number of cells that are required to cover at least the basic physics causing the transitional turbulence phenomena that govern heat sink thermal behavior. Assume a heat sink base area of 10 cm^2, with 20 fins of 10 mm length. Between the fins we need at least 10 cells, symmetrically nonequidistant, starting at 100 μm. We have 10 cells over the fin height, and 30 in the streamwise direction. This boils down to $200 \times 10 \times 30 = 60{,}000$ cells per heat sink. However, because of bleeding we are forced to add quite a number of cells around the heat sink, resulting in probably 200,000 cells. This is nowadays not a big deal when we have only one heat sink, but for sure we run into problems when we are dealing with tens of heat sinks, or a couple of big ones that are common in many LED-based systems, let alone the cooling of very large LED walls. Hence, full CFD is not a serious option in many cases in practice. We need to enter the domain of compact models, such as is the case for a long time in the integrated circuit (IC) world [7]. The additional difficulty is that we do not only need a compact model for the thermal part but also for the fluid flow part that should also contain information about the pressure losses.

There is one more difficulty associated with heat sinks designed for LED applications: the design freedom in the layout. For non-LED applications, the majority of heat sinks were of the type parallel-fin extruded, and a majority of compact models and design rules were developed for these types. The following Fig. 9.2 contains a number of LED heat sink designs showing the wide variety in shapes, and obviously many of them cannot be addressed using parallel-plate fin theory Fig. 9.2.

In summary, designers of LED systems need other approaches, in addition to the ones that are abundant in literature. Let us make a distinction between approaches that are useful for acquiring insight in the physics including the fluid dynamics, approaches that can be used when it comes to optimization of individual heat sinks, and finally approaches that are required for the final design stages.

This brings us the following list of potential approaches that will be discussed in more detail in subsequent sections.

- Approaches not recommended
- Approaches to acquire insight in the basic physics
- Approaches to acquire insight in optimization parameters
- Approaches to acquire insight in fluid dynamics
- Approaches suited for final design of heat sinks in LED applications.

9.4 Approaches not Recommended

9.4.1 Limits of Vendor Data

Around the mid-1990s, it was realized that there was a serious problem with the various methods of heat sink characterization by the vendors. The American Society of Mechanical Engineers (ASME) K16 committee took the initiative to try to solve this problem, and Belady [8] reported first results. Unfortunately, to our knowledge, no final report has been published, despite the impressive list of highly qualified

Fig. 9.2 A selection of heat sinks designs showing a wide variety

members of the committee. This seems to prove the fact that it is by no means simple to propose a standard that is satisfactory to everyone involved. We believe that not much has changed since its publication. One of the solutions proposed is to use a ducted wind tunnel test. While this solves many of the problems cited in the paper, one big disadvantage remains: when bypass is present the data are useless. Another serious shortcoming is that wind tunnel approach flow is significantly different from fan-driven approach flow. The conclusion must be that vendor data at best can be used for comparison purposes but should be treated with great care when these data are going to be used for design purposes.

9.4.2 Convective Resistance Models

The following sections are based on an article by Moffat [9]. Convective resistance approaches make use of a single thermal resistance invoking an effective surface area and/or an effective heat transfer coefficient:

$$R_{th} = 1/h_{\mathit{eff}}/A_{\mathit{eff}} \tag{9.9}$$

This is a legacy from the days of simple, low-density, finned heat sinks, but it does not accurately describe the thermal resistance of high fin-density heat sinks. In an age of searching for the most optimal design, this approach is not recommended except for very early design phases in order to get some idea about the heat transfer coefficient when the area is more or less fixed or vice versa, hence only for back-of-the-envelope calculations. A marginally better approach is to include the fin efficiency in the equation and add the surface of the prime area (the base area exposed to air):

$$R_{th} = \frac{(T_{base} - T_{air,in})}{q} = \frac{1}{hA_{eff}} = \frac{1}{h\left(A_{prime} + \eta_{fin}A_{fin}\right)} = \frac{1}{h\eta_0 A_{total}} \tag{9.10}$$

Often, the heat transfer coefficient is taken as the fully developed, laminar flow value for constant wall temperature. This value is independent of flow rate leading to the expectation that the thermal resistance of a heat sink should be independent of the flow through it. This is clearly not true, and hence many attempts are described in literature to find an acceptable value for the unknown h. However, what is the purpose of striving for better and better estimations for h, when at the same time the influence of the increasing air temperature is totally neglected? Therefore, a better "compact" model is the two-resistance model where the air temperature rise is taken into account.

9.4.3 *Two-Resistance Model*

Following Moffat, Azar and Tavassoli [10] proposed adding a "fluid resistance" term to the convective resistance to get the "total resistance".

$$R_{tot} = \frac{1}{h\eta A_{fin}} + \frac{1}{2\dot{m}c_p} \tag{9.11}$$

Simons [11] described this as a "caloric, or apparent fluid resistance, R_{fluid}" and showed its derivation in terms of the change in air temperature within the heat sink:

$$(T_{air,avg} - T_{air,in}) = \frac{q}{2\dot{m}c_p} \tag{9.12}$$

Hence,

$$R_{fluid} = \frac{1}{2\dot{m}c_p} \tag{9.13}$$

Simons points out that this equation is strictly speaking only valid for isoflux surfaces while in practice we are mostly dealing with (quasi) isothermal surfaces. Moffat shows clearly that the use of this equation leads to incorrect results in some limiting cases. The approach recommended is applying heat exchanger theory, discussed later. However, there is a more fundamental objection against using equation (2). Calling $1/(2\dot{m}c_p)$ a "resistance" simply violates physics. Analog elements such as resistors and capacitors must be based on the rate equations that describe the flow of a conserved property (one that is neither created nor destroyed within the component) in response to a potential drop that causes that flow. The physical laws of fluid mechanics, heat transfer, and mechanics identify the driving forces and flows. For example, Fourier's Law establishes that a temperature gradient is the driving force that forces the flow of heat by conduction. Regarding Eq. (9.11), the term $(T_{base} - T_{air,in})$ is clearly the driving force that drives the convective heat transfer rate and the convective resistance can be taken to be 1/hA. On the other hand, the term $(T_{air,out} - T_{air,in})$ is not a driving force in any recognized rate equation; it is simply the consequence of an energy balance on the fluid. Since we have no engineering rate equation that uses that as a driving force, we cannot use it as the basis for a resistance element. Maybe even more convincing, a resistance matrix is per definition symmetric: it does not matter in which direction the current flows. This is clearly not the case for the "fluid" resistor: the heat flows in one direction only, and even more serious, in the wrong one, because it flows from cold to hot. The only way to circumvent this problem from a circuit analysis point of view is to make the fluid resistance "voltage"-dependent. In summary, the process described by the rise in temperature of the air is not, itself, a heat transfer process, it is simply a mass-associated transport of thermal energy.

9.4.4 Empirical Correlations

First of all, we should make a distinction between (semi)analytical correlations and empirical correlations. While analytical correlations are more or less based on basic physics, most empirical correlations are of the type: There is a linear dependence between the number of churches and pubs. A thorough analysis of the problems associated with the use of correlations in real-life is provided in [12]. Regarding heat sinks, a plethora of papers are devoted to increasingly sophisticated correlations. However, one should always check if the correlations keep their validity outside the domain tested including nonwind tunnel and nonsingle source boundary conditions. Our advice for designers: stay away from empirical correlations, and focus on some other approaches from the list to follow.

9.5 Approaches to Acquire Insight in the Basic Physics

9.5.1 Heat Exchanger Model

Moffat again: Heat exchanger theory was developed to deal with two-fluid systems: a hot fluid transferring heat to a cold fluid. The nomenclature reflects this history. The three most important terms used in applying heat exchanger theory to heat sinks and cold plates are: Effectiveness (ξ), Capacity Rate ($\dot{m}c_p$), and number of transfer units (NTU) ($hA_{eff}/\dot{m}c_p$). The term NTU is sometimes referred to as the "dimensionless size" of the heat exchanger: the ratio of the convective conductance to the flow capacity. The higher the NTU, the more effective the heat exchanger.

The effectiveness of a heat exchanger is defined as "the ratio of the actual heat transfer rate to the maximum heat transfer rate that could be transferred for the same initial states and flow rates."

$$\xi \equiv \frac{q_{actual}}{q_{\max possible}} \tag{9.14}$$

Applying this definition to a heat sink, the actual heat transfer rate is:

$$q_{actual} = [\dot{m}c_p]_{air} (T_{air,out} - T_{air,in}) \tag{9.15}$$

The maximum possible heat transfer rate would be:

$$q_{actual} = \left[\dot{m}c_p\right]_{air} \left(T_{base} - T_{air,in}\right) \tag{9.16}$$

Hence, for a heat sink, the effectiveness equation is:

$$\xi \equiv \frac{(T_{air,out} - T_{air,in})}{(T_{base} - T_{cold,in})} \tag{9.17}$$

The actual heat transfer rate can then be written as:

$$q_{actual} = \left[\dot{m}c_p\right]_{air} \xi \left(T_{base} - T_{air,in}\right) \tag{9.18}$$

For the general heat exchanger case, the effectiveness is a function of three parameters: flow arrangement, C_{min}/C_{max}, and NTU. For the special case of uniform metal temperature, the effectiveness is independent of flow arrangement and the system behaves like a two-fluid situation with $C_{min}/C_{max} = 0$, in other words, one fluid has velocity zero. Only one parameter remains, *NTU*, and the effectiveness is given by:

$$\xi = 1 - e^{-NTU} = 1 - e^{-\left(\frac{hA_{eff}}{\dot{m}c_p}\right)} \tag{9.19}$$

The actual heat transfer rate is then:

$$q_{actual} = \left[\dot{m}c_p\right]\left[1 - e^{-NTU}\right]\left(T_{base} - T_{air,in}\right) \tag{9.20}$$

and the thermal resistance is:

$$R_{th} = \frac{1}{\dot{m}c_p \left(1 - e^{-\left(\frac{hA_{eff}}{\dot{m}c_p}\right)}\right)} \tag{9.21}$$

Equation (9.12) shows the complex way in which the flow affects heat sink performance. There is nothing new in the idea of using heat exchanger theory in electronics cooling. Moffat has reviewed this subject previously [13] and Webb [14] has also stressed the importance of using heat exchanger theory instead of simpler approximations.

In closure, Moffat shows a convincing table in which he compares the 1R and 2R models, and he concludes that in the region between 25 and 90 % effectiveness, where modern high-density heat sinks are designed, the 2R convective resistance model greatly overpredicts the heat transfer.

9.6 Approaches to Acquire Insight in Optimization Parameters

9.6.1 Literature Study

To get insight into heat sink behavior, or to start research to acquire this insight, the first step is of course to check the literature. In this section some general papers will be introduced, while more dedicated papers will be treated in other sections. All heat sink handbooks and review papers start with the treatment of parallel-plate channels, the basic thermal building block of the most common heat sinks in practice. The problem we face here is that, while this knowledge is mandatory for understanding what the parameters are that influence the performance of heat sinks, the designer tends to forget that optimization in practice might well lead to entirely different heat sinks concepts.

9.6.1.1 Natural Convection

Since the early 1960s there have been a number of pioneering publication suggesting heat-sink solutions for electronics cooling problems of their era [15–18], but these solutions do not prove to be applicable for today's solid-state lighting (SSL) thermal management needs unless we are dealing with large-size heat sinks (e.g., 20×20 cm).

Very recently, Kim et al. [19] published closed-form correlations that allow for thermal optimization of fully shrouded vertical plate-fin heat sinks in a fully developed-flow regime. The main purpose is that they can be used for optimizing the thermal performance of heat sinks in the early stage of thermal design. The work was inspired by the fact that the authors realized that existing least-material correlations focused on minimizing the weight and cost, while in some compact electronic devices the cooling space is severely limited and the prime consideration is not weight and cost but thermal performance only. The solutions were calibrated by comparing them with the results of the experimental data and direct numerical simulation. However, the assumption of fully developed flow is doubtful, and in addition, fully shrouded heat sinks are not the rule but rather the exception.

9.6.1.2 Forced Convection

Usually the treatment of forced convection heat transfer related to heat sinks follows a certain order: laminar fully developed flow, turbulent flow, the temperature field, the *Nu* correlations, developing flow, the analogy between heat transfer and fluid friction, etc.

A good introduction can be found in the book of Kraus and Bar-Cohen ([20], p. 275). Because of the dominance by the computer industry in electronics cooling, especially the cooling of central processing units (CPUs) and graphic cards, the emphasis is often on relatively high-speed flow and big heat sinks with small interfin spacing and relatively long fins. This approach leads to heat sink design equations based on fully developed turbulent channel flow. It should be realized that for many LED applications where forced convection is used we face often another situation: relatively short fins, and small fans. However, there are exceptions, for example, stage lighting.

In addition, formulating proper design equations is much more difficult than for natural convection because of the following reason. The heat transfer coefficient is a function of the air velocity, which is dependent on the available pumping power, which is dependent on the system air resistance. Hence, contrary to natural convection, it is not possible to define a single optimum fin array geometry. Optimization must be performed for each combination of parameters. Dedicated software can help considerably in finding an application-dependent optimum. This topic is also discussed in the paragraph on semi-empirical correlations where we present some simple design equations for first-order guesses.

To illustrate the complexity for the designer further, one of the attempts that have been described in literature to address the problem by some clever but complicated means will be discussed in some detail, namely the work by Holahan et al. [21]. The analytical methodology developed by these authors for calculating the thermal performance and pressure drop in fully shrouded, laminar, parallel plate heat sinks has been utilized to characterize the thermofluid performance of the side-inlet-side-exit (SISE) heat sink configurations.

This approach is unusual among the analytic approaches in dealing both with the impact of developing flow and axial (typically upstream) conduction in the fins (conjugate heat transfer), typical of thin fins combined with relatively low flow rates. They discretized the plate fin into a large number of "patches," and the local heat transfer coefficient was determined from the straight duct flow correlation for developing thermal and hydrodynamic laminar flow between parallel plates with uniform wall temperature, provided in Kakac et al. [22] and given by:

$$h_{fin,local} = \frac{k_{air}}{2s}\left[7.55 + \left(0.024X^{-1.14}\right)\frac{0.0179Pr^{0.17}X^{-0.64} - 0.14}{\left(1 + 0.0358Pr^{0.17}X^{-0.64}\right)^2}\right]$$

where h is the local heat transfer coefficient at position X.

The authors used a matrix approach to solve the many patch-related equations. They combined conduction and convection losses local to each patch, found by treating each patch as a circular fin, then applying the Bessel solution and using the convection coefficient from the parallel plate correlation. Heat dissipation rates obtained with this approach were shown to give good agreement with both experimental and CFD results. However, one should realize that this rather complex procedure is formally valid only for "fully shrouded, laminar, parallel plate heat sinks." When we are dealing with a practical situation, in most cases the fan is rather close to the fins. This simply means that the approach flow is highly turbulent, and thus violating any laminar-based correlation. In summary: an interesting academic exercise but of little practical value.

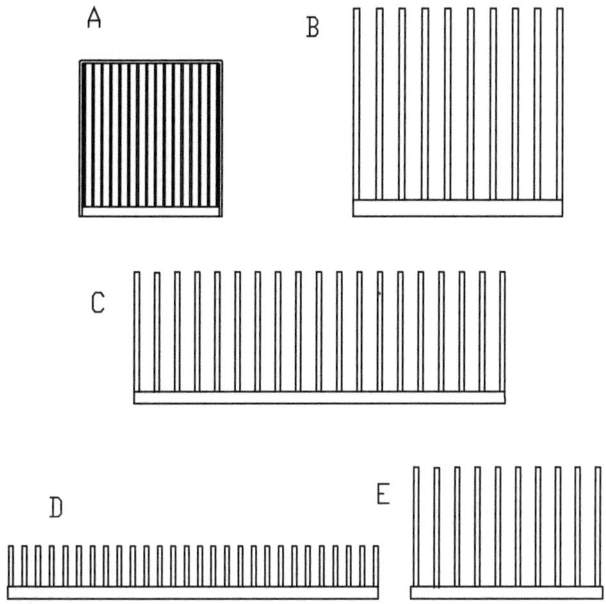

Fig. 9.3 Heat sinks tested

Biber and Belady [23] realized a long time ago that one of the most important parameters is the pressure drop through the heat sink, because it sets the velocity and hence the heat transfer. They compared correlations for fully developed flow and developing flow with CFD and test results, for various hat sink types, see Fig. 9.3.

For purpose of illustration, the pressure drop results for a typical heat sink is shown in Fig. 9.4.

Obviously, the results are not very encouraging. The authors concluded that not only the correlations were rather useless when it comes to accuracy but also that the prevailing test methods should be improved since no standardized test method did exist. Introductory texts using design equations for forced convection heat transfer can be found in Biber [24–26], and in an overview of Calculation Corner topics that appeared in the Electronics Cooling Magazine over the past 10 years or so [27].

9.6.2 Complex Shapes, Finned Arrays

When it comes to the analysis of heat sinks with complex geometries as often occur in modern LED-based systems with their freedom of design, we face a serious problem when we are after a first-order guess. The example shown in Fig. 9.5 proves the point.

Kraus et al. (see e.g., chapters in the book by Kraus and Bar-Cohen [5]) offered a solution based on matrix theory and published various algorithms to calculate finned arrays of the complexity shown Fig. 9.5. Unfortunately, the software is no longer offered, hence the current designer has only two options: measure or CFD.

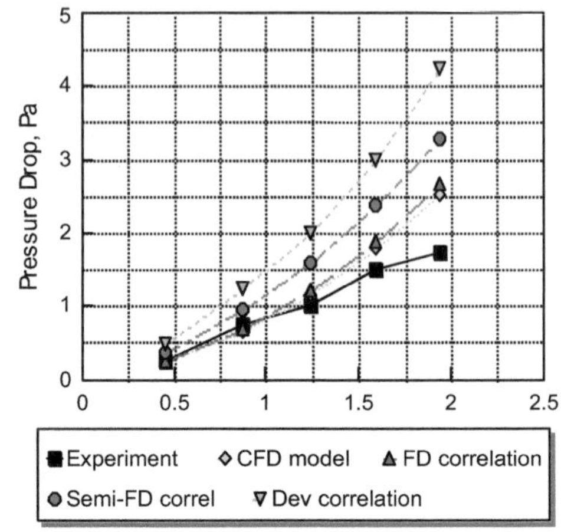

Fig. 9.4 Pressure drop results for typical heat sink

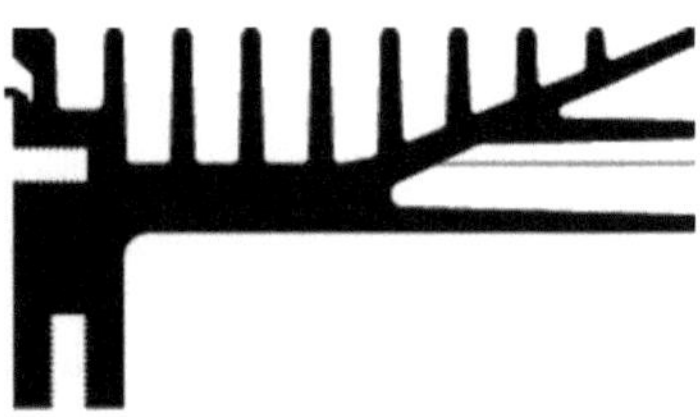

Fig. 9.5 Example of complex heat sink shape

9.6.3 *Pin Fin Heat Sinks*

Pin fin heat sinks are mostly used when the orientation of the system can vary in practice, for example, indoor floor lamps. The focus of this chapter is on parallel-plate heat sinks and space is lacking to treat pin fin heat sinks with the same attention. One of the problems is that they are considerably more difficult to treat analytically, complicated by the fact that literature data are sometimes contradictory when it comes to performance comparison. For details on this topic the interested reader may consult the following literature: Kraus and Bar-Cohen ([5], p. 320, natural convection), Bahadur [28] (natural and forced convection), Bar-Cohen et al. [29] (natural convection), and Khan et al. [30, 31] (forced convection).

9.6.4 *Inclined Plates*

Especially in application fields where it is not clear upfront how the end-user will position the product, such as in consumer lighting, care should be taken by designing the heat sink in natural-convection-driven cooling. In addition, it is not always obvious what the most optimal position is. When we are dealing with only one heat source, the direction of the airflow will be opposite to gravity. However, when more than one source is present, it might well be the case that some LEDs will experience so-called buoyancy induced forced convection, and the mean flow direction can be any direction. Positioning a plate fin heat sink in such a way that the air is parallel to the base plate then becomes a challenge. In those cases, preference could be given to pin fin heat sinks. Below the reader will find short descriptions of some relevant literature.

Papers by Starner/Marcus and Bilitzky have already been cited in Sect. 9.6.2. Both papers discuss different orientations and inclinations and are very useful because they show accurate test results for a variety of heat sinks under various circumstances and can be used to calibrate CFD models.

Ledezema and Bejan [32] performed an experimental and numerical study of natural convection and forced convection air cooling of plate finned heat sinks. The study documents quantitatively two effects: (1) the orientation of the fin array relative to the gravitational field in natural convection, and relative to a free stream in forced convection, and (2) the tilting of the crests of the plate fins relative to the approaching flow. In each configuration, the results based on complete three-dimensional (3D) numerical simulations of the flow and heat transfer confirm the validity of the results determined based on direct measurements. For example, it is shown that the overall thermal conductance of the heat sink increases when the fin crests are inclined so that they face the approaching fluid. This augmentation effect increases as the coolant velocity increases. The thermal performance of arrays positioned perpendicularly to the flow is also documented.

Naidu et al. [33] presented an experimental and theoretical investigation of natural convection heat transfer from two geometric orientations: (a) vertical base with vertical fin arrays and (b) horizontal base with vertical fin arrays, with five different inclinations. A comparison with the experimental data existing in literature indicated satisfactory agreement.

Malhammar [34] did observe an interesting phenomenon. In his own words:

> Fig. 9.6 shows a heat sink with vertical fins and another one with inclined fins. It is somewhat difficult to compare their performance on single unit bases but if they are stacked things get much clearer. For the vertical arrangement, it is apparent that the top heat sink is less effective than the bottom heat sink, because it receives heated air from below. For the inclined arrangement, this is not the case. Each stacked unit will in fact approximately contribute as much as the units below. My first idea was therefore that there must be a critical height above which inclined fins perform better than the vertical fins. That turned out to be wrong! What I found was that inclined fins always perform better than vertical fins.

For ratios of Height/Width >4 he shows that the gain can be 50 %, but some disadvantages are also discussed. It is clear that these results are of interest to many LED-based systems because of the freedom in design, especially when the casing is used as a heat sink/heat spreader.

Fig. 9.6 For stacked heat sinks, obviously the inclined stack experiences the smallest air temperature rise.

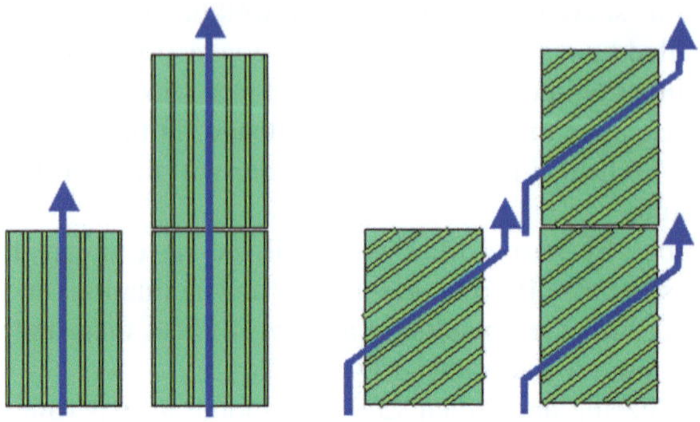

9.6.5 *Semi-analytical Correlations*

This section discusses approaches that may give some insight in the influence of common design parameters, such as optimal distance between fins, optimal fin cross-section, least-material fin, for both natural and forced convection. While the literature abounds with papers on how to optimize the shape of fins to ensure maximum heat transfer, in practice more mundane matters such as ease of manufacturing and cost usually result in the choice for a parallel plate-fin heat sink. In this section, we will limit the discussion to this type.

An important part of parallel plate fin heat sinks are the parallel plates, and especially when the fin height is large, heat sink fluid dynamics can be characterized by the flow between parallel plates only, hence without the base. The well-known book by Kraus and Bar-Cohen provides a good introduction into these matters.

Note Be aware that for many (but not all) LED applications this approach may lead to unacceptable simplifications because often the already relatively small fins are spaced widely, rendering the contribution of the base significant.

Much theoretical effort has been devoted to finding the optimal parameter settings for the so-called least-material fin. However, this topic is only of academic interest because the shape of the most optimal fin will significantly add to the costs of manufacturing. Fortunately, the next best optimal shape performance is not too far away from the mathematical optimum.

Kraus and Bar-Cohen ([5], p. 326) showed how to optimize the rate of heat transfer from a shrouded array of longitudinal rectangular fins, making a distinction between the fin average h and the base average h_w. They found iteratively:

$$(mb)_{opt} = 1.42 + 1.13\frac{h_w b}{k} \tag{9.22}$$

with m $= \sqrt{\frac{2h}{k\delta}}$, δ the fin thickness and b the fin height. Further analysis shows that the second term on the right hand side that describes the influence of the base only marginally influences the result for the optimum value for a single fin, $(mb)_{opt} = 1.42$.

Other useful equations to get insight in the parameters that determine the heat transfer performance of singe fins are the following.

For the longitudinal fin of rectangular profile, the optimum fin thickness and fin height are given by:

$$\delta_{opt} = \left(\frac{hA_p^2}{k}\right)^{1/3} \tag{9.23}$$

and

$$b_{opt} = \left(\frac{kA_p}{h}\right)^{1/3} \tag{9.24}$$

with the fin cross-sectional area $A_p = b \cdot \delta$. These equations are valid for specified A_p. Rewriting when the dissipation Q and the base temperature rise ΔT_b are given:

$$\delta_{opt} = \frac{0.63}{hk}\left(\frac{Q}{\Delta T_b}\right)^2 \tag{9.25}$$

and

$$b_{opt} = 0.8\frac{Q}{h\Delta T_b} \tag{9.26}$$

The optimum profile area is given by:

$$A_p = \frac{0.5Q^3}{h^2 k \Delta T_b^3} \tag{9.27}$$

It is interesting to compare the optimum profile areas for various cross-sections: rectangular, triangular, and parabolic. The conclusion is that the triangular fin requires only 60 % of the material of the rectangular fin for the same performance, while the concave profile is only marginally better.

From the equation, it becomes also obvious that the weight is proportional to the cube of dissipation Q. Hence, when the dissipation should be doubled, it is better to strive for two fins rather than extend a single fin.

Because area A_p is proportional to the volume V, the mass becomes proportional to the ratio of specific density over thermal conductivity ρ/k, meaning that the material with the lowest ρ/k gives the lightest fin.

Other optimum parameters discussed by Kraus and Bar-Cohen are the optimum aspect ratio and the optimum single-fin heat transfer rate, respectively:

$$\left(\frac{\delta}{b}\right)_{opt} = \frac{hb}{k} \tag{9.28}$$

and

$$\left(\frac{q}{L}\right)_{opt} = 1.26\left(h^2 A_p k\right)^{1/3} \tag{9.29}$$

The equations presented are valid for both natural and forced convection. In concluding this introductory section, it is worthwhile to mention that the book by Kraus and Bar-Cohen also treats pin and radial fins and fin arrays.

The next sections focus on finding the approximate values for the local heat transfer coefficient h.

9.6.5.1 Natural Convection

Let us first discuss parallel-plate channel flow only. Elenbaas, in a famous paper from 1942 [35], was the first to discuss natural convection between parallel plates. Elenbaas used the incompressible natural convection form of the Navier–Stokes equations and formulated semianalytically a Nu–Ra correlation based on bounding relations for the isolated-plate limit in the entrance region and for fully developed flow after the boundary layers merge. In recognition of his pioneering work, a modified gap-based Ra number, multiplied with the ratio of the gap width and the plate length, is referred to as the Elenbaas number, *El*, for an isothermal channel according to:

$$El = \frac{\rho^2 g \beta c_p z^4 (T_w - T_\infty)}{\mu k L} \tag{9.30}$$

with z the gap width, L the plate length, T_w the wall temperature and T_∞ the inlet air temperature. The Elenbaas correlation is:

$$Nu_0 = \frac{hz}{k} = \frac{El}{24}(1 - e^{-35/El})^{3/4} \tag{9.31}$$

with Nu_0 the gap-based *Nu* number.

When limiting expressions are known, often use is made of a formation proposed by Churchill and Usagi resulting in a function varying smoothly between the two bounds. Bar-Cohen and Rohsenow [36] proposed the following composite relation:

$$Nu_0 = \left[\frac{576}{(El)^2} + \frac{2.87}{(El)^{1/2}}\right]^{-1/2}$$

To give the reader an idea of what can be expected, let us take a temperature difference of 50 °C and a plate length of 15 cm (e.g., resembling a board). For a spacing of 20 mm we find the heat transfer coefficient h to be 6.1 W/m^2 K, for 5 mm this becomes 3.3 W/m^2 K, and for 2 mm it reduces to 0.2 W/m^2 K, showing the strong dependence of heat transfer coefficient on the gap width z. When we reduce the plate length to 1 cm, resembling a fin, these values become 12.1, 11.8, and 3.6 W/m^2 K, respectively. Notice the difference with $L = 15$ cm. The first two values clearly indicate developing flow, almost independent of the plate distance over a large range. The last value points again at fully developed flow. Kraus and Bar-Cohen also discuss the optimum fin spacing and the optimum Nusselt number, to result in:

$$h_{opt} = 1.25 \frac{k_g}{z_{opt}} \tag{9.32}$$

with

$$z_{opt} = \left(\frac{L\upsilon^2}{\eta g\beta\theta_b Pr}\right)^{(1/4)} = 3P \tag{9.33}$$

where η is the fin efficiency of the optimum single fin equal to 0.63, and P an array parameter:

$$P = \left(\frac{L\upsilon^2}{g\beta\theta_b Pr}\right)^{1/4} \tag{9.34}$$

Another useful optimum equation presented is the one that maximizes heat dissipation from a least-material rectangular fin with a prime area of given length L, width W, and specified temperature rise ΔT_b:

$$\left(\frac{Q}{LW\Delta T_b}\right) = 0.24\left(k\frac{k_{air}}{P^2}\right)^{1/2} \tag{9.35}$$

It is to be understood that results based on channel flow analysis can only be at best a reasonable approximation of heat sink heat transfer. This is especially true for LED-based applications because of their frequent use of short-wide fins. When we introduce a base, the problem becomes inherently asymmetric and 3D, leading to lateral inflow and/or outflow effects. These effects introduce often a significant reduction in the average interfin air temperature. Furthermore, near the base the local heat transfer coefficient will be reduced, violating the Murray–Gardner assumptions. Finally, because the fin temperature changes along its height, (semi)analytical calculations become impossible because of the coupling between temperature and flow. Usually the approach taken is to assume an average fin temperature using the fin efficiency as a correction factor. All these added complexities have resulted in many technical publications presenting empirically based and CFD-based results. An overview can be found in Kraus and Bar-Cohen ([5], p. 304). The main conclusion to remember is that the experimental Nusselt numbers are often substantially higher than those obtained from the Elenbaas correlation due to the strong impact of 3D flow, at least for the vertical-plate vertical-fin orientation. This observation has led several research groups to propose correction factors to the Elenbaas correlation, and other groups to propose correction factors to these correction factors, and so on, resulting in a set of increasingly complex equations. One research group examined 12 different heat sinks in various orientations, fin spacings, etc., to result in different *Nu* correlations for all 12 heat sinks! This is in our opinion not the path to walk for the average designer. To get some insight in the main parameters governing vertical parallel plate heat sinks it is better to use the Bar-Cohen-Rohsenow composite relation.

Further reading regarding the use of optimal individual plate and pin fins to compose multiple fin arrays: Iyengar and Bar-Cohen [37], Bar-Cohen et al. [38], and Bahadur and Bar-Cohen [39].

9.6.5.2 Forced Convection

Kraus and Bar-Cohen treat forced convection-driven heat sinks fin in a certain order starting with parallel-plate channels, the transition from laminar to turbulent flow, the transition from developing to fully developed flow, and the analogy between heat transfer and fluid flow. The interested reader is encouraged to consult this book for further details.

Open a heat transfer handbook and one will find a plethora of Nu–Re correlations. We are of the opinion that the differences between them are minor in relation to the differences that occur in real life. Hence, we will only present some typical and relatively simple equations to guide the designer in understanding the relative importance of various parameters.

For laminar developing flow along plates we recommend to use:

$$h = 0.66\left(\frac{k_{air}}{L}\right)Re^{0.5}Pr^{0.33} \tag{9.36}$$

For turbulent flow along plates:

$$h = 0.036\left(\frac{k_{air}}{L}\right)Re^{0.8}Pr^{0.33} \tag{9.37}$$

For fully developed channel flow we get:

$$h = \frac{4k_{air}}{d_h} \tag{9.38}$$

with d_h the hydraulic diameter: $d_h = 4A/P_w$ with A the flow area and P_w the wetted perimeter of the flow channel. For parallel plates the hydraulic diameter becomes equal to twice the plate spacing z and we find $h = 2k/z$.

For fully developed flow, the pressure drop per unit length is:

$$\frac{\Delta P}{L} = \frac{48\ \mu v}{d_h^2} \tag{9.39}$$

with v the average velocity.

The friction factor f is found to be:

$$f = \frac{24}{Re} \tag{9.40}$$

For turbulent flow, the friction factor can be written as:

$$f = 0.046Re^{-0.2} \tag{9.41}$$

To get an idea about the hydrodynamic entrance length x, for laminar flow this may be obtained from:

$$\frac{x}{d_h} = 0.05Re \tag{9.42}$$

with Re based on d_h. For turbulent flow, because there is no satisfactory general expression for the entry length, fully developed flow is usually assumed when:

$$\frac{x}{d_h} > 10 \tag{9.43}$$

According to Kraus and Bar-Cohen, when we are looking for an optimum for a forced convection array we meet a problem. Because the optimum profile area and hence the mass is inversely proportional to the square of the heat transfer coefficient, and because the heat transfer coefficient is dependent on the fluid velocity, the available pumping power limits the added mass. Hence, it is not possible to define a single optimum forced convection array. In other words, only in case the fan characteristics are given we are able to find the optimum fin array geometry. Within the framework of this chapter it is not possible to present even a limited overview of all correlations and optimization strategies that can be found in literature for various kinds of operating conditions. It is highly recommended to use a spreadsheet approach to be able to vary the multitude of parameters that are governing the world of forced convection-driven heat sinks. This way of working has been described by Biber and Fijol [40]. Following their description: spreadsheet-based thermal and pressure drop performance calculations are used, including developing flow, fin efficiency, and heat spreading in the base of the heat sink. The pressure drop for any given velocity between the fins is calculated using fully developed flow correlations in rectangular channels, varying with aspect ratio. Heat transfer calculations use developing flow correlations between parallel plates corrected for channel aspect ratio for the heat transfer coefficient. Air temperature rise is taken into account using the log–mean temperature difference method. Fin efficiency calculations are included to obtain the total fin-side thermal resistance. To obtain the total thermal resistance, an additional spreading resistance is added to the fin side thermal resistance. This spreading resistance captures the competing effects of heat spreading and conduction through the thickness of the base. To demonstrate this way of working, the authors discuss in detail the optimization of a certain volume occupied with a fan and a heat sink's given manufacturability and design constraints. They conclude that practical design considerations often place significant limitations on heat sink optimization, and that using a fan curve-based design method sometimes results in performance effects that run counter to conventional wisdom.

To get a flavor of what has been published, please consult papers by Knight et al. [41], Teerstra et al. [42], Iyengar and Bar-Cohen [43], and Copeland [44]. Optimization strategies based on the entropy generation minimization as proposed by Bejan [45] is treated by Shih and Liu [46].

Another popular trend is to take into consideration the impact heat sink design has on sustainability. Not only maximum heat transfer, but also the energy required to operate the heat sink and to manufacture it are part of the objective function to be optimized, see e.g., the paper by Bar-Cohen et al. [47]. This is especially true for LED-based lighting solutions where the energy spent on active cooling reduces the overall system efficacy. Worth mentioning is the attention paid to enhanced-conductivity polymers to replace aluminum as a possible material for natural convection-driven systems.

9.6.6 Constructal Theory

In a chapter on heat sinks as part of a book on LED-based systems thermal management, attention should be paid to potentially interesting solutions for the future. One of the solutions that could have an impact on the way the thermal community approaches heat sinks is given by Adrian Bejan through a theory he coined Constructal Theory. The author considers Bejan's approach of sufficient value to warrant the inclusion of an extended paragraph and he hopes it will challenge some readers to start walking along this promising path.

Bejan[1] developed a way of identifying and minimizing losses in the field of thermodynamics through the method of entropy generation minimization (EGM), which has its origins in his doctoral thesis [48]. The following introduction is a mix of what can be found on his very interesting web portal [49].

EGM is an aspect of thermodynamic optimization that sheds considerable light on the organization of the natural world. In the process of performing such analyses, engineers determine the entropy that a system generates as a function of its physical parameters, including size, shapes, and materials. After gaining this understanding, engineers—at least in theory—can go on to optimize the system's performance in terms of its constraints, which are responsible for its irreversible operation. Thermodynamic optimization can deliver not only the optimal distribution of material but also the optimal dimensions. For example, when considering heat transfer between two parallel plates the entropy generation is due to both heat transfer and fluid friction that compete against one another. The distance between the plates can be selected such that the sum of the two irreversibilities is minimal.

Around the mid-1990s he took this philosophy one step further, coining it as Constructal Theory. Usually, engineers begin the design of the device by first understanding its purpose. The device, always of finite size, must function subject to certain constraints. To analyze the device is not sufficient: to optimize it, to construct it, and to make it work is the real objective. All these features (purpose, finite size, constraints, optimization, and construction) can be seen in the animate and inanimate structures around us.

The Constructal Law is a first principle of physics that accounts for all design and evolution in nature. It holds that shape and structure arises to facilitate flow. The designs that arise spontaneously in nature reflect this tendency: they allow entities to flow more easily—to measurably move more current farther and faster for less unit of useful energy consumed. Rain drops, for example, coalesce and move together,

[1] Bejan is also the author of the book titled *Convection Heat Transfer* [71]. A.o, he derives common analytical correlations starting from simple scaling laws, and criticizes the common interpretation of numbers such as Re on the grounds that it is total nonsense.

generating rivulets, streams and the mighty river basins of the world because this design allows them to move more easily. The Constructal Law asks the question: Why does this design arise at all? Why can't the water just seep through the ground? The law provides this answer: Because the water flows better with design. It covers the tendency of nature to generate designs to facilitate flow.

Existing flow configurations are replaced by better flowing configurations, smoothly or stepwise, in animal design, river basin design, automobile design, and geopolitical design. The time arrow of the Constructal Law is not to be confused with the time arrow of the second law. The second law is the law of entropy generation, whereas the Constructal Law is the law of configuration generation. The concept defined by the second law is entropy. The concept defined by the Constructal Law is evolution of configuration (design, pattern, layout, drawing). Optimal flow organization minimizes entropy generation and thus maximizes the performance of systems.

While constructal theory gives rise to a fractal-like deterministic structure in space, constructal theory should not be confused with fractal theory. Fractals are mere descriptions but not explanations of natural phenomena.

The following principle is the essence of the Constructal Law:

> For a finite-size flow system to persist in time (to survive) its configuration must evolve such that it provides greater and greater access to the currents that flow through it.

It is a predictive theory of the geometry and rhythm of nature, and the principle according to which they can be deduced, is the optimized shape and structure obtained from engineering analysis.

Recent constructal theory predictions reported are the scaling laws of all animal locomotion (running, flying, swimming), the distribution of the sizes of human settlements, the scaling laws of river basins, the shape of all Egyptian pyramids, vegetation, and the evolution of science and civilization, which is pretty impressive.

Now, what has all this to do with the subject at hand? One of the many recently published papers [50] is devoted to maximizing the heat transfer density, by applying constructal theory, for natural convection between vertical plates. According to this method, the flow configuration is free to morph in the pursuit of maximal global performance under global constraints. The resulting optimal (constructal) configuration is deduced, not assumed. It is the winner of the competition in which all eligible configurations are simulated and compared.

What we especially like in this paper is the out-of-the-box thinking. Instead of optimizing only the distance between the plates, the authors sought for an optimized multiscale flow structure, under the same constraints as the single-scale structures optimized in the past. The key to higher performance is the increased design freedom, because more length scales (to be optimized) represent more degrees of freedom in the design. More length scales mean that the flow structures can morph in more design directions. In the end, the optimized multiple scales are distributed nonuniformly through the available volume. The basic idea is sketched in Fig. 9.7.

For developing flow, the area in between the boundary layers is more or less lost for heat transfer. Filling this area with fins of a certain optimized height, and repeating this process, leads to better performance. The optimization was done using scaling

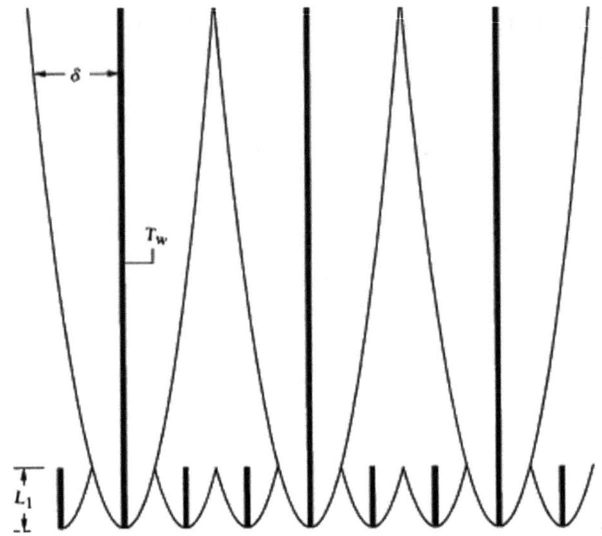

Fig. 9.7 Space between fins provided with much smaller fins

laws, and compared with numerical simulations. The authors reported improvements up to 20 %.

While it is not expected that constructal theory will become a mainstream method used by designers to optimize their heat sinks, we considered the method enough thought-provoking to decide to include the method in our list.

9.6.7 Web-Based Tools

While there are many heat sink calculators to be found on the web, the majority does not show background information. Providing this is exactly the major advantage of the tools generated by the University of Waterloo. An article in the Electronics Cooling Magazine [51] discusses a web-based tool for the calculation of natural-convection-driven heat sinks which is very easy to use and saves the designer from the burden to enter a lot of equations into a spreadsheet. The authors argue that the selection of the most appropriate heat sink for a particular application can be very difficult, given the many design options are available. While thermal analysis models, ranging from simple empirically derived correlations to powerful numerical simulation codes, can be used to analyze the thermal performance of heat sinks for a given set of design conditions, however, analytical models, based on the fundamental formulations for conservation of mass, momentum, and energy, provide distinct advantages. These include speed, accuracy, ease of use, and simplicity of coding, especially for web-based software tools. The authors present analytical models for calculating natural convection heat transfer as a function of the geometry, material properties, and the boundary and ambient conditions for three heat sink geometries a circular- and square-finned radial heat sink and a plate fin heat sink. Options are available to calculate both the total heat flow rate, Q, and the isothermal base plate temperature, T_s. through the independent control of the geometric parameters. Options for the solutions include nonisothermal fin temperatures to account for fin effects, as well as radiation heat transfer (radial fin) and an insulated back surface on the baseplate (plate fin). The interactive, real time solution tools described here, plus an extended set of complementary tools are available at the Microelectronics Heat Transfer Laboratory website [52]. Unfortunately, the website does not offer an equivalent tool for forced-convection-driven heat sinks. A first-order guess can be found at another website [53] that is based on research by Kays and London.

9.6.8 Dedicated Heat Sink Software

Apart from the free web-based tool discussed above, there exists also a commercial heat sink software code [54]. This code can address natural and forced convection, including bypass. One step further, something between this tool and CFD, there is software that can also handle fans, interface resistances, radiation, etc. [55].

The main objection in using these tools is that they exclude all heat sinks that deviate from the shape imposed by the extrusion process, neglecting many alternative design options, and they also fail in taking into account the significant 3D fluid flow phenomena that occur often in practice. Nevertheless, they can be fruitfully used in what-if studies.

9.7 Approaches to Acquire Insight in Fluid Dynamics

In the past 20 years, we have observed a significant increase in the use of CFD software to calculate the thermal behavior of electronic systems. The benefits are undisputed when it comes to performing parametric studies in early design phases. However, when the objective is accuracy, the discussion about what we can expect in practice becomes far from trivial. The heat transfer behavior of electronic systems is very complex when compared to the standard canonical cases usually treated in fluid dynamics classes. Section. 9.9.1 discusses the CFD analysis of single heat sinks, and the second section formulates answers to the question about the accuracy that can be expected.

9.7.1 Computational Fluid Dynamics Analysis of Single Heat Sinks, Boundary Conditions Assumed

When the purpose is to design the most optimal heat sink given a number of design constraints the best approach is to use a full-blown CFD code. The standard approach is to model a wind tunnel with uniform approach flow, set a certain bypass, and model a heat sink with any complexity that comes to mind, be it oblique fins or interrupted fins, fin arrays, pin fins, etc. Next, the best available turbulence model is chosen and a very fine mesh. Chances are high that one is able to find the best heat sink given the constraints, but clearly this the approach for the heat sink manufacturer, not the end-user. Another very useful outcome of such a CFD analysis could be insight in the factors that determine extrusion of air as a function of fin distance and fin dimensions. Finally, useful data are generated concerning pressure losses. However, the question remains: can these results been copied one-to-one to real-life situations? The answer is yes, provided real-life looks more or less as a wind tunnel. Usually, the answer is no.

In summary, while it is clear that this approach is very suited to acquire fundamental insight in many aspects of the fluid dynamics that govern heat sink heat transfer, it is only recommended for heat sink manufacturers, not for end-users. Please note that the discussion is about a **single** heat sink. A better approach is discussed in Sect. 9.8

9.7.2 *Pros and Cons of Computational Fluid Dynamics for Heat Sink Analysis*

9.7.2.1 Introduction

A natural question is how accurate numerical simulations are when compared to well-designed experiments. Many studies demonstrate amazing agreement, the conclusion often being that "validation of the numerical model" has been proven. It is demonstrated that these conclusions are subject to serious doubts. The major reason is the lack of sufficiently accurate input parameters and boundary conditions combined with complex geometries.

This section is based upon two papers [56, 57] and focuses on the following question: Can we predict the junction temperature of critical components in LED-based system provided with heat sinks with sufficient accuracy? The conclusion is; not without fitting!

9.7.2.2 The Difference Between "Validation" and "Calibration"

Regarding numerical analysis, the word "validation" is frequently used in different contexts. We should distinguish between verification, validation, and calibration (Table 9.2).

In this context, most authors mean "calibration," not "validation." In the following, we will mainly refer to calibration. The distinction is not only a matter of semantics: a "validated" code is much more general than a "calibrated" code, which is formally only useful for the test environment for which is was designed. When the test set up changes considerably, the calibration tests should be repeated.

Table 9.2 Various definitions

	Related to
Verification	Solving the equations correctly
Validation	Testing of predictive capabilities of physical models against detailed test data, including convergence
Calibration	Fitting a model in such a way that, e.g., temperatures are predicted to a sufficient degree of accuracy

9.7.2.3 Uncertainties

References [56, 57] summarize the various types of uncertainties that influence a comparison between numerical and experimental results. A distinction has been made between numerical and experimental uncertainties, often largely unknown input data, and a source of uncertainty that is called "comparison mismatch." An example is the mismatch between some cell-averaged critical temperature and the size of the thermocouple. Interested readers are encouraged to consult the quoted references. In this section, we will focus on numerical uncertainties only. Experimental and input data inaccuracies are not the dominant factors for heat sinks, usually the thermal conductivity of the metal is well-known but they are often the dominant factors for the LEDs themselves and especially the thermal interface materials (TIMs).

9.7.2.4 Numerical Uncertainties, Detailed Models

Accuracy (defined as *predicted* over *true* temperature rise) is influenced by the chosen discretization method, false diffusion, temperature-dependent physical properties, the Boussinesq approximation, the way radiation is treated, the convergence criteria, and the geometric representation of a real object in a database. Some of these uncertainties will certainly decrease in the future because almost all mentioned approximations can be relaxed by using a more powerful computer. However, the discretization error will stay for some time because it is still very difficult in practice to model every boundary layer in order to capture the local heat transfer correctly. This is especially the case for heat sinks because of the multiple fins, each requiring a very fine (order 100 μm) mesh near the fin surface. The radiation error suffers from another problem. All engineering calculations are based on the following assumptions:

- All surfaces are diffuse (or, sometimes, specular), gray, opaque, and uniformly irradiated.
- The medium between surfaces is transparent for the wavelength ranges of interest.

Especially the first generalization may cause unrealistic over-simplifications in applications dominated by natural convection (or vacuum). The ultimate challenge for an accurate numerical analysis, however, is posed by the modeling of a few complex physical phenomena that are inherently present in a real system. It is in this area that the largest errors are often made, caused partly by the discrepancy between reality and the modeling assumptions, and partly by the user's lack of awareness. Instabilities and "mini-jets" generated by a grille are never modeled but these effects persist long enough to cause a measurable influence on downstream local heat transfer, see Lasance and Rind [58]. For LED-based systems heat sinks are often located near luminaires with openings, causing significant interactions with the created turbulence near the walls. Transitional and turbulent convection pose another problem. Sophisticated turbulence models are not expected to be implemented in system-level analysis design tools for quite some time. To capture low-Re transitional flow requires a very fine grid and time-dependent analysis. Rodgers and Eveloy [59] discuss this topic in more detail and conclude that significant errors are unavoidable. A last complex issue that deserves attention in LED-based systems that often combine heat sinks and fans or synthetic jets is the modeling of fans and jets. For sure not every detail of a fan, both dimensionally and a velocity-wise, can be taken into account in a system level analysis in the near future. Synthetic jets pose even a more difficult problem, see a series of papers by our team [60–62]. Hence, accurate prediction of local heat transfer phenomena is considered problematic and requires often a lot of time in a research lab to formulate adequate conclusions, time that is normally lacking in a design environment.

9.7.2.5 Numerical Uncertainties, Compact Models

It is difficult to assess a meaningful value to this error, but contrary to boundary condition independent compact thermal models for ICs that show negligible errors as compared to detailed models [63] it is not to be expected that heat sink compact models will even come close to detailed CFD models. But then, detailed CFD models themselves neither come close to real-life test data. See also Sect. 9.8.3.

9.7.2.6 Errors When Comparing Numerical and Experimental Results

Let us consider the situation that substantial differences are found between the results of CFD analysis and experiments. Let us further assume that the experiment is well-designed, so that the results are correct to within 5 %, based on 20:1 odds. The question arises: How accurately are the real physics simulated by the numerical analysis and how accurately are the physical properties and input parameters known? Reference [59] jumps at the following conclusion.

Assume a brilliant designer with lots of time succeeded in reducing all errors to 5 %. Taking the root mean square of all error percentages, the final error is of the order of 20 %, notwithstanding the fact that the numerical and experimental analyses do meet current standards. The reader may wonder why these problems never show up in reports and literature. The reason that impressive results are very often claimed can simply be attributed to the fact that many parameters are available that can be used to match the results. In this way, it is relatively simple to reduce the errors to something between 5 and 10 %. It should be stressed that nothing is wrong with this practice; the problem is that it is often argued that the numerical code has been *validated*, while it is really *calibration* that we are talking about. As argued earlier, while it may seem only an academic distinction, it is not. Calibration does not guarantee extrapolation to other situations, while validation does.

9.7.2.7 How to Improve?

We may expect that on the numerical side some errors will decrease to more acceptable levels. The errors related to the treatment of complex fluid phenomena will not be reduced to the same extent unless Direct Numerical Simulation (DNS) or Large Eddy Simulation (LES) techniques become feasible for 3D complex systems. While it is true that some improvement could be gained by implementing more sophisticated turbulence models (see e.g., reference [59]) it is by no means a panacea for the problems mentioned. Suppose we want to model spray cooling. It makes a lot more sense to design appropriate experiments to estimate the local heat transfer instead of relying on very complex 3D highly turbulent two-phase CFD simulations. One way to enhance the predictability of CFD analyses is to use a pragmatic approach by employing correction factors. A recommended way to measure correction factors is by developing "ideal" experimental benchmarks for complex geometries where all boundary conditions and material properties are under control and are well known to within a few percent. Note: to prevent misunderstanding: the subject of the discussion is: how to arrive at the best heat sink for the application, not about the usefulness of CFD for system level design.

9.7.2.8 Conclusions

The continually shrinking time-to-market for LED-based systems will require an ever-increasing reliance on computational simulations. However, a number of issues still impede a greater reliance on predictive modeling capabilities. In particular:

- Computational resources for handling large, realistic problems,
- Databases on thermophysical properties of LED packaging materials,
- Accurate in-situ determination of physical properties,
- Assessment of interface thermal resistances,
- Wide availability of compact models supplied by vendors,
- Accurate benchmarks to assess and correct the influence of complex geometries.

The most important conclusion is: A numerical analysis of every electronic system (including the heat sinks of LED-based systems) may or may not be correct, and no one can tell. Suppose the calculated and measured junction temperatures differ by 20 %, then it is still possible that both analyses are correct to within 5 % or better, simply because sufficiently known input data are lacking. Another way to put it: An accurate match (± 10 %) is not possible without some kind of fitting.

Despite the conclusions, CFD tools are vital in design environments! However, do not expect or claim high accuracy in a direct comparison with experiments.

9.8 Approaches Suited for Final Design of Heat Sinks in Light-Emitting Diode Applications

9.8.1 Introduction

When it comes to business objectives such as cost reduction, time-to-market, optimization of weight, height and size from a sustainability point of view, and temperature reduction, it is obvious that the designer needs tools that allow prediction of real-life situations with sufficient accuracy. None of the approaches discussed above are suited to meet this goal. This section treats three approaches that may provide the designer with all relevant information that is required to make an optimal choice out of the thousands of heat sinks that are offered on the market. The first approach discusses a test dedicated to ranking real heat sinks under real-life boundary conditions. The second approach is purely numerical, applying user-friendly CFD codes at the system level using suitable heat sink compact models. This approach, as argued before, creates a lot of insight in the governing parameters, and lends itself for subsequent optimization, but high accuracy cannot be expected.

The final approach treats prototype testing, the proof of the pudding as it were. The combination with CFD modeling will bring expert experience to a higher level, useful for optimizing the next generation.

9.8.2 Dedicated Tests to Rank a Series of Heat Sinks

This section discusses a method to rank heat sinks given a certain application [64]. The measurement is based on the extraction of the average heat transfer coefficient from time-dependent temperature curves as a function of velocity and bypass. Scaling the measured effective heat transfer coefficient by mass, volume, weight or height provides several performance metrics allowing designers a novel way of ranking heat sinks in conditions that resemble the application.

When a heat sink has to be chosen that meets the required thermal criteria, the choice is often based on cost aspects and volume or weight restrictions. The only data that are usually available are the data from the vendor data sheets that are based upon closed-duct flow and with a source covering the heat sink area, see Sect. 9.4.1. From these sheets, the designer can get the following values: the thermal resistance from junction-to-ambient as a function of the fluid velocity and the pressure drop. Unfortunately, these data cannot be used in practice because we face bypass effects, complex inflow conditions, interface problems and spreading effects. Especially the last effect should not be underestimated.

In most cases, numerical modeling is not an option because reality is too complex. One of the conclusions of a panel on heat sink modeling held at Interpack 2003 [65] was that the errors using commercial codes could easily exceed 30 %. Because this problem is not related to CPU power we believe that not much has changed since 2003. What we need is a measurement method that mimics the flow conditions of the application and is relatively fast. To meet these objectives the heat sink performance tester has been developed. The basic idea is to measure the cooling curve of a heat sink and to extract an average heat transfer coefficient from this curve. Details are given in the paper cited. The measurements are performed in transient mode. In steady state, the losses of the heater into the substrate cannot be neglected. While this effect could be taken into account by elaborate calibration methods, the more serious problem is that the heated substrate will heat up the incoming air causing part of the heat losses to be coupled back, and this effect is much more difficult to account for. See also [58]. The described transient method reduces this effect to negligible proportions. The assumptions have been confirmed by 3D numerical modeling of a heat sink subject to natural convection on a substrate with the same conductivity as air. The results show that the heat losses in steady state are of the order of 10 % while for the duration of the transient measurements the losses are about an order of magnitude less. Another advantage is that the value of the heat dissipation is not required. Of course, the test equipment could also be used for steady state measurements. A further advantage is that heat spreading effects are measured directly without the need for a correction step. Especially for the case of forced convection and with the use of an aerogel for the substrate (with a thermal conductivity equal or lower than that of air) the losses can be kept to a minimum.

The measurement procedure is as follows. First, the heat sink is raised above the substrate and heated to say 60 °C. Next, the heater is switched off, the heat sink lowered onto the substrate and the fans are turned on. The resulting cooling curve is used to calculate a number of performance metrics. The curve can be described by:

$$\frac{T - T_\infty}{T_0 - T_\infty} = \exp\left[-\frac{hA}{\rho c_p V}(t - t_0)\right] \tag{9.44}$$

Let us define a time constant β to be equal to $\rho c_p V/hA$. From the slope of a logarithmic plot of the cooling curve β can be extracted. In practice, the performance metric hA is calculated that can be considered as an effective heat transfer coefficient. This parameter shows the effective performance of a heat sink, excluding for example, weight and volume influences. The repeatability and the reproducibility of the setup has been checked and stayed well within 5 %.

Performance Metrics The fundamental performance metric used to compare the heat sinks is hA, the heat transfer coefficient times the effective (wetted) area of the heat sink. To calculate hA, the thermal capacitance $\rho c_p V$ has to be known. ρV (the mass) can be measured by weighing. For the specific heat c_p, we need to know the material of the heat sink. The paper discusses the heat transfer behavior of 7 different heat sinks with the same footprint, but of different shapes, fin lengths, fin distance etc.. Heat sink 4 was made of magnesium, the others of aluminum. Usually the materials are known, and c_p values are well documented [66]. In case the material is not known, c_p can easily be found to a high accuracy using a Differential Scanning Calorimeter, standard equipment in every materials analysis lab. Here is a suggestion when this is not available: make a copy using a known material, the ratio of the cooling curves provides the required value.

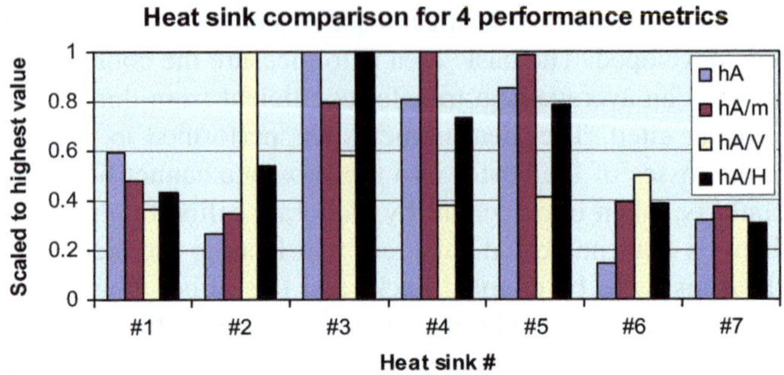

Fig. 9.8 Heat sink comparison for $v = 2.2$ m/s

The volume is calculated by multiplying the maximum values of the length, width and height of base plate and fins. By using the number of fins both area and volume of the heat sink can be obtained. For heat sinks 3, 4, 5, and 6 the height of the middle and outer fins are measured and then averaged to get one specific value. While *hA* can be considered as the most "fair" performance metric, in some cases other arguments to choose the heat sink play a role. For example, weight and height are selection criteria in portable applications, and volume in many others. By dividing *hA* by the mass, volume, or height one finds three additional performance metrics[2]. By way of illustration, Fig. 9.8 shows the results for the four performance metrics for the seven heat sinks for a velocity of 2.2 m/s and a relatively large bypass. The results for the other velocities tested show the same trend. *Note*: It is very well possible that a much smaller bypass results in a different ranking!

9.8.2.1 Observations for the Various Performance Metrics

hA (First Columns) Heat sink #3 performs best, followed by #4 and 5.

hA/m (Second Columns) Heat sink #4 scores best, directly followed by #5.

hA/V (Third Columns) Heat sink #2 has by far the best hA/V-ratio. Notice that this heat sink scored badly on the other metrics. On the other hand, # 4 and 5 score low on this metric while being among the best for the other two metrics.

hA/H (Fourth Columns) For example, notebook applications, the height of the heat sink should be as small as possible. Dividing *hA* by the total height *H* shows that heat sink #3 is the champion while #4 and #5 score also much better than the others.

Of course, it is up to the designer to invent other metrics, for example, based on performance and cost, or by combining several metrics using weighting factors. *Note*: When the heat source is (much) smaller than the base area of the heat sink, the effect of heat spreading will be included, but this measuring method cannot deal with the individual contribution of the spreading effect. This effect introduces an additional thermal resistance that can be easily addressed using the tricks outlined in [3].

[2] Iyengar and Bar-Cohen use the term "array heat transfer coefficient" for our "effective heat transfer coefficient," and they use thermal metrics such as the space-claim heat transfer coefficient for our volume–weighted h [38]

9.8.2.2 Suggestions for Additional Measurements

For particular industrial applications that define more or less the range of variables, a study towards the influence of the bypass on the choice of the heat sink makes sense. Pressure taps could be added to monitor pressure losses as function of a chosen set of variables, to provide a basis for an appropriate choice of the system fans. In addition, an optimal balance between the positive effects of a larger base area and the negative effects of increased spreading could be subject of study.

9.8.3 Computational Fluid Dynamics with Calibrated Heat Sink Compact Models

For the sake of completeness, this section shortly mentions compact models. Further details are discussed in Chap. 10.

Coetzer and Visser [67] introduced a compact model to predict the interfin velocity and the resulting pressure drop across a longitudinal fin heat sink with tip bypass. The compact model is based on results obtained from a comprehensive study into the behavior of both laminar and turbulent flow in longitudinal fin heat sinks with tip bypass using CFD analysis. The new compact flow prediction model is critically compared to existing compact models as well as to the results obtained from the CFD simulations. The results indicated a good match. The improved accuracy in velocity distribution between the fins also increased the accuracy of the calculated heat transfer coefficients applied to the heat sinks.

9.8.4 Computational Fluid Dynamics in General

For many applications where only a few LEDs and heat sinks are involved the use of compact models could be reconsidered. An example is a LED spot recessing in a ceiling, see Fig. 9.9.

Dong and Narendran [68] discuss the use of CFD in finding the optimum in conduction, radiation and convection for the recessed LED lamp. However, for more complex LED applications the design space becomes often too large to explore effectively. Hence, the use of appropriate correlations is recommended to narrow the choices. Note: In addition to what has been argued about the pros and cons using CFD for heat sink optimization, it should be stressed that using CFD at the system level is the recommended approach to support the designer in selecting the most suitable fan. For this choice, the designer needs a number for the total system air resistance in order to conclude about the most suitable operating point, and CFD is second to none in finding this number.

9.8.5 Prototype Testing

When a prototype becomes available, in many industrial cases a lot of verification tests are performed. Unfortunately, oftentimes there is no time for a thorough

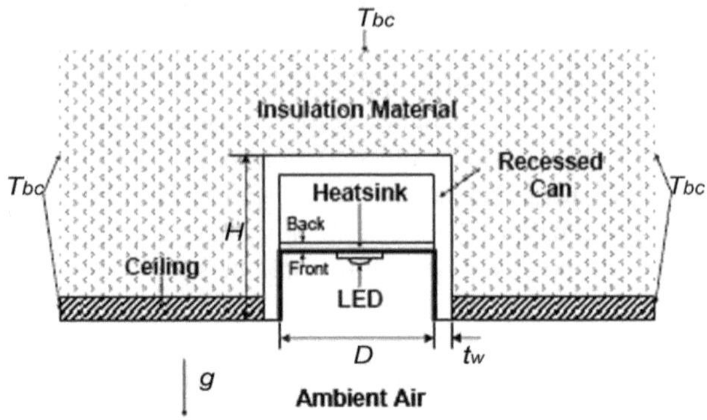

Fig. 9.9 Light-emitting diode (LED) lamp recessed in ceiling

thermal analysis except for the measurement of critical temperatures. One tends to forget that very useful insight can be obtained by comparing pressure losses and local air velocities and temperatures with the results of a CFD model, insight that can be fruitfully used to save considerable design time in developing the next generation. Whenever possible, measure a certain heat sink temperature, switch off (or reduce) the power of the component to which the heat sink is attached, measure the unpowered temperature, and deduce the adiabatic effective heat transfer coefficient for the component in question.

9.9 Pitfalls to Avoid

The author had some interesting discussions with Prof. Bar-Cohen from which some caveats popped up. In addition, when compiling this chapter, he found more nontrivial issues related to heat sink analysis that deserves highlighting.

9.9.1 Caveat 1. Know the Limits of the Elenbaas Correlations

The Elenbaas correlation was developed for isothermal plates and works very well for that configuration, as long as the Elenbaas number is not too small and the fins are much thinner than the gaps. When the Elenbaas number falls below 10 or so, the inflow from the open edge of the interfin gap keeps the channel from realizing the fully developed limit and *Nu* is much higher than predicted. Generally, however, this is not a desirable operating point, so this does not have much practical impact and does not affect the optimum spacing which is around Elenbaas = 50 and well above the values at which inflow is important. However, your design may force low Elenbaas numbers and then the use of these correlations becomes doubtful.

9.9.2 Caveat 2. Complex Composite Correlations Versus Simple Elenbaas Correlations

The composite equations were developed to deal with the effect of the plate/fin thickness and isoflux plates, as well as asymmetrically heated channels. For standard natural convection applications the general rule is to use the Elenbaas correlations when fin thickness can be neglected with respect to fin spacing. However, when you may decide what can be neglected? The recommendation is to acquire insight to try both to calculate the optimum spacings and heat transfer relations to quantify the effect of thickness.

9.9.3 Caveat 3. Be very Careful with Choosing the Right Temperature Difference in Calculating the Heat Transfer Coefficient h

The heat transfer coefficient h is always based on some temperature difference, and using published equations one should be very careful in selecting the right one. For example, heat exchanger theory is a simple and recommended way to get insight in many parameters that govern heat sink thermal behavior. However, the designer needs to get a value for h. Maximum dissipation is often calculated using the $T_{wall} - T_{inlet}$ difference, but then using h derived from a fully developed constant value that is based on $T_{wall} - T_{bulk}$ is incorrect. However, because h extracted from the Elenbaas equation does include the air temperature rise, the consequence is that when the heat exchanger concept is used this effect is taken into account twice. Our advice is to take an educated guess for h: 5 W/m^2 K.

9.9.4 Caveat 4. Nu = C Can Also Be Used for Natural Convection

It is not generally known that the fully developed constant *Nu* value is also valid for natural convection. A short reference can be found on p. 286 in the Kraus and Bar-Cohen book [5]. A more thorough discussion is provided by Moffat and Ortega in a chapter called *Direct Air-Cooling of Electronic Components*, especially, *Part II: Natural Convection* [69]. These authors not only state that the constant *Nu* assumption is valid for natural convection fully developed flow but also, within limits, there is almost no distinction between forced and natural convection for developing flow. It is informative to extend the principles discussed in the section called *Modeling Natural Convection Cooled Systems Using Forced Convection Principles* based on vertical channels with discrete components to smooth channels. The authors showed that the best way to treat the heat transfer for each individual component by defining the adiabatic heat transfer coefficient, in fact the only physically sound way. The argument could be extended to smooth channels. Consider the channel to consist of small horizontal strips divided from each other by infinitesimally thin layers with zero thermal conductivity. It is clear that the adiabatic temperature of each strip is caused by all upstream strips. If the strip's own contribution can be neglected, this could be viewed as buoyancy-driven forced convection heat transfer, explaining the fact that forced convection equations are also valid for natural convection channel flow. However, be aware of the temperature difference that is associated with the value for constant *Nu*! All other definitions state something like $Nu =$ C.Ra.(d/L^3) or equivalent, thereby obscuring its constant value, the reason being that this correlation is linked to another temperature difference. According to Bar-Cohen, this was one of Eckert's pet peeves, in later life, that the use of enclosure and channel aspect ratios in the various correlations obscured the underlying physics. For example, in some of the enclosure correlations, using the width as the length scale in the Nusselt number requires that you add an aspect ratio term in the correlation but, in fact, for a width that is greater than 2 $\times$ the boundary layer thickness, there is no dependence on width. On the other hand, one may question the physics behind a constant *Nu* number, hence a constant h, hence independent of the velocity. This topic is discussed in [70].

9.9.5 Caveat 5. Be Aware of the Useful Range of Your Correlations

An example is given in Caveat 4. If you use a correlation based on the fin spacing, this correlation is not valid beyond a certain width.

9.10 Conclusions

Whenever cooling a certain electronic component or system asks for the addition of a heat sink, the designer is confronted with the question: which one? Apart from the plethora of choices in shapes, sizes, materials, volumes etc., most handbooks and papers focus on parallel plate heat sinks and wind tunnel flow. The objective of this chapter is to provide designers with the right mindset to enable an optimal choice for their application, with a focus on LED applications. After discussing some heat sink thermal basics, the body of this chapter is devoted to heat sink modeling approaches from a designer's perspective. A distinction is made between approaches that are not recommended, approaches that are useful to acquire either insight in the physics, or in optimization parameters, or in fluid dynamics, and finally approaches that are suited for the final design stage in LED applications. The final conclusion boils down to the following: the only way that leads to optimization of the whole system is to use CFD modeling supported by dedicated tests. The use of correlations is only recommended in the first stages of a design.

References

1. Lasance C (2010) How to estimate heat spreading effects in practice. J Electron Packag 132:031004
2. Lasance C (2004) The influence of various common assumptions on the boundary-condition-independence of compact thermal models. IEEE CPT 27:523–529
3. Lee S (1995) Calculating spreading resistance in heat sinks. Electron Cool 1(January Issue)
4. Lasance C (2008) Heat spreading: not a trivial problem. Electron Cool 14(May issue): 24–30
5. Kraus A, Bar-Cohen A (1995) Design and analysis of heat sinks. Wiley, New York p 61
6. Maheswaran U, Sekhar S (2006) Evaluation of concepts of fin efficiency using monte carlo simulation method. Int J Heat Mass Transf 49:1643–1646
7. Lasance C (2008) Ten years of boundary-condition-independent compact thermal modeling of electronic parts: a review. Heat Transf Eng 29:149–169
8. http://www.electronics-cooling.com/1997/09/standardizing-heat-sink-characterization-for-forced-convection/
9. Moffat R (2008) Modeling air-cooled heat sinks at heat exchangers. Electron Cool (February)
10. Azar K, Tavassoli B (2003) How much heat can be extracted from a heat sink? Electron Cool (May)
11. Simons R (2006) A simple thermal resistance model—isoflux versus Isothermal. Electron Cool (February)
12. Lasance C (2005) Sense and nonsense of heat transfer correlations applied to electronics cooling. Proceedings of EUROSIME pp 8–16
13. Moffat R (2007) Modeling Air-Cooled heat sinks as heat exchangers. In: Proceedings of the 23rd IEEE SEMI-THERM Conference 20–22 March 2007
14. Webb R (2007) Heat exchanger design methodology for electronic heat sinks. Transactions of the ASME. J Heat Transf 129(7):899–901
15. Starner K, MacManus H (1963) An experimental investigation of free convection heat transfer from rectangular fin arrays. J Heat Transf 85:273–278
16. Welling W, Wooldrigde C (1965) Free convection heat transfer coefficients from rectangular vertical fins. J Heat Transf 87:439–444
17. Van de Pol DW, Tierney JK (1974) Free convection heat transfer from vertical FinArrays. IEEE Trans Part Hybrids Packag PHP 10(4):267–271
18. Bilitzky A (1986) The effect of geometry on heat transfer by free convection from a fin array, MS Thesis, Dept. of Mechanical Engineering, Ben-Gurion University of The Negev, Beer Sheva, Israel
19. Kim T, Do K, Kim D (2011) Closed form correlations for thermal optimization of plate-fin heat sinks under natural convection. Int J Heat Mass Tran 54:1210–1216
20. Kraus A, Bar-Cohen A (1995) Design and analysis of heat sinks. John Wiley, p. 275
21. Holahan MF, Kang SS, Bar-Cohen A (1996) A Flowstream based analytical model for design of parallel plate heat sinks. In: ASME Proceedings of the 31st National Heat Transfer Conference, HTD-Vol. 329, Volume 7, pp. 63–71
22. Kakac S, Shah RK, Aung W (1987) Handbook of single-phase convective heat transfer, 2nd edn. Wiley, New York, pp 3–35, 3–42
23. Biber C, Belady C (1997) Pressure drop prediction for heat sinks: what is the best method? In: Proceedings of Interpack' 97 Conference, ASME, Mauna Lani, Hawaii
24. http://www.coolingzone.com/library.php?read=489
25. http://www.coolingzone.com/library.php?read=490
26. http://www.coolingzone.com/library.php?read=491
27. Simons B (2011) Calculation corner: a useful catalog of calculation corner articles. ElectronicsCooling (September)
28. Bahadur R (2005) Characterization, modeling, and optimization of polymer composite pin fins, Thesis University Of Maryland (http://drum.lib.umd.edu/handle/1903/2821)
29. Bar-Cohen A, Bahadur R, Iyengar M (2006) Least-energy optimization of air-cooled heat sinks for sustainability-theory, geometry and material selection. Energy 31:579–619

30. Khan WA, Culham JR, Yovanovich MM (2005) Optimization of pin-fin heat sinks using entropy generation minimization. IEEE Trans Compon Packag Technol 28(2):247–254
31. Khan WA, Culham JR, Yovanovich MM (2008) Modeling of cylindrical pin-fin heat sinks for electronic packaging. IEEE Trans Compon Packag Technol 31(3):536–545
32. Ledezma G, Bejan A (1996) Heat sinks with sloped plate fins in natural and forced convection. Int J Heat Mass Trans 39(9):1773–1783
33. Naidu SV, Dharma Rao V, Govinda Rao B, Sombabu A, Sreenivasulu B (2010) Natural convection heat transfer from fin arrays: experimental and theoretical study on effect of inclination of base on heat transfer. ARPN J Eng Appl Sci 5(9)
34. http://frigprim.x10.mx/articels2/parallel_pl_Inc.html
35. Elenbaas W (1942) Heat dissipation of parallel plates by free convection. Physica 9:1–28
36. Bar-Cohen A, Rohsenow W (1983) Thermally-optimum arrays of cards and fins in natural convection. IEEE CHMT CHMT 6:154–158
37. Iyengar M, Bar-Cohen A (1998) Least-material optimization of vertical pin-fin, plate-fin, and triangular-fin heat sinks in natural convective heat transfer. In: Proceedings of the ITHERM Conference, Seattle, USA, pp 295–302
38. Bar-Cohen A, Iyengar M, Kraus AD (2003) Design of optimum plate fin natural convection heat sinks. ASME Trans—J Electron Packag 125(2):208–216
39. Bahadur R, Bar-Cohen A (2005) Thermal design and optimization of polymer-based pin fin natural convection heat sinks. IEEE Trans Compon Packag Technol 28(2):238–246
40. Biber C, Fijol S (1999) Fan-plus-Heatsink "optimization"—mechanical and thermal design with reality. In: Proceedings of the international systems packaging symposium, pp 285–289
41. Knight R, Goodling J, Hall D (1991) Optimal thermal design of forced convection heat sinks-analytical. JEP 113:313–321
42. Teertstra P, Yovanovich MM, Culham JR, Lemczyk T (1999) Analytical forced convection modeling of plate fin heat sinks. In: Proceedings of 15th IEEE SEMI–THERM
43. Iyengar M, Bar-Cohen A (2001) Design for manufacturability of SISE parallel plate forced convection heat sinks. Trans of the IEEE-CPT 24(2):150–158
44. Copeland D (2003) Fundamental performance limits of heat sinks. JEP 125:221
45. Bejan A (1996) Entropy generation minimization. CRC, Orlando
46. Shih C, C Liu G (2004) Optimal design methodology of plate-fin heat sinks for electronic cooling using entropy generation strategy. IEEE trans compon packag technol 27(3):551–559
47. Bar-Cohen A, Bahadur R, Iyengar M (2006) Least-energy optimization of air-cooled heat sinks for sustainability-theory, geometry and material selection. Energy 31:579–619
48. Bejan A (1975) Improved thermal design of the cryogenic cooling system for a superconducting synchronous generator. Thesis. Ph.D.–Massachusetts Institute of Technology. Dept. of Mechanical Engineering
49. http://www.constructal.org/
50. Silva AK da, Bejan A (2005) Constructal multi-scale structure for maximal heat transfer density in natural convection, Int. J Heat Fluid Flow 26:34–44
51. Teerstra P, Yovanovich M, Culham J (2000) Natural convection modeling of heat sinks using web-based tools, ElectronicsCooling (September)
52. http://www.mhtl.uwaterloo.ca
53. http://www.novelconceptsinc.com/calculators-forced-convection-heat-sink-thermal-resistance.cgi
54. http://www.heatsinkdesigner.com/
55. http://www.qfin.net/drupal/qfin47
56. Lasance C (2002) The conceivable accuracy of experimental and numerical thermal analyzes of electronic systems. IEEE CPT 25:366–382
57. Lasance C (2005) CFD simulations in electronic systems: a lot of pitfalls and a few remedies. Electron Cool 11(2):20–27
58. Lasance C, Rindt C (2007) Accuracy comparison of a standard CFD code for the thermal analysis of non-simple geometries with baseline experiments. In: Proceedings of 23rd SEMI-THERM, San Jose, CA

59. Rodgers P, Eveloy V (2004) CFD prediction of electronic component operational temperature on PCBs. Electron Cool 10(2):22–28
60. Lasance C, Aarts R (2008) Synthetic jet cooling Part I: overview of heat transfer and acoustics. In: Proceedings of SEMI-THERM 24: San Jose
61. Lasance C, Aarts R, Ouweltjes O (2008) Synthetic jet cooling Part II: experimental results of an acoustic dipole cooler. In: Proceedings of SEMI-THERM 24
62. Lasance C, Nicole C, Aarts R, Ouweltjes O, Kooijman G, Nieuwendijk J (2009) Synthetic jet cooling using asymmetric dipoles. In: Proceedings of SEMI-THERM 25 San Jose
63. Lasance C (2008) Ten years of boundary-condition-independent compact thermal modeling of electronic parts: a review. Heat Transf Eng 29:149–169
64. Lasance C, Eggink H, (2005) A method to rank heat sinks in practice: the heat sink performance tester. In: Proceedings of 21st SEMI-THERM, pp 141–145, San Jose, CA
65. Copeland D (moderator) (2003) Modeling, optimization and performance of heat sinks with non-traditional flow and fin geometries. Panel Session at Interpack, Maui
66. Lasance C (2003) Thermal capacitance. Electron Cool 9 (May)
67. Coetzer C, Visser J (2003) Compact modeling of forced flow in longitudinal fin heat sinks with tip bypass. J Electron Packag 125:319–324
68. Dong T, Narendran N (2009) Understanding heat transfer mechanisms in recessed LED luminaires. In: Ninth International Conference on Solid State Lighting, August 3–5, 2009, San Diego, Proceedings of SPIE 7422: 74220V.
69. Moffat R, Ortega A (1988) Chapter 3 in advances in thermal modeling of electronic components and systems, vol 1, ed. Bar-Cohen and Kraus, Hemisphere
70. Lasance C (2010) Thermal facts and fairy tales: fully-developed channel flow: why is Nu constant? Electron Cool (September 2010)
71. Bejan A (2004) Convection heat transfer, Wiley Fourth Edition, 2013

Part IV
Applications

Chapter 10
Considerations for an Optimal Choice of Heat Sinks for LED Applications

Norbert P. Engelberts

Abstract This chapter on heat sinks is devoted to examples of industrial LED applications and offers ways to arrive at an optimal choice for a heat sink given a certain application. The selection of cooling solutions for LED applications should be based on an overall system-board-component level investigation rather than on solving individual component problems. From a total system perspective, this approach will result in cheaper or better thermal management solutions. Computational fluid dynamics (CFD) analyses are a very valuable tool to arrive at the best possible solution at the system, board, and component levels. Within this process, compact heat sink modeling is required and a method is presented how to realize this for various geometries.

10.1 Examples of Industrial LED Applications

Nowadays, LED lighting is one of the fastest growing market sectors across the globe. There are three primary areas that make LEDs attractive as a replacement for the standard lighting products we enjoyed for decades. These are (1) energy savings, (2) easier control of managing the response of light (freedom of application), and (3) long lifetime. These advantages are all extensively discussed in other chapters and hence the discussion will not be repeated here. We will only present in this section an overview of industrial LED applications highlighting the role of heat sinks therein.

1. Automotive lighting
   a. Taillights: break lights, turning lights
   b. Day light saving
   c. Low beam/high beam
   d. Internal lighting within the cabin
2. Backlighting/flat screen displays (extensively discussed in Chap. 14, will not be treated further)
3. Social activities and screen adverting

N. P. Engelberts (✉)
Optimal Thermal Solutions BV, Bussum, The Netherlands
e-mail: nengelberts@ots-eu.com

C. J. M. Lasance, A. Poppe (eds.), *Thermal Management for LED Applications*, 391
Solid State Lighting Technology and Application Series 2,
DOI 10.1007/978-1-4614-5091-7_10, © Springer Science+Business Media New York 2014

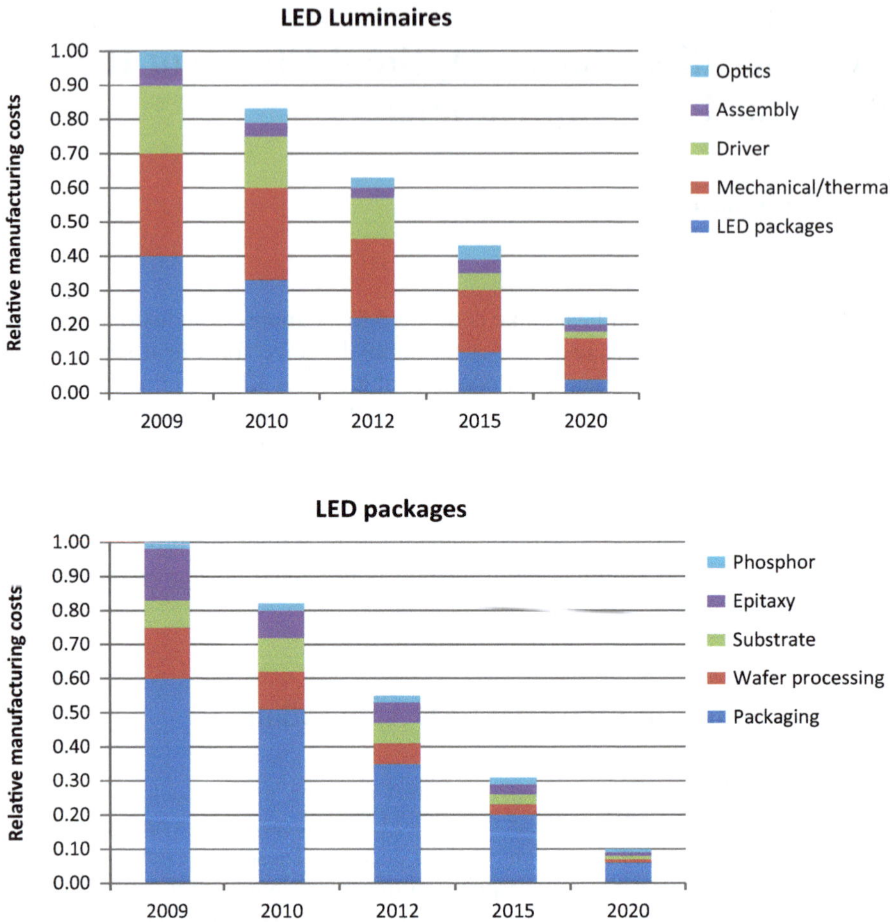

Fig. 10.1 Cost reduction roadmap. (Source: Department of Energy, SSL Manufacturing Roadmap, 2010)

The application in the opening ceremony of Beijing Olympics is a typical example: a screen of length 147 m and width 22 m employing 44,000 LEDs

4. General lighting
   Street lighting, retrofits

All LED-based applications have the following parts in common next to the LED: a heat-spreading submount on a PCB or a Metal Core PCB (MCPCB) to hold and interconnect the LED, additional optics, driver electronics to power the LED, mechanics to hold the whole together, and a thermal management solution to cool both the LED and driver electronics. The cost impact of each part is shown in Fig. 10.1, both for the LEDs themselves and the LED applications. Concentrating on the applications, we see that in the future, the mechanical/thermal part is expected to become the major

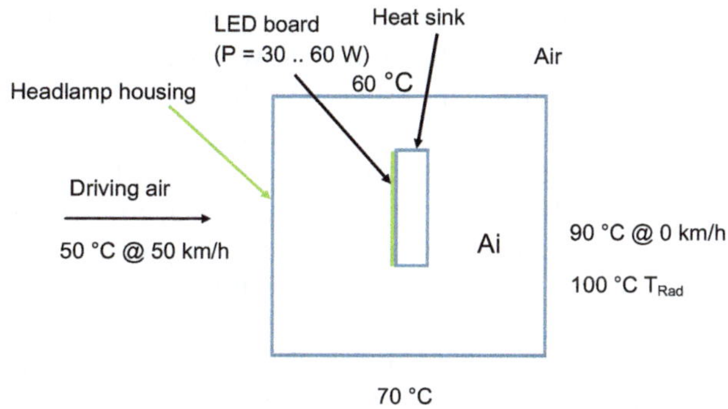

Fig. 10.2 Boundary conditions of a headlamp of a car

part of the total price of the application. Obviously, this trend suggests that it pays off to have a critical look at the factors that determine the cost of a cooling solution. In the following examples of industrial LED applications with their specific design challenges are discussed.

10.1.1 Automotive Lighting

LEDs are used since many years within cars mainly for the breaking light, turning light, taillight, signaling, and interior illumination. These applications do not ask for sophisticated thermal management solutions, but cost savings by optimization of heat sink material should certainly be explored. This is quite different for front lights. One of the biggest advantages of using LEDs in front lights is styling flexibility, which enables manufactures to introduce distinctive designs with unique freedom and variety in styling for the development of brand characteristic styling elements. Other advantages are as follows: the color gamut of the LEDs nearly achieves daylight quality, a lifetime more than the vehicle's life, and significant lower power consumption.

At this moment in time, the front side uses LEDs mainly for daylight saving, whereas the headlight is made available as an option for some high-end cars, such as the Lexus LS600h (2007–2008), Audi V10 R8 (2008), Cadillac Escalade (2009), and Toyota Prius (2010), that are using LEDs in headlamps for both high and low beam. The taillight, daylight saving lights, and signaling lights are in most cases high-brightness LEDs that use 1–2 W input electric power. This power makes it rather easy to cool by using either a metal core PCB combined with a heat sink or an enhanced FR4 board with a heat sink. This is different if we look at the headlights where a number of LEDs, each with an input power of 5–11 W, are required to guarantee sufficient light output. Together with the high ambient temperature requirement (typically up to 100 °C), cooling becomes really a challenge (see Fig. 10.2).

With a maximum temperature of 125–150 °C for the LED junction, we have a maximum temperature rise with respect to the ambient of 55–80 °C for a total power dissipation of 30–50 W within the lamp. Currently, high-brightness LEDs are used with an input power between 5 and 11 W. Assume the thermal resistance from the LED junction to the slug to be 2 K/W, about 20 °C out of the allowed temperature drop is already inside the package, so the slug temperature needs to be at 105–130 °C.

With some spreading and interfacing losses, the internal heat sink temperature needs to be kept at 95–120 °C, which means a thermal resistance from the heat sink to outside ambient of less than 1 K/W, which poses a real challenge. Cooling is done by internal fans and/or heat pipes that transport the heat to an external heat sink.

To cool 50 W with liquid is straightforward and several options have been proposed. For example, one option is to make use of a liquid cooling circuit that makes an efficient connection between the internal LED heat sink to an external heat exchanger, see for example [1]. When complete car fronts are offered by vendors, it becomes attractive to integrate the LED headlamp cooling with the liquid cooling circuitry of the motor, thereby getting rid of heat sinks and fans altogether.

Dimming is recommended at lower driving speeds or in traffic jams to compensate for the reduced external cooling. It should be added that fans are not only used within the headlamps to generate sufficient cooling but also for deicing and defrosting [2], which was not a problem with the significant infrared radiation leaving the front of halogen and HID lamps.

10.1.2 Large Screens and Video Walls

Large LED-based screens are used for events, advertisement, branding, and sports. They are mounted in free air or in front of buildings. Uniform brightness and color stability is important here. Because both parameters are influenced by temperature, it is mandatory to control this temperature within certain limits. Outdoor displays are used during day and night and need to be designed for all environmental conditions including solar load. In the market, we see different options: systems cooled by fans, systems with integrated air-conditioning such as large displays for use at events in stadiums as well as passive cooled systems using extruded or plate heat sinks together with solar shields to reduce the effect of solar load during daytime use.

10.1.3 Street Lighting

Street lighting luminaires with embedded LEDs are becoming increasingly popular because of the lower energy consumption and reduced maintenance need. The enclosure is usually made of metal that serves a number of goals: it holds all parts together, makes it water tight, and is used for heat spreading. A variety of designs can be found on the market (see, for example, Figs. 10.3 and 10.4).

Depending on the location on earth, the ambient conditions will differ. The maximum ambient temperature will change depending on the climate. See the different climates that are found on earth in Fig. 10.5. The ranges show statistical probabilities; the limits of range 1 are within approximately 1 %, range 2 within 0.1 % at the lower end and 0.05 % at the higher end, and range 3 shows the absolute maximum

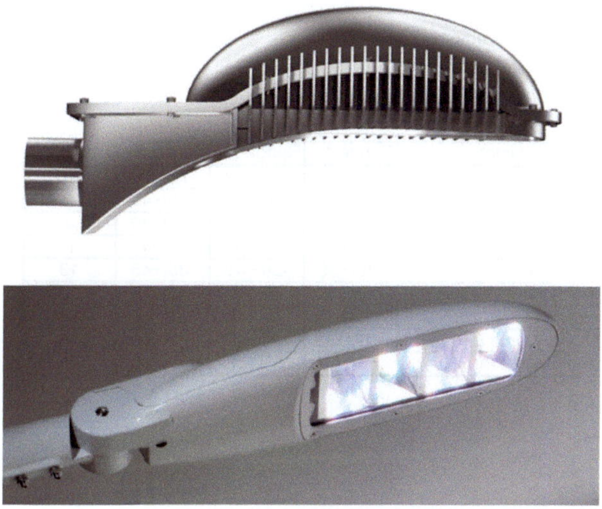

Fig. 10.3 LED based streetlights. (Courtesy of Arianna (*upper*) and Innolumis (*lower*))

Fig. 10.4 T-line LED tunnel lighting luminaire, Vlake Tunnel, The Netherlands (June 2011)

and minimum values ever measured. Another important effect is the solar load. In most cases, lamps are used during night time, but can be switched on during daytime for several reasons, for example, maintenance and weather conditions. Wind will also have a major effect on the cooling of the lamp. In any case, a strategy for optimal cooling is by no means straightforward, and a choice has to be made for street lamps optimized for certain regions or for the whole planet. No general rules can

Tambient : -30/+45°C
Wind?
Solar load ?

Type of climate	Range 1	Range 2	Range 3	Tmean
Extremely cold	-55/+26	-65/+32	-75/+26	-15
Cold	-45/+25	-50/+32	-60/+40	-10
Cold temperate	-29/+29	-33/+34	-45/+40	0
Warm temperate	-15/+30	-20/+35	-30/+40	8
Warm dry	-10/+35	-20/+40	-30/+45	13
Mild Warm Dry	0/+35	-5/+40	-15/+45	20
Warm Damp	+12/+35	+5/+40	+0/+45	24
Warm Damp Equable	+17/+33	+13/+35	+4/+40	25
Extremely Warm Dry	+8/+43	+3/+55	-10/+60	26

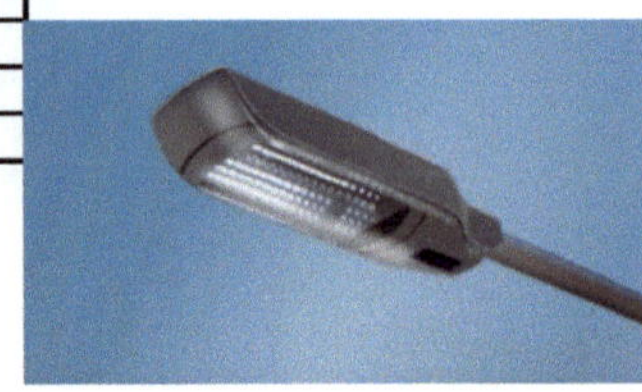

IEC 721-2-1

Fig. 10.5 Ambient conditions for a street lamp. The temperatures are extrapolated from IEC721-2-1

be given, but it is recommended to apply smart designs that provide enhanced heat transfer/heat spreading inside the luminaire and yet maintain cost effectiveness. It is mandatory that the lifetime of these options should at least be comparable with that of the LEDs and electronics. When using thermal interface materials, their reliability and ageing properties should also be considered.

The cooling of street lighting luminaires is in most cases done by the housing itself improved with additional finning and in some cases with heat pipes to better spread the heat. The use of heat pipes connected to efficient plate heat sinks can give a major reduction in the weight of the full luminaire assembly. Lowering the spreading resistance is key in this kind of design. There is plenty of surface area in most cases, but the temperature gradients can be large, hence the use of heat pipes. Another option is to add internal ribs to reduce the spreading resistance and keep the weight acceptable. In summary, because of the complexity of optimizing a streetlamp from a thermal point of view, it is strongly recommended to use computational 3D analyses together with some dedicated experiments as discussed in Chap. 9.

10.1.4 Retrofit Solutions

Figure 10.6 shows some typical examples of LED-based retrofit lamps. The thermal challenges of these lamps are extensively covered in Chap. 12 and will not be repeated here.

10.1.5 Down Lighters

Down lighters as well as spot lights placed within the wall and floors are increasingly replaced by LED-based solutions. These may pose severe problems with mounting

Fig. 10.6 LED-based retrofit lamp examples. (Courtesy of Philips, Lemnis Lighting, Osram, and GE)

Fig. 10.7 Typical mounting of a down lighter in the ceiling

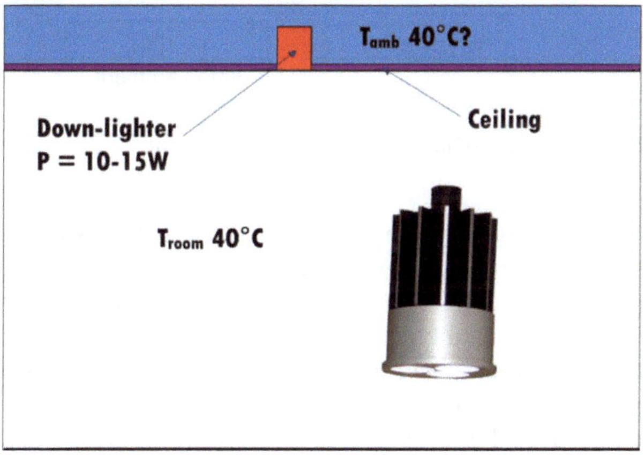

because most of the heat goes backward in contrast to standard lighting. Mounting can block the free airflow around the lamp and the effective ambient temperature may be quite different from the one that is measured near the surface where the lamp resides (see Figs. 10.7 and 10.8).

Mounting the LED lamp in a hole in the ceiling covered with insulation material shows a temperature increase of 16 °C for the lamp. The same will happen if no insulation material is used but the free airflow is blocked. Some lamps use openings to improve ventilation.

10.2 Industrial Examples Showing How More Expensive Heat Sinks Can Save Money

The question arises why to pay more for a more efficient heat sink that is also smaller and lighter in weight, while a simple and cheap extruded heat sink can also do the job? In most cases, the single-piece part cost price is the main driver why engineers and purchasers stay with lesser effective solutions because they are of the opinion that they save money, but is this really true?

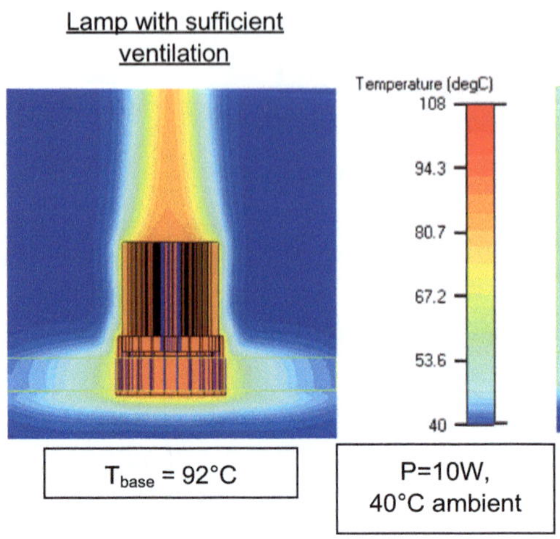

Fig. 10.8 Effect of mounting on lamp temperature

Let us compare the price of two single heat sinks, a highly efficient type vs. a standard extruded type with the same thermal performance, but not with the same flow performance, effect on nearby and downstream components, weight, volume usage etc. The standard type will always be cheaper if the quoted thermal performance is the single metric and other issues are accounted less important. However, from a system's point of view, the conclusion may be different. A more efficient heat sink may reduce the requirements on weight, volume, height, and/or fans, thereby saving money on the system level.

What makes a heat sink an efficient heat sink you may wonder?

Such a heat sink is optimized for both flow and heat transfer and hence makes much better use of the volume available for cooling and the available cooling air.

The result of the comparison test as shown in the Sect. 9.8.2 of Chap. 9 is proving this. The flow is optimized by using thin fins and a special design to lower the air resistance, resulting in the highest possible velocity between the fins, minimizing the effect of bypassing flow. Because of this, it also has a lesser intrusive effect to the flow resulting in a positive effect on the neighboring and downstream flow and in lower pressure drops over the board and through the system.

To arrive at a better comparison, let us look at other related effects that are affecting the system price:

1. Effect on the flow
2. Heat sink design and weight
3. Attachment to heat sink
4. Raw material usage to manufacture the cooling solution
5. Required fan performance
6. Product reliability
7. Transportation cost
8. End-of-life cost

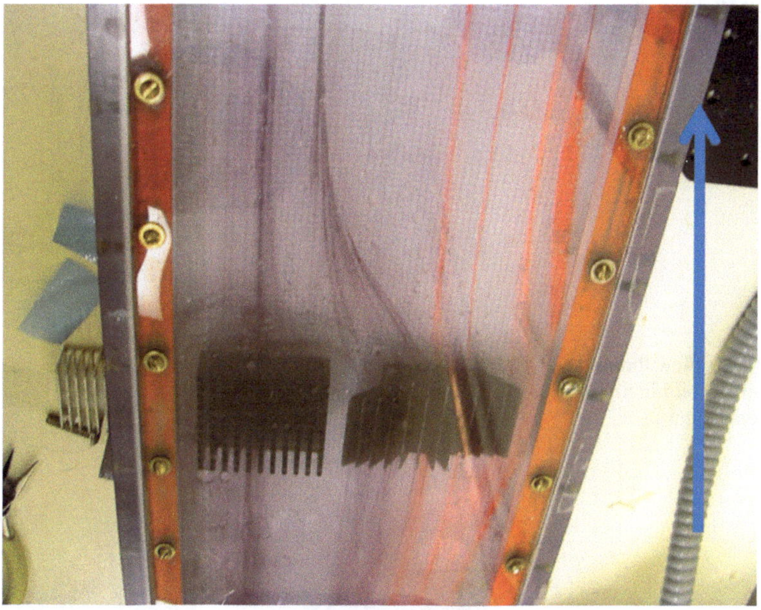

Fig. 10.9 Flow test in water tunnel of pin fin (*right*) and DUT. Flow is from bottom to top. (Courtesy of Advanced Thermal Solutions, Inc)

10.2.1 Ad 1: Effect on the Flow

Figure 10.9 is a picture of a fluid flow visualization test through a pin fin and the device under test (DUT) in a water tunnel. Compare the amount of flow entering the pin fin and the DUT. Most of the water "hitting" the pin fin is bypassing it because of the high air resistance of the fin area. Watch the flow that is left over downstream of the pin fin, being much less. Imagine what will happen if we put multiple pin fins in a row downstream of each other because we have to cool multiple devices in a row.

The first pin fin will have sufficient cooling but the further we get downstream the lesser effective the heat sinks become. Limited air will be available for cooling because most of the air is bypassing the heat sinks. What is the first reaction without knowing anything about the flow structure? We need a larger heat sink downstream to get the same cooling effect or, alternatively, we need to consider a more powerful fan system to drive more air through the system. However, if we had started from a system level point of view instead of concentrating on a single heat sink, we would have studied the flow field and the interaction between the heat sink and the flow more closely; we could have come up with better solutions.

In many datasheets of LED suppliers, an estimate is given for the total amount of surface area required for the heat sink. What is often seen is a very narrow fin spacing, which generates a large surface area but at the cost of a significant pressure drop. Furthermore, when there is bypass, the flow will rather favor the bypass path.

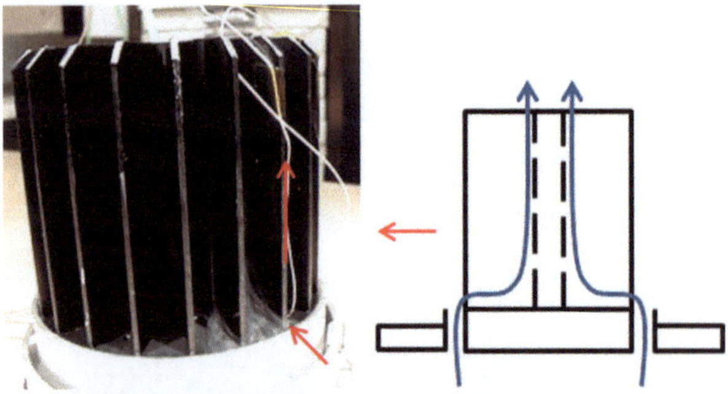

Fig. 10.10 Flow through down lighter heat sink

Fig. 10.11 Flow through two same-diameter pin fins with different height

While the problem can be solved using a high-pressure fan, we run into problems with natural convection. Especially for this type of cooling, we need to pay attention to optimize the fin spacing of the heat sink. A flow test on the first prototype will show if the design is performing as expected.

Flow visualization in general will help to understand if a heat sink is working in an optimal way or if a lot of air is bypassing. In Figs. 10.10 and 10.11, the benefits of a simple flow visualization test are demonstrated. Figure 10.10 shows the air passing the fin field of a star-shaped heat sink used for an LED-based down lighter. Openings were made in the lamp housing to allow lair to enter from below the ceiling. In the picture, the air nicely enters at the bottom of the heat sink near the base and goes to the center to rise, resulting in a full "wetting" of the heat sink. Figure 10.11 shows the flow through two pin fin heat sinks differing only in fin height. Both versions show a correct behavior regarding proper flow through the heat sink. What they also show is that it could help to have the fins in the center longer than at the sides to ensure a more uniform local heat transfer coefficient over the whole length of the pins.

10.2.2 Ad 2: Heat Sink Weight

Standard extruded heat sink profiles, but also die-cast heat sinks, normally have a thick base and thick fins and are made of lesser thermally conductive aluminum alloys. (The lower conductivity is a result of the additives that improve the manufacturing of the product.) The base and the fins need to be thicker than required from a thermal point of view because it is more difficult to extrude thin and high fins. Especially for natural convection, an optimal heat sink can easily be an order of magnitude thinner than what is offered.

The main design driver for these types of heat sinks is the ease of manufacturing and not the overall thermal performance. The end result is a bulkier and heavier heat sink that makes bad use of the available volume and has a negative effect on the flow.

The same is valid for those LED designs that use the housing as a heat sink. The housing is often made by extrusion or die casting processes, which limit freedom of design. The thermal conductivity is much lower than that of pure aluminum; for instance zinc aluminum (Zamac) has a thermal conductivity around 115 W/mK, Aluminum used for molding of 100–150 W/mK, brass annealed of 60 W/mK. Compare this with standard aluminum alloys with a high thermal conductivity such as 6063 (200 W/mK) or 1050A (230 W/mK).

In most LED applications, heat spreading is an important factor, which drives the thermal design. If analysis shows that heat spreading is important, the consequence may be that for lower conductivity materials, the only option is to either increase the wall thickness or embed heat pipes or vapor chambers, adding to cost and weight.

Forged heat sinks are made of highly conductive aluminum, but the manufacturing method itself is very limited regarding design freedom and size. Hence, in general, to get a better performance, a larger heat sink is required.

Especially for natural convection, the use of thermally conductive plastic could be of interest because of its lower weight and design freedom. Plastic enhanced with carbon fibers could also be used but needs special attention because of its anisotropic conduction behavior. Other options are designs that are a combination of highly efficient heat sinks and heat pipes to either improve heat spreading when the heat sink is much larger than the source or to transport the heat from the source to a remote heat sink. Counting the weight of a standard heat sink, the overall product weight could be much higher than for a design based on more efficient cooling solutions.

10.2.3 Ad 3: Attachment to Heat Sink

The attachment of the CoB module (Chip on Board)/Light Engine (PCB with LEDs) is important to guarantee a good interface between the board and the heat sink/lamp housing. The pressure on the interface, the surface roughness, and flatness are key here. Making a heat sink housing of die-cast material is good from a cost standpoint. However, be aware of the interface condition. In many cases, this surface needs to be machined afterwards to guarantee a thermally good interface. Using thin PCBs enhanced with thermal vias will be good from a thermal perspective, but proper attachment to the heat sink base/housing is as important. Figure 10.12 shows the differences in temperature between an LED board made by using 0.8 mm FR4 with thermal vias below the LED and 1.6 mm MCPCB and a thin (0.15 mm) interface material. The mounting surface within the die-cast housing was not machined, and the problem was further increased because the FR4 board was rather flexible because of its low thickness. The LEDs close to the screw location are showing a much lower temperature than in between the screws. Also see Fig. 10.12c were we put additional pressure on the surface between the screws, causing the temperature to drop. For the MCPCB, the effect is less pronounced; while the heat is first spread over a larger surface area, before going through the interface, resulting in a lower heat flux, the interface is becoming less critical. For LEDs dissipating 1 W, there is not a big risk in applying FR4 enhanced with thermal vias, but for higher power dissipations, the interface needs to be carefully chosen.

10.2.4 Ad 4: Raw Material Usage to Manufacture the Cooling Solution

As discussed in the other points, a more bulky heat sink solution requires more raw materials. However, the initial manufacturing process of aluminum is energy intensive, something we like to reduce in a world where reduction of energy consumption is key (see [3]). Fortunately, it can be recycled without the loss of its properties, and the recycling process uses only a fraction of the energy of the initial manufacturing process. Finally, there are manufacturing techniques, such as bonding, folding, and skiving, that do not suffer from these sustainability issues.

10.2.5 Ad 5: Required Fan Performance

As already discussed in Ad 1, a lesser efficient heat sink such as a pin fin or standard extruded type will lead to higher air resistance and lesser optimized flow over the LEDs and through the system. To overcome the higher air resistance and allow for more flow to compensate for the not optimized airflow, a more powerful fan or multiple fans are required. A more powerful fan can mean either a larger fan type or

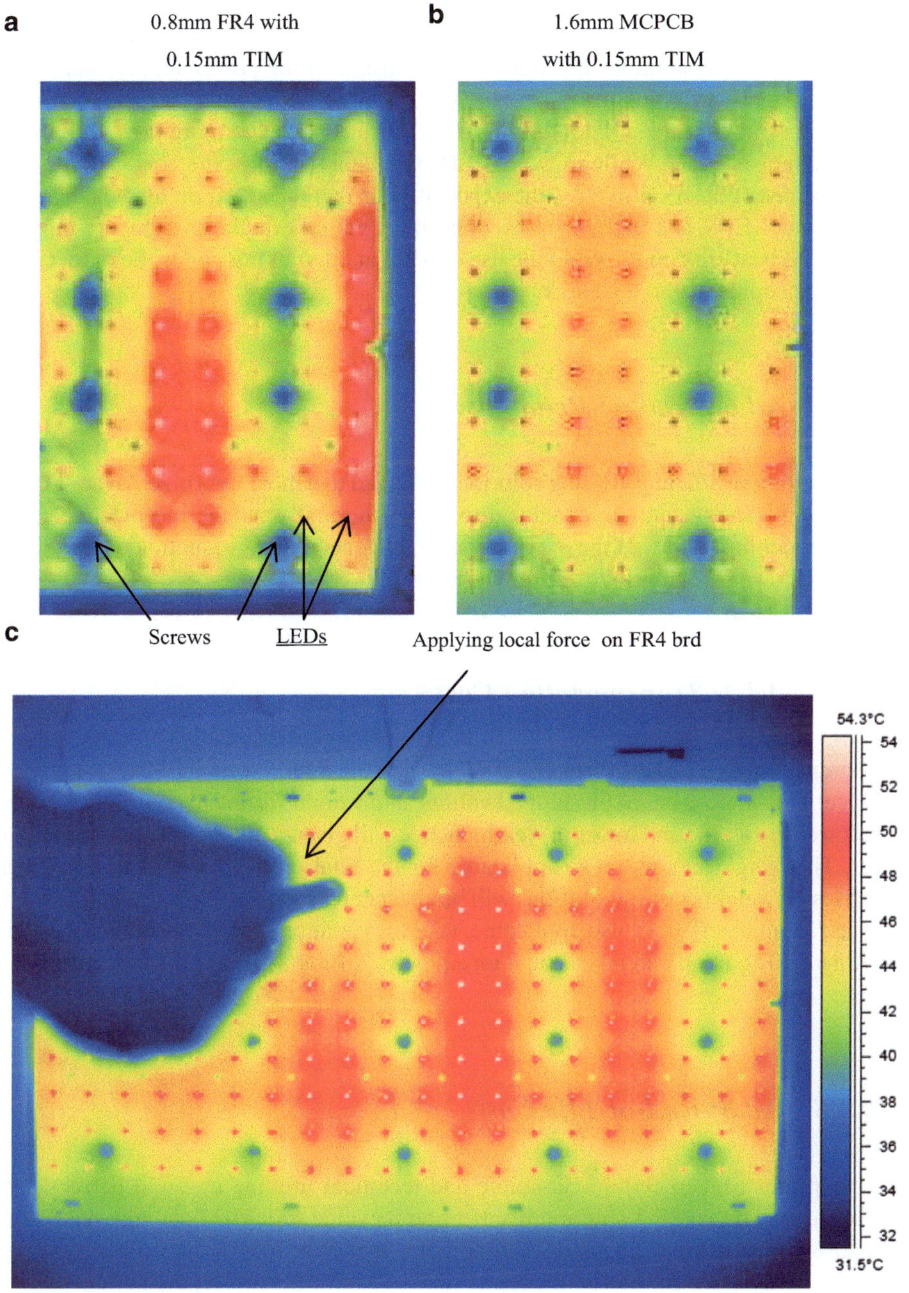

Fig. 10.12 Infrared picture of FR4 and MCPCB LED board

a higher rotational speed of the current fan. However, doubling the fan speed means increasing the input power to the fan by a factor of 8. As a result, we will pump more heat in the system, have a higher current usage of the power supply, and more power dissipation in the fan itself, which will lead to a higher fan temperature and reduction of its lifetime. On top of this, a higher fan speed and more flow will result in higher noise levels. While in many LED-based applications, the goal is to rely on natural convection cooling only, the expectation is that the situation may change for high-power LEDs, and in these cases, optimizing your thermal design by optimizing around the heat sink could avoid the use of a fan at all, making up for the extra costs of a more sophisticated heat sink.

10.2.6 Ad 6: Product Reliability

The use of more efficient cooling solutions will lead to a more optimized overall thermal design of the system, influencing directly the thermally and thermo-mechanically related reliability issues of the overall system. In general, lower junction temperature results in lower probability of different failures.

10.2.7 Ad 7: Transportation Cost

The transportation cost of the cooling solution to the manufacturer of the system has a price. This price is based on shipped volume and weight. Efficient cooling solutions are lower in volume and lower in weight, hence will give a reduction in transportation cost. The weight factor is also applicable for the final product, a product equipped with lesser weight cooling solutions will be cheaper to transport. Apart from transport issues, a human effect is also applicable, take for instance an LED-based streetlamp. Lifting a 30 kg lamp and installing it on a pole vs. lifting a 20–25 kg lamp speaks for itself.

10.2.8 Ad 8: End-of-Life Cost

Every product has a certain economic and technical lifetime and will be recycled afterwards. The cooling solution needs to be recycled too. The heat sink and lamp enclosure in most cases are made of aluminum and can be recycled in cost and energy effective ways, but obviously, the lesser mass we have to recycle, the better.

In summary, the conclusion must be that it pays off to focus on the costs of the total system, and not only on the costs of the individual parts. In times hopefully long gone by it was standard practice that project leaders got bonuses for buying parts as cheap as possible. Needless to say that such an attitude cannot survive in a world where end-users buy total systems, not a collection of parts. However, in the case of heat sinks we still notice a suboptimal purchasing policy, often based on lack of knowledge and outdated protocols.

10.3 Computational Fluid Dynamics with Calibrated Heat Sink Compact Models, Including Optimization

This section contains a detailed description of how to build and use compact models in a popular computational fluid dynamics (CFD)- based thermal simulation tool dedicated to electronics cooling. This modeling method is generic and hence can also be used for LED-specific cooling.

10.3.1 Compact Models for Use in CFD Analysis; Practical Example

Slanted fin heat sink types (see. for example, [4]) are often complex heat sink designs and require a compact way of modeling to reduce the number of grid cells. CFD packages such as FloTHERM® (from Mentor Graphics) and 6SigmaET (from Future Facilities) are "limited" to an orthogonal hexahedral mesh (however, the advantages over unstructured grid generation outweigh the disadvantages by far). This makes the modeling and meshing of non-parallel-plate heat sinks more problematic, although rotation and slanting are better solved within 6SigmaET at this moment in time. To capture the complex flow correctly, the resulting very fine mesh drastically increases the total mesh size and therefore increases the simulation time. At the system level, detailed heat sink modeling is often not an option. Henceforth, what is required are compact models of heat sinks. Usually, these compact models represent the total volume and the flow and pressure drop characteristics. This is rather straightforward for parallel-plate heat sinks, but not for slanted heat sinks. The next section proposes a way of working to arrive at a compact model for more complex heat sinks. The method is described for a particular CFD code but may be easily adapted to other CFD codes.

10.3.2 Structure of the Compact Model

This section will detail the compact modeling of a complex heat sink (referred to as device under test (DUT)) as an example. The structure of a detailed DUT is shown in Fig. 10.13. Figure 10.13 does not show the heat sink base length, *L*. The fin length of the heat sink is the same as the heat sink base.

A compact model of a heat sink in CFD package consists of the following (Fig. 10.14):

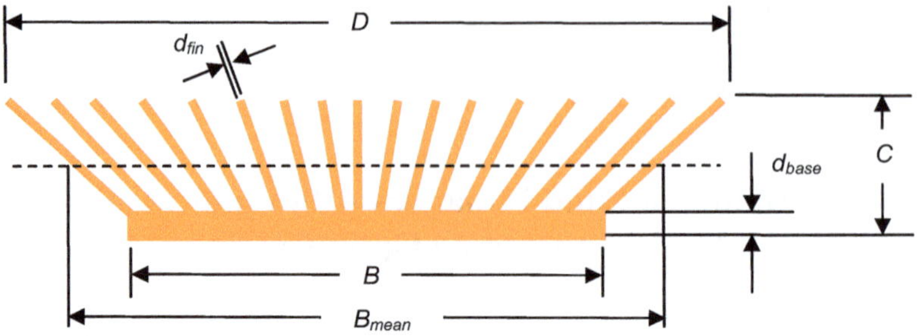

Fig. 10.13 Front view sketch of a DUT with the dimensional variables indicated

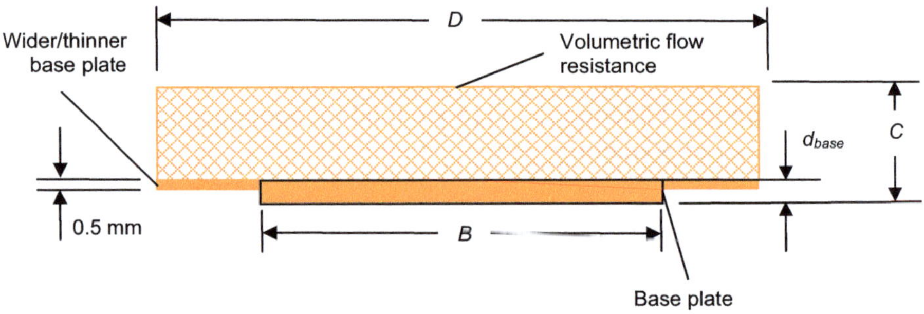

Fig. 10.14 Front view of a heat sink compact model

- A base plate
- A wider and thinner base plate
- Volumetric flow resistance
- Planar flow resistances on all volumetric flow resistance surfaces
- Heat transfer coefficient from the base to the air surrounding it

The compact model base plate has the same dimensions as the original base plate. The base plate's conductivity in the lateral direction is the same as the base material conductivity, k_{HS}. In the flow direction, the base plate has a higher conductivity value, k_{base}, because of the fins enhancing the conductivity in the flow direction. This thermal conductivity can be calculated by the following equations:

$$n = \text{odd}: H = \frac{2}{n-1} \sum_{i=1}^{(n-1)/2} \sqrt{\left(i\frac{D-B}{n-1}\right)^2 + (C - d_{base})^2} \tag{10.1}$$

$$n = \text{even}: H = \frac{2}{n} \sum_{i=1}^{n/2} \sqrt{\left((i-1)\frac{D-B}{n-1}\right)^2 + (C - d_{base})^2} \tag{10.2}$$

$$k_{base} = k_{HS} \frac{n \times H \times d_{fin} + d_{base} B}{d_{base} B} \tag{10.3}$$

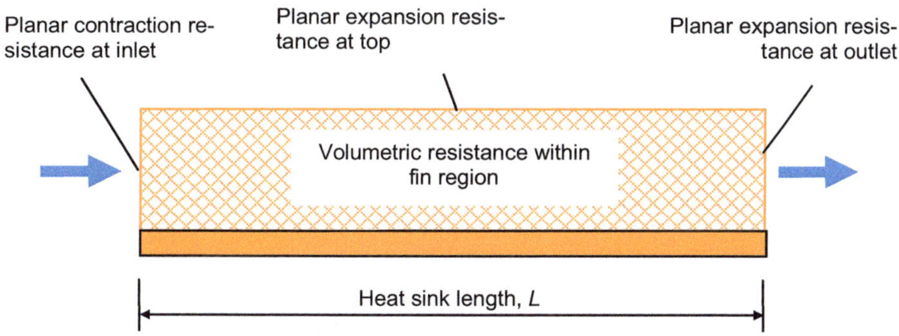

Fig. 10.15 Sketch of the planar and volumetric flow resistances of a compact model

Please note that the following description is code-dependent. The author is of the opinion that it is more instructive this way rather than trying to describe the procedure in a general sense.

The compact model base plate is overwritten by a wider but thinner base plate to allow for the slant effect with attached volumetric *Surface Heat Exchange*. The *Surface Heat Exchange* is determined in Sect. 10.3.7 further on. The thin plate has a conductivity of 600 W/mK in lateral direction. In the flow direction, it has the same effective conductivity as the base plate, k_{base}.

In the fin region, a volumetric *Flow Resistance* is modeled. The flow resistance values are calculated in Sect. 10.3.6. The fin region has the same height and overall width as the original heat sink. The flow region for the compact model is larger than the real trapezoidal region occupied by the DUT. The settings for the flow resistance and heat transfer are done using the real fin surface area, real average fin pitch, etc. The planar flow resistance for the planar regions is determined in the Sect. 10.3.4.

10.3.3 Flow Resistance

This section discusses the calculation of the flow resistance values for the planar and volume resistance of the compact model. There are two flow resistances for the planar regions. The inlet, outlet, and top planar flow resistances are determined in the next section, whereas in the section after the next, side planar regions flow resistances are determined. The volumetric flow resistance is derived in the Sect. 10.3.6 (Fig. 10.15).

10.3.4 Inlet, Outlet, and Top Planar Flow Resistances

For electronic equipment applications, the air flow between the fins of the DUT can be assumed to be laminar in almost all cases. Therefore, the compact model is restricted to laminar flow between the fins. A typical DUT has thin fins in the order of 0.5 mm to 0.8 mm thick. Contraction and expansion losses for heat sink inlet and

Fig. 10.16 Sketch illustrating the hydraulic diameter calculation

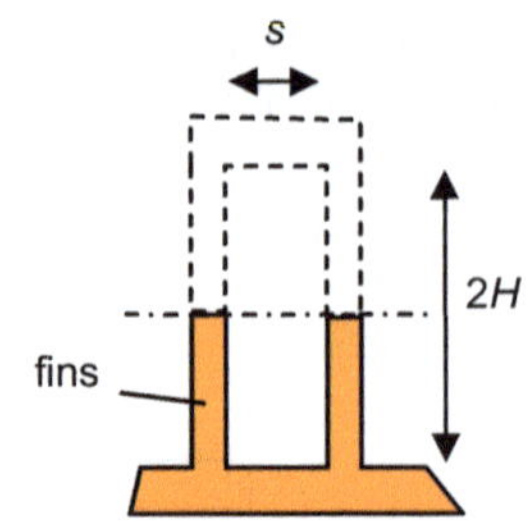

outlet with thin fins are not very important. Therefore, the losses are treated in a simplified way.

The planar contraction/expansion resistances are treated, similarly to the code's compact heat sink model, by applying a constant loss coefficient, based on the approach velocity. Using the same coefficient for inflow and outflow avoids distinguishing for the flow direction, which would not be possible in the used code. Hence, the heat sink model can be placed in the calculation domain without regard for the flow direction.

Contraction/expansion pressure drop:

$$\Delta p = \xi_{c/e} \frac{\rho}{2} v_{appr}^2 \tag{10.4}$$

where

$$\xi_{c/e} = 0.09 \tag{10.5}$$

In the *Resistance dialog* the *Standard* formula is selected, and *Approach Velocity* is selected.

10.3.5 Side Planar Flow Resistance

In lateral direction, the flow is blocked by the fins, and an appropriate very high resistance is to be set.

10.3.6 Volumetric Flow Resistance

For calculating the hydraulic diameter d_h, the space between the fins is treated as one half of a closed duct. For slanted fins, the average fin height, *H*, can be calculated by using Eq. (10.1) for odd number of fins and Eq. (10.2) for even number of fins. The fin spacing, *s*, is the average distance between the fins (Fig. 10.16).

$$B_{mean} = \frac{(B + D)}{2} \tag{10.6}$$

$$s = \frac{B_{mean} - d_{fin}}{n - 1} - d_{fin} \tag{10.7}$$

$$d_h = \frac{4x \text{ cross sectional area}}{\text{perimeter}} = \frac{4(s \times 2H)}{2(s + 2H)} = \frac{4sH}{s + 2H} \tag{10.8}$$

The friction factor for round or square ducts and developed laminar flow is:

$$f_{sq,dev} = \frac{64}{\text{Re}} \tag{10.9}$$

For ducted flow with entry length, the friction factor is augmented by an entry-length term:

$$f_{sq,entry} = f_{sq,dev} + 2\left(\frac{L}{d_h}\right)^{-0.68} \text{Re}^{-0.25} \tag{10.10}$$

For rectangular instead of square ducts, a shape factor is introduced:

$$f_{rect,entry} = C_s f_{sq,dev} + 2\left(\frac{L}{d_h}\right)^{-0.68} \text{Re}^{-0.25} \tag{10.11}$$

where

$$C_s = 1 + 0.35\left(1 - \frac{s}{2H}\right)^6 \tag{10.12}$$

The entry-length term and the shape factor are deduced from detailed simulations of air flow in a single duct. It has been found that the results correlate best if C_s is applied only to $f_{sq,dev}$ and not to the entry-length term as well.

The above formulas can readily be implemented using *Volume Resistance–Advanced Treatment*. The loss coefficients are based on *Device Velocity*, which uses the *Free Area Ratio* (FAR) for calculating the accelerated velocity between the fins.

The friction loss is calculated from the general formula for pressure drop in a duct of length L:

$$\Delta p = f \frac{\rho}{2} v^2 \frac{L}{d_h} \tag{10.13}$$

Thus, pressure drop per length is:

$$\frac{dp}{dL} = f \frac{\rho}{2} v^2 \frac{1}{d_h} \tag{10.14}$$

The formula used in the code is:

$$\frac{dp}{dL} = \tilde{f} \frac{\rho}{2} v^2 \tag{10.15}$$

where

$$\tilde{f} = \frac{C_A}{\text{Re}} + \frac{C_B}{\text{Re}^{Index}} \tag{10.16}$$

From Eq. (10.4), it can be seen that

$$\tilde{f} = \frac{f}{d_h} \tag{10.17}$$

From Eqs. (10.9) and (10.11), C_A and C_B parameters in Eq. (10.16) can be determined.

$$C_A = \frac{64C_s}{d_h} \tag{10.18}$$

$$C_B = 2\frac{1}{d_h}\left(\frac{L}{d_h}\right)^{-0.68} = 2L^{-0.68}d_h^{-0.32} \tag{10.19}$$

$$Index = 0.25 \tag{10.20}$$

Further parameters to be set are:

$$Length\ ratio = d_h \tag{10.21}$$

$$FAR = \frac{v}{v_{app}} = \frac{B_{mean} - nd_{fin}}{B_{mean}} \tag{10.22}$$

The *FAR* is built with the cross-sectional area of the fin region belonging to the real heat sink. The coefficients in Eqs. (10.18)–(10.22) are set for the length and vertical direction.

10.3.7 Heat Transfer

The heat transfer in the compact model is described by the appropriate setting of the Surface Exchange parameter in the code. This section will describe the derivation of the heat sink thermal resistance vs. velocity relation. The reader is encouraged to read the sections on "Literature Study" and "Empirical Correlations" in Chap. 9 first to get an impression about the accuracy of the following simplifications.

The Nusselt number for developed laminar flow in a square duct, for isothermal walls (Incropera and DeWitt [5]), has a constant value

$$\text{Nu}_{sq,dev} = 2.98 \tag{10.23}$$

Similarly, to the flow resistance formula, an additive term for the entry-length effect and a shape factor for rectangular ducts is introduced:

$$\text{Nu}_{rect,entry} = C_{s,\text{Nu}}\text{Nu}_{sq,dev} + 0.26\left(\frac{d_h \text{Re}_{d_h} \text{Pr}}{L}\right)^{0.65} \tag{10.24}$$

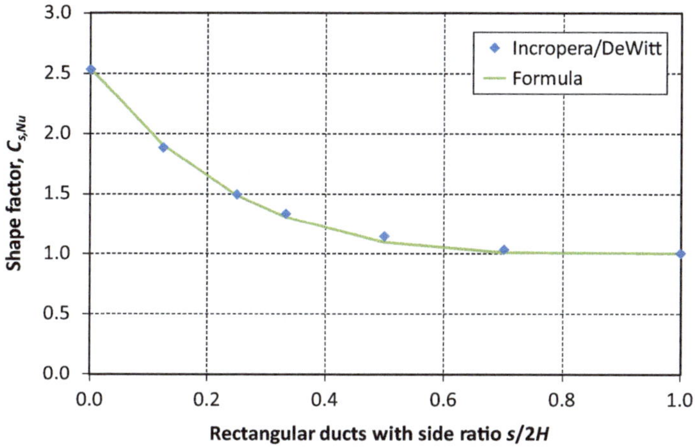

Fig. 10.17 The shape factor $C_{s,\mathrm{Nu}}$ for discrete rectangular ducts as a function of the side ration $s/2H$

where

$$C_{s,\mathrm{Nu}} = 1 + 1.53\left(1 - \frac{s}{2H}\right)^4 \quad (10.25)$$

The shape factor is a close approximation to the discrete values for rectangular ducts with side ratio $s/2H$, as given in Incropera and DeWitt [5] (Fig. 10.17).

The entry-length term is deduced from detailed simulations of air flow in single rectangular ducts. It is also based on a Nu-law discussed by Renz [6]. As with the flow resistance, it fits the simulation results best if the shape factor is applied only to $\mathrm{Nu}_{sq,dev}$, and not to the entry-length term as well.

For deriving the overall heat sink resistance, the Effectiveness–Number-of-Transfer-Units (NTU) method for heat exchangers is used. A slight difficulty is that FloTHERM does not allow temperature dependency of this quantity. Fortunately, the final resistance relationship is to a large extent independent of the chosen temperature.

Starting from an arbitrary, nonaccelerated velocity in the fin region (denoted as approach velocity), the mean air velocity between the fins is calculated with the *Free Area Ratio* as given in Eq. (10.22):

$$v = v_{app}\frac{B_{mean}}{B_{mean} - nd_{fin}} \quad (10.26)$$

The Reynolds number, with d_h as characteristic length, is calculated as:

$$\mathrm{Re} = \frac{\rho_{air} v d_h}{\mu_{air}} \quad (10.27)$$

From Eqs. (10.24) and (10.25), the Nu-number for the specific approach velocity can then be calculated. The mean heat transfer coefficient between fins and air flow is:

$$\bar{h} = \frac{\text{Nu}\, k_{air}}{d_h} \tag{10.28}$$

For thin fins, as used in the DUT, the fin efficiency has an effect on the overall heat transfer rate. Therefore, the fin efficiency is calculated and included in the heat transfer calculation:

$$\eta_{fin} = \frac{\tanh(M \times H)}{M \times H} \tag{10.29}$$

where

$$M = \sqrt{\frac{2\bar{h}}{k_{fin} d_{fin}}} \tag{10.30}$$

The NTU can be calculated by Eq. (10.33) using the total heat-transfer surface area A_W and cross-sectional area of the fin region A_C.

$$A_w = 2nHL + BL \tag{10.31}$$

$$A_c = B_{mean}(C - d_{base}) \tag{10.32}$$

$$\text{NTU} = \frac{\eta_{fin}\bar{h} A_w}{c_{p,air}\rho_{air} A_c v_{app}} \tag{10.33}$$

The effectiveness, η_b, is readily calculated from:

$$\eta_b = 1 - e^{-\text{NTU}} \tag{10.34}$$

The overall thermal resistance, as used in the *Surface Exchange* feature, is based on the temperature difference between base plate and inlet into the heat sink:

$$R = \frac{(T_{base} - T_{in})}{P} \tag{10.35}$$

where P is the total heat transferred.

The total power transferred to the air can be rewritten with the effectiveness term.

$$P = \dot{m} c_{p,air} (T_{out} - T_{in}) = \rho_{air} A_c v_{app} c_{p,air} \eta_b (T_{base} - T_{in}) \tag{10.36}$$

Rearranging Eq. (10.37), the final equation for the thermal resistance can be deduced:

$$R = \frac{1}{\rho_{air} A_c v_{app} c_{p,air} \eta_b} \tag{10.37}$$

For any approach velocity, the overall thermal resistance can thus be calculated.

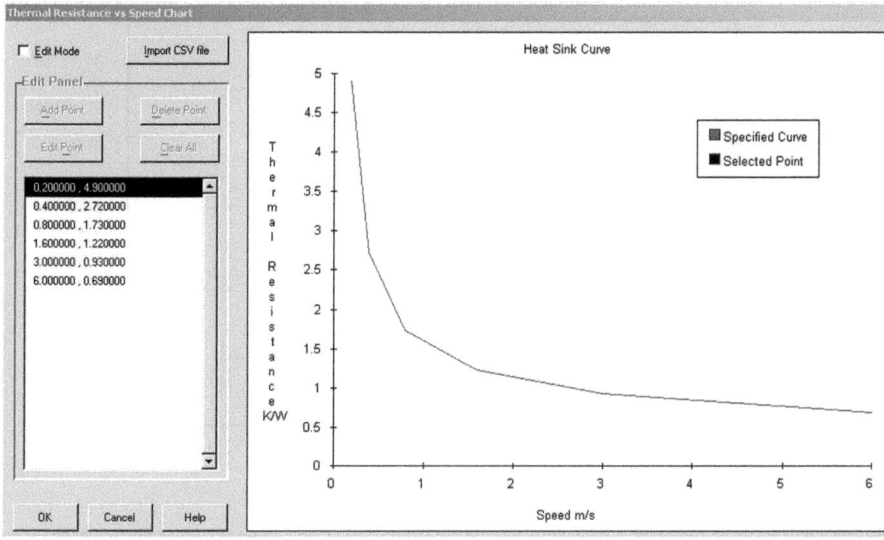

Fig. 10.18 The heat sink thermal resistance applied to the *Surface Exchange* dialog in FloTHERM

Table 10.1 Inlet temperature dependence on the overall thermal resistance as a function of approach velocity

Approach velocity, v_{app} (m/s)	Ambient temperature (°C)					Average	Standard deviation
	25	5	35	50	75		
0.2	4.89	4.61	5.03	5.25	5.62	5.08	0.38
0.3	3.41	3.26	3.49	3.62	3.84	3.52	0.22
0.4	2.71	2.62	2.77	2.85	2.99	2.79	0.14
0.6	2.05	2.00	2.07	2.11	2.19	2.08	0.07
0.8	1.72	1.70	1.74	1.76	1.80	1.74	0.04
1	1.53	1.51	1.53	1.55	1.58	1.54	0.02
1.3	1.34	1.33	1.34	1.35	1.36	1.34	0.01
1.6	1.21	1.21	1.22	1.22	1.23	1.22	0.01
2	1.10	1.09	1.10	1.10	1.10	1.10	0.00
3	0.92	0.92	0.92	0.92	0.92	0.92	0.00
4.5	0.78	0.78	0.78	0.78	0.78	0.78	0.00
6	0.69	0.69	0.69	0.69	0.69	0.69	0.00

The usage of the overall thermal resistance of the heat sink is straightforward for a heat sink in ducted flow, where the velocity does not change. For unducted flow, it is not immediately clear why this should work as well, because a great part of the air flow escapes through the open top of the heat sink, such that the velocity drops from inlet to outlet. However, since FloTHERM uses the average velocity within the fin region, this effect is fairly well compensated. Hence, the same resistance profile gives accurate results for both ducted and unducted flow.

Implementing the resistance vs. inlet velocity profile in the *Surface Exchange* dialog requires setting the correct fin height (Extent of Heat Transfer), and selecting Specified Profile. The resistance profile should be constructed with an appropriate number of points. Figure 10.18 shows an example (Table 10.1).

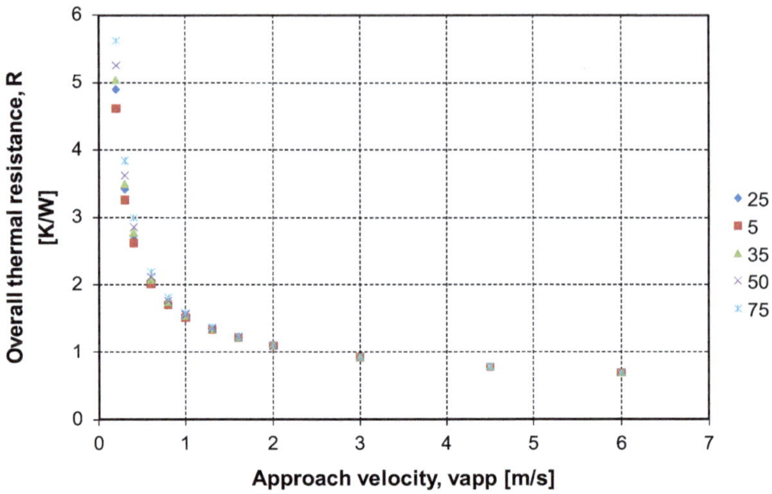

Fig. 10.19 The effect of inlet temperature on the calculated heat sink thermal resistance as a function of temperature

The procedure for calculating the resistance vs. velocity profile has been implemented as an Excel worksheet, which requires only the geometry data of the heat sink and the material properties. If the user wants to include radiation in the simulation, radiation can easily be added. This can be accomplished by setting a radiation attribute to the upside surface of the thin plate. The effect of air temperature on the calculated resistance vs. velocity profile can be shown to be reasonably small, especially for higher velocities. Figure 10.19 shows the profile for a DUT, following the described procedure, at three different temperatures.

It should be realized that the resistance profile that is set in FloTHERM as *Surface Exchange* attribute is not identical with the heat sink thermal resistance that would result from simulations in either ducted or unducted flow. This is due to the fact that the resistance definition is typically based on the center point base plate temperature and not on the mean temperature in the "thin plate." Also, the velocity changes for unducted flow, so that the air velocity approaching the heat sink is generally different from the average velocity within the fin region.

10.4 Simulations Comparing Detailed and Compact Models

The DUT was used in order to compare detailed and compact models at various approach velocities. A detailed model of this heat sink already existed, and also a comparison to measurements has previously been performed, showing good agreement, provided the effect of radiation in the measurement is compensated for.

The simulations were done for ducted and unducted flow. A very basic numerical wind tunnel was modeled. In order to save computation time, only one half of the system is taken into account. On the center plane and on all outside borders, symmetry

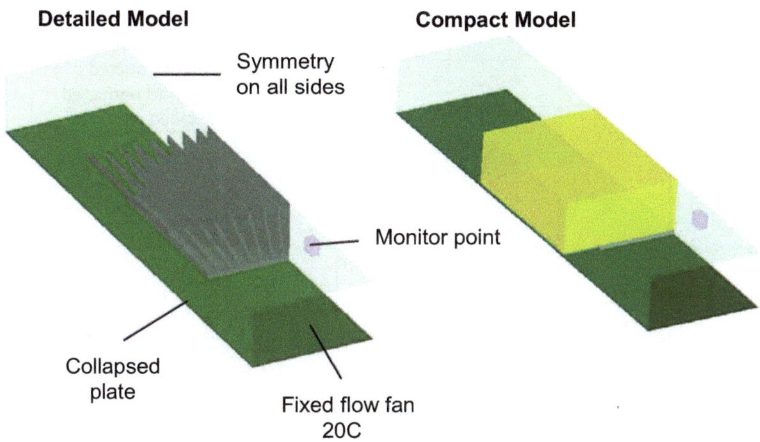

Fig. 10.20 3D view of the detailed (*left*) and compact (*right*) ducted flow models

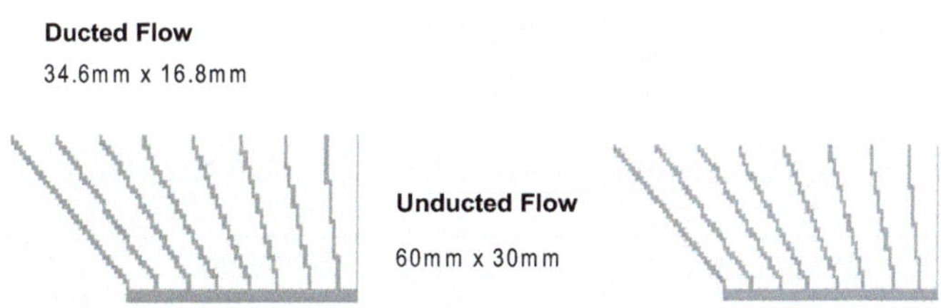

Fig. 10.21 Front view of the detailed models showing the duct size for the ducted (*left*) and unducted (*right*) simulations

planes are set. The ground plane consists of a nonconductive collapsed plate with symmetry condition. At the inlet, a fixed flow fan is modeled. The approach velocity is monitored at a distance of 15 mm in front of the heat sink and 10 mm above the ground plane. The heat source is located at the bottom of the base plate and has, in both cases, the same size as the base plate (Figs. 10.20 and 10.21).

The results of the simulations are shown in Fig. 10.22 and Table 10.2. The resulting thermal resistance values obtained with the compact model are consistently higher than the results obtained with the detailed model, as shown in Table 10.2. However, it is generally desirable that the compact model gives somewhat conservative results.

10.5 Conclusions

The industry uses passive and active solutions, including air-conditioning, to cool LED-based systems. The mechanics including the cooling solution is the major part of the cost at the present and in the future; hence, smart and cost effective thermal management is required.

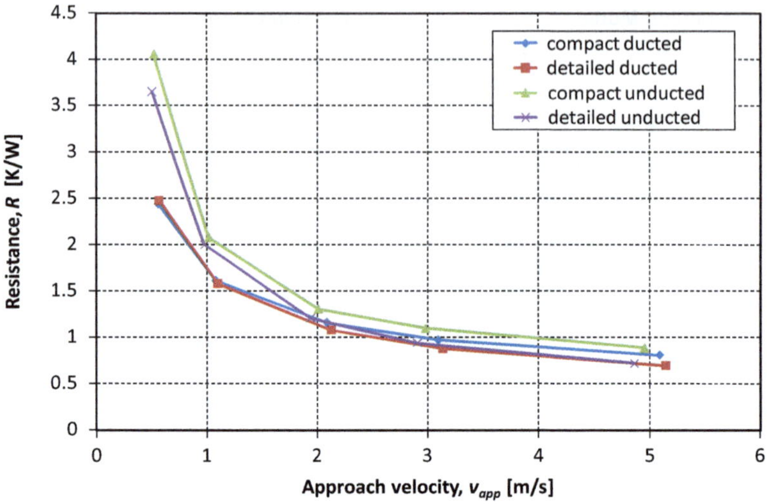

Fig. 10.22 Results of detailed and compact ducted and unducted simulations of the ATS97-09-034B heat sink

Table 10.2 Results of detailed and compact ducted and unducted simulations of the ATS97-09-034B heat sink

Heat sink condition Value	Ducted			Unducted		
	v_{app} (m/s)	R (K/W)	dR (%)	v_{app} (m/s)	R (K/W)	dR (%)
Detailed	0.56	2.44	1.6 %	0.52	4.06	−10.1 %
	1.08	1.61	−1.9 %	1.02	2.08	−3.8 %
	2.09	1.16	−6.9 %	2.01	1.31	−7.6 %
	3.1	0.97	−9.3 %	2.98	1.1	−14.5 %
	5.09	0.81	−13.6 %	4.95	0.89	−19.1 %
Compact	0.57	2.48		0.5	3.65	
	1.1	1.58		0.98	2	
	2.13	1.08		1.94	1.21	
	3.14	0.88		2.91	0.94	
	5.14	0.7		4.86	0.72	

The selection of cooling solutions for LED applications needs to be based on an overall system–board–component level investigation rather than on solving individual component problems. Computational (CFD) analyses are a very valuable tool to arrive at the best possible solution at the system, board, and component levels. Within this process, compact heat sink modeling is needed and a method is presented how to realize this for various geometries.

References

1. Lai Y, Cordero N, Barthel F, Tebbe F, Kuhn J, Apfelbeck R, Würtenberger D (2009) Liquid cooling of bright LEDs for automotive applications. Appl Thermal Eng 29(5–6):1239
2. Whitaker T (2007) LED Headlamp illuminate the way forward. LEDs Magazine Issue 17, November/December, pp 18–19
3. Aluminium recycling in Europe: the road to high quality products of the EAA/OEA Recycling Division, see http://www.oea-alurecycling.org/de/verband/oea_eaa_aluminium_recycling.pdf
4. http://www.qats.com
5. Incropera F, DeWitt D, Bergman T, Lavine A (2006) Introduction to heat transfer, 5th edn. Wiley, New York
6. Renz U (2002/2003) Vorlesung Wärme und Stoffübertragung, Technische Hochschule Aachen, Ausgabe WS 2002/2003 (http://www.wsa.rwth-aachen.de/)

Chapter 11
Testing Issues in LED Manufacturing

Richard Young

Abstract Power to a light-emitting diode (LED) is converted to light and heat. However, these are not independent and affect each other in complex ways. Knowing the thermal state of the LED chip is essential to understanding the light output properties of the LEDs. Controlling the temperature of thermal contacts is used in the testing of LEDs, where possible. Changes to some LED optical and electrical properties with temperature are described with examples.

The setups used in testing LEDs for total flux and averaged LED intensity are discussed with practical equipment illustrations. The configurations and conditions for laboratory and production lines are compared. The relationships are explored so that the correction of production test results to laboratory quantities can be made.

The effects of measurement equipment on the results are examined; in particular, the requirements of nonequilibrium testing with short pulses. Short pulses are routinely employed in production testing and results "corrected" to equilibrium conditions.

The role of uncertainties and tolerances and their differences are discussed. Measurement precision, result distribution, and their relationship to actions such as rejection and binning protocols are explored. In particular, the effect of traceability paths on the observed differences between production lines is explained. Comparison of results depends on the number of variables involved in tests and the scope of the intercomparison.

11.1 Thermal Effects on LED Properties

11.1.1 Light and Heat

The power supplied to the light-emitting diode (LED) would ideally be converted to light, but the process is not 100 % efficient and so some of the power appears as heat. As the temperature increases, this then decreases the efficiency and more of the

R. Young (✉)
Instrument Systems GmbH, Munich, Germany
e-mail: young@instrumentsystems.com

C. J. M. Lasance, A. Poppe (eds.), *Thermal Management for LED Applications*, Solid State Lighting Technology and Application Series 2,
DOI 10.1007/978-1-4614-5091-7_11,

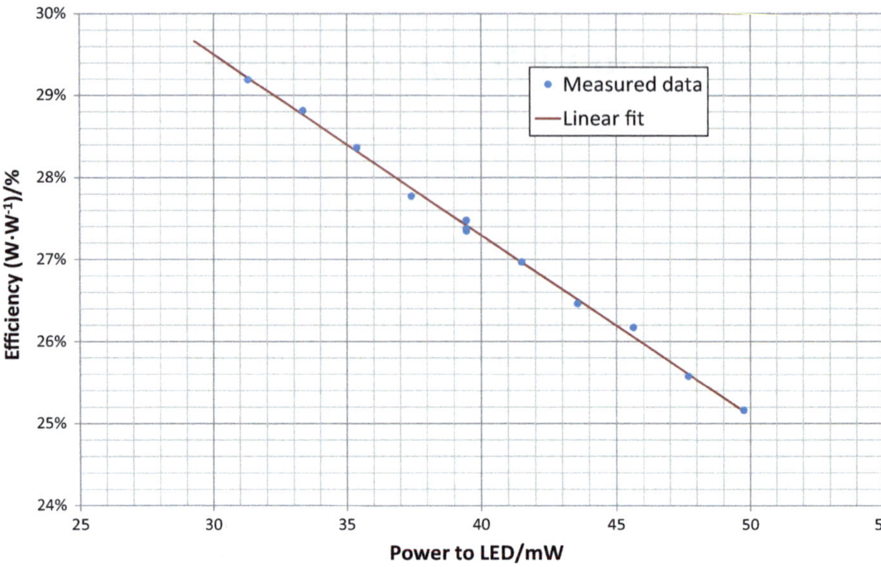

Fig. 11.1 Changes to LED efficiency with input power for an orange LED

input power is converted to heat. This process will continue until the LED fails or the heat generated is dissipated somehow to reach equilibrium at some temperature.

A similar process happens if one pumps more power into an LED. The amount of light increases, but the extra heat causes the efficiency to drop, as shown in Fig. 11.1, so the light out does not keep pace with the power in. The LED is most efficient at lower power, but that may not provide sufficient light for the application. The thermal aspects of this are detailed in Chap. 4 on thermal testing.

Often with LEDs, the light output of interest is visible and measurements of the light will be photometric quantities. Efficacy[1] of the light source, which is the light output in lumens (luminous flux) divided by the electrical power input in watts, is extremely important to most applications when manufacturing LEDs. Efficacy is a performance metric which is used to compare different light sources. Efficacy depends on both the emitted light power and the spectral power distribution—popularly referred to as *spectrum*. The spectrum also changes with temperature, so both need to be considered. Producing an optimal product involves compromises between desired light output, input power, and thermal management.

[1] One needs to make the distinction between two kinds of efficacy. ILV, the *International Lighting Vocabulary* published by CIE [1] defines two terms. The one used to compare electric light sources (term 17-729 of ILV) is the usual (luminous) *efficacy of light source* defined as the ratio of the emitted luminous flux and the input electrical power. Throughout this chapter, the word efficacy refers to this quantity. The second type of efficacy is called (luminous) *efficacy of source of radiation* (term 17-730 of ILV) defined as the ratio of the emitted luminous flux and the emitted radiant flux. Both quantities are measured in lm/W. The third term related to efficiency to characterize LEDs is called *radiant efficiency* (term 17-1018 of ILV) defined as emitted radiant power divided by the supplied input electric power. See also the JEDEC JESD51-53 [2] document for some of these definitions.

11.1.2 Defining the Reference Temperature

Testing in industry can take place at each and every stage of integration: wafer, die, package, module, light engine, and luminaire. Each stage will employ its own techniques and procedures in testing. Often the finished product of one stage is the starting point of another, so there are some relationships between testing methods, but these are not necessarily fixed between manufacturers or applications.

However, before continuing, we should recall what we mean by LED temperature.[2] In any powered system, the temperature will vary continuously from the hottest point, usually the LED chip junction, to the coldest point which is ambient air or other means of heat dissipation. Each point in that sequence of temperatures could be used based on accessibility and convenience. In practice, just a few locations are chosen for temperature measurement, but the location chosen often varies with the level of integration.

The temperature reference point defined for LED end-users is normally decided by the manufacturer and depends on the stage of integration and type of tests to be performed.

T_J is the *junction temperature* of the LED chip. It is not directly accessible, but can be inferred from measurements of forward voltage under constant current conditions (for details refer to Chap. 4). The temperature of the junction is the best indicator of the thermal state of the LED in all environments.

T_c is the *case temperature*, normally of the heat slug of an LED package. In light measurement standards, this is called *contact temperature*. According to classical thermal testing methods, a simple way of measuring this is to attach a thermocouple to it (or any other device can be used to measure its temperature directly).[3]

T_p is a *test point temperature*. For modules involving more than one LED chip, there is no longer a single junction or heat slug, so the reference temperature defined for end-users is more likely a location on a thermal conductor, e.g., copper block, metal PCB, etc., ($*T_p$) or on the heat sink ($**T_p$). Generally, the former is a more reliable indicator, but the latter is more accessible.

T_s *the solder point temperature*— similar in use to T_p. This location is typically at one of the electrical pins of the LED, it is very close to the PCB holding the LED package (not indicated in Fig. 11.2).

T_a is the *ambient temperature*. Some modules and all luminaires require the measurement of ambient temperature, but some rapid testing methods also use an ambient temperature of 25 °C as their reference value. As illustrated in Fig. 11.2.), as $*T_a$, $**T_a$ (referred to as T_q in some standards), and $***T_a$, there are several

[2] Chapter 4 on thermal testing and Chap. 6 on standardization provided definitions of LED temperature. These definitions are based on repeatability considerations of measurements. In daily practice, convenience of LED users is the major factor which determines how LED vendors define thermal reference points for their products.

[3] Usually, the purpose of measuring exactly the case temperature is to determine the R_{thJC} junction-to-case thermal resistance.

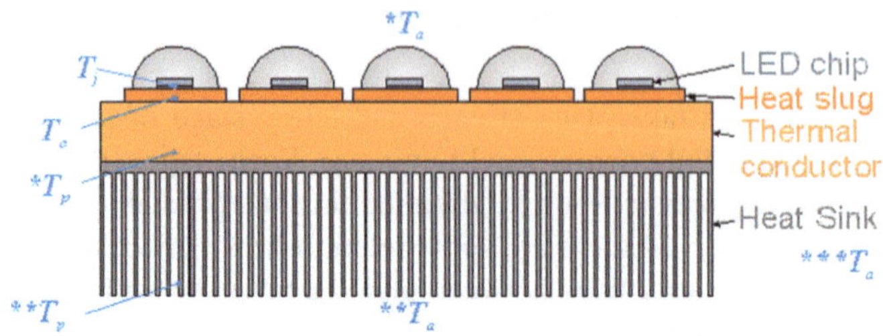

Fig. 11.2 Diagram showing possible definitions and locations of temperature reference points

possibilities for the location of ambient temperature measurements, all of which would be expected to give different values. The most common recommendation[4] for T_a is *shielded from direct illumination in a horizontal plane containing the device and within a radius of 1 m*, but that still leaves scope for variations. (Note that, in thermal testing, there is less flexibility related to the definition of the ambient temperature.)

As a general principle, the smaller the thermal resistance between the junction and the reference temperature point,[5] the more reliable it is in indicating the thermal state of the LED. The nature of the so-called thermal transfer impedance (see, e.g., Sect. 6.1.4.2 in Chap. 6) between the junction and the reference point (of which the temperature is controlled during tests) may strongly influence the optical measurements. Measurements soon after changes to electrical power to the LED or temperature changes at the reference point should be avoided; especially, in systems with long thermal time-constants.[6]

11.1.3 Active Temperature Control

Temperature can not only be measured, it can also be controlled. Often in applications, the cooling of the LED will use passive components such as heat sinks. This may be adequate for general use, but for testing, it is better to use active control of the temperature.[7] As the heat flow path from the LED junction to the active control element (typically a cold plate) is kept short, the LED temperature can be stabilized at some new value in a short time—seconds or minutes instead of hours.

[4] In testing standards of light sources and luminaries.

[5] For a physically sound definition of thermal resistance between two locations along a heat-flow path refer to Sect. 6.1.4 of Chap. 6.

[6] One should wait with starting the optical measurement at least as much after the electrical switching or temperature change as the thermal time-constant of the system containing the test LED, see, e.g., the timing diagram shown in Fig. 6.20 of Chap. 6.

[7] According to the JEDEC LED thermal testing standards (JESD51-51 [3] and JESD 51-52 [4]), the temperature of the LED under test has to be controlled. For details refer to Chap. 4 and Chap. 6.

Fig. 11.3 A system for active thermal control. The LED slug is held in thermal contact against the metal heat sink that can be both heated and cooled to maintain a constant temperature. Four electrical contacts to either side allow power and sense connections to the LED

A wide range of LED packages and modules are available and the physical shape of the temperature control device should match the range. There are some common features however, so one design may accommodate several packages as illustrated in Fig. 11.3.

Temperature control is not just about matching physical characteristics of the LED, it is also a matter of matching power. The cooling mechanism can only remove a certain amount of heat and if the thermal power generated by the LED is greater than that of the cooling capacity, full regulation cannot be achieved. Notice the term "thermal power" is used; part of the input power will be converted to light and this will not contribute to heating. The greater the optical efficiency of the LED, the lower the heat dissipation requirements for the same input electrical power.

A Peltier element is normally used for temperature control, as this reacts quickly (see Fig. 11.4) and can be used for both cooling and heating. The LED reference temperature can therefore be set and held both below and above room temperature. The short settling time indicated by optical parameters in Fig. 11.4 means that the characteristic thermal time-constant of the junction-to-heat sink thermal transfer impedance is small.

11.1.4 Forward Voltage

As described in Chap. 2 about the fundamental physics of LED operation under constant forward current conditions, the V_F forward voltage of an LED is sensitive to the thermal state of the LED junction and hence may be used to indicate the junction temperature once the calibration curve has been established, see Chap. 4 about K-factor calibration.

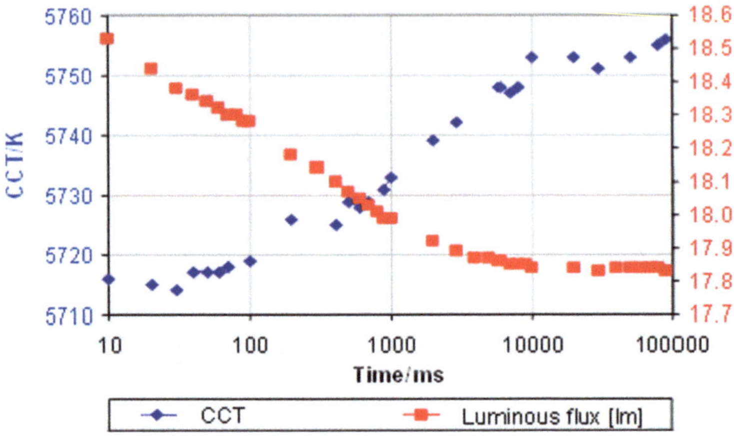

Fig. 11.4 Changes of optical parameters after changing the temperature of the contact point. For this white LED attached to the temperature controlled DUT holder the parameters stabilize in around 10 s

The biggest problem associated with the measurement of forward voltage at a given temperature is that applying the current abruptly heats up the chip and changes the temperature.

If one allows an unpowered LED to come to equilibrium with a stable environmental temperature, then the LED and associated components must be at that temperature. If the current is then applied and measurements of forward voltage are made rapidly (<1 μs per measurement), then the measurements may be extrapolated back to zero time (switch-on) to give the forward voltage at the equilibrium temperature [5]. This of course requires some time to complete a calibration curve since between each set of measurements, which are quick, there may be a long wait for equilibrium to be established at each new temperature.

Generally, a graph of forward voltage against junction temperature (T_J) would look very similar to that of Fig. 11.5, but shifted slightly to higher temperatures.

The relationship revealed in Fig. 11.5 is a gentle curve. This is typical of measurements made over wide temperature ranges. For narrower temperature ranges, a linear relationship is often assumed.

11.1.5 LED Spectra

The spectrum of a simple LED often depends on the band structure of the semiconductor materials used (see Chap. 2) and their conditions of use. LEDs change their light output and spectral distribution with temperature, as previously discussed, but a white LED may also use phosphors with very different temperature characteristics.

Typically, a phosphor-based white LED is a combination of a blue "pump" LED and a yellow phosphor. The ratio of the blue LED peak and the yellow phosphor peak is generally responsible for color changes with temperature. The exact details of the phosphor and LED vary between manufacturers.

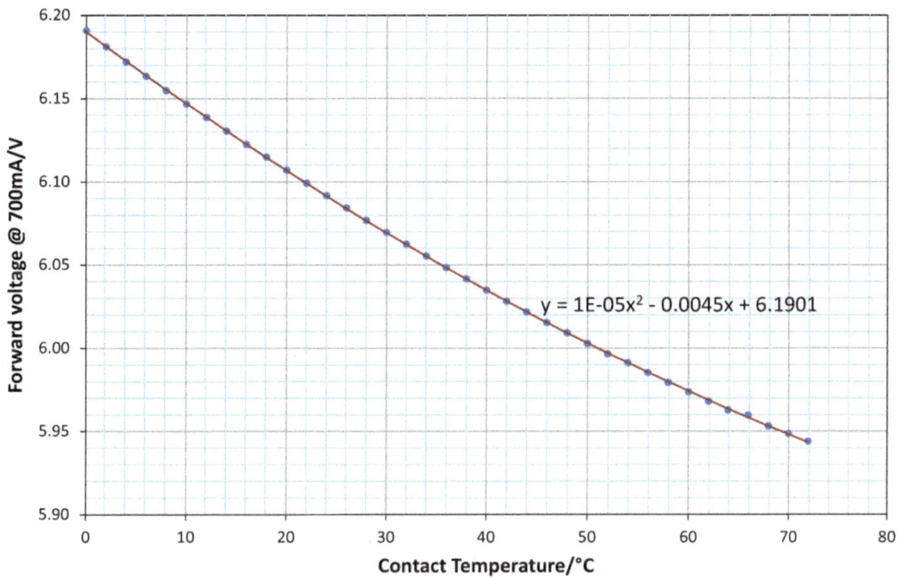

Fig. 11.5 Measurements of forward voltage with contact temperature for a 5 W white LED operated at 700 mA

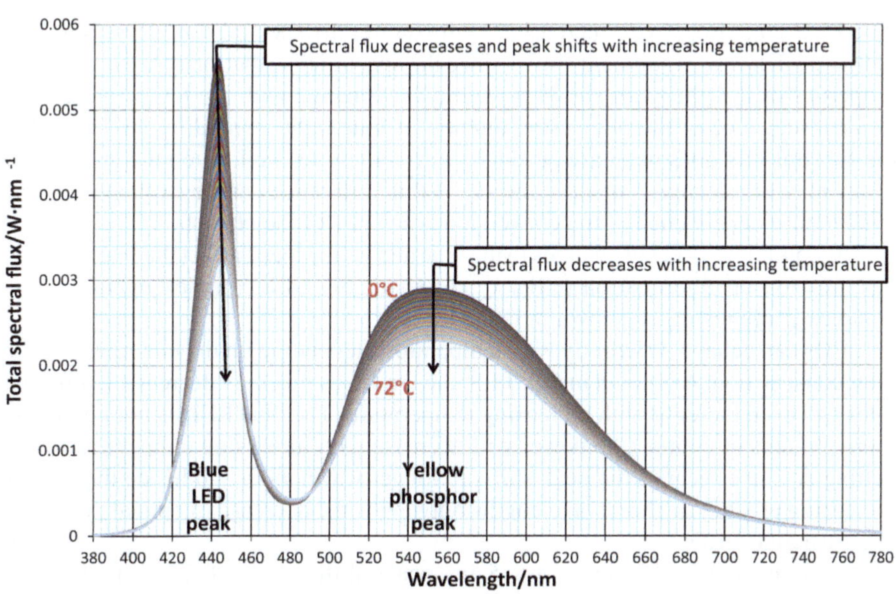

Fig. 11.6 Changes in spectra of a white LED with contact temperature

Figure 11.6 shows the spectral changes for a white LED as the reference temperature is varied. The blue peak moves to longer wavelengths with increasing temperature as described in Sect. 5.1 of Chap. 5. As the radiant efficiency of the

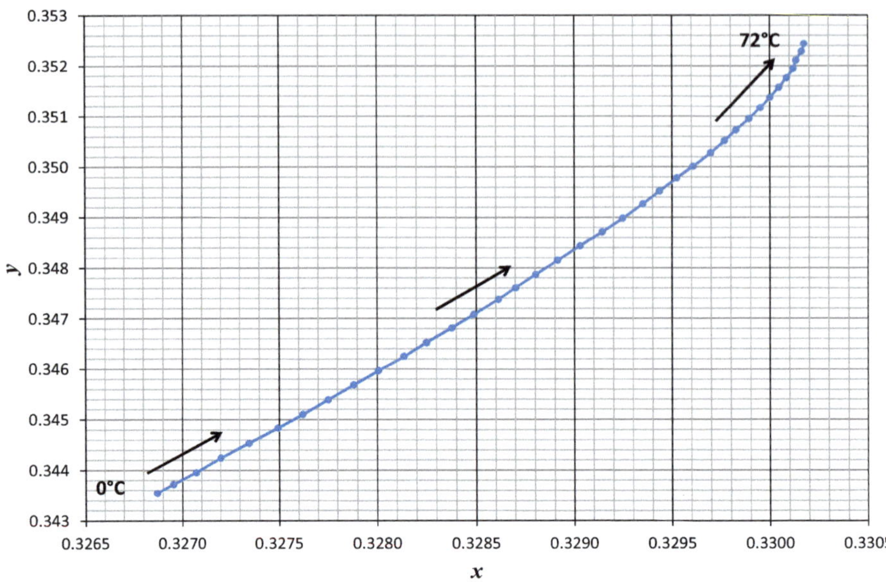

Fig. 11.7 CIE *xy* chromaticity changes for the white LED shown in Fig. 11.6

blue LED chip drops with increasing temperature, the spectral flux decreases. The yellow peak has no significant shifts (since the spectral distribution of the conversion properties of the phosphor shows no significant change with the temperature), but as a consequence of the decreasing amount of blue light emitted by the LED chip the spectral flux does decrease at longer wavelengths as well. The blue and yellow peaks may not decrease as the same rate—this difference can be explained by the temperature dependence of the conversion efficiency of the phosphor.

The combined effect of changes to the blue and yellow peak affect the International Commission on Illumination (CIE) *xy* chromaticities in a complex manner and a complex curve results as shown in Fig. 11.7.

LEDs can behave differently with temperature depending on the materials used in their construction. As shown in Fig. 11.8, the luminous flux may decrease at different rates and even increase with increasing temperature. The blue LED is not giving out more photons of course (the power will decrease with increasing temperature); it is getting brighter because the peak shifts to longer wavelength and hence the rapidly increasing efficacy at short wavelengths outweighs the decrease in optical power for this particular LED.

As the maximum efficacy is at 555 nm, peaks shifts in blue LEDs can offset the loss of optical power output somewhat. However, the light must also be absorbed by the phosphor and in general less is absorbed with peak shifts to longer wavelength. The spectral behavior of phosphor-based white LEDs is not simply a mix of two independent emitters.

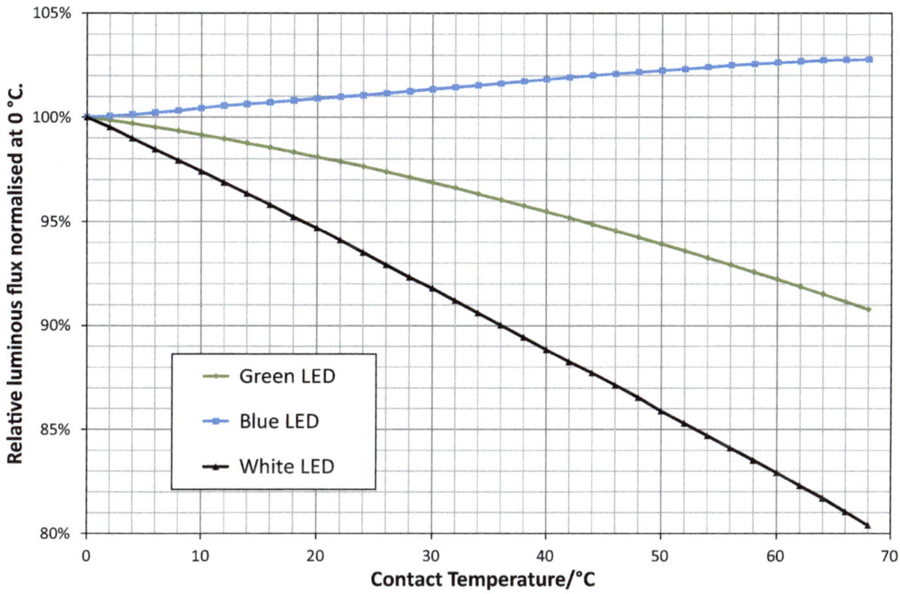

Fig. 11.8 Luminous flux changes with temperature for blue, green, and white LEDs

For 3- and 4-color white LEDs, which are a mixture of single-color LEDs (RGB- or RGBA-based white), each individual chip is of a different material and, as discussed earlier, behaves differently with temperature. Intensity and color changes of the complete package or module can be even more complex than the phosphor-based white LEDs.

Remote phosphor LEDs are perhaps the most complex to characterize. The phosphor is no longer in physical contact with the LED chip, so the temperature of the chip and phosphor may differ considerably in use. Therefore, the spectral power distribution of such an LED-based light source is generally related to both chip temperature and ambient temperature.

11.2 Nonsteady-State Conditions

Steady-state may take a long time to establish making such tests impractical in a production environment. The question then becomes whether one can make tests in nonsteady-state conditions that give meaningful results.

As the forward voltage is a good indicator of the thermal state of the LED chip, we can measure the forward voltage with respect to time to see how fast the junction heats up. As shown in Fig. 11.9, the junction heats immediately. There is no delay even on a submicrosecond timescale. The forward voltage decreases (increasing temperature) very fast, typically up to the first 20–50 μs or so, followed by an

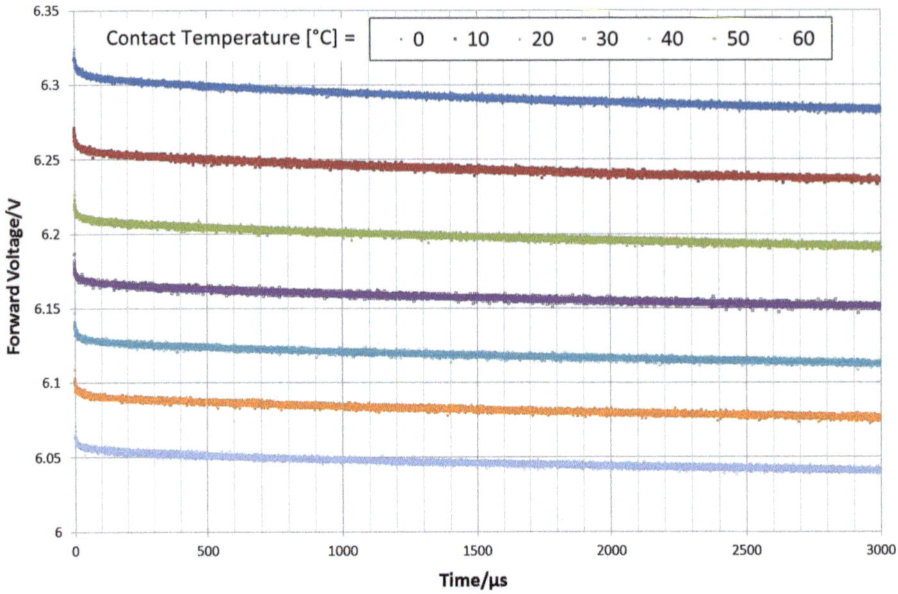

Fig. 11.9 Measurements of forward voltage with respect to time at various case temperatures for a white phosphor-based LED at 700 mA. The data interval is 0.1 μs

approximately linear decrease with time up to typically ~3 ms. This corresponds to an initial heating of the junction itself, after which the heat would be transferred to the structure following the active LED chip in the heat flow-path structure such as die attach, submount, heat-slug, etc., which can be determined with the help of structure functions described in Chap. 4 dealing with thermal testing.

On these time scales, the temperature increase is within the chip and no loss of heat to external components has yet occurred. The time required for heat flow to external components depends on the size and materials of the semiconductor chip, but generally external components such as heat sinks have no effect on the temperature of the junction at these time scales, which can last tens of milliseconds.

If we have the calibration curve for the LED, we can transform the forward voltage curves in Fig. 11.9 to temperatures. In particular, we can look at the temperature differences between the junction temperature and the contact temperature.

A generally similar behavior is seen at all case temperatures, confirming a rapid initial increase of 2 or 3 degrees in the first 100 μs followed by a slower, but sustained continued increase.

11.3 Optimization of Testing in Production Environments

It is clear from Fig. 11.10 that whether the external temperature is controlled or not, it will have little effect on testing at short times. The temperature is changing constantly, so a test at a specific temperature is not possible. However, it can be

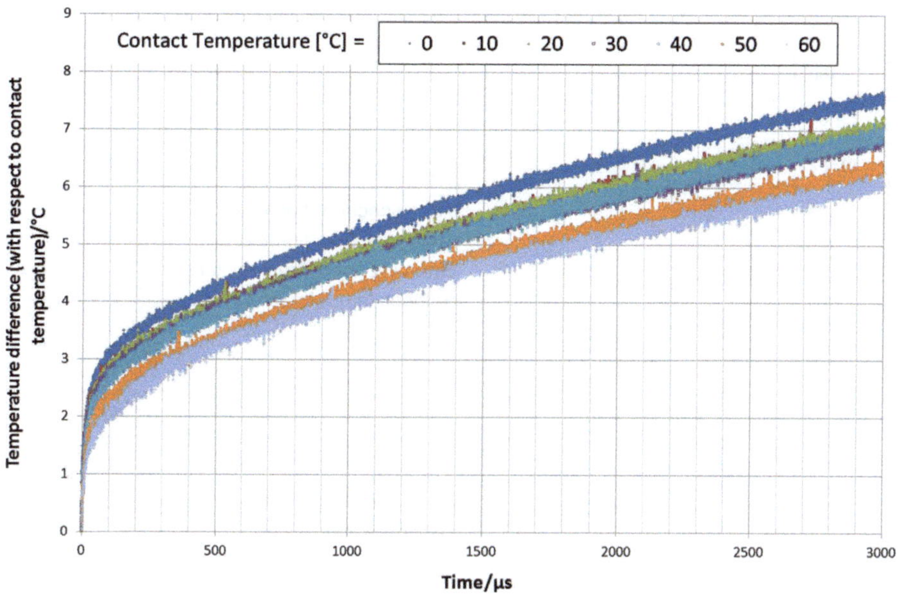

Fig. 11.10 Temperature difference with respect to contact temperature for the LED tested in Fig. 11.9

seen that if one first has a stable temperature before applying power, then a fairly repeatable temperature profile will result. High-speed testing ambient environments are normally very well-regulated for exactly this reason, usually to a temperature of 25 °C.

In production, testing compromises are made on the conditions of measurements, but no compromises are made on the accuracy. Making measurements very quickly necessitates operation under transient conditions, but if these test results correlate highly to steady-state condition tests, then one can be used to indicate expected result in the other. Similarly, it may not be possible (due to time considerations) to place an LED in the center of an integrating sphere for instance, but if a test condition can correlate to results inside a sphere, it can be used instead.

Often production tests will include compromises in both timing and exact test conditions, and test results will be "corrected" to the standard measurement conditions.

One implication of fast measurements is the use of array spectroradiometers in testing. Scanning spectroradiometers measure wavelength by wavelength and require the source to be stable throughout the scan. Clearly, this is untenable in production testing. Some test stations use a combination of photometers and colorimeters to provide the speed of measurement required. Increasingly, array spectroradiometers are used as they can be operated at the speeds required while providing higher accuracies and more information than photometer/colorimeter approaches.

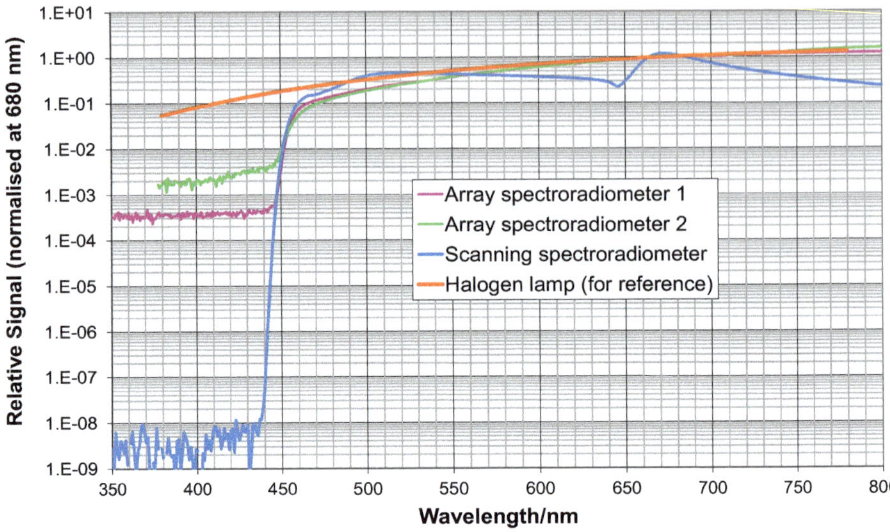

Fig. 11.11 Differences in stray light between systems

11.4 Measurement Equipment

In laboratory environments, both scanning and array spectroradiometers may be used as well as broadband photometers and colorimeters (see Chap. 5 for details on laboratory testing of LEDs). The highest accuracy is generally obtained using scanning spectroradiometers, but these require significant time to complete measurements and may not be suited to some forms of LED control such as AC or modulated. Array spectroradiometers offer shorter scan times, are unaffected by the LED control method, and can be used in place of scanning systems without much loss in accuracy. However, the quality of array spectroradiometers varies greatly, so careful selection is required to maintain accuracy. Photometers, colorimeters, and lower quality array spectroradiometers are used where the accuracy requirements are not so stringent.

When considering such characteristics as stray light, essential to accurate measurements, the differences between systems is evident. Stray light is the contribution at some wavelength from light which is actually at another wavelength. It is an unwanted cross-talk between wavelengths due to scattering from gratings and other components; and it would not be present in an ideal spectroradiometer. Figure 11.11 shows the results of a simple stray light test: a halogen lamp is measured with a 455 nm long-pass filter so that all shorter wavelengths are completely blocked. As there is no light at these shorter wavelengths, values are then a direct measure of stray light. Low stray light is essential for accurate measurements, so array spectroradiometer 2 is not suitable for many applications. High stray light will tend to move measured chromaticity values toward the equal energy point ($x = 0.3333$, $y = 0.3333$); the higher the stray light, the greater the distance moved.

Sensitivity is also essential for fast measurements. In measurements at equilibrium, high brightness LEDs are often too bright, requiring neutral density filters to attenuate the light, so that they can be measured. If the LED is only on for microseconds, even high brightness LEDs may not achieve optimum measurement conditions. Low signals mean low signal-to-noise and poor results. High sensitivity is essential to obtain high signal-to-noise ratio (SNR) in short time scales. Low SNR will tend to decrease the precision of measurements, giving randomness to repeated tests.

Wavelength accuracy is essential to most LED applications as parameters such as peak position and chromaticities critically depend on the wavelength scale. Scanning spectroradiometers generally have sufficient accuracy for most applications, but the wavelength accuracy of array spectroradiometers depends very much on their design and the calibration techniques. A synopsis of the various techniques is available [6] and will not be repeated here. However, for most LED applications, wavelength accuracies of 0.5 nm or better are required. It should also be noted that the wavelength scale should be based on centroid positions of measured line spectra (atomic emission lamps, lasers, etc.) rather than the position of the maximum. This is because the centroid represents the effective wavelength of the energy measured and hence chromaticity and other parameters that depend on the spectral distribution of energy will be correct.

An optical bandwidth of around 3 nm or less should be used if possible to prevent broadening of LED peaks. Asymmetric bandpass shapes can distort LED spectra in some cases, so bandpasses should be as symmetric as possible. Ideally, a spectrum would be corrected for bandpass, so all broadening or distortion is removed.

Linearity is the change of signals measured with incoming light. Ideally, if the amount of light is doubled, the value of the signal also doubles. Being close to this ideal is the basis of all radiometry. Most systems have a dark signal, i.e., a signal that is present even with no light, which must be measured and subtracted in order to give the signal that is only due to light. The dark signal depends on temperature and integration time, so a new dark signal should be taken when either changes. Cooled, temperature-regulated CCD systems or temperature-compensated CCD systems are the only ones to offer sufficient stability of dark signals for high-accuracy measurements.

Linearity relates an output quantity (x_o), e.g., signal, to an input quantity (x_i), e.g., light, using $x_o = k \cdot x_i$ for all x. Nonlinearity, n, is thus:

$$n = \frac{x_o}{k \cdot x_i} - 1$$

Here, the input quantity is light (power through the entrance slit) and the output quantity is electrical signal (current) on the array detector. If the system is completely linear ($n = 0$), then doubling the input power will double the output current. However, the nonlinearity should be determined over the useable dynamic range of the instrument and this may be several decades.

Even with suitable dark signal subtraction techniques, the signal may still not be completely proportional to light. The characterization and correction of any deviations from ideal should be an important feature of the system, so the user is not required to make compensations. Typically, a system should have nonlinearities of 0.5 % or less.

11.5 Quantities Measured

11.5.1 Total Luminous Flux

Total luminous flux is a quantity required for most modern LEDs in all applications. Details of the definitions and measurement methods are given in Chap. 5. Basically, the total flux is measured:

1. In the center of an integrating sphere.
2. At the wall of an integrating sphere (forward-emitting LEDs only).
3. With a goniometer.

Measurements in the center of an integrating sphere should be made with a large sphere (0.5 m diameter or above) to prevent the size of the source being measured adversely affecting results. The use of an auxiliary lamp to correct for self-absorption differences between the standard lamp and LEDs is essential. To conform to LM-79 [7], the sphere should also include measurement of ambient temperature. Measurements in the center of a sphere (properly corrected for self-absorption effects, angular response variations, and radiation patterns of the LEDs and lamps) are considered to be the reference values for the total flux of the LEDs, since they apply to all types of lamps or LEDs.

However, measurements in the center of the sphere may not be the most accurate for high-power LEDs. The cooling system for high-power LEDs may be large and if this goes into the sphere, it will add large self-absorption corrections and make the ambient temperature inside the sphere rise steeply.

If the light is emitted in the forward direction, as is the case with most high-power LEDs since heat sinks tend to be opaque, then measurements can be made more accurately and conveniently at the sphere wall.[8] An example of such a system is shown in Fig. 11.12.

Goniometers are important for measuring the radiation pattern of LEDs, and results may then be used to calculate total flux. However, if total flux is the prime parameter, then a sphere system, which gives results quickly, is usually preferred.

The above systems all apply to laboratory environments and are often not possible in production testing requiring large throughput. If one makes the assumption of constant spectral and angular properties of the LEDs to be tested, a measurement at any angle would correlate to total luminous flux. Naturally, real production lines do not have constant spectral and angular properties between individual LEDs or

[8] In case of forward emitting LEDs attached to a cold plate at the side of an integrating sphere it is sufficient to control the cold plate temperature which in such a setup is used as reference temperature. This is the requirement of the JEDEC JESD51-52 standard [4].

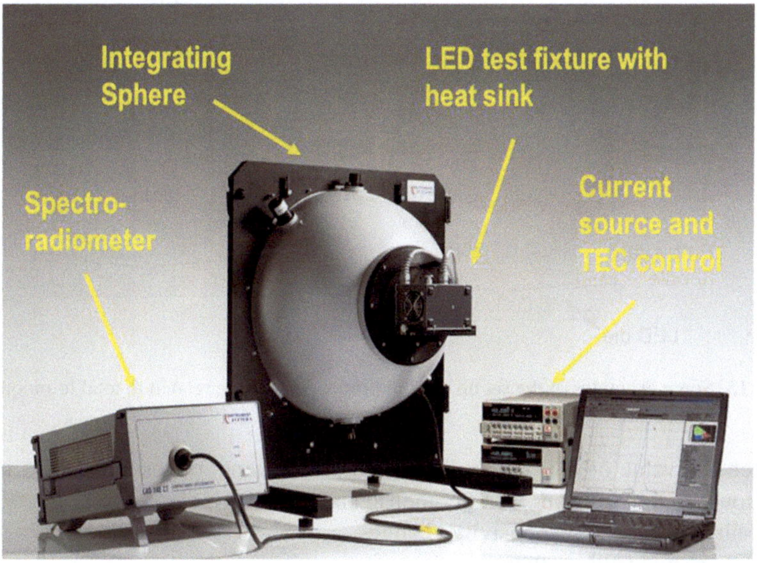

Fig. 11.12 A possible setup for measuring the total flux of high-power LEDs

batches, so the key is to find a test condition that is close enough to the standard methods to correlate highly enough among the LED variations expected.

Often a variation on method (2) is used for production measurements relating to total luminous flux. The LED is placed as close to the open sphere port as possible given the requirements of physical electrical connections and other tests to be performed on the LED at the same time. Also, contamination of spheres is a major issue in maintaining calibrations, so an "inverted" dome or flat window may be used to minimize dust and other particulates from entering the sphere (Fig. 11.13).

In a specific test set up, the angle, θ, for the LED being tested and the distance to the edge of the input port, r, can be defined and the relationship:

$$\Phi = r^2 \int_{\varepsilon=0^\circ}^{180^\circ} \int_{\eta=0^\circ}^{360^\circ} E(\varepsilon,\eta)\sin\varepsilon \, \mathrm{d}\eta \, \mathrm{d}\varepsilon = \alpha r^2 \int_{\varepsilon=0^\circ}^{\theta^\circ} \int_{\eta=0^\circ}^{360^\circ} E(\varepsilon,\eta)\sin\varepsilon \, \mathrm{d}\eta \, \mathrm{d}\varepsilon \qquad (11.1)$$

is assumed.[9] The correction factor, α, represents the factor between the measured flux and the total flux. This should ideally be a constant factor for all production tests of that type of LED and the uncertainty in the value of α determines whether or not this measure of "partial flux"[10] can be successfully used to indicate the total flux of the LED.

[9] See Sect. 5.1.2 and Eq. 5.9 in Chap. 5.

[10] Note: this is not the "Partial LED flux" quantity defined by CIE, but is instead a simple expression that not all the flux is measured.

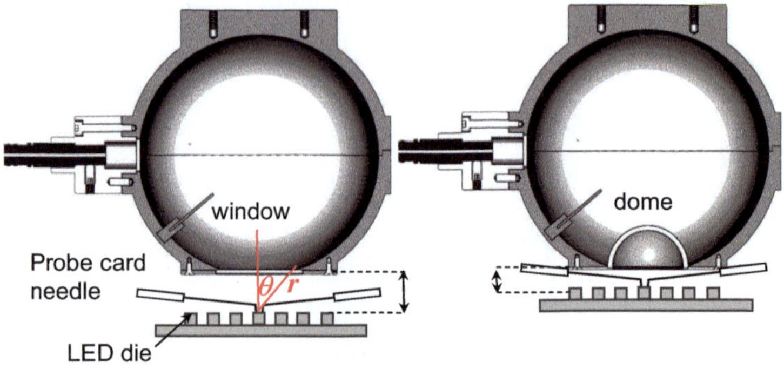

Fig. 11.13 Some examples of the set up used in production testing related to total luminous flux

Often the value of α is determined by measurements of a set of LEDs in both conditions, as this is easier than geometric analysis and includes other factors such as self-absorption and responsivity changes.

The corrected value is then:

$$\Phi = \alpha \Phi_m \tag{11.2}$$

where Φ_m is the measured flux.

The relative uncertainty in the final flux value is:

$$\frac{u(\Phi)}{\Phi} = \sqrt{\frac{u^2(\alpha)}{\alpha} + \frac{u^2(\Phi_m)}{\Phi_m}} \tag{11.3}$$

The uncertainty in α should include: the uncertainty in the total flux value of the reference LEDs; the precision in measurements using both the test and laboratory conditions; the variations between LEDs tested; variations in power supplied to the LEDs; variations in environmental conditions; and any variations in alignments that are possible when moving the LEDs into position prior to testing.

11.5.2 Averaged LED Luminous Intensity

Averaged LED intensity (referred to as ALI in short) is particularly of interest when the LED is used in applications of direct viewing, e.g., indicators or signs, rather than general lighting. It is defined by the CIE in Publication 127:2007 [8] and summarized in Fig. 11.14. Essentially, it is the flux (luminous flux, radiometric flux, or spectro-radiometric flux) received via a 1 cm^2 circular aperture at a given distance from the LED tip along the mechanical axis. Two conditions are defined corresponding to: condition A is a 0.001 sr solid angle, which corresponds to a distance of 31.6 cm; condition B is a 0.01 sr solid angle, which corresponds to a distance of 10 cm.

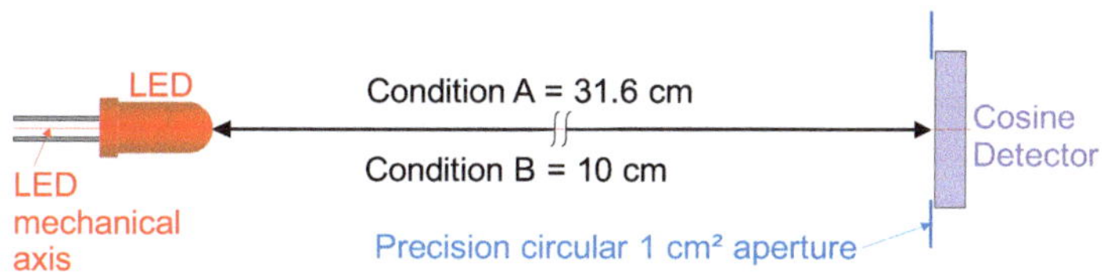

Fig. 11.14 Drawing illustrating the key features of the averaged LED intensity definition

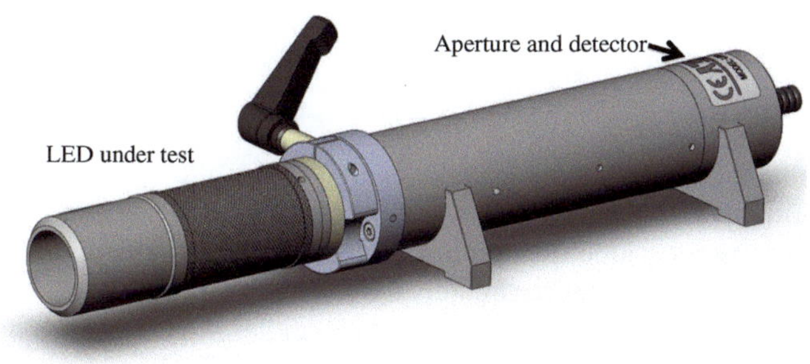

Fig. 11.15 Example of a baffle tube for measuring averaged LED intensity according to condition B

As these conditions prescribe an exact physical geometry rather than a fundamental quantity, adherence to these conditions is essential to ensure the correct value is obtained.

Measurements are normally made using a fixed baffle-tube that ensures the geometry and provides a light-tight condition, so a dark laboratory is generally not required.

Although the accessory in Fig. 11.15 is quick and easy to use, it is still far too slow for many production line requirements. Also, only a small fraction of the light emitted is within the 0.001 or 0.01 sr solid angles and the cosine collector loses much of that light. So, generally the problems are to measure a value that is a tiny fraction of the emitted light and to measure it rapidly.

Clearly, the arrangement in Fig. 11.16 does not conform to the definition of ALI condition B. However, the lack of a diffuser or integrating sphere greatly enhances the sensitivity of the system. This arrangement, if operated at a slight angle to the wafer normal, can even allow other equipment such as inspection microscopes to be used at the same time.

Instead of the 11.3 mm diameter uniform response detector that is part of the averaged LED intensity definition, the fiber optic bundle is small, e.g., 1.5 mm diameter, and the individual fibers within it mean the response across the fiber is very nonuniform (light hitting between the individual fibers is not transmitted). Therefore, it is essential that the light hitting the fiber is uniform; hence this technique would not be used in testing narrow-beam LEDs. The distance is also not necessarily 10 cm, but this is not essential if the LED tested is small. So, for small LEDs with a

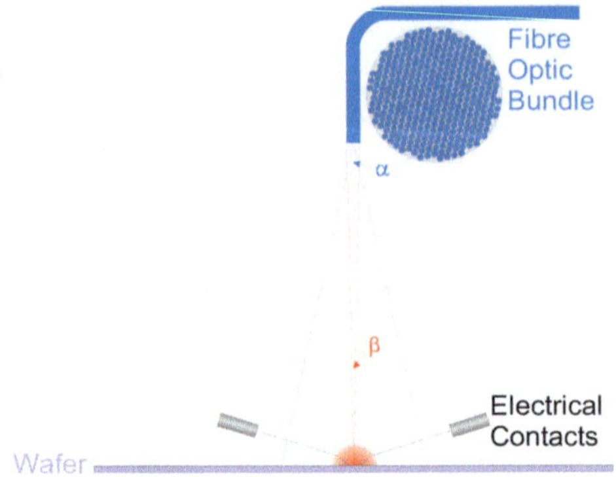

Fig. 11.16 Example of a production set up for tests related to averaged LED luminous intensity, condition B

broad radiation pattern this test ought to correlate reasonably well to averaged LED luminous intensity provided the condition B angle of 0.01 sr is approximated. This corresponds to angle β of approximately 2°.

The full acceptance angle of the fiber optic bundle, β, is related to the numerical aperture (*NA*) of the fiber by:

$$\beta = 2\sin^{-1}(NA) \tag{11.4}$$

Within this acceptance angle, the fiber transmittance varies with the cosine of the angle, but outside it there is no transmittance as the rays are no longer totally internally reflected within the fibers. Bends in the fiber can decrease the angle α, as higher angle rays are lost. Also, short fibers may maintain some of the angular properties of the spectroradiometer or other measurement device, giving less uniform response within the field of view.

Ideally, the angle α should be large enough to completely contain the LED, so as to prevent alignment difficulties, but small enough to exclude any other LEDs or devices that may be operating at the same time. The averaged LED intensity can be expressed as:

$$I_{LED,B} = \frac{\Phi_A}{10^2} = \frac{E}{100} \tag{11.5}$$

where Φ_A is the flux in a circular 1 cm^2 area and hence the same as the irradiance, E, expressed in W · cm^{-2}. The measured "intensity" is given by:

$$I_m = \frac{E'}{d^2} \tag{11.6}$$

where d is the distance between the LED and fiber optic bundle and E' is the irradiance measured at the fiber optic bundle tip.

Again, a conversion factor α is determined by comparative measurements on reference LEDs with both methods so that:

$$I_{LED,B} = \alpha I_m \tag{11.7}$$

The relative uncertainty in the final averaged LED intensity value is:

$$\frac{u(I_{LED,B})}{I_{LED,B}} = \sqrt{\frac{u^2(\alpha)}{\alpha} + \frac{u^2(I_m)}{I_m}} \tag{11.8}$$

The uncertainty in α should include: the uncertainty in the $I_{LED,B}$ value of the reference LEDs; the precision in measurements using both the test and laboratory conditions; the variations between LEDs tested; variations in power supplied to the LEDs; variations in environmental conditions; and any variation in alignments that are possible when moving the LEDs into position prior to testing.

11.5.3 Chromaticity

Binning of LEDs based on color has been done since early commercialization of LEDs. Applications such as traffic lights require a close match of colors within the LED array to ensure they look uniform. In modern applications such as solid-state lighting (SSL), LEDs are binned in terms of correlated color temperature (CCT) based on ANSI C78.377 [9]. This is based on seven-step MacAdam ellipses as originally defined for compact fluorescent lamps. One step of an ellipse refers to the color shift required to produce a "just noticeable difference" (JND) in color of the source, so seven steps (the ellipse is seven times larger) are easily perceived. Programs such as Energy Star® have adopted this [10] as a criterion of eligibility for SSL, but express it in CIE *xy* color space, as shown in Table 11.1. Although the color spaces in Fig. 11.17 and Table 11.1 are different, they are essentially the same specifications and the color spaces are easily transposed.

Some manufacturers have binning procedures based on much finer gradations than this and current discussions within the standardizing bodies indicate that possibly smaller step sizes of the MacAdam ellipses will emerge in future. Chromaticity is therefore a critical parameter to be measured and the requirements of the measurement instrumentation, already stringent, are likely to be pushed to the limits of accuracy. As chromaticity also changes with LED temperature, a properly chosen LED reference temperature as parameter will become important in this regard.

Many LEDs and SSL lamps have chromaticities that change with the viewing angle, as seen by gonio-spectroradiometric measurements (Fig. 11.18).

The chromaticity quoted for an LED is generally that of the entire chip rather than a specific direction. The measurements at different angles must be therefore combined correctly to give a single chromaticity value. The most accurate method is to form the integral, shown in the first part of Eq. 11.1, in terms of spectral radiant intensity to give total spectral flux and then calculate the chromaticities from this.

Goniometry is not usually employed routinely in a production environment however, since it requires steady-state conditions for the LED and each angle is measured separately, giving long measurement times. In such circumstances, it is much more common to employ integrating spheres as described in Sect. 5.1. This not

Table 11.1 Energy Star® chromaticity recommendations for integral LED lamps

Energy Star®	2700 K		3000 K		3500 K		4000 K		4500 K		5000 K		5700 K		6500 K	
	x	*y*	*x*	*y*	*x*	*y*	*x*	*y*	*x*	*y*	*x*	*y*	*x*	*y*	*x*	*y*
Center point	0.4578	0.4101	0.4338	0.403	0.4073	0.3917	0.3818	0.3797	0.3611	0.3658	0.3447	0.3553	0.3287	0.3417	0.3123	0.3282
Tolerance quadrangle	0.4813	0.4319	0.4562	0.4260	0.4299	0.4165	0.4006	0.4044	0.3736	0.3874	0.3551	0.3760	0.3376	0.3616	0.3205	0.3481
	0.4562	0.4260	0.4299	0.4165	0.3996	0.4015	0.3736	0.3874	0.3548	0.3736	0.3376	0.3616	0.3207	0.3462	0.3028	0.3304
	0.4373	0.3893	0.4147	0.3814	0.3889	0.3690	0.3670	0.3578	0.3512	0.3465	0.3366	0.3369	0.3222	0.3243	0.3068	0.3113
	0.4593	0.3944	0.4373	0.3893	0.4147	0.3814	0.3898	0.3716	0.3670	0.3578	0.3515	0.3487	0.3366	0.3369	0.3221	0.3261

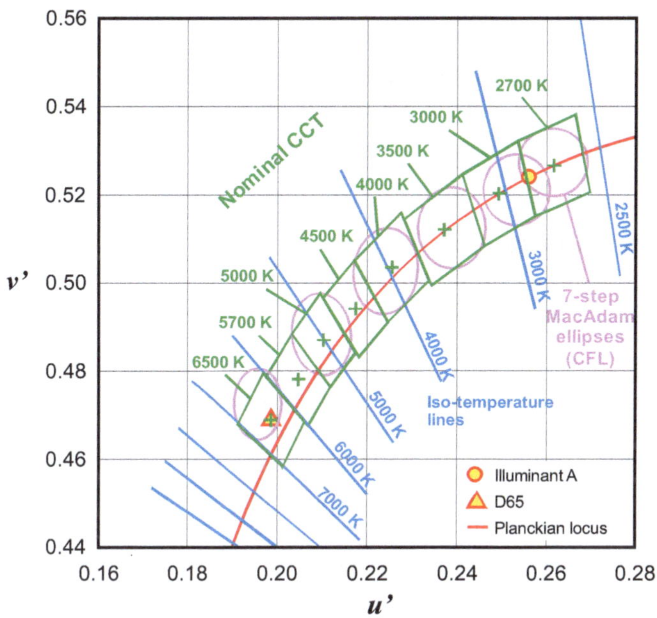

Fig. 11.17 CIE 1976 u′v′ chromaticity diagram with ANSI C78.377 color boundaries (*green* quadrangles) for SSL products

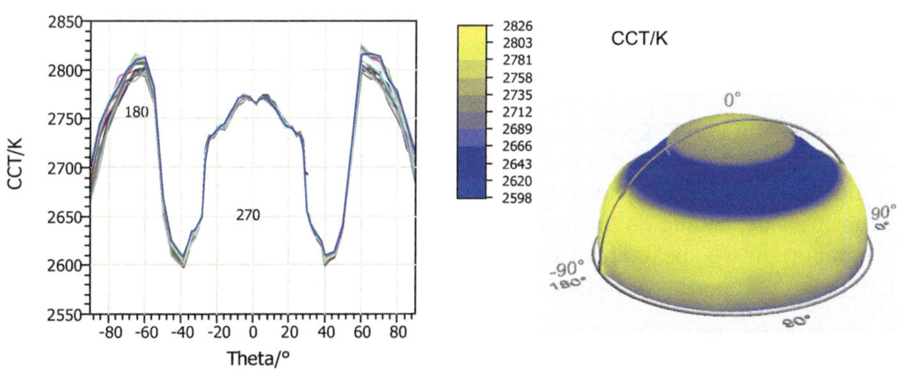

Fig. 11.18 Variation in CCT with angle for a white LED example

only provides relatively quick measurements, it also gives the correct averaging to chromaticity values.

Of course, if production methods such as shown in Fig. 11.13 are used, then any chromaticity changes due to unmeasured parts of the LED emission should be considered and corrected. If the LED is also operated with short pulses, chromaticity corrections due to nonequilibrium conditions during testing may also need to be applied.

SSL products are used in general lighting and task-specific lighting, so their color rendering is also important. Two sources with the same chromaticity and CCT may not have the same color-rendering properties. Questions on color fidelity and preference are discussed in Sect. 5.5.3 of Chap. 5; however, the general color-rendering index (Ra) along with some special indices remains the industry standard metric [11]. Spectral values are required for calculation of color rendering and hence only spectroradiometers can be used in measurements.

11.6 Uncertainties and Tolerances

11.6.1 Calibrations

In order to make absolute measurements, the measurement equipment must be calibrated. Unlike quantities such as length or mass a physical standard cannot be made. Light cannot be stored in any practical sense and must be constantly generated during calibrations. Standard lamps are used that convert electrical power into light and their power supplies must be extremely stable. For the most part incandescent and deuterium lamps form the sources used in calibration. Of these, only specific designs are suitable and they must be operated under specific conditions in order to give stable and consistent light output. Moreover, the calibration sources should be traceable to a National Metrology Institute (NMI) with a short traceability chain to ensure international recognition of measurement results.

11.6.2 Units

Light is a form of energy, so the basic unit of light is the joule [J]. As light must be constantly generated the energy per unit time, or power in watts [W = J/s], appears in most commonly encountered applications. The calibration lamp is used in specific geometries depending on the quantity being calibrated. When the energetic aspects are important we speak about radiometry. The most relevant quantities related to radiometric properties of light sources are:

- *Radiant intensity* in W sr^{-1}.
- *Irradiance* in W m^{-2}.
- *Radiance* in W m^{-2} sr^{-1}.
- *Radiant flux* in W.

When humans see the light from lamps, they do not see the energy, except those operated in short flashes, and the visual stimulus is related to the power. It is also related to the spectrum of the lamp as all wavelengths are not perceived equally.[11] Therefore, all radiometric quantities have their so-called photometric counterparts

[11] The spectral distribution of the sensitivity of human eye under photopic conditions is described by the V(λ) function—see also Chap. 5.

in which the spectral distribution of the sensitivity of the human eye is considered. For example, the human visual equivalent of power is *luminous flux*[12] measured in lumens [lm]. The photometric quantities of interest are:

- *Luminous intensity* in lm sr^{-1} (this is an SI unit called the candela [cd]).
- *Illuminance* in lm m^{-2}.
- *Luminance* in lm m^{-2} $sr^{-1} = cd\ m^{-2}$.
- *Luminous flux* in lm.

Calibration light sources can be calibrated both for radiometric or photometric quantities. Calibrations are performed in the geometry of the desired measurement, which must agree with the unit in the calibration certificate of the calibration lamp. In this way, measurements are effectively made by substitution of a test light source for a calibration light source, both of which are measured using the same equipment. The separation into calibration and measurement phases of the procedure is purely for convenience.

11.6.3 Uncertainty and Precision

Uncertainty estimation is a requirement of modern measurements. However, it is a difficult subject to master and misconceptions are common. If many measurements are made, each result is slightly different and sometimes it is felt that this variability is the uncertainty. This variability is actually a component of uncertainty called precision and can be expressed as the standard deviation of a set of measurements under specific test conditions.

A standard lamp is chosen because it is stable and its uncertainty is therefore not an estimation of its variability; it is instead a statement of how exactly we know its value.

This idea may be illustrated with a simple example: measuring a distance between two objects. We can take a calibrated ruler, which has an uncertainty, and measure the distance several times. The overall uncertainty is a combination of the uncertainty in the ruler calibration and our ability to read the distance and to define the reference points on the objects. We combine all of these and arrive at an uncertainty in the distance. This does not mean the distance is varying, it only means we cannot know the exact distance better than this.

When a standard is calibrated, it will have an uncertainty. Again this is not variability in the standard, but reflects knowledge of its value. The ruler has an uncertainty, but it is not varying in its scale. Generally, values are quoted as $X \pm u(X)$, where $u(X)$ is the standard uncertainty (equivalent to a standard deviation of our knowledge in the value X). Alternatively, the value might be quoted at some expanded uncertainty, e.g., $X \pm U(X)$ where $U(X) = k \cdot u(X)$. Many NMIs recommend quoting at $k = 2$ which corresponds to 95 % confidence.

[12] See the definition of luminous flux in Chap. 5.

This concept of confidence level is an important one. The above statements give a value for X, which is important in enumeration of measurements, but it does not say the quantity value *equals* X. Instead it says that we are 95 % confident that the real value of X is somewhere in the range of $X - U(X)$ to $X + U(X)$. This is a very important distinction and one that many people find confusing, but thinking of X as the true value of the quantity is the most common misconception in the field of measurement uncertainty. Instead, X is just the central value of a range of possible values.

11.6.4 Traceability

A NMI will generally have access to some *absolute standard* where the light output can be calculated from fundamental physics, e.g., cryogenic blackbody radiator or synchrotron radiation source. Here, the uncertainty is from the possible values of input variables, e.g., temperature and emissivity, or from the reliability (i.e., lack of assumptions) in the fundamental physics. Each of these is estimated, measured, or simulated to provide a range of possible values for this absolute standard. The central "expectation" value, X_{AS}, is the one assigned for evaluation purposes in the full knowledge that this carries an uncertainty that reflects the range of possible values. Again, this does not reflect an actual fluctuation in the output of the source, though that may be component, but is an estimate of uncertainty associated with our knowledge of the value.

Linear accelerators are hard to fit in laboratories, so this absolute standard is used to calibrate a *primary standard* that can be moved to another location and used there. Primary standards, e.g., high-temperature blackbodies, often have different characteristics to the absolute standard and estimates of systematic effects, e.g., geometry, and measurement variations due to setup and equipment give a transfer uncertainty that is combined with that of the absolute standard to give an overall uncertainty. As the "expectation" value of the absolute standard is used in the evaluation, then a result, X_{PS}, follows an assumed evaluation path.

The primary standard is then used to create a *working standard*, usually of the type that will be sold, e.g., an FEL lamp. Again, there will be a transfer uncertainty and this increases the combined uncertainty associated with the value X_{WS}.

When we buy a calibration standard from an NMI to be used as a *laboratory primary standard*, it generally comes with a certificate giving values and uncertainties in the form $X_{LPS} \pm U(X_{LPS})$. We use the value of X_{LPS} in our evaluations when we use this standard to create in-house, *laboratory working standards* and hence continue the chain of assumed evaluation path to give X_{LWS}.

Figure 11.19 shows a diagrammatic illustration of this process. The different standards in the *traceability chain* are shown horizontally. Each standard has an expectation value (shown as a black filled circle) and uncertainty (shown as a black error bar). Note: the value scale is not necessarily the same for each standard in the chain and is shown purely for illustration purposes as if mapped to a common scale. Uncertainties increase at each stage since these must be a combination of the uncertainty in the prior standard and transfer uncertainty to the current standard.

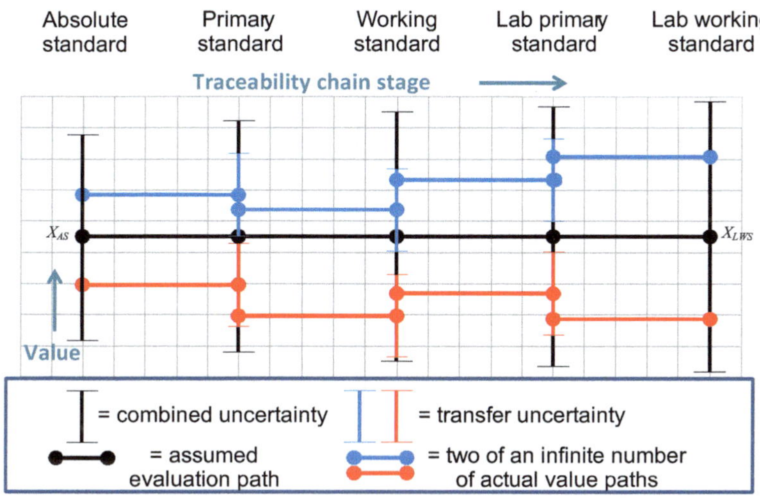

Fig. 11.19 Propagation of uncertainties through a traceability chain. The assumed evaluation path is based on "expectation" value, but actual value paths may be different. The actual value path is unknowable and an infinite number of paths are possible (two of these are shown for illustration)

At any stage in this process, the "true" value may not have been the "expectation" value; in fact, it is very unlikely that they are the same. So, two standards that nominally have identical values and uncertainties do not necessarily agree. It is not uncommon to send a lamp for recalibration and have new values even if the lamp has not changed.

Uncertainties comprise both random and systematic components. However, once a value is assigned to a standard, even random components are "frozen" into a systematic uncertainty. If the calibration of the standard was repeated, it would have the same systematic components, but different random components and hence would be assigned a different "frozen" value. These systematic offsets are however unknowable and we can only give a range of possible values that they could be relative to the assumed evaluation path. As they are unknowable, and hence cannot be corrected, the range of possible values is treated in the same way as random uncertainties in propagation calculations.

The fact that all new lamps from an NMI are probably calibrated with the same working standard(s) gives some consistency, but if lamps are calibrated with different working standards, such as those obtained from different NMIs, larger differences may be expected. This does not mean one is right and another wrong—both agree within their overall uncertainties—but again it is a common misconception that differences must mean that if one is right the other must be wrong. In Fig. 11.19, two possible evaluation paths are shown that give different values to a laboratory working standard. However, these values lie within the uncertainties and hence within the limit of our knowledge of the value. Although we can compare them and see they are different, we cannot say which is right and which is wrong and in that sense they are both right.

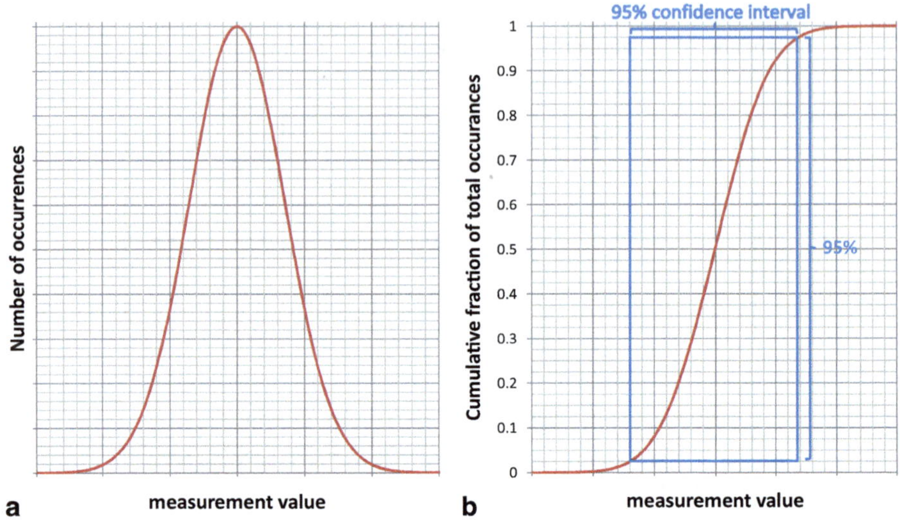

Fig. 11.20 **a** Distribution of results from repeated measurements. **b** Cumulative distribution of results

11.6.5 Measurements

We have calibrated the measurement equipment with a stable calibration source and now we measure a stable LED with our stable measurement equipment. If we make repeat measurements over a short period of time, there will still be some variation due to noise, stability of power supplies, etc., but those variations will be generally small compared with the overall uncertainty of the measurement.

If we plot the number of times a particular measurement result is obtained against the measurement value, a familiar bell-shaped distribution is usually seen. This is the Gaussian distribution often encountered in statistics. It indicates that some values are more likely to occur than others and the tendency is to group around some central mean value. Generally, it is more useful to plot the cumulative fraction of the occurrences against the measurement value. The cumulative fraction is then the probability of a measurement result being less than or equal to the value. As shown in Fig. 11.20b, the values corresponding to fractions of 2.5 and 97.5 % are easily identified, giving the 95 % confidence interval values directly. In the case of the Gaussian distribution, this is equivalent to calculating the standard deviation of results and multiplying by 2 (the expansion factor k).

It is a common misconception to see the close agreement between repeated measurements like this as evidence of accuracy; it is not. This is the measurement precision, not the uncertainty. To obtain the uncertainty, a budget of all possible contributions is drawn up.

The first stage of creating a budget is the model. This is simply the equation used in calculating the value complete with any corrections. This model describes variables used in calculation, so the next stage is to quantify the value of those variables with estimates or measurements of their uncertainty. For instance, if correction is made for temperature, then the temperature at the time of measurement needs to be known and the thermometer used to measure it will have an associated uncertainty provided by the manufacturer. The variables are listed individually in a budget table. Propagating each variable value and uncertainty through the model will then give the value and uncertainty in the result. For further details see the GUM [12] which is now freely available from the BIPM website.

11.6.6 Tolerances

Most people involved in production testing are familiar with tolerances. When talking of tolerances we often use similar language to when talking about uncertainties so naturally they get confused at times. Tolerances are not uncertainties. Measurement uncertainty is defined as a "non-negative parameter characterizing the dispersion of the quantity values being attributed to a measurand, based on the information used" [13]. Tolerance is defined almost on an individual company basis, but the NIST handbook 44 [14] defines it as "A value fixing the limit of allowable error or departure from true performance or value." Essentially, an uncertainty is associated with a measurement value and no judgment is implied whether the measured value is "good" or "bad." Conversely, the word *allowable* in the definition of tolerance implies that the measurement result triggers an action, e.g., rejection or adjustment.

Tolerances are often linked to specific test procedures and results outside trigger actions that cost money. This economic incentive means it is very important that tolerances are set at sensible values to catch defective products, but not reject good products. For instance, setting tolerances at the 95 % confidence values shown in Fig. 11.20 would reject 5 % of products purely on statistical grounds. Manufacturers decide what a defective product value is and set the tolerances based on that rather than an arbitrary limit based on what a good product should be. The greater the difference in values between a good and bad product is, the easier it is to set the tolerance limits.

Most production-testing concentrates on consistency of products and often uses the small measurement precision to assess products and set tolerances rather than the uncertainty. This is quite acceptable for comparing a series of products under tightly controlled conditions, e.g., same calibration standard, same equipment, and same environmental conditions. However, it should be realized that this cannot then easily be compared with results from another line that used a different standard, illustrated in Fig. 11.21, equipment or conditions.

As described earlier, the "true" value for each calibration standard is likely to differ from the central value given in the report. That difference will be within the uncertainty, but will not be known exactly; so, each standard may give a different result for the same LED. Measurements on a set of LEDs will therefore give distributions that are offset, depending on which standard was used.

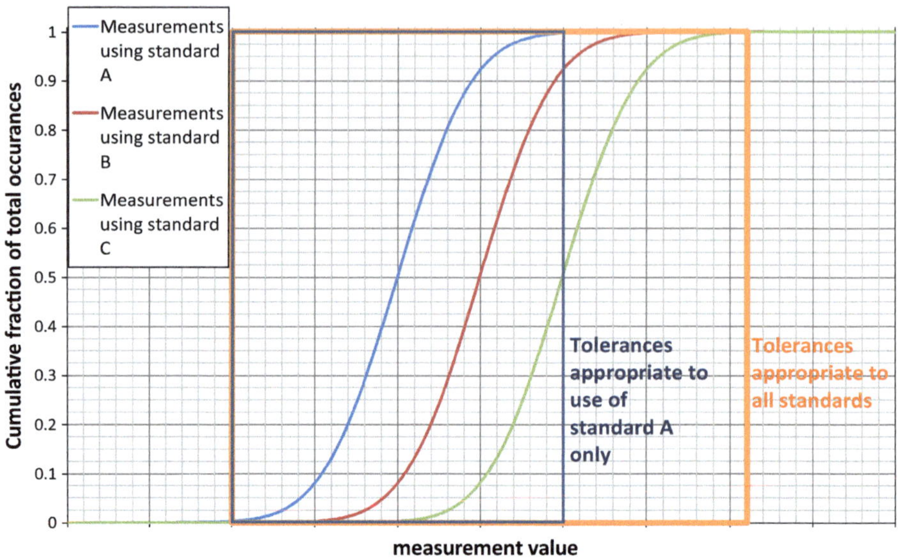

Fig. 11.21 Simulated measurement distributions for measurements of the same set of LEDs, but with equipment calibrated using different standards

Therefore, tolerances should be set according to the scope of comparison. The more variables involved (e.g., standards, measurement equipment, control equipment, environmental conditions, procedures, etc.), the greater the potential shift in measurement precision distributions. Naturally, when comparing across all variables, both within the company and between companies, they should all agree within the measurement uncertainty at some prescribed level of confidence.

Consistency requirements placed on production are constantly shifting to tighter agreement and this leads in turn to tighter control over all variables. However, as companies introduce multiple production lines, often in multiple countries, maintaining tight consistency between them becomes a mammoth task.

11.6.7 Binning

The issues involved in binning are obvious analogues to those of tolerances. Here the limits are used in sorting the product into groups based on intensity, color, etc. For this to work, the variations in the product must be large compared with the measurement precision for repeat measurements of a single LED. Even so, the measurement precision places a limit on the "sharpness" of the sorting process. An LED that lies exactly on a bin boundary has a 50 % chance of being assigned to either bin purely on statistical grounds. As the "true" value of the LED moves further from the bin boundary, it becomes more and more likely to be assigned to the correct bin. The narrower the measurement precision distribution, the greater the proportion of LEDs that will be assigned correctly. *Bins that are narrow compared with the measurement precision should be avoided.*

Correct assignment may be purely relative; a bin may contain LEDs whose chromaticities agree within 0.0002 in x and y chromaticity for instance, but that does not mean the "true" values are known within 0.0002. Another production line may place these same LEDs in a different chromaticity bin, but the sequence order of the sorting would generally be the same. Therefore, the absolute values only become important when comparing different lines or company comparisons.

11.7 Summary

The power to LEDs is converted to light and heat. The temperature rise caused by the heat component affects the quantity and spectrum of the emitted light. The efficiency of light emission is also affected, so heat removal is the key to operation of LEDs at their optimum.

Suitable equipment is needed to make measurements, but different setups are often required in laboratory measurements or in production line testing. An array spectroradiometer is often used in both setups and high-quality types are generally preferred to maintain accuracy requirements.

Compromises on exact conditions of measurement, but not on accuracy, are made in production lines. They may use related geometries rather than the ideal conditions and determine the factor that relates the test to the ideal. This factor can then be used to correct test results during production. Short pulses of current are routinely used instead of steady-state conditions. These short pulses mean that LEDs are measured in nonequilibrium conditions, but careful control over the testing environment means results can be corrected to the steady-state condition.

The distribution of test results follows the measurement precision, which should not be confused with uncertainty. Uncertainty does not imply variability, but a limit to knowledge of a value. Measurement precision is a variation and limits how precisely a value can be expressed or compared. The parameter to use in comparisons depends on the scope of the test results being compared.

All testing in laboratory or production environments requires high-quality instruments. Array spectroradiometers provide the required accuracies as well as the correct quantities for all optical parameters while being fast enough for production line requirements.

References

1. CIE S 017/E:2011 ILV: International lighting vocabulary
2. JEDEC Standard JESD51-53 (2012) Terms, definitions and units glossary for LED thermal testing. http://www.jedec.org/sites/default/files/docs/jesd51-53.pdf

3. JEDEC Standard JESD51-51 (2012) Implementation of the electrical test method for the measurement of the real thermal resistance and impedance of light-emitting diodes with exposed cooling surface. www.jedec.org/sites/default/files/docs/JESD51-51.pdf
4. JEDEC Standard JESD51-52 (2012) Guidelines for combining CIE 127-2007 Total flux measurements with thermal measurements of LEDs with exposed cooling surface. www.jedec.org/sites/default/files/docs/JESD51-52.pdf
5. Zong Y, Ohno Y (2008) New practical method for measurement of high-power LEDs, Proceedings of CIE Expert Symposium 2008, July 2008, Turin, Italy, CIE x033:2008, 1002-106
6. Young R, Häring R (2011) Wavelength calibration of array spectrometers, Proceedings of CIE 27th Session, South Africa, vol 1, Part 1
7. IES LM-79-08 (2008) Approved method: electrical and photometric measurements of solid-state lighting products
8. CIE 127:2007 (2007) Measurement of LEDs
9. ANSI_ANSLG C78.377-2011 (2011) American national standard for electric lamps—specifications for the chromaticity of solid state lighting (SSL) products
10. ENERGY STAR® (2010) Program requirements for integral LED lamps, eligibility criteria—version 1.1, amended—March 22. Energy Star® website. http://www.energystar.gov/
11. CIE 013.3-1995 (1995) Method of measuring and specifying colour rendering properties of light sources
12. JCGM 100:2008 (2008) Evaluation of measurement data—guide to the expression of uncertainty in measurement. BIPM website. http://www.bipm.org/en/publications/guides/gum
13. JCGM 200:2012 (2012) International vocabulary of metrology—basic and general concepts and associated terms (VIM), 3rd ed. BIPM website. http://www.bipm.org/en/publications/guides/vim.html
14. NIST Handbook 44 (2012) Specifications, tolerances, and other technical requirements for weighing and measuring devices, 2012 ed.

Chapter 12
Thermal Management of Sophisticated LED Solutions

Theo Treurniet

Abstract Light-emitting diodes (LEDs) are already widely applied in general lighting applications. For almost all lamps and luminaires based on conventional technologies, an LED equivalent has been developed and put on the market. The application space of LEDs ranges from small systems, using a few watts of power to systems that generate more than 10,000 lumens and dissipate more than 200 W of power. Over this wide application area, many different system architectures are used and many different thermal solutions are applied. This chapter focuses on the application of LEDs; it is split-up in sections that cover different application areas. First, it describes the application of LEDs in lamps. These retrofit LED lamps have a prescribed form factor and replace existing lamps, based on conventional technologies (incandescent, halogen, or compact fluorescent). Next, the application of LEDs in new systems is described for several application fields. Finally, LEDs in some special applications are discussed. These applications have special requirements that need more advanced thermal solutions. This chapter concludes with a section on remaining challenges.

12.1 Retrofit Lamp Applications

In light-emitting diode (LED) lighting applications, we can make a distinction between retrofit lamps and systems that are developed explicitly to profit from the advantages LEDs are offering. Retrofit LED lamps are designed to be a replacement for a conventional lamp and have the advantage that they can be applied in existing luminaires. This allows users to upgrade their lighting system to LED technology, without having to replace the whole luminaire. The requirement that a retrofit lamp has to replace an existing conventional light source puts many limitations on the design of the lamp. The electrical, mechanical, and optical interfaces are prescribed. A lamp is equipped with a certain socket. It has to fit in a certain form, has to provide a certain amount of light preferably with a comparable light distribution and quality. It will be supplied with the same voltage as the conventional lamp. A retrofit LED lamp is applied in a variety of different fixtures that often predates the LED technology.

T. Treurniet (✉)
LED Platform Development, Philips Lighting, 80020, 5600 JM,
Eindhoven, The Netherlands
e-mail: theo.treurniet@philips.com

C. J. M. Lasance, A. Poppe (eds.), *Thermal Management for LED Applications,*
Solid State Lighting Technology and Application Series 2,
DOI 10.1007/978-1-4614-5091-7_12, © Springer Science+Business Media New York 2014

A number of these requirements have a direct impact on the thermal design of these lamps. The fact that the form factor is prescribed limits the possibility to optimize the heat transfer from the lamp to the ambient air. In addition, one has to take into account that the lamp is applied in a variety of different fixtures. Especially in the case of conventional incandescent and halogen lamps, there is no provision in the conventional system to remove heat from the lamp to the ambient air. Almost all the excess heat from conventional lamps radiates to the environment, due to the high-operating temperatures of these lamps. Since LED lamps operate at much more moderate temperatures, the generated heat is transferred mainly via conduction and convection. Although radiation still plays an important role, in LED lighting it is not the dominant heat transfer mechanism. An overview of different lamp types and their system properties is given in Table 12.1.

Next to the limited form factor, the fact that the fixtures are in general not designed to contain lamps that are cooled via conduction and convection means that all the provisions to remove the heat have to be designed into the lamp. Also, even for highly optimized LED lamps, one has to realize that there will be fixtures—especially compact closed ones—that will block almost all convective cooling and are not suited to be equipped with LED lamps at all.

All retrofit LED lamps are equipped with integrated electronics to convert the supply voltage—typically the mains AC voltage or 12 V AC—to a voltage and current that the LED light source can handle. Next to this conversion function, the electronics can have additional functionality like the possibility to dim the lamp. The electronics is typically 80–90 % efficient, thus adding a thermal load of 10–20 % of the input power to the lamp system.

There are numerous different lamp types in the market. Most of the lamp types are standardized in IEC documents. For example, IEC 60064:2005 [1] describes the performance of tungsten filament lamps for general lighting applications while, e.g., IEC 60630:2005 [2] describes the maximum outlines of these lamps. For other retrofit lamps like the PAR38 or MR16 reflector lamps, similar standards are available [3]. The most regular light bulbs are the so-called A-size bulbs. In Europe, these are indicated as A55 and A60 bulbs, where the number indicated the bulb diameter in millimeters. In the United States, bulbs are often indicated as an A19 bulb, which is comparable in size to the European A60 form factor. Other commonly used lamps are reflector lamps that provide collimated beams. These are available as large PAR38 lamps but also as smaller GU10 and MR16 lamps. The GU10 and MR16 lamps have a similar reflector size. The GU10 lamps are powered by the mains AC voltage, while the MR16 lamp requires a 12 V AC supply, see Table 12.2.

All these lamp types are available in a large number of different power and luminous flux levels. Due to the limitation in form factor, it remains a challenge to replace the higher power levels of the conventional lamps with LED lamps. The mentioned IEC standards describe the required performance levels that should be realized. In the market, a lot of LED lamps could be found that claimed performance levels that where equivalent with high power conventional lamps. However, in practice, many of these lamps could not fulfill their performance claims, mainly due to the thermal limitation that is imposed by the prescribed form factor.

Table 12.1 An overview of the difference between conventional and LED lamps

Lamp type		Conventional lamps			Equivalent LED replacement lamps	
		Typical level of luminous flux	Level of electrical power	Efficacy	Level of electrical power	Efficacy[a]
A-bulbs[b]		720 lm (@60 W)	25–100 W	12 lm/W	5–12 W	55–70 lm/W
MR-16		550 lm (@50 W)	20–50 W	10–15 lm/W	4–10 W	50–55 lm/W
PAR38		1340 lm (@90 W)	45–90 W	12–15 lm/W	13–18 W	45–55 lm/W

[a] These efficacy values are LED system level values, including losses in the driver and the optics for products that are available in the market in 2012
[b] In many countries, conventional incandescent bulbs above a given electrical power level are already banned. For details on legislation, see e.g. [4]

There is a risk that this leads to consumers that become disappointed in LED technology. A similar situation occurred with compact fluorescent lamps, which were often perceived as lamps that provide a poor light quality (some more information about the lessons learned from mistakes in the past can be found in Chap. 6 on standardization). In order to avoid that consumers become disappointed with LED technology and continue to use more energy consuming alternatives, various governments implemented additional legislation. European regulations [4] provide a number of requirements that a lamp has to fulfill before it can claim to be equivalent with a conventional lamp. In the United States, the same is achieved via the Energy Star program. In order to carry the Energy Star logo, a lamp has to fulfill a number of requirements [5] that ensure a minimum luminous flux, light quality, light distribution pattern, and lifetime.

Table 12.2 A comparison of MR16 12 V and GU10 mains voltage lamps

MR16 LED Lamp (12V AC)	GU10 LED Lamp (mains voltage)

12.1.1 Thermal Performance Metric for Lamps

A first high-level assessment of the thermal performance of an LED system can be done by evaluating the global energy balance of the system. We can determine the thermal dissipation P_H by:

$$P_H = P_{el} - \Phi_e \tag{12.1}$$

where P_{el} is the electrical input power and Φ_e is the optical power that leaves the system in the form of visible light (also known as the *radiant flux*). The user of a light source is interested in the perceived light that is characterized by the *luminous flux* (denoted by Φ_V)[1] Since the optical radiation is often indicated as the luminous flux, one has to convert the luminous flux to the optical power with

$$\Phi_e = \Phi_V / LE \tag{12.2}$$

LE is the *luminous equivalent*[2] in lm/W. The relation of the radiant and luminous flux and thus determining the value of the luminous equivalent is quite complex. Details are provided in Chap. 5.

The *efficacy* (donated η_V) relates the emitted luminous flux to the electrical input power. Thus the luminous flux Φ_v can be determined by:

$$\Phi_V = \eta_V \cdot P_{el} \tag{12.3}$$

Efficacy is also defined on system level—in this case the luminous flux obtained from the lamp or luminaire is divided by the total electrical input power (including

[1] Luminous flux is an important photometric quantity used to characterize a light source. Each radiometric quantity has a photometric counterpart. The subscript V is used to distinguish the photometric quantities from their radiometric counterparts. See further details in Chap. 5.

[2] According to IES RP-16-10 "Nomenclature and Definitions for Illuminating Engineering" this quantity denoted by LE here is officially known as "luminous efficacy of radiant flux". CIE S017/E:2011 "ILV: International Lighting Vocabulary" defines the same quantity under the name "luminous efficacy (of radiation)".

the power consumed by the electrical ballast or driver circuitry). Substitution of (2) and (3) in (1) gives the following expression for the thermal dissipation P_H as a function of the input power P_{el} and the (system) efficacy η_V:

$$P_H = P_{el} \cdot \left(1 - \frac{\eta_V}{LE}\right) \tag{12.4}$$

The definition of the thermal dissipation as provided by Eq. (12.4) based on the global energy balance does not distinguish between losses in the driver, the LEDs or the optical part of the system, for example caused by absorption in the phosphor, if present. However, since all the dissipated heat has to be transferred to the ambient air, it gives a reasonable first-order estimate of the thermal load of the system.

For such system-level approximations, one may calculate with typical values for *LE*: 280 lm/W for 2,700 K warm white light and 320 lm/W for 4,000 K neutral white light. If one does not know the exact luminous equivalent, 300 lm/W is a safe first-order estimate. However, since improving the luminous equivalent is one of the ways to improve the overall LED efficiency, the mentioned values might increase over time.

If we define the maximum temperature of the heat sink—that is typically the temperature underneath the LEDs—as $T_{heatsink}$ we can define the thermal resistance of the LED lamp or LED system as:

$$R_{th} = \frac{T_{heatsink} - T_{ambient}}{P_H} \tag{12.5}$$

We will use this definition in the remainder of the chapter to indicate the global thermal performance of the various LED systems.

12.1.2 Retrofit Bulb Applications

Incandescent light bulbs are available in a number of power levels, ranging from 15–200 W. The most common power levels are 60 and 100 W. In order to claim that an LED lamp can replace a conventional lamp of, e.g., 60 W both in the United States and in Europe, the LED lamp has to fulfill a number of requirements related to luminous flux levels, light distribution, and color quality.

In Table 12.3, an overview of flux levels that are required by the European DIM regulations is given. The Energy Star requirements for the US market are very similar.

The power levels that can be dissipated in this form factor depend on many things like the system efficacy of the lamp, the maximum allowable operating temperature and the design of the heat sink. Next to the required flux levels, Table 12.3 shows an estimate of the system efficacy that is needed to realize this flux level. For this estimate, we assume a system thermal resistance defined in Eq. (12.5) of 8 K/W, a maximum heat sink temperature of 85 °C in an ambient of 25 °C and a luminous equivalent of 300 lm/W. This allows a maximum thermal dissipation of 6.5 W for the total system, including driver and optical losses.

Table 12.3 Required flux and system efficacy (assuming a 6.5 W thermal dissipation and a lumen equivalence of 300 lm/W)

Incandescent electrical input power	25	40	60	100
Required luminous flux (lm)	249	470	806	1521
Required system efficacy (lm/W)	30	52	79	121

With the 2011/2012 state of the technology, 60 W bulb equivalent replacement lamps can be realized and are indeed on the market. However, in order to realize a 100 W bulb equivalent LED lamp, a system efficacy of 121 lm/W is required.

Figure 12.1 shows the *LED Efficacy Roadmap* as published by the Department of Energy [6]. Note that this predicts LED luminous efficacies at 25 °C at LED levels. This means that losses in the driver and the optics are not included in this figure. Due to these losses and the higher-operating temperature in a system, the resulting system efficacy is typically 25–50 % lower than the indicated LED efficacy at 25 °C. This means that a system efficacy of 121 lm/W with warm white LEDs will not be realized shortly.

Therefore, other improvements are needed to realize a 100 W equivalent LED lamp. Potential improvements are increasing the operating temperature at comparable performance and efficacy levels, improving the heat sink design and/or breakthroughs in system architecture that lead to very low losses. One of the obvious directions is to increase the form factor. By increasing the form factor, a big improvement in thermal performance can be realized. Of course, there is a risk that larger lamps do not fit in a number of luminaires.

In Table 12.4, a comparison is made between conventional and LED-based bulbs. The LED bulbs show a heat sink that is integrated in the lamp. The first generation of LED bulbs had a straightforward architecture of a heat sink with an LED board mounted on top of the heat sink. This results in a light distribution that differs from conventional lamps. Conventional lamps radiate light in all directions (omnidirectional), while the first generation of LED lamps mainly emitted light upwards. This causes undesired effects in some applications, especially when the lamp is mounted in a lampshade.

The latest generation on LED lamps resolves this issue by mounting the LEDs sideways instead of on the top of the heat sink. The result is a much more omnidirectional light distribution. This optical design also has a strong impact on the heat sink design and the appearance of the lamp. The yellow caps on the second-generation lamp contain phosphor that is mounted remotely from the LEDs. This gives an efficiency gain compared to the situation where the phosphor is mounted directly on the LEDs. Additionally, the Stokes losses that are generated in the phosphor can be cooled directly to the ambient air. With phosphor directly on the LEDs, the Stokes losses are transferred via the LED and the heat sink.

Next to the challenge of increasing the thermal performance, there is a strong drive to lower the LED system cost. In order to become a success in the market, the price

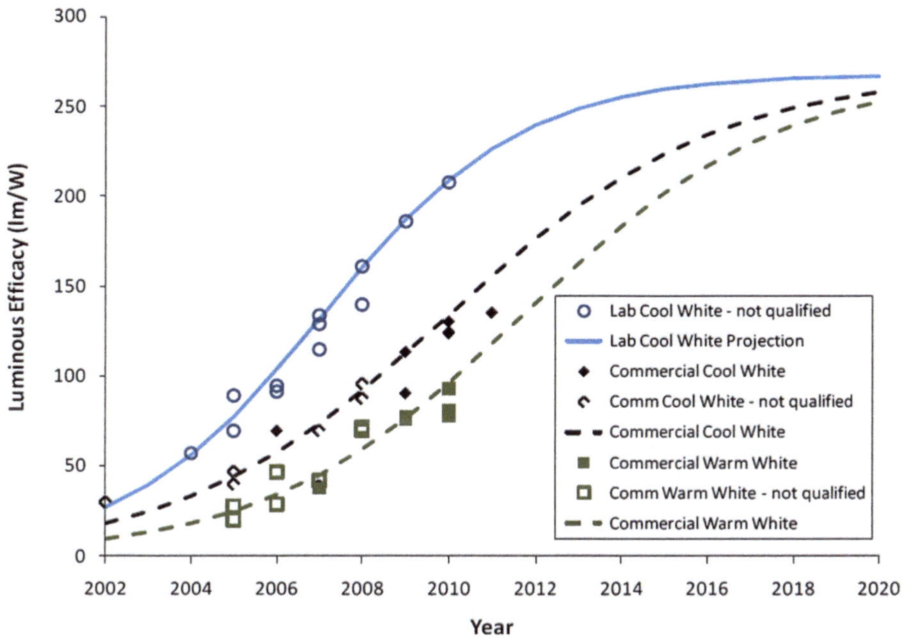

Fig. 12.1 LED efficacy roadmap at 25 °C and 35 A/cm^2 current density. (Solid-state lighting research and development: multi year program [6])

Table 12.4 A comparison of conventional and LED-based bulbs

Conventional	LED (1st generation)	LED (2nd generation)

should come down to levels comparable to the benchmark products: conventional incandescent and compact fluorescent lamps. Currently, the LED costs are dominant in the total cost prize of an LED lamp. However, with the current strong price erosion of the LEDs, the costs of the remainder of the system, including the heat sink, become relatively more important. Therefore, low-cost heat sink solutions with comparable performance are needed. See Chap. 9 for a more detailed discussion on heat sinks.

Table 12.5 A conventional and an LED-based PAR38 lamp

Conventional	LED

12.1.3 Retrofit Spot Applications

Compared to bulbs, spot lamps direct the light in a beam. These lamps are available in numerous sizes, power levels, beam angles, and supply voltages. In this chapter, as examples, we describe two commonly used lamp types, the PAR38 and the MR16. The performance of spot lamps is indicated with the *luminous intensity* measured in candelas (cd). This quantity describes the amount of flux in a certain direction in a certain solid angle. This quantity of course depends on the amount of luminous flux that is generated by the lamp, but also on the beam angle. A smaller beam angle typically results in a higher luminous intensity. Since different lamp types are usually available in different beam angles, we have to take care when comparing the performance of different lamps.

12.1.3.1 PAR38

The PAR38 lamps are relatively big and have a diameter of 120 mm. In conventional technology, they are available in various power levels. The most common power levels are 60 and 90 W in the United States, which is their main market. They are supplied with the mains voltage. A conventional and an LED-based PAR38 lamp is shown in Table 12.5.

Due to their relatively large form factor, we can dissipate a relatively large amount of heat in these lamps. The system thermal resistance is typically 3.0 K/W or lower, compared to 8 K/W for the 60 W bulb replacement. With this performance, the equivalent luminous flux levels are quite straightforward to realize.

12.1.3.2 MR16

On the other size of the range of retrofit spot lamps, we find the MR16 halogen lamps. They have a small form factor with a diameter of 50 mm and have a supply

Table 12.6 Conventional and LED-based MR16 lamps

Conventional	LED (passive cooling)	LED (active cooling)

voltage of 12 V AC. Low voltage halogen lamps are typically more efficient than mains voltage lamps. Available power levels for conventional halogen are 20, 35, and 50 W. Table 12.6 provides pictures of three different MR16 lamps.

Given the compact form factor and the better efficiency of halogen lamps, compared to standard incandescent lamps, the thermal challenges are much bigger for MR16 lamps. Next to the thermal challenge, one has to accommodate electronics in the compact housing in order to ensure compatibility with a large range of 12 V transformers in the market.

LEDs can provide the high brightness source that is needed to make an energy saving version of the MR16. Other energy saving light sources like compact fluorescent do not provide the brightness that is needed. Therefore, for MR16 lamps, there is no other energy saving technology available that could be integrated in the MR16 form factor. Hence, next to the technical challenges that the MR16 form factor provides, the opportunity for an LED-based MR16 is significant.

With the current state of the LED technology, a 20 W equivalent MR16 product has been developed and is available in the market. With the 2012 LED efficiencies, a 35 and 50 W equivalent lamps uses 7 and 10 W of electrical power respectively, resulting in a thermal power dissipation of ca. 5 and 8 W. These power levels are very challenging to dissipate in the MR16 form factor. In the market, this is solved in several ways. Some vendors choose to extend the form factor and produce MR16-like lamps that do not comply with the MR16 outline and thus do not fit in a number of fixtures. Another way to dissipate this power level in the compact MR16 form factor is by actively cooling it with an integrated fan. Both a passively cooled and an actively cooled MR16 are shown in Table 12.6.

Although integrating a fan in the MR16 form factor is a challenge in itself, it enables much higher power dissipation rates and flux levels equivalent with 50 W halogen lamps. Besides the physical integration of the fan, there are a number of additional challenges that have to be addressed. One has to ensure that the noise generated by the fan is kept to an absolute minimum in order to make it acceptable for customers. In addition, a lot of attention is needed to ensure a reliable operation of the fan over the lifetime of the system. Although these issues can be resolved, there is a strong market drive toward passively cooled lamps. With the 2012 LED roadmap, we can expect passively cooled 35 and 50 W equivalent lamps in the market shortly.

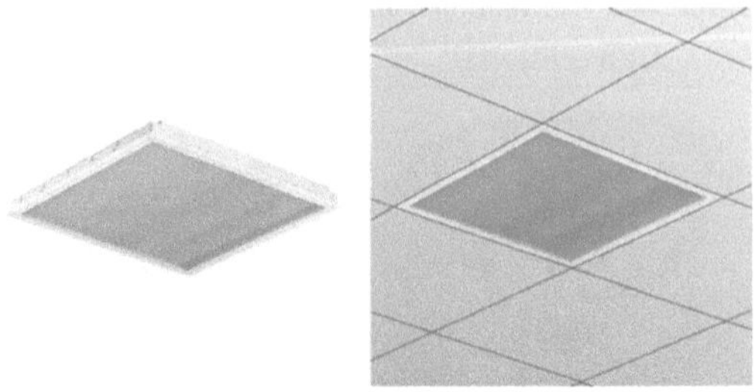

Fig. 12.2 Example of an LED-based office luminaire

12.2 Application in New Systems

Compared to retrofit LED lamps, new systems offer much more opportunity to optimize the system for the application of LEDs. Since the housing is usually exposed to the ambient air, often a heat sinking function is added to the LED System housing. This often allows for a thermally much more effective solution compared to retrofit lamps, where the fixture housing in most cases is not optimized for cooling.

Additionally, most of the time, there is no legacy that prescribes the exact form factor of the system. This results in much more design freedom to optimize the LED system.

12.2.1 General Indoor Illumination

The benchmark for general indoor illumination is lighting using fluorescent tubes. Fluorescent lighting provides high luminous flux level and high efficiency at a low cost. For this application, high brightness sources are not required. As a result, the systems have a relatively low power density. Therefore, the thermal challenges in this application field are limited. However, since the benchmark technology is very cost effective, there is a high pressure on cost. Figure 12.2 shows an example of an LED-based office luminaire. The luminaire consist of a large array of LEDs and is recessed in the ceiling. The recessed nature of this type of luminaires in some cases caused additional thermal issues. Some false ceilings are covered with insulation material that is seriously interfering with the cooling provisions in the luminaire. In this case, even the large, low power density recessed luminaires might run at elevated temperatures, resulting in lower lifetimes.

12.2.2 Indoor Spot Illumination

Indoor spot illumination is mainly applied in retail environment. The benchmark technologies are halogen and high-pressure discharge lamps. LEDs provide some clear advantages over these technologies. Compared to halogen, LED technology is

Table 12.7 Active cooling solutions for LED spot systems

Synthetic jet-based cooling system (Nuventix)	fan-based system (Sunon)

much more energy efficient. High-pressure discharge technology that has also been introduced as an energy efficient alternative for halogen has some other drawbacks. High-pressure discharge lamps have a relatively long run-up time and are difficult to dim. LED technology provides much more opportunity for control. For example, it is easy to switch or dim. These features can be used to generate dynamic lighting effects.

Spot illumination is a so-called etendue critical application. This means that application requires a high brightness at source level. This high brightness is required to generate a beam with a small beam angle. In principle, LED technology is very well suited to provide this high brightness. However, a high power density, combined with a high total system power, provides a serious challenge from a thermal point of view.

Another application requirement is compact design. The compact design, combined with high power requirement leads to the application of active cooling in this application domain. A number of vendors offer dedicated active cooling solutions for this application. Some pictures are shown in Table 12.7.

Next to fan-based systems, synthetic jet cooling is offered as an active cooling solution for LED lighting systems [7]. The main advantage of synthetic jet cooling is its lifetime. The most critical part of a fan is the bearing. A synthetic jet-based system has components that are less sensitive to wear out and therefore has a longer lifetime.

When active cooling is applied, noise is always a concern. One has to evaluate carefully what noise levels are acceptable in which application. Lasance et. al. [8] and [9] put considerable effort in reducing the noise levels of synthetic jet cooling by using the acoustical properties of the device. Up until now, fan cooling is often a bit more compact than synthetic jet cooling at comparable performance levels. For more details on active passive cooling solutions for LED systems refer to Chap. 7.

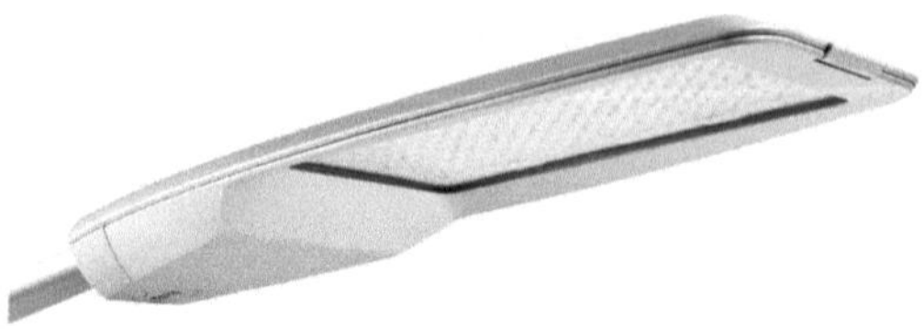

Fig. 12.3 Example of an LED-based outdoor luminaire for road lighting

12.2.3 Outdoor Illumination

Outdoor illumination is characterized by the combination of high luminous flux and tight spatial distribution requirements, combined with high efficiency and high reliability. Systems, based on high-pressure sodium lamps, are the benchmark in this area, although recently high-pressure metal halide finds its inroads as well. This technology combines the high efficiency of High-Pressure Sodium lamps (HPS) with the high lighting quality of Metal Halide lamps (MH). Both technologies reach lifetimes of 12,500 hrs for MH till 17,000 hrs for HPS (L5 life). Since maintenance for outdoor systems is expensive and can lead to disrupted traffic or other inconvenience, the big advantage of LED systems is the additional reliability it could provide. In order to ensure this high lifetime, the operating temperature of outdoor systems has to be managed very well and is often lower than other LED-based systems. An example of an LED based outdoor luminaire is shown in Fig. 12.3.

The luminaire in Fig. 12.3 provides a luminous flux of 20,000 lm, which is equivalent to a 250 W HPS lamp at a comparable efficiency. The main advantage of this system, compared to a conventional system, is its lifetime of 60,000 h. In order to realize this lifetime, the solder temperature, which is critical in this system, is limited to 75 °C. The cooling system is actually quite straightforward. The die cast aluminum housing provides sufficient cooling to maintain this temperature, provided that the thermal interfaces along the LED's junction-to-luminaire heat-flow path (as in any other LED application) are of good quality. Harsh environments may impose special requirements in such applications (see Chap. 15 on LEDs used in harsh environments). Of course, for good thermal design, numerical codes should be used to optimize the heat spreading capabilities of the housing, such as the optimum thickness given the area.

12.3 LEDs in Special Illumination Applications

The applications described below are not considered to be general illumination. However, they are important markets and interesting from a thermal point of view. Automotive lighting has very strict reliability requirements in combination with high ambient temperatures, especially for automotive front lighting.

Entertainment lighting requires very high flux and power levels, in combination with a very high power density, due to the optical requirements.

12.3.1 Automotive Lighting

Automotive lighting has very high reliability requirements in harsh environments. Especially for under-the-hood applications like front lighting, one has to take into account that the ambient in which the lighting system has to operate can become as high as 105 °C. Furthermore, automotive front lighting has high optical requirements, resulting in relatively high power densities at the LED die level.

As a result, automotive front lighting is one of the most challenging LED applications in terms of thermal management. In order to face these challenges, the LED packaging technologies for automotive front lighting often differs from packaging technologies for general lighting. For example, the Lumileds Luxeon Altilon package, which is developed for this application, has a maximum case temperature of 130 °C [10]. See also the Chap. 15 on LEDs in harsh environments.

Fig. 12.4 Example of an LED-based entertainment luminaire

12.3.2 Entertainment Lighting

Entertainment lighting combines high flux and power levels with high power density requirements, again due to the optical constraints of the systems. Figure 12.4 shows an example of an LED-based entertainment luminaire.

The current LED-based entertainment luminaires contain active cooling. Future generations probably require advanced heat spreading solutions like heat pipes, vapor chambers, or even liquid cooling to spread the heat from the highly concentrated sources, in combination with active cooling to transfer the heat to the ambient air.

12.4 Challenges in LED Applications

The main thermal challenges in LED applications are low-cost heat sink solutions, active cooling solutions and finding the optimal stack of the LED PCB (Printed Circuit Board), the thermal interface material and the heat sink.

The current LED heat sinks are often die cast solutions. Due to the need of the processing of die cast solutions, like machining and applying a coating, there are limited opportunities for cost cutting. Since there is a strong cost pressure, especially on the retrofit LED lamps, there is a need for low-cost heat sink solutions.

Additionally, still part of the lighting applications cannot be served with LEDs, mainly due to thermal constraints. These applications are typically a combination of compact system with a high flux level, like the 70 W compact HID spot luminaires for retail applications. High flux entertainment and stadium lighting also cannot be covered with LED technology yet. For this, advanced active cooling system could provide a solution. However, the current active cooling systems are often not optimal. Fans are a potential reliability risk, especially in harsh environments, while other technologies like synthetic jet cooling are still relatively expensive and large. If these applications are going to be replaced with LED technology, one has to integrate an advanced cooling solution in the system that required the replacement of the whole luminaire. Due to thermal constrains, it is very unlikely that we will be able to develop an LED-based replacement for an HID lamp.

Another challenge is to find the optimal stack of the right LEDs, the LED PCB, a thermal interface material to connect the PCB to the heat sink. Currently, LEDs are often applied on insulated metal substrates. However, these are relatively expensive. Lower cost PCB solutions like FR4 with thermal vias often can be applied as well. However, the thermal spreading capability of FR4 and the low stiffness of the PCB results in high requirement for a thermal interface material, possibly combined with a more complex mechanical fixation of the LED PCB to the heat sink. See also the Chap. 9 on heat sinks.

Although a large part of the lighting applications can be equipped with LEDs, there is a continuous challenge to develop better and more cost-effective solutions.

References

1. International Standard IEC 60064, Edition 6.3 2005–05, Tungsten filament lamps for domestic and similar general lighting purposes—Performance requirements
2. International Standard IEC 60630, Edition 2.5 2005–04, Maximum lamp outlines for general lighting lamps
3. International Standard IEC 60357, Third edition 2002–11, Tungsten halogen lamps (non-vehicle)—Performance specifications
4. Commission Regulation (EC) No 244/2009 of 18 March 2009 implementing Directive 2005/32/EC of the European Parliament and of the Council with regard to ecodesign requirements for non-directional household lamps
5. ENERGY STAR® program requirements for integral LED lamps—Eligibility criteria-Version 1.4

6. Solid-state lighting research and development: multiyear program plan, March 2011 (Updated May 2011), Department of Energy
7. Mahalingam R, Heffington S, Jones L, Williams R (2007) Synthetic jets for forced air cooling of electronics. Electronics Cooling 13(2)
8. Lasance C, Aarts RM (2008) Synthetic jet cooling part I: overview of heat transfer and acoustics. Proc. SEMI-THERM 24, San Jose, pp 20–25
9. Lasance C, Aarts RM, Ouweltjes O (2008) Synthetic jet cooling part II: experimental results of an acoustic dipole cooler, Proc. SEMI-THERM 24, San Jose, pp 26–31
10. Luxeon Altilon automotive forward lighting source datasheet, DS 66, 2009

Chapter 13
Thermal Challenges in LED-Driven Display Technologies: The Early Days

Kazuaki Yazawa

Abstract This section discusses a variety of thermal challenges in LED-driven displays, especially in early development phase of the LED displays. The unique fundamentals of the displays are their wide screen size, flat and thin envelop, light weight for a wall mount, and the requirements of bright and high contrast on screens. A historical background in the introduction is followed by the technological challenges of the system primarily based on a passive air convection cooling for LED displays. Even with active air cooling solution, this is still important for the initial design aimed for energy efficient and low acoustic noise systems. The following two sections "Thermal Spreading and Transient for Design" and "Active Cooling of Laser LED for Projection Displays" discuss the technologies to handle the device-focused thermal issues in conjunction with thermal imaging characterization of the LED device.

13.1 Introduction

The plasma flat display was invented in 1964 [1]. The definition of a flat panel display is a bit obscure, but it is getting popular to name it "Wall-mount TV" in consumer electronics. Television sets started their long history as a symbol of modernization among electronics with the cathode ray tube (CRT), also known as the Braun tube, invented in 1895 by Karl Ferdinand Braun [2]. The television set symbolized modern life in the 1950s and 1960s. Wall-mounted televisions were already visible in comics and science fiction of the 1960s, but it took more than three decades to become a reality for ordinary life. Nowadays, it is common to see large screen TVs on the walls of electronics stores, public houses, coffee shops, stations/airports, etc. Before modern liquid crystal display (LCD) [3] screens, some systems used wall-embedded projectors with optical magnification to create a flat screen image, called a rear projection TV [4]. Others used a direct projection on a flat wall [5].

K. Yazawa (✉)
Birck Nanotechnology Center, Purdue University, 1205 W. State Street,
West Lafayette, IN 47907-2057, USA
e-mail: kyazawa@purdue.edu, kaz@soe.ucsc.edu

University of Santa Cruz, Santa Cruz County, CA, USA

C. J. M. Lasance, A. Poppe (eds.), *Thermal Management for LED Applications*, Solid State Lighting Technology and Application Series 2,
DOI 10.1007/978-1-4614-5091-7_13, © Springer Science+Business Media New York 2014

LCD itself is a matrix of pixel-level shutters without illumination, hence LCD requires back lighting to generate a bright and beautiful screen. The replacement of the plasma displays (themselves a matrix of very small CRT-like structures to emit electrons) by LED back-lit LCD displays [6] seems like a natural progression, since the components of plasma displays have larger dimensions and require a more power to drive. Another method of lighting with LEDs is to use edge-lit technology that uses concentrated in-line arrays of LED with optics to spread the light across the entire screen.

13.2 Flat Panel Displays and Thermal Challenges

LEDs serve as controlled light sources in terms of color (wavelength) and intensity of luminescence. Using this unique characteristic, the color gamut has been enhanced [7], which was difficult using the older cold-cathode fluorescent lamps (CCFL) [6] that only adjusted the wavelength range by changing the phosphor characteristics. LEDs, however, cover the entire spectrum, providing richer colors.

The challenge of using LEDs for display applications is the thermal drift of the diode characteristic and the color spectrum. This means that the device temperature shift is shown directly on the screen. As a consequence, the quality of LED back-lit LCD displays depends highly on proper thermal management.

Since the material is based on band gap emission, it is wavelength dependent, using GaAsP for red, GaP for green, and InGaN for blue [8]. Fortunately, these junctions can be small, since the quantity of photons created at the junction is such that human eyes can easily perceive the information. This is of interest because the materials used in the LED junction are still expensive. The illumination at the junction starts out quite small, but then optically spreads out on the screen. However, the concentrated energy conversion causes a loss that transforms to heat, causing temperature-related concerns such as a color drift and reliability issues of both active materials and interconnects between the device and the substrate (Fig. 13.1).

Back-lit LCD displays started out using CCFLs as their light source, but LED lighting is quickly getting popular in the marketplace. Many of the recent consumer LCD TVs are using white LEDs, see Fig. 13.1b for more details. In the early days, to attain high-quality color, white light was not made by a single device but rather three or more, so that red (R), green (G), and blue (B) are mixed together to generate a flat spectrum. Since a single wavelength originates from the band gap of the active layer in the LED devices, the intensities of each R, G, and B are not the same due to the differences in the materials. Therefore, sometimes three (RGB) are used, but alternately four (RGGB) or five (RRGGB) devices can work together as an array [7]. Considering the different power consumptions and different energy conversion from electron-to-photons, the losses, i.e., heat generation from the junctions, are different. Consequentially, the junction temperatures are also different. The challenge becomes how to balance the temperature across the array, or how to spread heat to the heat sink. A typical solution could be a thermal spreader attached to the structure where the

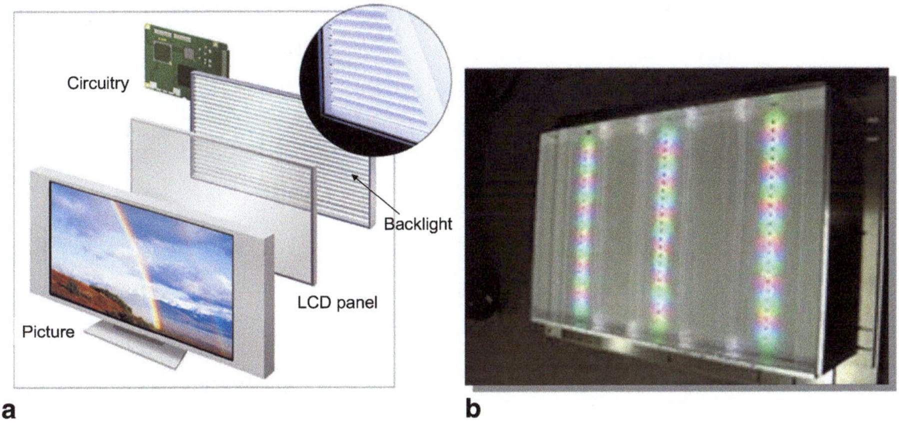

Fig. 13.1 **a** LCD + back-lit and **b** a backlight assembly for high-quality LCD backlighting. (Reproduced from [9] with permission of the author)

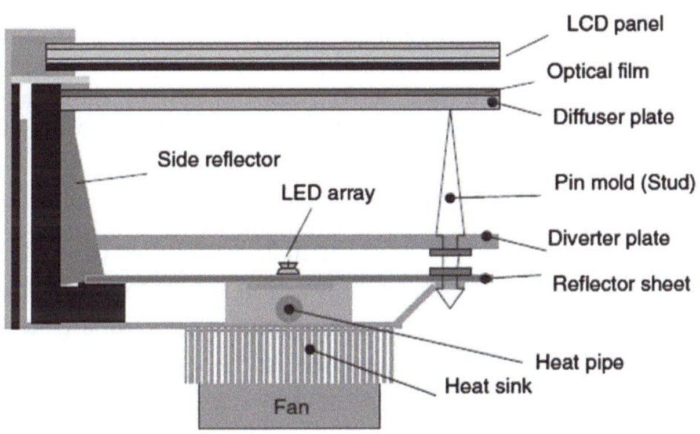

Fig. 13.2 Mount example of LED array. (Reproduced from [7] with permission of the author)

LEDs are mounted. The following figure shows one example of a high-performance backlit, see [7] using a heat pipe as a passive heat spreader (Fig. 13.2).

In a large-screen display, such as 60 in or larger, the thermal impact of the dispersed heat sources becomes a serious issue. The heat generated from the LEDs needs to dissipate to the ambient temperature. Before the heat can be transferred to the outside of the enclosure, the heat generated at the LED junction should conduct to the substrate. Unfortunately, the substrate is usually made of organic materials with some copper tracks. Since the volumetric ratio of copper remains relatively low, the effective thermal conductivity of the substrate is not very high. To calculate the effective thermal conductivity, various methods can be found in the literature, see e.g., [10]. In practice, a heat spreader plate is necessary in order to properly transfer the heat toward the ambient. Sometimes, such plates can take on another function as part of the mechanical structure.

Convection from the heat spreader plate to the local air is the final step in the heat transfer chain. Forced convection using cooling fans or other air-moving devices is

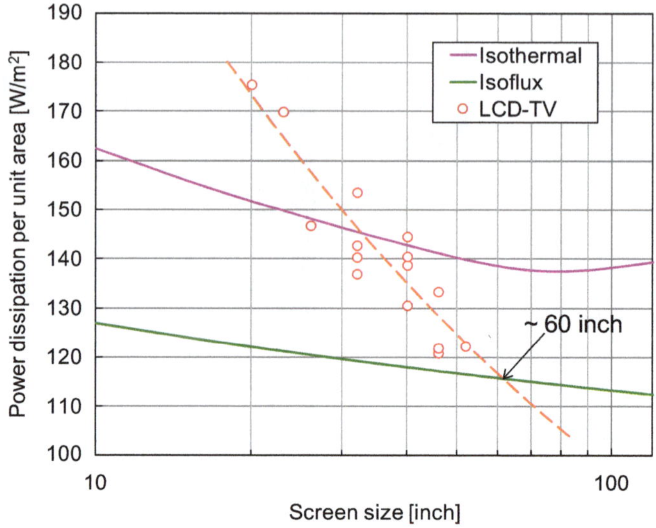

Fig. 13.3 Power dissipation limits by screen size (maintaining an aspect ratio of 16:9) based on analytic prediction considering natural convection with assuming black body radiation to the ambient temperature wall. The mean surface temperature is 18 °C over ambient. Data points and trends are based on the catalog specs of LCD TVs sold in 2006. The size of 60 in screen display crosses the passive cooling limit

the best method to transfer the heat, at the cost of noise, reliability, and price. In the majority of cases, natural convection is employed. Natural convection theory shows that a large screen system could have sufficient surface and height to create enough buoyancy to maintain an average surface temperature of 18 °C above the ambient as seen in Fig. 13.3. This large screen suggests a size of 60 in for wide screen or larger sizes.

Reference [11] presents a unique passive heat transport mechanism by introducing vertically oriented channels. The optimum design for a particular system of 1 m height might result in around 10 mm × 10 mm of space for each channel. The channels create a fully developed flow, for which the Nusselt number is much higher than in a vertically oriented single wall. The heat source wall and the opposite facing wall have different heat fluxes. In total, this configuration nearly doubles the heat dissipation performance for natural convection. Equation (13.1) may be used to get an impression of the power that can be dissipated (Fig. 13.4):

$$Q = N\left(\frac{1}{\psi_{ch1}} + \frac{1}{\psi_{pw} + 1/(1/\psi_{ex} + 1/\psi_{ch2})}\right)(T_b - T_a) \tag{13.1}$$

where Q is total heat dissipation, N is the number of channels, T_b is heat source temperature, T_a is ambient temperature, ψ_{ch1} is a channel convective thermal resistance, which shows the convection from the interim wall surfaces to the air flow in the channel. ψ_{pw} is passive wall convective thermal resistance, and ψ_{ex} and ψ_{ch2} are

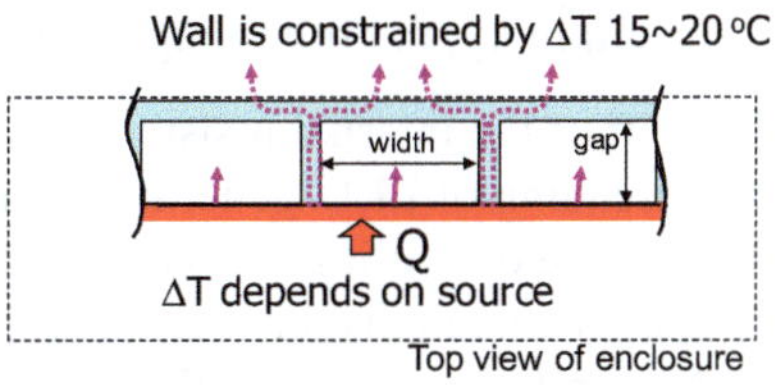

Fig. 13.4 The multichannel wall and thermal resistance network

the thermal resistances conducting the heat through the exterior wall and the bridge wall across the channels, respectively.

The uniformity of color and local brightness on the screen are important for the quality of the display [12]. Thermal nonuniformity occurs naturally due to the above phenomena, and the problem intensifies with larger screens. Also, due to the edge cooling effect, the upper/center section of the screen tends to become a hotspot, even if all the LEDs are powered equally. The temperature difference causes a difference in drift in terms of the color and intensity of light emission (due to the internal resistance change). This cannot be compensated with a fixed rate weighting of illumination since the load is always changing. A sophisticated solution of this passive nonuniformity may be given by local temperature sensing and local feedback drive for back-lit LEDs. Figure 13.5 shows the temperature nonuniformity prediction of a vertically oriented screen.

New technologies enable much stronger light emission from solid-state devices based on the optical pumping or oscillation structures. Using this laser LED technology, building laser LED projection displays [13] becomes feasible. Considering the small form factor of the laser LED, the heat flux generated by the device is nearly 10 W/mm^2 [14]. A challenge for the display is to maintain the wave lengths, so that color does not visibly change due to self-heating, while simultaneously maintaining the spectrum. For this particular case, active refrigeration cooling might be necessary. Thermoelectric devices are typically used to adapt to the local and the very quick temporal thermal load changes. However, at this moment in time, this technology has no mainstream applications.

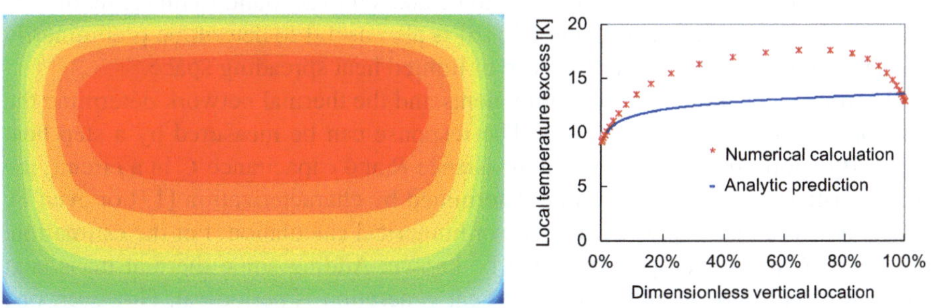

Fig. 13.5 **a** Thermal map based on a full 3D computational fluid dynamic (CFD) numerical simulation and **b** vertical local temperature plot with 2D analytical prediction for an example of a 1 m-tall wide screen. The discrepancy between numerical calculation and analytical prediction on this plot is due to the 3D nature of the airflow nature near the edges

In summary, the following thermal challenges need to be considered:

1. Drift of the internal resistance and wavelength of the devices.
2. Spreading of LED-generated heat.
3. Balance of cooling for back-lit RGB LEDs.
4. Larger screen displays dissipates more heat.
5. Nonuniformity due to the natural convection and edge effect for large screen displays.
6. Thermoelectric cooling for high-power LEDs for future displays.

13.3 Thermal Spreading and Transient for Design

Spreading thermal transport contributes to the system thermal resistance in an electronics system design. Due to the complexity, the spreading thermal resistances are neglected or not properly designed in some cases. Moreover, it is important to know that the thermal resistance changes by the relation of internal thermal conductivity to the external heat transfer coefficient, i.e., Biot number. Often, a closed formula [15] is used to determine the spreading thermal resistances. The model itself is proven to be accurate for an engineering prediction, but the boundary conditions need to be accurate to use this model for proper design. For example, the model is only valid for one-sided heat transfer, and there are limits to the source area/heat sink area ratio.

$$\psi_s = \frac{\sqrt{A_P} - \sqrt{A_s}}{\lambda\sqrt{\pi A_P A_S}} \frac{\beta\lambda A_P \psi_E + \tanh(\beta L)}{1 + \beta\lambda A_P \psi_E \tanh(\beta L)} \qquad \beta = \frac{\pi^{1.5}}{\sqrt{A_P}} + \frac{1}{\sqrt{A_S}} \tag{13.2}$$

where ψ_S is spreading thermal resistance, ψ_E is substrate bulk thermal resistance, λ is the thermal conductivity of the spreader. L is the thickness of the spreader, and A_P and A_S are the areas shown in Fig. 13.6. Designing sparse heat spots such as LEDs can be considered as a lateral 2D thermal diffusion problem to solve. However, spreading thermal resistance still needs to be taken into account. In this case, the 45-degree rule or a more accurate model [16] is also useful to determine preliminarily the spreading contribution for the case of limited heat spreading space.

Thermal capacitances of the components and the thermal network determine the temporal system thermal response. The response can be measured by a step heat input. By approximation to a single resistance R and capacitance C in a circuit, the time constant of the system can be determined by characterization [17] or even by an RC network model [18] by analytic or numerical calculation. For the estimation, the R and C of the components must be known. Adding some thermal mass next to the hot spot may be effective if the heat generation is periodical. However, one needs to remember that this approach to reduce peak temperature by adding mass also creates additional design trade-offs such as extra space, additional material, and so on (Fig. 13.7).

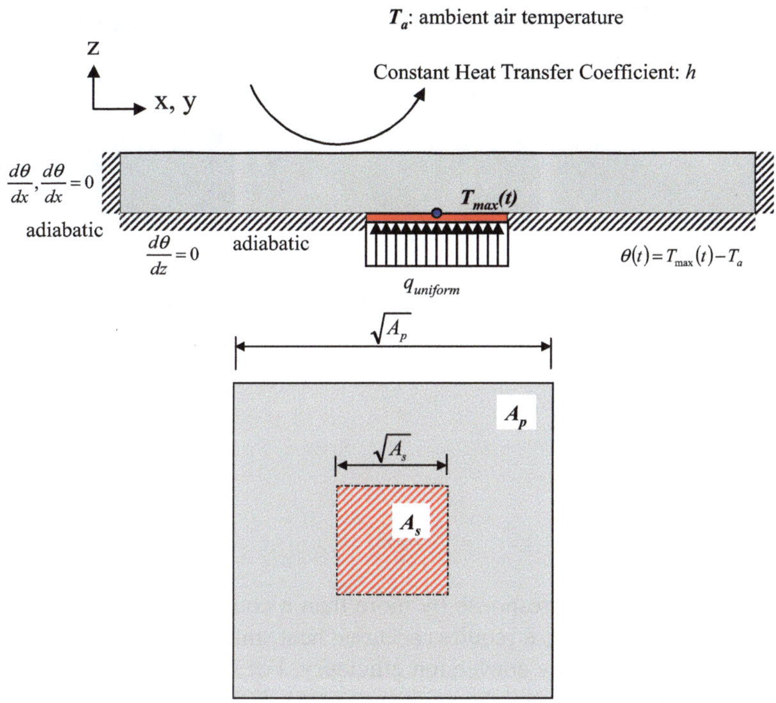

Fig. 13.6 Example of a transient thermal spreading in a low-profile solid

Fig. 13.7 Thermal time responses of high-heat flux devices with appropriate thermal spreaders

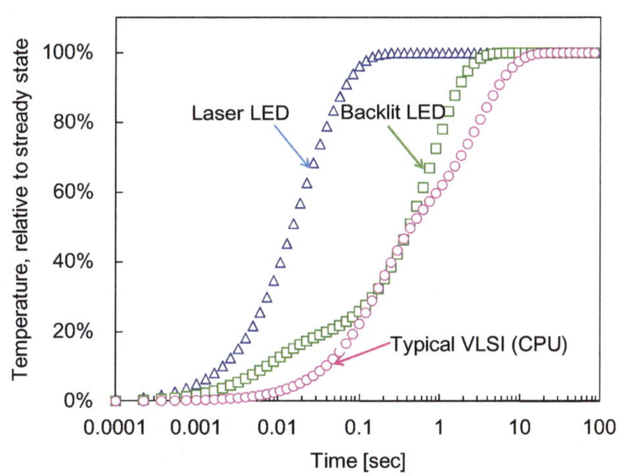

13.4 Active Cooling of Laser LED for Projection Displays

Super heat flux devices dissipating heat in the range of 10 W/mm^2 requires pumping a lot of heat to the heat sink. In these cases, thermoelectric devices may be considered for thermal management. Pumping electrons to release the heat energy allows very

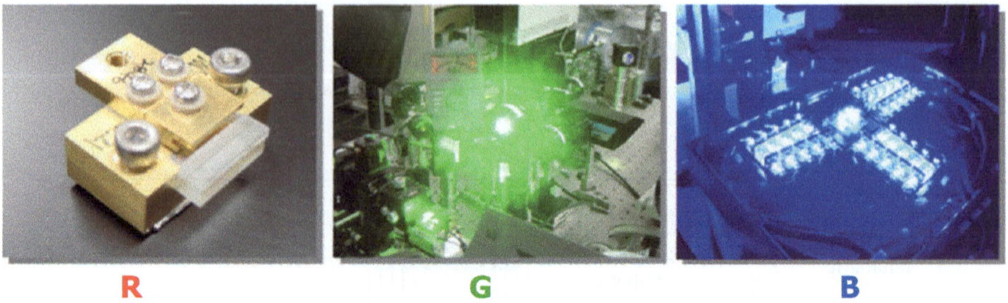

Fig. 13.8 Laser LED devices. (Reproduce from [10] with permission of the author)

Fig. 13.9 Example of a multistage thermelectric module – right. (Reproduced from permission of Komatsu Electronics Ltd. [20])

precise control and quick response by more than a couple of orders of magnitude in time. On the other hand, it requires accurate heat sink design and module design due to the moderate energy conversion efficiency. For such higher power fluxes, a multistage thermoelectric module can be an option. Figure 13.8 shows a two-stage thermoelectric module. Thermoelectric cooling can be fast as the Peltier effect [19] is a surface effect. Thus, thermoelectric cooling could be the preferred solution for quick temperature control. On the other hand, it should be realized that Joule heating [19] and thermal conduction lower the efficiency quite a bit, in such a way that in some cases water cooling by microchannel heat sinks becomes a viable solution. Of course, this is not an option for consumer LED–LCD displays (Fig. 13.9).

13.5 Thermal Characterization of the Devices and Systems

Thermal characterization is necessary to ensure whether the system is properly built from a thermal management point of view, not only for the LED or light sources, but also for the other electronic components, as well as for the enclosure surface (e.g., touch temperatures).

The preferred measurement technique to acquire the temperature data in space and time is using thermocouples. Sometimes, infra-red (IR) imaging is used to find the hot spots in the system or in a subsystem. IR imaging can also be useful at the package level or even device level. For these smaller features, special optics such as a microscope and high-sensitivity imaging sensors are required.

At the package level, understanding of the thermal structure is helpful to verify the thermal resistances and thermal capacitances of the components as well as the contact/interface thermal resistances. The methods discussed in [21] and [22] were

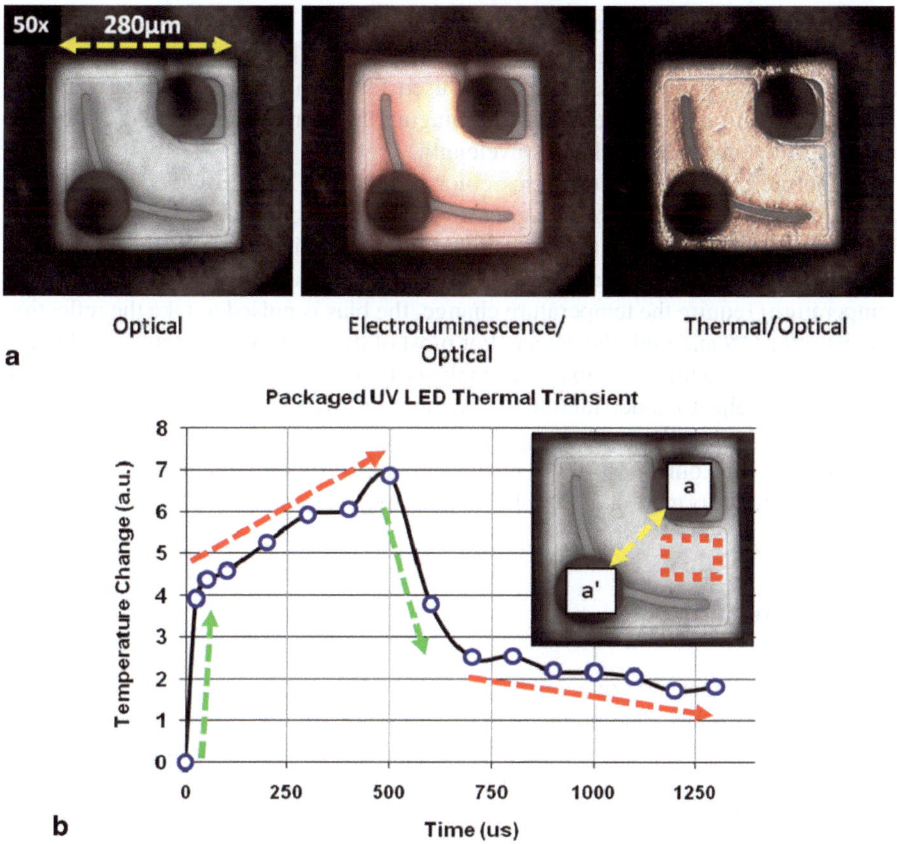

Fig. 13.10 Image example of LED device. **a** Optical image, **b** light emission from the device, **c** thermal signal intensity map on optical image, and **d** thermal time response of LED. (Reproduced from [27] with permission of the authors)

developed based on deconvolution of the time series of the temperature data and subsequent fitting of the data with series of R-s and C-s in a Cauer-ladder, which is an electrical filter topology. An example at the device level is shown in [22].

These system or package-level thermal characterizations are not unique to displays. However, the device thermal characterization requires special treatment. Unlike other electric components, an LED emits the photon with a high energy conversion efficiency. Some significant work related to thermal characterization has been described in literature [23, 24]. Further interest in the thermal profile in the device itself led to new developments in the field of thermal characterization methodologies. With IR imaging, filtering the emitting wave length allows one to extract the thermal emission only. Because the amount of photons emitted from the device is not enough to obtain a good signal-to-noise ratio at near room temperature, a certain time of accumulation is necessary. Because of this, IR imaging is often not fast enough to capture the device response. Thermoreflectance imaging methods have been developed for this type of device characterization. This technique is based on the fact that light reflectance changes with temperature. This method was discovered already half a century ago and has been used for sensing point thermal response. However, recent enhancement enabled imaging the thermal profile. The relation is unique for a certain material and wavelength.

A thermoreflectance system uses color-filtered LED and a CCD camera with a shutter. These lighting and imaging devices are synchronized with a signal for biasing the device under test (DUT). This "lock-in" technique enables high-resolution and high-speed transient characterization [25]. Since the reflectance coefficient (by temperature) require the temperature change, the bias is pulsed to take the reflection intensity for ON and OFF the device. For most of the cases, such electronics devices contain layer structures. Numerical analysis based on characterizing the surface temperature helps to understand the temperature distribution in the layers [26].

The interested reader is encouraged to consult Kendig and Shakouri [27] and Vermeersch and Shakouri [28] for the separation of electro luminescence and reflective thermal signal information. Examples of the images are shown in Fig. 13.10.

13.6 Summary

In this chapter, thermal concerns and solutions have been discussed with a focus on LEDs in displays. The uniqueness of LED back-lit displays is explained from the needs and history of display systems. Due to the tight coupling of color and brightness of LED, the thermal management contributes significantly to the quality of the screen. Passive thermal spreading in the device mounts and in the system is a major design challenge. For very high-heat fluxes as occur on LED laser-based applications, thermoelectric cooling might offer a potential solution to ensure proper thermal management, but it also requires additional technical considerations. Finally, these designs can be investigated using recently developed thermal characterization technologies such as thermoreflectance imaging.

References

1. http://www.life123.com/technology/home-electronics/plasma-tv/when-was-the-first-plasma-tv-invented.shtml. Accessed 29 May 2013
2. Kurylo F, Susskind C (1981) Ferdinand Braun: a life of the nobel prizewinner and inventor of the cathode-ray oscilloscope. The MIT Press, Cambridge
3. Kaneko E (1987) Liquid crystal TV displays (advances in opto-electronics). Springer, Tokyo
4. Walker BH (1998) Optical engineering fundamentals. The International Society of Optical Engineering (SPIE) Press, Bellingham
5. Brennesholtz MS, Stupp EH (2008) Projection displays, 2nd edn. Wiley, New York
6. Kobayashi S, Mikoshiba S, Lim S (2009) LCD backlights. Wiley, West Sussex
7. Kakinuma K (2006) Technology of wide color gamut backlight with light-emitting diode for liquid crystal display television. Jpn J Appl Phys 45(5B):4330–4334

8. Takahashi K, Yoshikawa A, Sandhu A (eds) (2010) Wide bandgap semiconductors (fundamental properties and modern photonic and electronic devices). Springer, Berlin
9. Kubota SR (2007) Thermal challenges deriving from the advances of display. In: Gariimella SV, Fleischer AS (eds) Proceedings of thermal challenges in next generation electronics systems. Millpress, Rotterdam, pp 23–24
10. Lemczyk TF, Mack BL, Culham JR, Yovanovich MM (1992) PCB trace thermal analysis and effective conductivity. ASME J Electron Packag 114:413–419
11. Yazawa K, Nishino Y, Nakagawa S, Ishizuka M (2006) Modeling of a 3-dimensional conjugate heat transfer on the channel-composite-wall for electronics cooling. Proceedings of 2007 ASME international mechanical engineering congress and exposition, IMECE2006-13524
12. West RC, Takagi Y, Eberle S, Harbers G, Tan TW, Chan CE (2003) High brightness direct LED backlit for LCD-TV. SID Symp Dig 34:1262
13. Chellappan KV, Erden E, Urey H (2010) Laser-based displays: a review. Appl Opt 49(25): F79–F98
14. Beach R, Benett WJ, Freitas BL, Mundinger D, Comaskey BJ, Solarz RW, Emanuel MA (1992) Modular microchannel cooled heatsinks for high average power laser diode arrays. IEEE J Quantum Electron 28(4):966–976
15. Lee S, Song S, Au V, Moran KP (1995) Constriction/spreading resistance model for electronic packaging. Proc 4th ASME/JSME Therm Eng Joint Conf 4:199–206
16. Vermeersch B, De Mey G (2007) A fixed-angle heat spreading model for dynamic thermal characterization of rear-cooled substrates. Proceedings of 23rd SEMI-THERM symposium, pp 95–101
17. Mentor graphics. http://www.mentor.com/products/mechanical/products/t3ster/. Accessed 29 May 2013
18. Yazawa K, Yamashita T, Kuroda H (2007) A study of transient thermal spreading of VLSI packaging with multi scaling. Proceedings of ASME international mechanical engineering congress and exposition, IMCECE2007–41658
19. Rowe DM (ed) (2006) Thermoelectrics handbook: macro to nano. CRC Press, Wales
20. http://www.kelk.co.jp/english/thermo/index.html. Accessed 29 May 2013
21. Farkas G, Haque S, Wall F, Martin SS, Poppe A, Vader QV, Bognar G (2004) Electric and thermal transient effects in high power optical devices. Proceedings of 20th SEMI-THERM symposium, pp 168–176
22. Ezzahri Y, Shakouri A (2009) Application of network identification by deconvolution method to the thermal analysis of the pump-probe transient thermoreflectance signal. Rev Sci Instrum 80:074903
23. Lasance CJM, Poppe A (2009) LED thermal standardization a hot topic. Electronics Cooling Issue May, 2009. on-line article, http://www.electronics-cooling.com/2009/05/led-thermal-standardization-a-hot-topic/. Accessed 29 May 2013
24. LED thermal characterization committee in Euro therm
25. Yazawa K, Shakouri A (2011) Ultrafast submicron thermoreflectance imaging. Electronics Cooling Magazine, Spring Issue 2011, ITEM Publications, pp 10, 12–15
26. Raad PE, Komarov PL, Burzo MG (2009) Numerical simulation of complex submicron devices. Electron Cool 15:18–22.
27. Kenig D, Shakouri A (2011) Thermal imaging of encapsulated LEDs. 27th IEEE SEMI-THERM symposium
28. Vermeersch B, Shakouri A (2010) Simultaneous thermal imaging of Peltier and Joule effects. Proceedings of IMAPS advanced technology workshop on thermal management

Chapter 14
Thermal Challenges in LED-Driven Display Technologies: State-of-the-Art

G. A. Luiten

Abstract Recent years have shown an explosion in different LED–LCD displays, and a corresponding proliferation of LED TVs in the consumers' living rooms. They offer a great picture, a great styling with a thin-form factor, increased functionality such as 3D and internet access, and great value for money. Likewise, the technical product design of LED–LCD TVs also made great strides in the past years. One of the key aspects of LED TV design is cooling of the LEDs, which also has evolved to cater for steadily increasing power densities in ever thinner product enclosures. In the lighting and computer industry, large finned heat sinks are used to address comparable thermal challenges. In the thin-form factor of a modern TV set, this is not feasible, and the only large surface areas available for cooling are located on the front and the back of the TV set. The role of in-plane heat spreading in LED–LCD thermal management has not been addressed so far. The scope of this subchapter is to clarify the thermal significance of heat spreading in the in-plane direction in the set, especially in relation to cost down initiatives using less LED packages at equal total light output.

14.1 The Need for Thermal Management

The need for system-level LED thermal management is well documented. Van Driel et al. [1] describe the reliability chain of an LED lighting product, mentioning temperature-related failure mechanisms along the chain from LED junction, LED package, LED substrate, and LED module to LED lighting system.

For an LED–LCD TV, the LED junction, LED package, and LED substrate levels are similar to the lighting world, and have similar reliability and performance concerns: At die- (junction) level, temperature affects both performance and lifetime. At high junction temperature, LED lumen efficacy is lowered and the color temperature is shifted, see e.g., Liu et al. [2] and Biber [3]. Furthermore, as shown in Lumileds WP15 [4], LED lifetimes in terms of lumen depreciation are dependent on junction temperature. On package level, Koh et al. [5] describe the degradation of

G. A. Luiten (✉)
Philips Research, Eindhoven, Netherlands
e-mail: wendy.luiten@philips.com

C. J. M. Lasance, A. Poppe (eds.), *Thermal Management for LED Applications*, Solid State Lighting Technology and Application Series 2,
DOI 10.1007/978-1-4614-5091-7_14, © Springer Science+Business Media New York 2014

epoxy materials in the LED optical system as a function of temperature, and on the substrate, the thermomechanical failure due to low-cycle solder fatigue in the LED solder joints is also well documented as mentioned by van Driel et al. [1].

In an LCD–LED TV, the LED module is the LCD module with the LED backlight. On module level, both the absolute temperature and the temperature distribution over the screen can lead to performance and lifetime issues. Performance issues are directly related to picture quality: Barnhoefer [6] mentions that the LED thermal management constraints are so severe, that brightness levels are primarily limited by LED power dissipation and cooling issues. Kim et al. [7] mention that nonuniform LED temperatures lead to unwanted spatial nonuniform lighting in terms of color uniformity and brightness, and also mention thermomechanical considerations. These are important in view of unwanted effects on the optical geometry. Furthermore, the LC temperatures are performance related to loss of the LC switching property at too high temperature (typically 60–70 °C), and to sluggish switching at too low temperatures, causing fast-changing pictures to display blurred, which is very performance degrading in modern high-resolution 3D TVs. Finally, aging of the polarizer material over lifetime is a temperature dependent issue (typically above $\sim$ 60 °C).

At TV system level, one of the important thermal management issues is the safety issue of touch temperatures or what the user experiences in touching the TV. Metal surfaces especially can be unpleasantly hot to the touch. This is reflected in the safety requirement to touch temperatures, which is justifiably much lower for metal touchable surfaces. According to safety standard IEC60065 [8], the safety requirement for TV products is 75 °C for metal and 95 °C for nonmetal surfaces [9]. This is well above the pain threshold of around 60 °C.

Thermal management of flat TV products and thermal management of LED systems are discussed in numerous papers. Luiten [9, 10] describes thermal management of a flat TV plasma monitor, including CFD simulations from component to system level. Sooshtari et al. [11] treat simulations and measurements on a plasma TV, with smeared electronics, and imposed external heat transfer coefficient, thereby ignoring the effect of boundary layer heating on the outside surfaces of the set. Biber [3] discusses spreadsheet calculations including the thermal effect on light output for a general LED case, but without deriving the heat dissipation in the LED from the light output and the electrical input power. Wilcoxon et al. [12] treat thermal management of a light engine for airborne applications. FEM calculations were performed with fixed LED heat dissipation, including geometrical details and heat spreading. Kim et al. [7] describe computational fluid dynamics (CFD) simulations on an edge-lit LED backlight. They address the effect of LED pitch, part thickness and thermal conductivity of the LED board and the LED heat sink, and conclude that a proper combination is crucial for a cost-effective backlight design.

Both Biber [3] and Wilcoxon [12] discuss the use of a finned heat sink to obtain a large surface area for the LED cooling. This is not feasible within the form factor of a thin flat display. In a thin TV, the only large surface areas available for cooling are the front and the back. The role of the chassis, or back of the backlight unit, in enabling thermal access to these cooling surfaces has not been addressed so far. First, the derivation of the LED heat dissipation from electrical input power and light output is demonstrated, and the effect of junction temperature-related efficacy loss is assessed. Next LED thermal management and heat spreading are discussed. Module (LED–LCD display) and System-(TV set) level CFD simulations are shown. It is shown that a cost benefit in terms of a larger LED pitch, or alternatively less LED packages with the same total heat dissipation, will be partially offset by an increased cost for heat spreading in the in-plane direction of the display.

14.2 LED-Related Thermal Issues

14.2.1 LED Power Balance

Only a fraction of the LED electrical input power will be converted into light, and the remaining majority of the input power will become heat. LED suppliers specify the LED efficacy of LED devices in lumens per Watt electrical input power (η_V). The power content (Watts) of this light is obtained through the luminous efficiency (LE).[1] LE is not a constant, but dependent on the wavelength or the color of the light. If the light has a color that is less visible to the human eye, the lumen efficiency is lower since more energy is needed to have the same effect on the human visual system. For white light typically $LE = 300$ lm/W.

To determine how much power (Watts) is converted into light (lumens) for an LED device, first one has to determine the generated luminous flux by multiplying the electrical input power by the η_V efficacy of the device, and then determine the energy content of the light, by dividing by the luminous efficacy LE. This leads to:

$$P_{opt} = P_{el} \cdot \frac{\eta_V}{LE} \tag{14.1}$$

The remainder of the electrical input power is heat, or:

$$P_H = P_{el} \cdot \left(1 - \frac{\eta_V}{LE}\right) \tag{14.2}$$

With a typical white LED device of efficacy $\eta_V = 75$ lm/W, and a typical white light luminous efficacy of 300 lm/W, this means $\eta_V/LE = 75/300 = 0.25$, or 25 % of the electrical input power is converted into light. The remaining 75 % of the electrical input power is heat. A schematic view of the power balance for a white LED is shown in Fig. 14.1.

The power balance shows that as the efficacy increases, the electrical input power is converted into more light and less heat. As stated earlier, the LED energy conversion efficiency as well as efficacy are dependent on the junction temperature. For reasons

[1] According to IES RP-16-10 *Nomenclature and Definitions for Illuminating Engineering*, the quantity denoted by *LE* here is officially known as "luminous efficacy of radiant flux." CIE S017/E:2011 *ILV: International Lighting Vocabulary* defines the same quantity under the name "luminous efficacy (of radiation)."

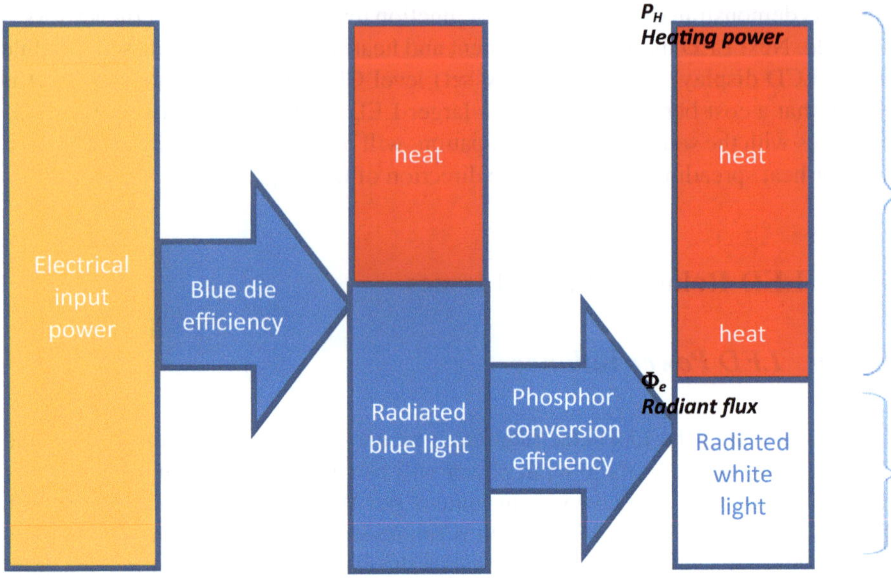

Fig. 14.1 Power balance of a white LED

of easy in-line measurements in production, LED suppliers prefer to specify the LED efficacy at a low junction temperature such as 25 °C room temperature. In practice, this is not a realistic specification condition and it is advisable to ascertain that efficacy/efficiency data for LED devices relate to a realistic operating junction temperature such as 95 °C. (In addition, as LED chips react thermally very fast, during in-line testing significant change of the junction temperature may take place.)

Liu et al. [2] show that the energy conversion efficiency

$$\eta_e = \frac{\eta_V}{LE} \tag{14.3}$$

as a function of junction temperature is nearly linear. This is confirmed by Biber in [3]. Liu measured the effect of junction temperature for three different samples, A, B, and C. All three samples were white high-power LEDs, but different in internal construction and obtained from different suppliers. A conversion of the values of Liu's samples A, B, and C, using $LE = 300$ lm/W, is shown in Fig. 14.2.

Figure 14.2 shows that with a junction operating temperature around 95 °C, typical for display LEDs, the difference in luminous efficacy at room temperature and at operating temperature can be very significant, dependent on the LED. A 10 °C excursions above or below the operating temperature are much less significant: for sample A, the most temperature dependent sample, the conversion efficiency of electrical input power to light is 27 % at 85 °C junction temperature, and 25 % at 105 °C junction temperature, and for sample B, the least temperature-dependant sample, the conversion efficiency remained stable at 23 %.

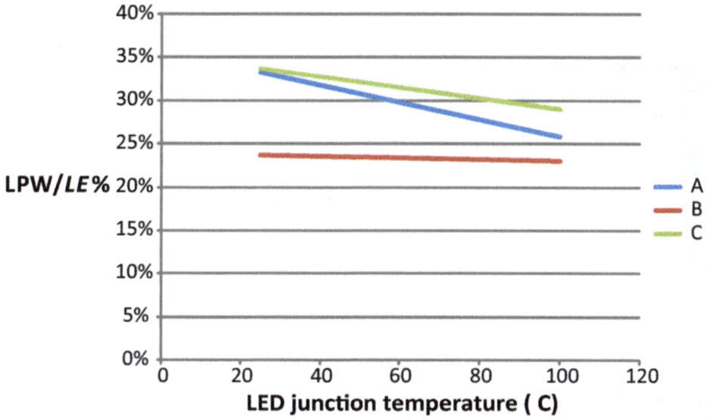

Fig. 14.2 Electrical input power to light power conversion, based on data by Liu et al. [2]

14.3 Thermal Management of an LED–LCD TV

14.3.1 LEDs in a TV Set

LEDs in a TV set can be used in two different manners: they can be the prime light sources in the backlight, or they can be used to create an ambient light setting supportive of the TV picture, referred to by the trade name *Philips Ambilight*®. The most common use of LEDs is in the backlight of an LED–LCD TV. In a direct-lit LED–LCD, the LEDs are placed in a matrix over the entire display area, and light falls directly through the optical foils and on to the TFT display. In an edge-lit display, the LEDs are located on one or more display edges and the LED light is coupled into the side of an optical light guide plate. The light guide plate is provided with an optical pattern, which reflects the light at 90 degrees angle. The light exits in the direction normal to the light guide plate onto the TFT display panel. Grading of the optical pattern provides uniform lighting over the display area. Schematic views of the direct-lit and edge-lit displays are shown in Fig. 14.3.

14.3.2 Dimming and Boosting

An important difference between the use of LEDs in a TV and in a lighting application is that they are used in very different time domains. In lighting applications, LEDs are typically used in a steady-state manner, with timescales in the order of hours or longer. In a TV, two different timescales prevail. Heating and steady-state behavior of the set as a whole is governed by the average power consumption, and this has a long timescale similar to domestic lighting applications. However, the moment-to-moment changing of the video content happens at a much higher frequency, and this creates an additional much shorter timescale.

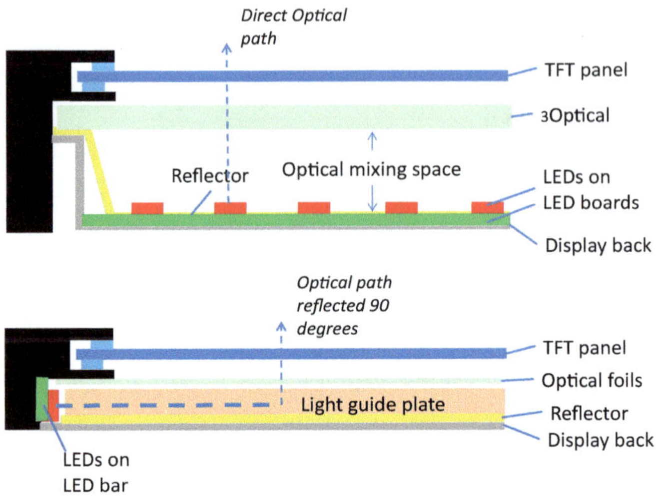

Fig. 14.3 Principle of direct-lit (*top*) and edge-lit (*bottom*) LCD–LED display

Figure 14.4 illustrates the shorter time constant for video material by means of the average shot length (ASL). ASL is a measure of importance for the cinematographic character of video material. It is the average time length of a cinematographic shot, in seconds. The Cinemetrix database [13] contains ASL data for more than 4,000 items of video material. The scatter plot of ASL data shows that most ASL are below 30's. More specific, the cumulative probability plot shows that 80 % of the ASL is below 15's, and 95 % is below 30's. In a plasma display panel (PDP), there is a direct one-on-one correlation between the displayed video content and the instantaneous power consumption. Sooshtari et al. [11] measured instantaneous power consumption of a PDP displaying an arbitrary movie, and showed a large difference between the instantaneous peak power consumption and the long-time averaged value.

In an LED–LCD panel, the light emitted from the display is a combination of the light emitted by the backlight and the state of the pixels in the TFT panel. If a TFT pixel is open, light goes through to reach the viewer. If a TFT pixel is closed, light is blocked. In display with a static light-emitting backlight, pixels are predominantly open in a bright image, and pixels are predominantly in the "closed" state if the image is dark. However, from a picture quality and from an energy consumption point of view, a static light output from a backlight is not optimal. In order to improve the visual experience, the backlight is usually set to emit as much light as possible. In a dark scene, the pixels in the LCD are shut to prevent this abundance of light reaching the viewer. Light leakage, however, causes black to be perceived as less black.

A common scenario is dimming the LEDs in dark scenes. This improves both the picture quality (deeper black) and reduces the energy consumption of the display. Instead of leaving the light on at the brightest setting and closing the TFT pixels to block it from the viewer, the light is turned down during darker scenes. Apart from improving the picture quality and this also lowers the average power consumption

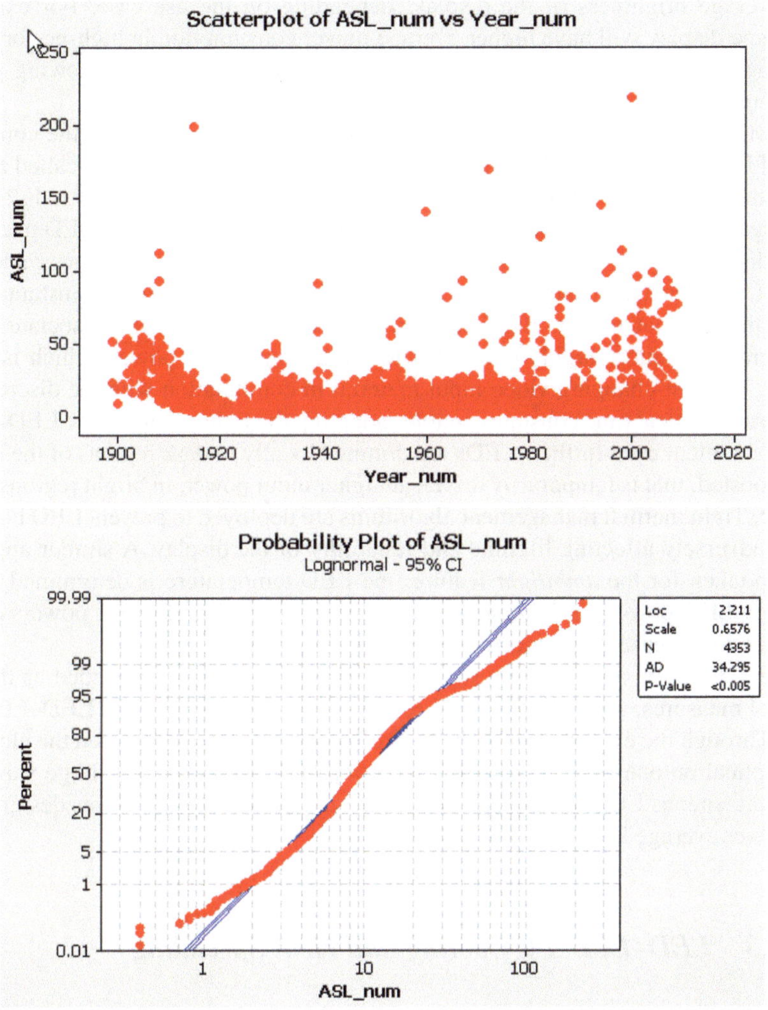

Fig. 14.4 Scatter plot and cumulative probability plot of average shot length (s) data from cinemetrix database [13]

and therefore has a positive effect on both the thermal management and the energy efficiency of the display. In direct-lit displays, a further enhancement is obtained through 2D (= local) dimming. In 2D dimming, LEDs are dimmed not only in time, but also depending on location: LEDs in dark parts of the moving picture are dimmed, providing a further improvement both in picture quality (higher contrast over the screen area) and in energy consumption. Barnhoefer [6] describes spatially adaptive local LED driving, allowing for a high-brightness range at less system power dissipation through local dimming of LEDs. In practice, the thermal room created by the lower time average system power consumption can be partially offset by raising the average brightness of the display, depending on the use case. For example, the same display will have higher average power consumption in high-performance gaming mode, and lower average power consumption in normal viewing or eco viewing mode.

Further expansion of the dynamic brightness range is possible by the combined use of both dimming and boosting of LEDs, for example as in the so-called *bright-pro feature* [14]. The "seconds" range time constants for video material is 1–2 orders of magnitude smaller than typical thermal time constants for display LED packages. This discrepancy in time means that fast oscillations in LED input power and light output will be dampened considerably in the thermal domain; the instantaneous LED light output reflects the instantaneous LED power, but the associated LED temperature is more highly correlated to the average LED power, which is much lower. The fast-changing video input material in combination with the discrepancy in thermal/optical time constants enables adapting the light output of the LEDs to the picture content even further: LEDs are dimmed locally in dark regions of the image, and boosted, that is temporarily driven at higher input power, in bright regions of the image. Tight thermal management algorithms are deployed to prevent LED boosting from adversely affecting lifetime and reliability of the display. A similar approach is also taken for the *ambilight* feature: the LED temperature is determined by the average LED power dissipation, and large instantaneous LED peak powers can be allowed to increase the immersive experience.

LED dimming and boosting scenarios, while not primarily intended as thermal control measures, are very beneficial to the thermal management of LED–LCD TV sets. Through the exploitation of the time constant differences between the electrical and optical on one side, and the thermal domain on the other side, large variations in instantaneous LED driving can be tolerated in thermal architectures designed for the lower average LED heat dissipation.

14.3.3 LED–LCD TV Cooling and Heat Spreading

In TV cooling temperature ranges, the heat flow is approximately proportional to the temperature difference and hence flow of heat can be described by a linear thermal resistance network. Figure 14.5 shows thermal circuits for the cooling of an LED junction in an LED TV in both a direct-lit and edge-lit display.

The thin-form factor of the TV limits internal airflow and the use of internal finned heat sinks. With limited internal cooling surfaces, cooling has to be predominantly on external surfaces: the front of the screen and the back of the TV set.

The thermal resistances can be derived for both conductive and convective/radiative heat transfer. In ***conduction***, thermal resistance is given by: $R_{th} = \frac{L}{Ak}$ and the heat flow is equal to $q = \frac{Ak}{\mathrm{L}}(T_1 - T_0)$ where L (m) stands for the distance between the hot (T_1) and cold (T_0) faces in the conductive path, A for the cross-sectional surface area (m^2), and k for the thermal conductivity [W/(mK)]. The heat flow for ***combined radiation and convection*** is equal to $q = Ah(T_1 - T_0)$, and the thermal

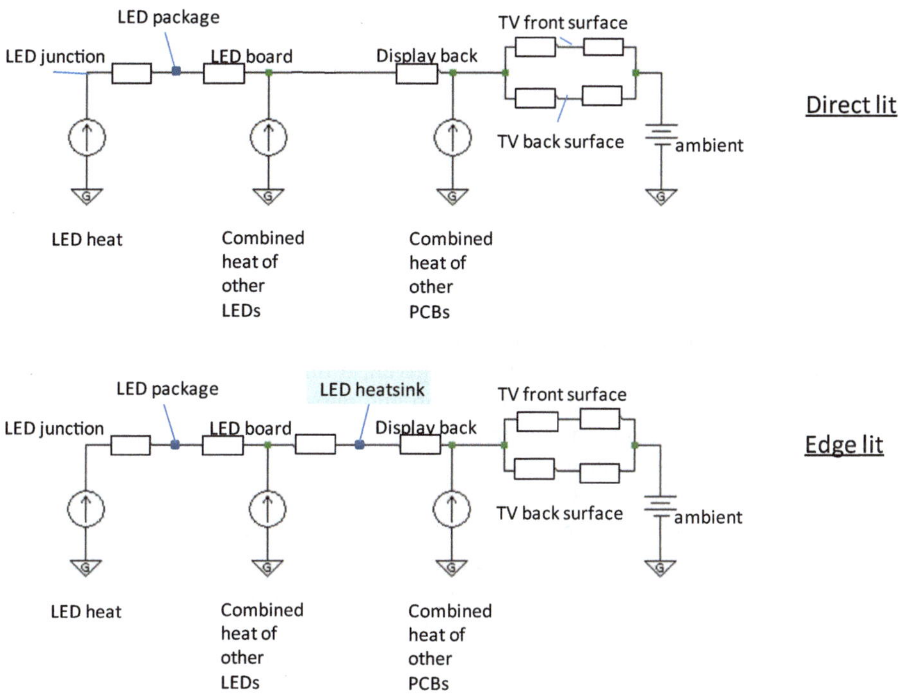

Fig. 14.5 Thermal circuit for a direct-lit and edge-lit TV

resistance by $R_{th} = \frac{1}{Ah}$ where A is the heated area in m^2, and h the heat transfer coefficient in $W/(m^2K)$. T_1 and T_0 are the hot surface temperature and the ambient temperature respectively. As a first estimate, the combined radiation-convection heat transfer coefficient for the outside of the enclosure is about 10 $W/(m^2K)$. For more details please consult Chaps. 3 and 9.

From a physical point of view, the one-dimensional formulas for conduction and convection/radiation cited above are only correct if the temperature refers to an isothermal area. In other words, in the resistance formulas, the entire area A must be uniform in temperature.

If the area is not isothermal, then the total heat transfer is given by: $q = \iint f(T(x,y) - T_0)d_x d_y$ with $f = k/t$ in case of conduction, and $f = 1/h$ in case of convection/radiation. Thus, in case of a nonisothermal surface, the total amount of heat transfer is proportional to the volume below the temperature surface or, in a 1 dimensional case, the area below the temperature curve. In electronics cooling, the object of interest is usually not the amount of heat transfer at a specific source temperature, but rather the source temperature related to a given value of power dissipation. In other words, when looking at the temperature distributions for different cooling options, we are comparing cases at a given power dissipation, so an equal area below the temperature curve. The curve with the lowest top represents the coolest heat

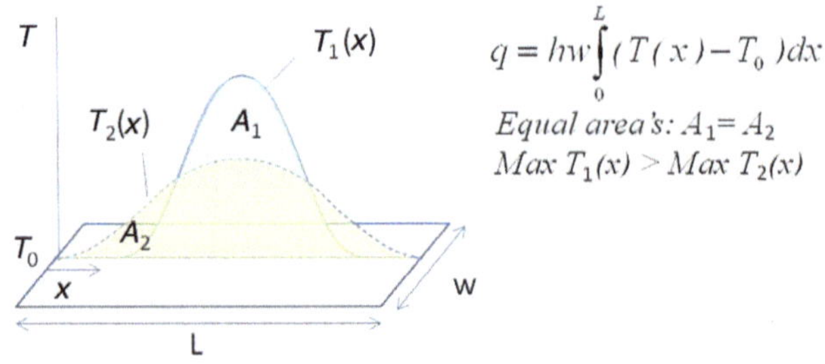

Fig. 14.6 Heat transferred from a plate with non uniform temperature

source and is thus the preferred choice. Since the area below the curve is directly related to the power dissipation, and this is given, it immediately becomes obvious that a wide, low curve is preferable to a high and narrow shape, and thus that heat spreading, effectively flattening the temperature curve, is an important mechanism to keep temperatures in control.

This is illustrated in Fig. 14.6. The area below both curves is equal, $A_1 = A_2$, so an equal amount of heat is transferred. Consequently, the narrower shape curve T_1 has a higher peak, representing a higher source temperature. The wider shape curve T_2 has a much lower peak value, so a lower source temperature, and is preferable. Spreading the heat over a larger area has lowered the temperature of the source.

Luiten [15] describes cooling of a heat source on a thin plate in terms of a cooling circle: heat spreads in radial direction from the source over a specific distance, and the source cools as if it has the surface of a cooling circle with this spreading distance added to the radius of the source itself. For large sources, this distance is equal to the characteristic length, Lc, defined in the same paper. For small sources, the spreading distance is smaller and a correction factor needs to be applied. The characteristic length is determined by the balance between how well heat is conducted in-plane, and how well heat is removed from the plate through the heat transfer coefficient boundary condition. At equal cooling heat transfer coefficients, Lc is proportional to the square root of the product of plate thickness and plate thermal conductivity.

A numerical example of the effect of heat-spreading is given in Fig. 14.7. A $3 \times 3 \times 2$ mm well-conducting heat source is placed on a $100 \times 100 \times 1$ mm vertical plate, cooled by radiation and natural convection. On the left side of the picture, the plate is aluminum, $k = 160$ W/mK, and on the right side glass, $k = 1$ W/mK. The calculated temperature fields show that in case of the aluminum plate, the entire plate surface is heated. The characteristic length for this case is much larger than the distance between the source and the rim of the plate. The cooling circle is much larger than the plate, and the cooling area available to the source is the front and the back of the plate area. In contrast, on the glass plate, only a small circle around the source is heated. The characteristic length for the glass is $\sqrt{160} = 13$ times smaller, and the cooling area available to the source is restricted to its cooling circle, which is much smaller than the plate area. The results of the calculation show that the source thermal resistance is lowered considerably by the effect of better heat spreading.

In a TV, most thicknesses are in the same 1 to 5 mm order of magnitude, so the thermal conductivity of materials is a good indicator of how well heat will spread.

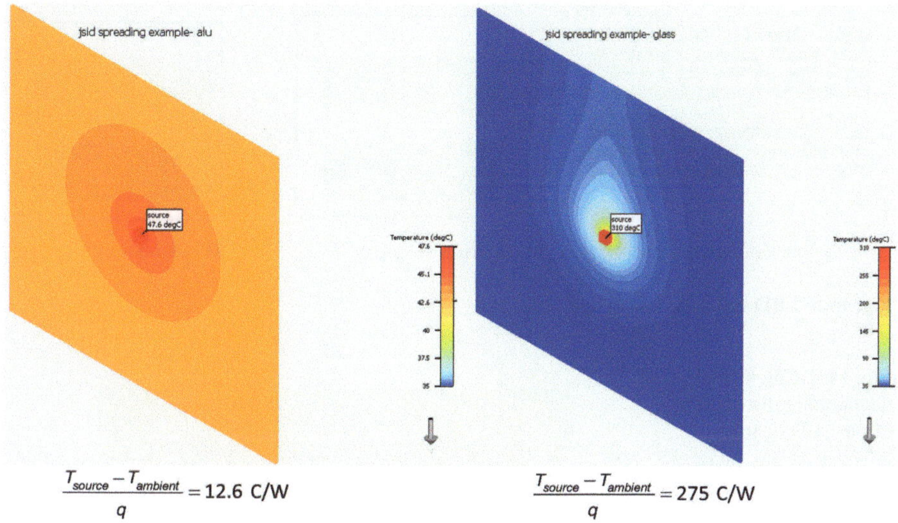

Fig. 14.7 Heat-spreading example

Table 14.1 Thermal conductivities of common materials at room temperature

Material	*k*
Still air	0.03
Plastic	0.2
Glass	1
Mild steel	50
Aluminum	160
Copper	390
Graphite, in-plane	300–600
Heatpipe, in-plane	10,000

An overview of the thermal conductivity of common materials in electronics cooling is found in Table 14.1, including the in-plane conductivity of graphite heat spreaders and a literature value by Ryoson for the equivalent conductivity of a flat heat pipe [16].

14.4 CFD Simulations of Display Heat Transfer

14.4.1 Direct-Lit Displays

FloTHERM® [17] CFD simulations were performed on a stand-alone direct-lit display, cooled by a heat transfer coefficient typical of natural convection including radiation on the front and on the back. The display is in vertical orientation, in a 35 °C ambient. The display build up, from the back to the front, consists of the metal back, or chassis, the LED boards, the optical mixing space (filled with air), optical

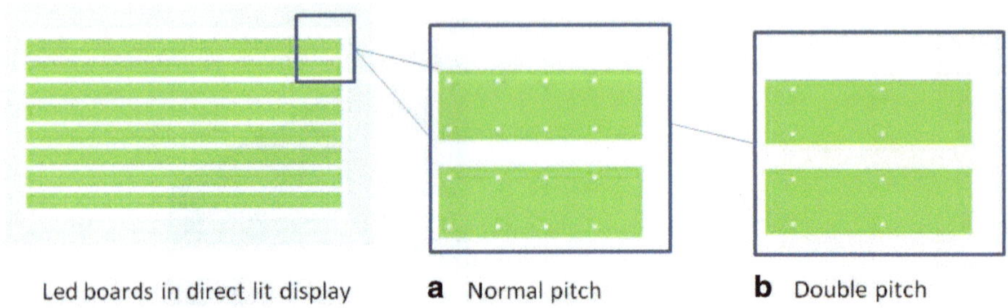

Fig. 14.8 LED board geometry

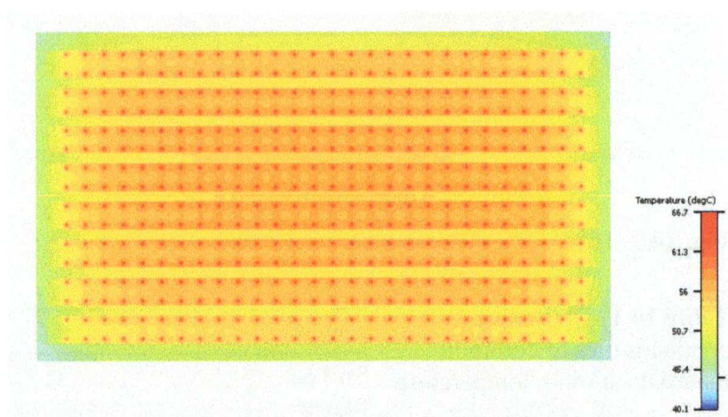

Fig. 14.9 Calculated temperature distribution of the LED boards

foils, and TFT panel on front. A typical interface resistance equivalent to 0.5 mm air was included between the display back and the LED boards.

Figure 14.8 shows the simulated LED board geometry and Fig. 14.9 the calculated temperature field, which shows that even though individual LEDs act as a heat source within their own hotspot, due to the specific combination of display geometry, LED heat dissipation and LED pitch, the heat spreading in the LED boards and the metal back of the backlight unit is sufficient to merge all individual LED heat sources into one more or less uniform heat source as far as the effect on the temperature distribution on the front of the TFT screen and on the metal chassis back is concerned. The image illustrates an important trade-off between the number of LEDs and thermal issues. Using less LEDs is attractive from a cost point of view, but less LEDs for the same light output of the total display means more power per LED and also more distance between the LEDs. This means more heat has to be spread over a larger distance, and hence in-plane heat spreading needs to be improved in this case.

Table 14.2 shows the result of display variations with a double pitch at the same total LED heat dissipation, i.e., double LED dissipation per LED package, as compared to the case treated above. The results show that with half the number of LEDs at the same total heat dissipation, the maximum LED package temperature in the display is increased. When the in-plane conduction of the LED board is increased,

Table 14.2 Direct-lit display results for various pitches and in-plane conduction

			In-plane conduction			Max-LED
Case	Pitch (%)	Led heat (%)	Display back (%)	LED board (%)	Interface resistance	°C
a	100	100	100	100	Yes	67
b	200	200	100	100	Yes	79
c	200	200	100	400	Yes	63
d	200	200	400	100	Yes	78
e	200	200	400	100	No	60

this effect is counteracted. Increased conduction in the display back was only effective when the interface resistance between the led board and the display back was removed (case e).

14.4.2 Direct-Lit LED TV

Figure 14.10 shows the set geometry of a TV with a direct-lit LED–LCD display. The set has three boards mounted on the display back. The display LED heat source is distributed uniformly over the entire display area. The set is cooled by natural convection including radiation, and is in vertical orientation in 35 °C ambient. Figure 14.11 shows the calculated temperature distribution in the horizontal and vertical cross-sections (top) and on the front (bottom). Heat flows perpendicular to the isotherms, so the cross-sectional temperature distributions confirm that also at set-level, heat spreads in-plane, and is lost predominantly from the front and the back surface of the set. The temperature distribution at the front of the screen shows that in this display geometry, the optical mixing space inside the display (typically in the order of 10–20 mm) is sufficiently wide to allow hot air to rise. Since the display is closed, a hot air bubble is trapped in the top of the display. The boards mounted on the back of the display contribute to the thermal load that has to be cooled locally. Since the thermal resistances remain the same while the thermal load is locally higher, this shows up as locally higher temperatures, both on the board itself and on the outer surface of the set. The results show that the screen temperatures are well below the aging limit for the polarizer ~60 °C, and that temperature nonuniformity has the shape of a vertical temperature gradient with a top to bottom difference of about 8 °C.

The calculated temperature field compares well to the measured temperatures on the front of a direct-lit TV set, as shown in Fig. 14.12.

In the direct-lit TV, the effect of hot air rising is visible as higher temperatures at the top of the display. Also, the positions of the three boards are visible as locally higher screen temperatures. The infrared picture confirms that screen temperatures are well below the aging limit ~60 °C in 35 °C ambient) and that the temperature difference over the screen is around 8 °C.

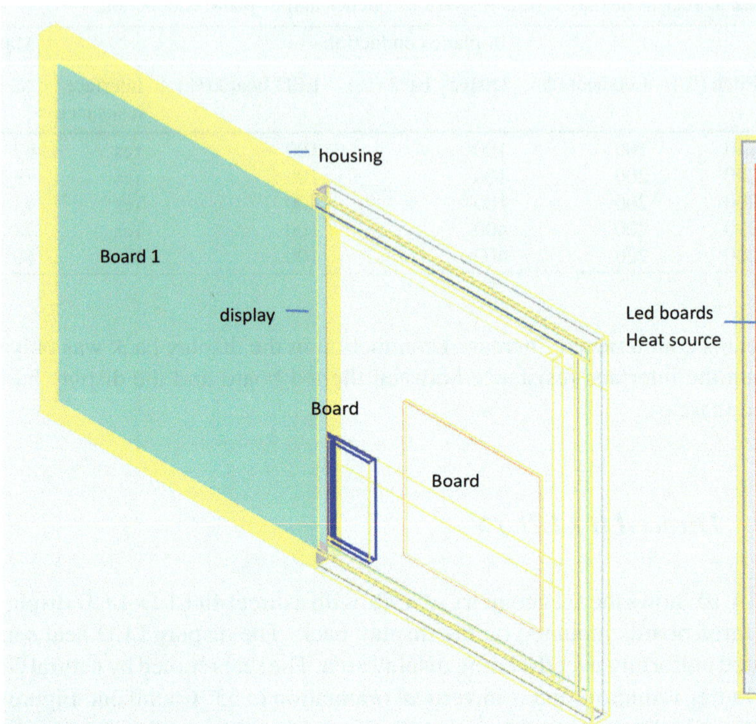

Fig. 14.10 Direct-lit LED TV

14.4.3 Edge-Lit Display

FloTHERM® CFD simulations were also performed on a stand-alone left/right edge-lit display, in vertical orientation, in 35 °C ambient, cooled by natural convection and radiation. Figure 14.13a shows the cross-section of the display edge, Fig. 14.13b shows the front view of the display, 14.13c–e show variations in the in-plane conduction of the display back, namely the use of graphite heat spreaders (14.13c), and the use of two or four heat pipes per side (14.13d and e). The heat pipes were modeled as orthotropic conducting strips, with a horizontal thermal conductivity of 10,000 W/mK, as per the paper by Ryoson [16]. The thermal conductivity in through plane and in vertical direction was set to 200 W/mK. Simulations were performed for the default case, and for the case of double LED bar heat dissipation, to simulate the thermal consequences of replacing two LR LED bars with one LED bar at double heat dissipation.

The calculated temperatures at the front of the display are shown in Fig. 14.14. The results show that the heat of the LED bars is transported through the LED heat sinks to the back of the display, and is lost from the front and the back surface. Increasing the in-plane conduction in the display back, by using better conducting material (not shown), graphite heat spreaders (Fig. 14.14b), or heat pipes (Fig. 14.14c, d) increases

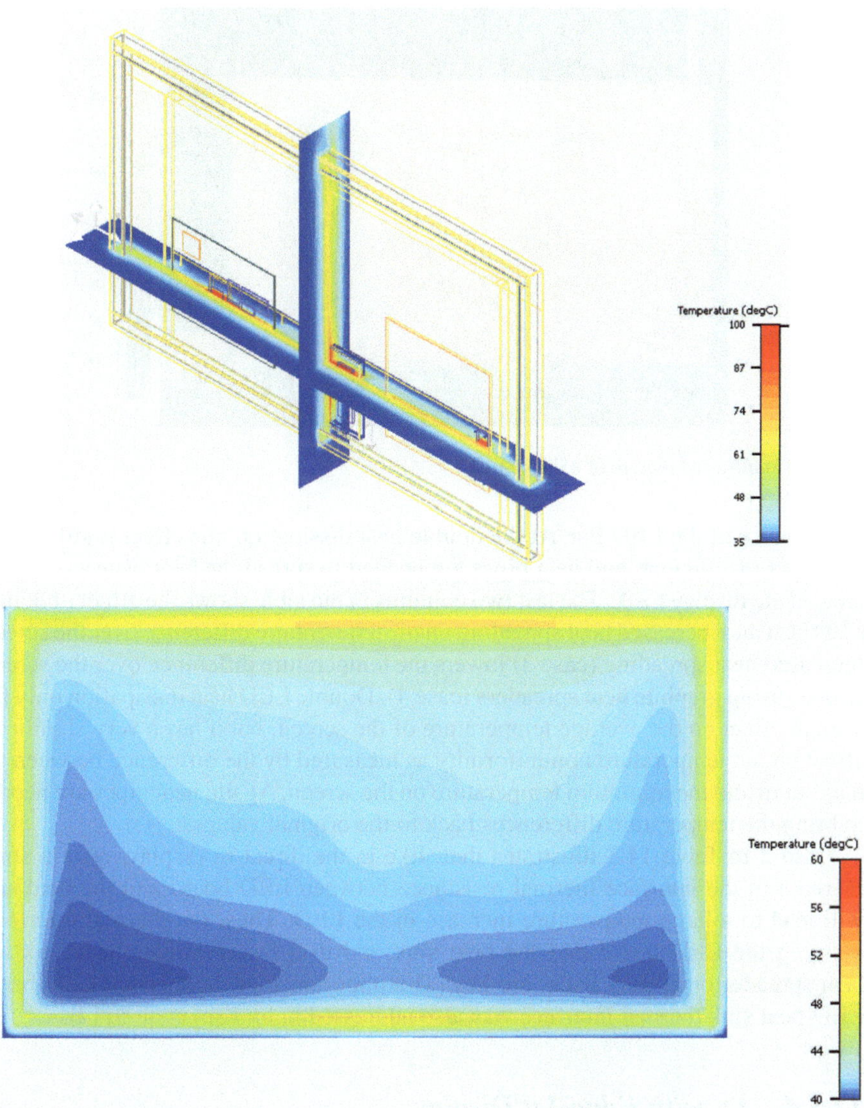

Fig. 14.11 Calculated temperature distributions for a direct-lit TV

the heated area and lowers maximum temperature on the LED bar. The combined thermal design of the LED heat sink and the display back, including the thermal interface between LED board and heat sink, determines how much area is accessible to lose the LED heat, and thus how well the LED bars are cooled.

Table 14.3 shows the calculated maximum temperatures at the LED bar. The results show that enhancing the in-plane conduction of the display back by using graphite spreaders next to the LED heat sinks is effective to lower the maximum

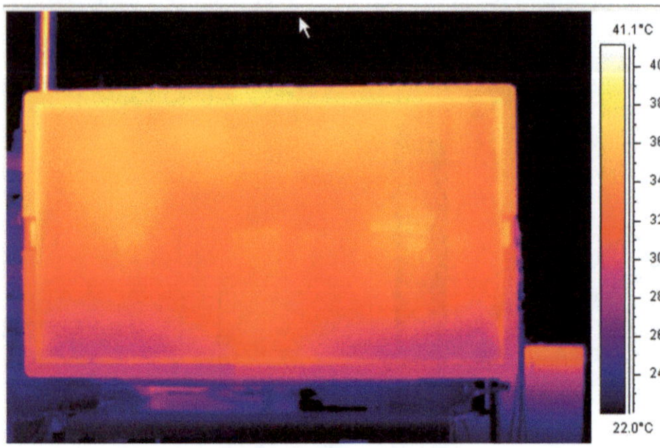

Fig. 14.12 Infra-red picture of a direct-lit TV

temperature of the LED-bar. At the double heat dissipation this effect is still significant, but insufficient, and heat pipes are needed to spread the heat over a sufficient area on the display back. The last two columns in the table shows the effect of double LED heat and increased heat spreading on the temperature difference over the screen. Increased heat spreading (case 3) lowers the temperature difference over the screen, as does using graphite heat spreaders (case 4). Double LED heat dissipation only has a small effect on the average temperature of the screen, but it has a very significant effect on the temperature nonuniformity as measured by the difference between the maximum and the minimum temperature on the screen. Again, heat pipes are needed to bring the temperature differences back to the original values.

Case 2 in Table 14.3 illustrates that, like in the direct-lit display, even a small increase in the interface thermal resistance between LED board and the heat sink will lead to a large temperature increase in the LED. Thus, the thermal interfaces between the LED board and the heat sink and to a lesser extent, heat sink and heat spreader or display back, are key. The importance of the thermal LED-bar to LED-heat sink thermal interface was also addressed in by Kim et al. in [7].

14.4.4 TV with Edge-Lit Display

FloTHERM® CFD simulations were performed on the L/R edge-lit TV shown in Fig. 14.15. The TV has 3 boards mounted at the back of the display, and the display is equipped with graphite heat spreaders. The TV is in vertical orientation and cooled by natural convection and radiation to an ambient of 35 °C.

The calculated temperature cross-sections (top) and the front of the screen (bottom) are shown in Fig. 14.16 Calculated screen temperatures at the left and right edges of the screen, next to the led bars, are critical with respect to the aging criterion

Table 14.3 Results for edge-lit variations

		In-plane conduction		Geometry	Max-LED bar	Average screen T	Max–min screen T
Case	Led heat (%)	Display back (%)			[°C]	[°C]	[°C]
1	100	100		b	81	42	33
2	100	100	0.1 mm air LED board-heat sink	b	131	42	33
3	100	300		b	72	43	25
4	100	100	Graphite spreader	c	72	43	25
5	200	100		b	116	47	56
6	200	300		b	102	48	44
7	200	100	Graphite spreader	c	101	49	44
8	200	100	2 heat-pipes/ side	d	96	48	39
9	200	100	4 heat-pipes/ side	e	82	48	28

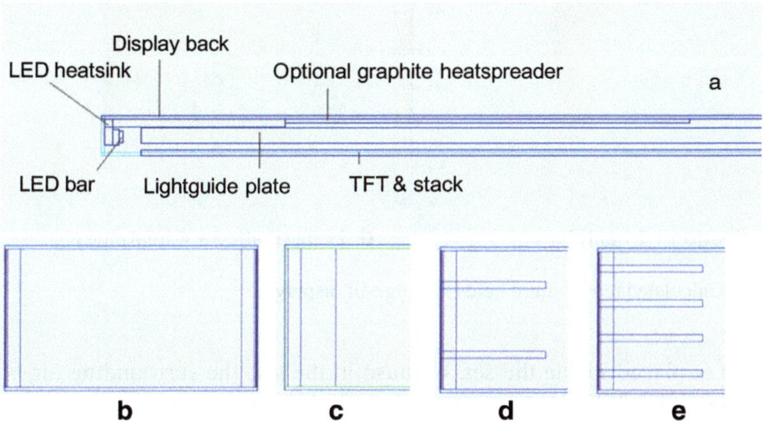

Fig. 14.13 Edge-lit display geometry and thermal management variations

(60–70 °C). Temperature nonuniformity is about 20 °C between edge and central temperatures. The results show that the three boards locally add to the heat load, and show up as hotter areas at the front of the screen.

The calculated maximum LED-bar temperature is 79 °C, a 7 °C increase from the bare display (Table 14.3, case 4). This illustrates the thermal interaction at TV system level, which has important implications for single parts measurements. While it is entirely feasible to do electrical and optical tests in a standalone setting and directly translate the results to the system situation, for thermal this is not the case. A temperature measured on a standalone board is not representative of the temperature once

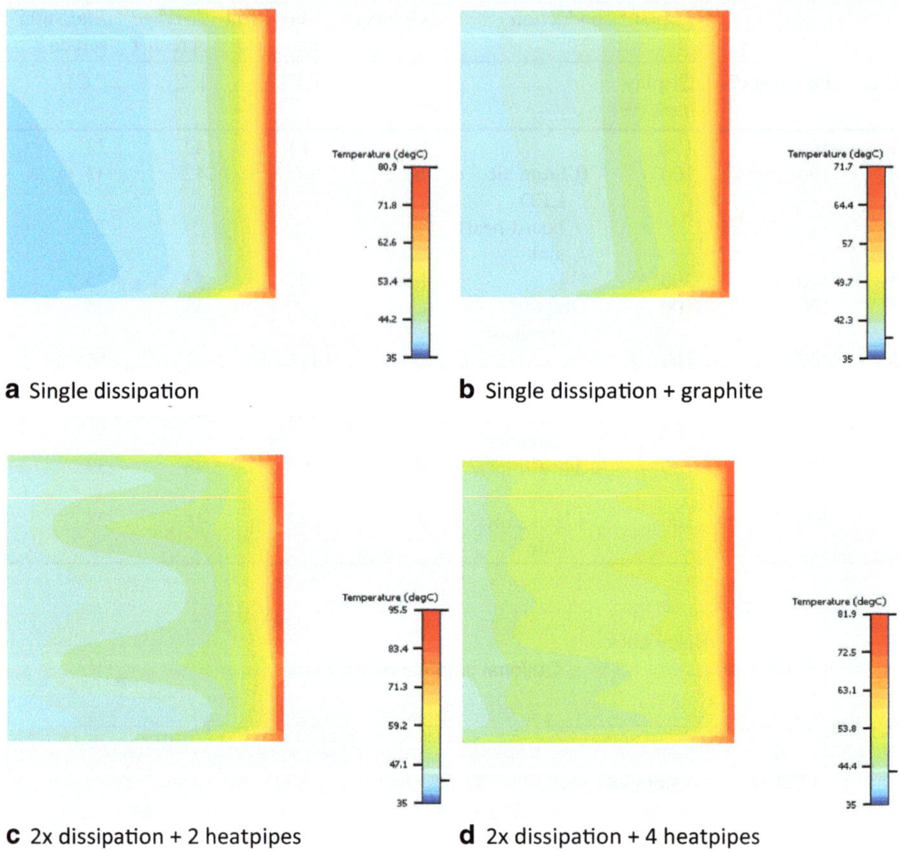

Fig. 14.14 Calculated temperature fields for edge-lit display

that board is placed inside the set, because in the set, the surrounding air is above room temperature. Likewise, measured temperatures for a bare display are not valid for the display once it is inside the TV set. There are two causes for this difference: first, the presence of the set back cover, which causes an additional thermal resistance from the display back to the room temperature, and second, the direct thermal interaction between the display and boards, especially if the boards are mounted on or at close proximity to the display back. This illustrates the difficulties encountered in co-design, translating from supplier measurements to expected behavior in the finished product and vice versa.

Figure 14.17 shows the infrared images of the front of a direct-lit TV, equipped with internal graphite heat spreaders. Comparison of the infrared images with the simulation results, Fig. 14.16, shows good agreement. In the image of the side-lit TV, the locally high temperatures due to the left and right LED bars are visible, as well as the zones heated from the graphite heat spreaders, and a hot zone caused by

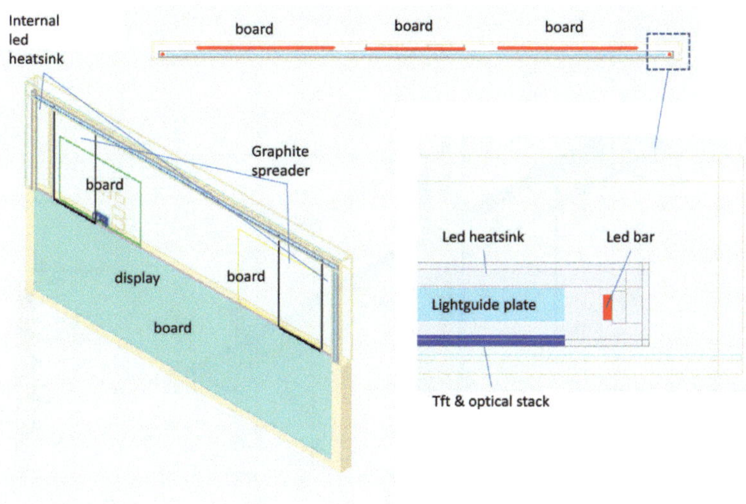

Fig. 14.15 Edge-lit TV

the central board. The measurement confirms that the screen is critical with respect to the aging criterion in the zone directly adjacent to the LED bars, and that there is roughly 20 °C temperature difference between the high temperatures at the side and the temperatures in the center.

14.5 Summary and Conclusions

The cooling of an LED–LCD display in an LED TV set was discussed. The derivation of the LED heat dissipation from its electrical input power was demonstrated, and it was shown that, for a typical white LED with a luminous efficacy of 75 lm/W roughly 25 % of the electrical input power is converted into light, while the remainder is dissipated as heat. Literature data on the temperature dependence of the lumen efficacy show that this temperature dependency can be relevant in relation to room temperature versus operating temperature of the LED junction, but is less significant for a 10 °C temperature range around the operating point.

In contrast to lighting applications, the fast-changing video material displayed on a TV allows for exploitation of the time constant differences between the optical and the thermal domain. Larger instantaneous peak light values can be tolerated in thermal architectures designed for the lower average LED heat dissipation. Dimming and boosting scenarios, primarily intended to improve picture quality, are very beneficial to power efficiency and thermal management of LED–LCD TV sets.

It was shown that heat spreading enlarges the cooling area that is accessible to a heat source, and thereby reduces source temperatures. FloTHERM® CFD simulations were shown at module (LED–LCD display) level and at system (TV) level,

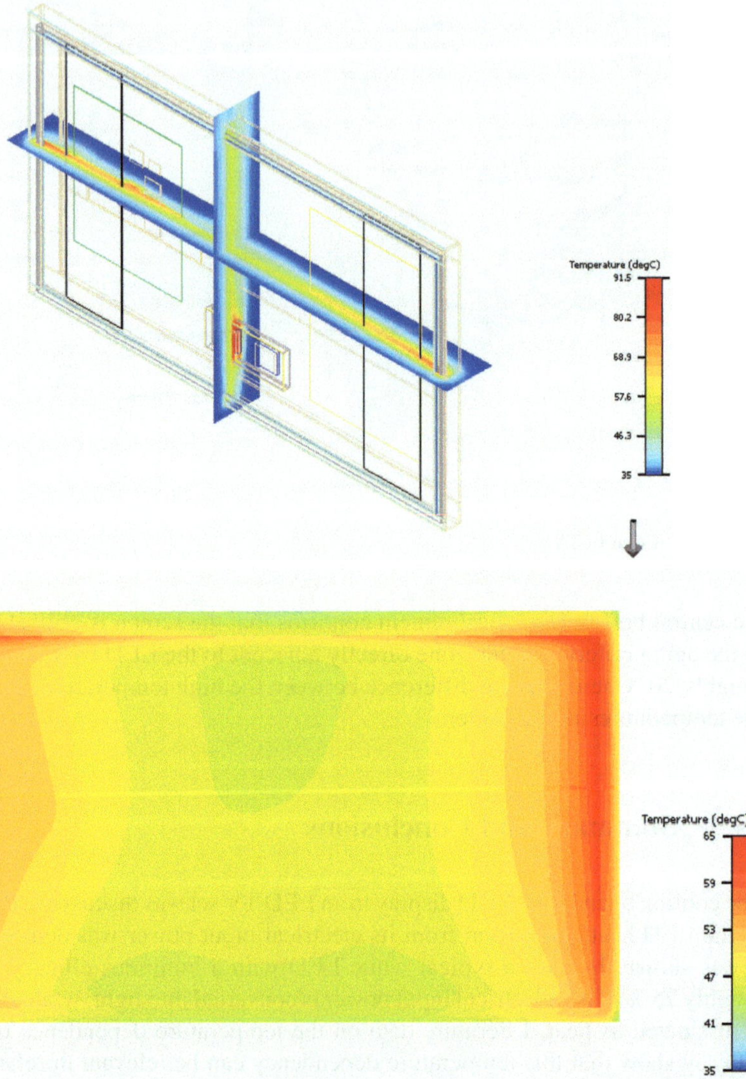

Fig. 14.16 Calculated temperature distributions in a left/right edge-lit TV

showing that in-plane heat spreading in the display is an important mechanism in LED–LCD display thermal management, because it enables the heat to access the only large surface areas available for cooling, that is the front and the back of the display or, at system level, the front and the back of the TV.

In a LCD display with direct-lit LED backlight this means the thermal contact and heat spreading of the LED boards and the back of the backlight is key, while for an edge-lit configuration accessing sufficient area through heat spreading from the edge to the center of the back of the backlight is more important. For the future, further

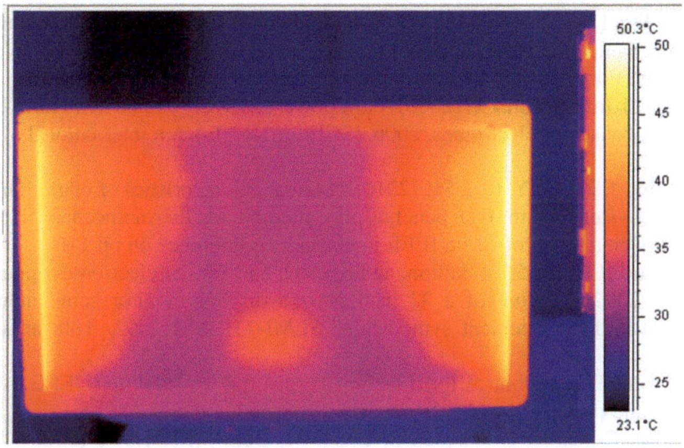

Fig. 14.17 Infra-red picture of a side-lit TV

on the path of higher efficacy, cost down and thinness, this means the cost benefit of using less LED packages will be partly compensated by increased cost in better conducting boards or backlight units backs in case of direct-lit LED backlights, and by increased cost for graphite heat spreaders and heat-pipes in case of edge-lit.

The temperature distribution in the LED–LCD display was shown to be a system issue, the result of both display properties and set interaction consisting of the presence of the housing and the interaction with the electrical boards. This means that, unlike in the optical and electrical domain, thermal measurements on standalone displays and boards are not one-to-one transferable to the situation of a completed TV set.

Acknowledgments A Thank-you to Philips TV innovation laboratory Eindhoven and Philips TV development site Bruges, whose ongoing support in TV-cooling activities is gratefully acknowledged.

References

1. van Driel WD, Yuan CA, Koh S, Zhang GQ (2011) LED system reliability, thermal, mechanical and multi-physics simulation and experiments in microelectronics and microsystems (EuroSimE), 12th international conference on pp 1/5–5/5
2. Liu L, Yang D, Zhang GQ, You Z, Hou F, Liu D (2011) Thermal performance analysis of photoelectric parameters on high power led packaging modules, thermal, mechanical and multi-physics simulation and experiments in microelectronics and microsystems (EuroSimE), 12th international conference on, pp 1/5–5/5
3. Biber C (2008) LED light emission as a function of thermal conditions, semiconductor thermal measurement and management symposium, SEMI-THERM 2008. Twenty-fourth annual IEEE, pp 180–184
4. Lifetime behavior of LED systems (2011) Lumileds white paper WP15, www.philipslumileds.com/uploads/167/WP15-pdf. Accessed 24 Jan 2011

5. Koh S, Van Driel W, Zhang GQ (2011) Degradation of epoxy lens materials in led systems, thermal, mechanical and multi-physics simulation and experiments in microelectronics and microsystems (EuroSimE), 12th international conference on, pp 1/5–5/5
6. Barnhoefer U, Kim M-J, Erez M (2006) A low power passively cooled 2000 cd/m2 hybrid LED-LCD display, consumer electronics, 2006. ISCE '06. IEEE tenth international symposium on, pp 1–4
7. Kim SK, Kang J, Kan WY, Kim SH (2010) Thermal characteristics of edge-illumination type led backlight system in a thin flat panel display, thermal and thermomechanical phenomena in electronic systems (ITherm), 12th IEEE intersociety conference on pp 1–5
8. Safety standard IEC60065 7th edition, audio, video and similar electronics apparatus
9. Luiten GA (2002) Cooling of a 32 inch plasma display monitor semiconductor thermal measurement and management symposium, SEMI-THERM 2002. 18th annual IEEE, pp 119–124
10. Luiten GA (2005) Fan control for a ceiling mount flat TV monitor, IEEE transactions on components and packaging technologies, vol 28, issue 4, pp 680–685
11. J Kahn S, Bar-Cohen A, Dessiatoun S, Ohadi M, Getz M, Norley J (2006) The impact of a thermal spreader on the temperature distribution in a plasma panel display, thermal and thermomechanical phenomena in electronics systems, ITHERM '06. The tenth intersociety conference on, pp 395–401
12. Wilcoxon R, Cornelius D (2006) Thermal management of a LED light engine for airborne applications, semiconductor thermal measurement and management symposium, SEMI-THERM 2006 IEEE twenty-second annual IEEE, pp 178–185
13. http://www.cinemetrics.lv/database.php. Accessed 30 March 2010
14. http://live.philips.com/index.php/nl_be/video/bright-pro-2010-english/73378776001. Accessed 15 Oct. 2011
15. Luiten GA (2010) Characteristic length and cooling circle semiconductor thermal measurement and management symposium, SEMI-THERM 2010. 26th annual IEEE, pp 7–13
16. Ryoson H, Yajima T, Goto K, Hirata K, Oniki K (2010) Thermal performance of novel thin heatpipe, thermal and thermomechanical phenomena in electronics systems, ITHERM 2010 12th IEEE intersociety conference on, pp 1–7
17. FloTHERM® Electronics cooling CFD software. http://www.mentor.com/products/mechanical/products/flotherm. Accessed 15 Oct. 2011

Chapter 15
LEDs in Harsh Environments

Ross Wilcoxon and Jim Petroski

Abstract Light-emitting diodes (LEDs) are increasingly used in applications with extreme temperatures, pressures, dynamic loading, and conditions, which can degrade materials, such as solar loading and fluids. This chapter discusses a number of these harsh environment LED applications that include military/aerospace vehicles, automotive lighting, and outdoor lighting. It also describes test methods used to assess the reliability of LEDs and system integration issues, with an emphasis on thermal management, related to harsh environments.

15.1 Introduction

A wide variety of electronics are used in less-than-optimal environments and there is no single application or definition that could be universally agreed upon as what constitutes a harsh environment. For example, while military electronics are generally considered to fall into the category of harsh environment equipment, a number of these systems spend the majority of their service in an air-conditioned office environment. Outdoor lighting may be used in installations that could be subjected to extreme temperatures, severe solar loading, debris, high humidity, and/or driving rain. In other cases, harsh conditions may be defined by regulatory or other specifications that impose conditions rarely seen, but which nonetheless must be passed in order for a product to be sold to end users.

This chapter considers a range of light-emitting diode (LED) systems used in environments that meet a relatively broad definition of 'harsh'. These environments may include one or more extreme conditions such as ambient temperature extremes with maximum values $> \sim 50$–$70\,^{\circ}$C and/or minimum ambient temperatures below -20 to $-40\,^{\circ}$C. Exposure to fluids that can cause material degradation and/or corrosion can also be considered to be a harsh environment. These fluids encompass a number of materials that may range from 'conventional' fluids such as humidity, salt fog, sea water, solvents, fuels, and combustion by-products to less obvious fluids

R. Wilcoxon (✉)
Cedar Rapids, IA 52402, USA
e-mail: ross.wilcoxon@gmail.com

J. Petroski
Rambus, Brecksville, OH, USA
e-mail: jpetroski@rambus.com

C. J. M. Lasance, A. Poppe (eds.), *Thermal Management for LED Applications*,
Solid State Lighting Technology and Application Series 2,
DOI 10.1007/978-1-4614-5091-7_15,

such as spilled drinks and the 'blue water' found in commercial aircraft lavatories. Harsh environments may also be a factor of exposure to high dynamic loading due to shock and/or vibration. Other factors that constitute harsh environments include extended exposure to severe solar loading, extremely low pressures at high altitudes and in spacecraft, high pressures that occur in deep sea and downhole drilling applications, etc. In addition to these extreme environmental factors, harsh environment equipment tends to experience severe variations in these environments (such as ambient temperature or relative humidity) that create cyclical stresses that can cause fatigue failures. These varying environmental conditions may be the result of daily and yearly weather fluctuations (such as in outdoor lighting), platform induced changes in the environment (such as temperature, pressure, and vibration fluctuations encountered during the flight of an aircraft), and severe changes in environments during storage and/or transport of equipment.

Another consideration for lighting equipment receiving the harsh environment label is whether it is part of a safety critical system. In the case of lighting, systems such as traffic/pedestrian/rail signals, navigation lights, or general lighting for personal safety (e.g., area lighting) are often safety-critical [1], not to mention automotive lighting such as tail lights or head lights. These safety-critical lighting systems are often subjected to harsher requirements at the system level to ensure greater reliability and functionality at the extreme ranges of operating conditions. Other lighting tasks may be deemed safety-critical but have no such formal control document, and for liability reasons are more strictly designed and manufactured. In aerospace/military applications, more severe requirements are typically imposed on systems that are deemed flight- or mission-critical.

As alluded earlier, regulatory agencies play an important role in determining specifications of design and test for lighting systems. Testing specifications used to evaluate systems and provide a safety factor for their use environment can lead to worst case conditions that are more severe than an LED or LED product may see in actual service.

This chapter focuses specifically on the use of LEDs in this variety of harsh environments. It describes a number of applications that use LEDs in harsh environments, briefly discusses how these environments can impact the reliability of the LEDs, presents some examples of test methods used to assess the suitability of LEDs and systems that include them for use in harsh environments, and finally discusses some system integration issues for LEDs in harsh environments with an emphasis on thermal management approaches for those applications.

15.2 Applications of Light-Emitting Diodes in Harsh Environments

15.2.1 Military and Aerospace

LEDs are seeing widespread use across military and aerospace applications due to a number of factors including reliability, efficiency, size, and capabilities to provide flexibility in system design and integration. LEDs have been included in military equipment for more than 2 decades as light sources for fiber-optic interconnect in optical data links on commercial airliners as well as military aircraft [2–4]. These systems generally required hermetic packaging and the LEDs dissipated relatively little power. Over the past decade, improvements in LED technology have led to their use as the primary light source for display technologies such as backlights for liquid crystal diode (LCD) displays [5], as well as light engines for projectors and heads-up displays [6]. These types of displays are used in virtually all commercial and military aircraft as well as many military and law enforcement ground vehicles. LEDs are also widely used in aircraft lighting applications ranging from anticollision lights, flight deck lighting, reading lights, and runway/taxiway lighting. The transition to LEDs in these applications has been driven both by their reduced power consumption and the order of magnitude increase in mean time between failures (MTBF) relative to traditional lighting [7].

In parallel to the growing use of LEDs in aircraft applications, ground-based military is also employing the technology in a variety of applications. For example, the standard lighting in a portable, expandable military shelter as a remote operational center [8] has transitioned to LEDs for all lighting. The advantages of the technology in this application include the very low physical profile of the light fixtures (which must fit within a minimum volume when the shelter is not expanded and essentially three rooms are packed into the volume of one) and the ability to easily define multiple lighting scenarios that include night vision compatibility. Other military LED lighting applications include vehicle lighting, check point lights and high output beacons, and night vision enhancement [9]. LEDs have also been used for optical communication links for military convoys and Identification Friend or Foe (IFF) systems [9].

Applications of LEDs in space include the crew lighting for a proposed inflatable space station [10] and future systems to provide light for growing plants in space [11].

15.2.2 Automotive

LEDs are seeing widespread use in automotive applications, both in external lighting applications for headlights, fog lights, indicator lamps (turning signals and brake lights) as well as interior lighting for instrument cluster illumination, courtesy lights, map lights, and display backlights [12]. The Center High Mount Stop Lamp (CHMSL), i.e., the "third" brake light indicator, was among the first applications for automotive due to a need for small lamp designs [13]. While the LEDs in this application are not typically high power dissipators, their location provides limited opportunities for effective thermal management.

The growing use of LEDs in automotive applications has been enabled by the availability of higher output power devices with improved semiconductor technologies, primarily aluminum gallium indium phosphide (AlInGaP) for reds and yellows or the indium gallium nitride (InGaN) for greens, blues, and white (with phosphors), and increasingly gallium nitride (GaN) LEDs [14] that can survive high temperatures in the higher power white applications. High temperature compatibility is critical in automotive applications since the internal (under hood, etc.) ambient temperature for external lighting is typically at least 25 °C higher than the outside air temperature [13] due to engine waste heat and solar loading. Significant advantages of LEDs in automotive applications include the improved lighting, the consistent brightness and chromaticity over the life of the product, the rapid turn on time for the devices [13], vibration resistance due to no bulb filaments, and the possibility for variable directional lighting. The rapid response of LEDs can lead directly to safety improvements; incandescent bulbs in automotive tail lights have a thermal start-up time of approximately 250 ms. LEDs have a nearly instantaneous start-up time, which can increase stopping distance by 8 m for a vehicle traveling 125 km/h (78 mph) [12]. In addition, the ability to control the colors emitted from the LED can reduce the need for colored lenses in the light assembly; the higher efficiency reduces power consumption and lowers housing and lens temperatures, which reduces the need for high temperature materials. The smaller size of the lighting devices directly leads to smaller reflectors and lenses in the light fixture [13].

15.2.3 Outdoor Lighting Applications

LEDs are used in a very wide range of outdoor lighting applications such as street lighting, traffic control, etc. More severe applications include lighting in harsh environments such as underwater [15], and in downhole drilling, in which commercially available cameras that use LEDs, can be subjected to temperatures of 125 °C and pressures of over 1,000 atm [16].

Environmental conditions can add to severe thermal requirements in outdoor applications. A luminaire installed outdoors is not likely to remain in pristine condition over time. Debris accumulated on such luminaires can degrade system's thermal performance and frequently must be accounted for in thermal design. Debris sources can include material shed from trees (such as leaves, needles, and tree sap) or animals (most frequently, bird nests and/or fecal matter build-up, though other small animals may pose problems).

Another issue for outdoor lights is the reduced performance through light blockage. This may force a design to increase heat dissipation by increasing LED light output either with the initial design or with progressively higher light levels over time. One ongoing study by the US Department of Energy noted that the degradation of outdoor luminaires due to the build-up of dust and dirt varied from approximately 4–10 % after a 1–2-year period [17]. This drop was in part due to the optical elements being on the underside of the luminaires where they were not subjected to cleaning by rain or snow.

While the higher efficiency of LEDs does provide significant advantages, in outdoor lighting it can introduce a maintenance concern in cold environments. The thermal dissipation from less efficient lighting will more effectively melt snow and ice that accumulates on them during the winter. LEDs may require additional maintenance to remove this material [18].

15.2.4 Critical Performance Metrics and Failure Mechanisms

An extensive review article [19] has thoroughly documented the failure mechanisms historically observed in LEDs. These mechanisms are summarized in Table 15.1, which lists the failure mechanisms and identifies environmental and operational conditions that typically lead to these failures. The harsh environment systems described in the previous section are due to environments that include high ambient temperatures, extreme temperature excursions between minimum and maximum operating and/or storage temperatures, exposure to water in the form of both, liquid and humidity, high currents and voltages associated with high power densities, and exposure to severe solar and other radiation effects. In other words, all of the factors that tend to exacerbate the failure mechanisms in LEDs tend to be more severe in the classes of systems described in the previous section as 'harsh environment'.

Table 15.1 summarizes the failure mechanisms at the device and package levels. Harsh environment can degrade LED packaging component and the systems in which they are used that in turn affect the performance of LEDs. For example, the optically clear encapsulant that is typically used in the packaging of LEDs must be able to withstand environmental effects without degrading. In particular, the effects of temperature, radiation, and chemical exposure on this encapsulant, which is typically silicone-based, must be recognized. Extreme temperatures can cause outgassing of volatile components in the encapsulant that may change its properties and have a severe impact in applications such as spacecraft. High radiation can cause cross linking in the encapsulant, which increases its stiffness and leads to higher stresses from thermal expansion mismatch. The fuel used in aircraft and other vehicles can cause the silicone to significantly expand, which creates stresses and affects optical performance of the lens. Formulations of specific silicones can be used to address these environmental effects if necessary [20].

The thermal characteristics of LEDs introduce constraints for their system level thermal management. Traditional lighting sources have typically operated at higher temperatures and therefore rely on radiation to transfer much of their dissipated heat to ambient; with their lower operating temperatures, radiation is not a dominant node for LEDs. Since LED thermal management is often accomplished by conduction, issues may arise when multiple parts are in an assembly and the heat energy must move across interfaces between parts. Traditional luminaire designs have made extensive use of sheet metal parts (for cost control); this can lead to luminaire constructions with poor contact between parts if not carefully accounted for in design. Joints may also degrade at thermal interfaces over time and is another item to consider for harsh environments. In lieu of ensuring sound contact at a joint, using a highly thermally conductive material across a contact joint provides another option to reduce thermal resistance across the interface [21].

Table 15.1 LED failure mechanisms. (Adapted from reference [19])

Failure site	Failure mechanism	Extreme high temperatures	Temperature cycling	Moisture	Electrical overstress	Photo-degradation
Semiconductor	Defect/dislocation generation and movement	X	X			
	Die cracking	X	X			
	Dopant diffusion				X	
	Electromigration				X	
Interconnect	Bond wire fracture	X			X	
	Wire ball bond fatigue	X	X	X	X	
	Electrical contact metallurgical interdiffusion	X			X	
	Electrostatic discharge				X	
Package	Carbonization of encapsulant	X			X	
	Delamination	X	X	X		
	Encapsulant yellowing	X				X
	Lens cracking	X	X	X		
	Phosphor thermal quenching	X	X			
	Solder joint fatigue	X	X			
Military and aerospace		X	X	X	X	X
Automotive		X	X	X	X	X
Outdoor lighting and signs			X	X		X
Industrial applications		X		X	X	

LED light-emitting diode
X indicates that that failure mechanism applies to that application

Table 15.2 Energy Star® proposed reliability testing requirements. (Adapted from reference [25])

Test	Requirement
Color maintenance	Lamp change in chromaticity from initial/0-hour measurement, at any measurement point during the first 6,000 h of lamp operation, shall be within 0.004 on the CIE 1976 u′v′ diagram
Life time	All tested units shall be operational at 3,000 h. ≥ 90 % of the tested units shall be operational at 6,000 h. Lumen maintenance as per Fig. 15.1
Rapid cycle stress	Lamp shall survive cycling once for every hour of rated life (minimum of 10,000 cycles). Each cycle shall be 5 min On, 5 min Off
Transient protection	Lamp shall survive seven strikes of a 100 kHz ring wave, 2.5 kV level, for both common mode and differential mode
Typical ambient test temperature	45 °C

CIE International Commission on Illumination

15.3 Testing Requirements for Harsh Environment Light-Emitting Diodes

LEDs are components within a larger electronics system and in many cases only the overall system level testing requirements are specified for a given application. Individual component level testing methods used for particular product requirements are not necessarily uniform among various manufacturers or even publically available. The following section provides some representative examples of LED component and/or subsystem testing to provide insight into the types and levels of testing that may be required of these devices. These examples are not meant to be comprehensive requirements for the types of testing that may be needed for LEDs used in harsh environments, but instead are meant as guidance that must be adapted before being applied to other products.

New standards are currently being developed to provide guidance and requirements for LED testing; these may be helpful for defining system requirements [22–24], see also Chap. 6. Testing to these types of standards for what may seem normal products used in a normal manner may also constitute a harsh testing condition. As shown in Table 15.2, the requirements for an Energy Star bulb in the United States [25] (as of this writing, a draft standard) can impose severe conditions.

Figure 15.1 shows the performance requirements for solid state lighting as a function of their rated operating life. The solid line shows the minimum lumen maintenance (starting to final) after 6,000 h of operation while the dashed line indicates additional testing required for lights rated as 'Extended Life'; these must complete the indicated level of additional testing and maintain ∼91.5 % (depending somewhat on the rated operating life) of the emitted luminous flux.

Table 15.3 summarizes a set of testing that was performed in the evaluation of a new LED component for use in an LED array backlight in an avionics display

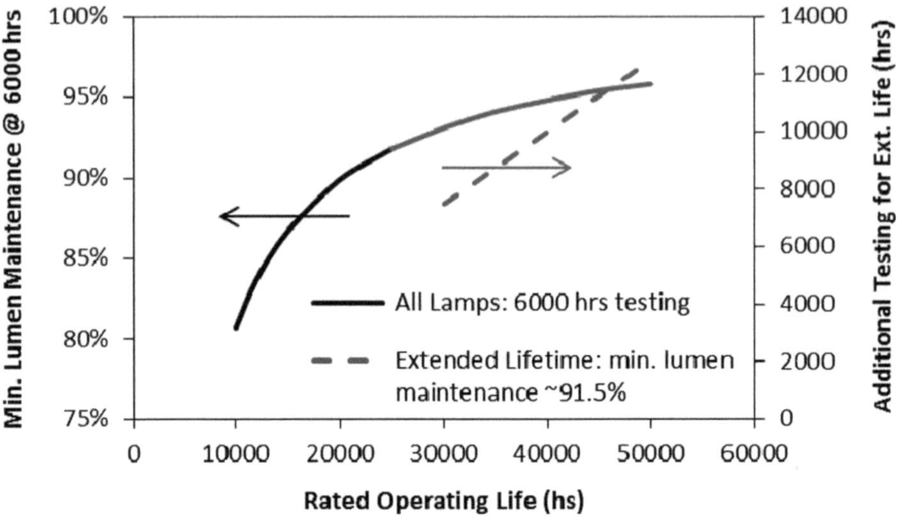

Fig. 15.1 Proposed life time testing requirements for Energy Star® lights

[26]. This testing was conducted on individual LEDs as well as arrays of LEDs in a backlight to compare their component level and subsystem level performance as a product line transitioned to a new generation of higher current, higher efficiency LEDs.

The LEDs were tested for performance as well as reliability over a range of operating current conditions. This provided the necessary information to conduct design trade-offs to assess the impacts of reducing the number of LEDs in an array to ensure that the design operating current met optical requirements with sufficient margin for any severe operating conditions as well as any performance degradation. Based on these results as well as other system-level constraints, the nominal operating current for the devices was ultimately set to approximately 15 % of the maximum device current.

Table 15.4 shows another example of a set of subsystem reliability assessment tests. These tests were conducted on an LED module for internal crew lighting in a proposed inflatable space station [10]. After each test, the electrical performance was evaluated and the module was visually inspected to ensure that the testing had not caused significant damage. For this particular application, the most severe tests were primarily targeted at the extreme conditions of getting into orbit, i.e., dynamic shock and vibration, and worst-case scenarios such as complete loss of pressure.

Synthetic jets, which use either electromagnetic actuators or piezoceramic elements to oscillate a diaphragm to improve heat transfer from a surface by generating local turbulence, have been used to cool high brightness LEDs with a target application of lighting applications in harsh environments [27]. Table 15.5 summarizes a set of qualification testing procedures that were reported by a synthetic jet manufacturer to characterize the suitability of these devices for LED lighting applications. The

Table 15.3 Test criteria for LED devices and backlight for an avionics display backlight. [26]

Test	Conditions
LED efficacy	LEDs mounted to a printed wiring board and operated at a range of currents (5 mA to ~ 80 % of max current) and efficacy was measured using a large integrating sphere
LED chromaticity	LED average chromaticity measured as a function of currents ranging from ~ 10–80 % max current
LED backlight evaluations as a function of current and temperature	Luminance under range of bias over temperatures ranging from − 40 to + 80 °C in 20 °C increments
LED voltage characteristics	Forward voltage as a function of current for ambient temperatures from − 40 to + 90°C
LED temperature cycling	− 55–85 °C thermal cycling with 15 min dwells and >15 min ramps. LED on during cold temperatures. Measured luminous flux and chromaticity after 500 cycles
LED life testing	Measured luminous flux during continuous operation with a 45 °C baseplate for 4,000 h at multiple current levels (~ 0.5 × , 3 × , and 7 × normal operating current)
LED fault analysis	Room temperature operation; current was applied to the LED and increased in steps of 100 mA until the device failed. (~ 25 × normal operating current)
LED humidity testing	10 day humidity testing
Backlight luminance	Measured backlight luminance over − 50 to + 50° viewing angle in increments of 10°
Backlight uniformity	Localized luminance was measured across the backlight. Nonuniformity (maximum luminance − minimum luminance)/mean luminance was required to be <40 % (in some cases this could be as high as 50 % depending on customer requirements)

LED light-emitting diode

topic of synthetic jets that use piezoelectric or electromagnetic actuation is further discussed later in this chapter.

15.4 Implementation in Harsh Environments

15.4.1 System Integration and Manufacturing Issues

Rapid improvements in commercially available LEDs over the past decade have led to a changing landscape for system designers. New generations of devices with higher light output, greater efficiency, and a wider spectrum of colors introduce opportunities for system improvements. But a clear understanding of system design trade-offs is needed if these opportunities are to be effectively exploited. For example, the higher light output of new LEDs allowed the designers of an avionics LCD display to significantly reduce the number of LEDs in the backlight [5]. This resulting larger

Table 15.4 LED light fixture testing for space vehicle. [10]

Test	Condition	MIL-STD
Performance baseline	Luminance measurements	n/a
Steady-State-Life	Powered at + 125 °C for 1,000 h with periodic performance verification tests	883, Method 1005
Temperature cycling	25 cycles, − 55 to + 85 °C	883, Method 1010, Condition A
Thermal shock	5 cycles, − 65 to + 125 °C. 15 min dwells at each temperature extreme with <1 min transition time between cycles	202, Method 107, Condition B
Thermal vacuum, reduced pressure (space/altitude)	While under power, external pressure reduced from 760 Tor (1 atm) to 1 mTor. First 24 h at − 55 °C, second 24 h at + 85 °C	STD-883, Method 1001 & MIL-HDBK-340, Method 6.4.3
Radiation hardness	30 krad (Si) total dose from Cobalt-60 gamma ray source	883, Method 1019, Condition C
Moisture resistance	24 pre bake at 125 °C followed by 24 h humidity cycling between 95 and 0 % RH	883, Method 1004
Acoustic noise	Uniform intensity shaped spectrum of acoustic noise that impacts all the exposed materiel surfaces. Frequency range 30–10,000 Hz with 127 dB maximum at ~ 200 Hz. 60 s per axis	810, Method 515, Procedure I
Vibration: random	Power spectral density (11.95 Grms), frequency range of 20–2,000 Hz. (0.1 G^2/Hz from 100–1,000 Hz)	202, Method 214, Condition 1D
Vibration: sinusoidal	10 g, frequency range of 10–2,000 Hz	202, Method 204, Condition A
Mechanical shock: specified pulse	50 g, half sine pulse, 11 ms. Applied on each of six mutually perpendicular axes	202, Method 204, Condition A
Assessments conducted after each test		
Mechanical and visual inspection	Dimensional inspection (60 × microscope) per design specifications. Solder joint inspection per NASA-STD-8739	
Electrical test	Electrical verification of performance per design specification. Device output in Lumens/m^2 measured	

LED light-emitting diode, *MIL-STD* military standard, *MIL-HDBK* military handbook, *RH* relative humidity, *NASA-STD* National Aeronautics and Space Administration standard

spacing between the LEDs required an increase in the distance from the display to backlight emitter plane to ensure uniform light output. Because the LEDs were

Table 15.5 Test methods for electromagnetic fans for cooling LEDs in harsh environments. [27]

Test	Power On	Description
HAST/Storage	No	60 h at 123 °C, 96 % RH, 2 atm
Low temperature/storage	No	– 40°C
Accelerated life testing	Yes	3,300 h at 105 °C
Life testing	Yes	9 h at 85 °C
Low temperature	Yes	– 40 °C
Humidity	Yes	3,250 h, 85 °C/85 % RH
Thermal cycling	Yes	200 cycles, – 40 to + 105 °C, 5 °C/min ramp rate
Vibration: sine sweep	No	5–150 Hz, 2G, 3 sweeps/axis
Vibration: sine dwell	No	5G at first 3 resonances, 10 min dwell
Random vibration	Yes	2.2 Grms, 20 min each axis
Shock	No	Six half sine shocks, 40 G, 11 ms
Bump test		1,000 shocks, 25G 10 ms
Dust	Yes	12 h IEC 60529 (IP5X)
Cyclic humidity	Yes	200 cycles between + 50 °C/95 % and + 5 °C/5 % RH
Freezing rain	Yes	100 cycles between + 25 °C/95 % RH and – 10°C

HAST highly available storage, *RH* relative humidity, *IEC* International Electrotechnical Commission

also more efficient than their predecessors, the size of the backlight finned heat sink could be reduced. This depth reduction component was used to allow the backlight to meet its optical requirements by expanding the depth of the backlight cavity. This expansion offsets the negative effects of reduced LED counts on display uniformity. This is but one example of how a clear system level understanding is needed to adapt products that rely on the ever changing (for now at least) landscape of LED technology. LED technology is still sufficiently immature enough that designs are changing at a much faster pace than the equipment and platforms in which harsh environment electronics are used. It is particularly important for harsh environment equipment, which tends to be constrained by environment and volumetric requirements, that designers have a 'big picture' understanding of system design when using an evolving technology such as LEDs.

The equipment used in many harsh environment applications, such as airborne, military, naval, outdoor lighting, etc., typically have service life requirements that may be measured in decades. Therefore, obsolescence can be a critical issue since new and improved devices might not work the same in the platform [28]. New aircraft are being designed for LEDs but there is a desire to retrofit older systems too. This can require extra cost and effort, such as if the old aircraft used incandescent lights operating on alternating current (AC) current that needs to be converted to direct current (DC) for LED applications.

Most packaged LED devices are produced for commercial applications. When used in military or aerospace environments, the equipment manufacturer must treat them as Commercial Off-The-Shelf (COTS) and may need to adapt design and manufacturing processes to address the limitations of COTS. For example, while commercial circuit board assembly extensively utilizes No Clean processes, the longer service life and more severe environments experienced by Harsh Environment

Electronics generally dictate that multiple cleaning processes are used in assembling circuit boards. The importance of these cleaning processes for extended life in harsh environments is growing as electronics manufacturers, even those producing military and avionics equipment, must utilize lead-free components due to regulations and part availability. The increased solderability challenges of some lead-free solders dictate the use of more aggressive fluxes that will increase the likelihood of electrochemical failures if circuit boards are not adequately cleaned. However, due to material compatibility issues, manufacturers of LEDs targeted at commercial applications may limit the cleaning processes that may be used on circuit boards that include packaged LEDs. It has been shown that a 'reduced clean' approach can be successfully utilized in the production of an LED backlight used in an avionics display, but only with a thorough understanding of the solder, assembly processes, and product use environment [29].

15.4.2 Thermal Technologies Applied to Light-Emitting Diode Systems in Harsh Environments

Thermal management of LEDs in harsh environments is generally similar to those used in other LED applications. There are three primary factors that tend to increase the challenges encountered by LEDs that are used in these applications. First, the junction temperature limits of harsh environment LEDs tend to be similar to those used in less rugged applications. Since ambient temperatures are generally higher in harsh environments, there is typically a smaller thermal budget in the design. This may push the system to utilize more exotic cooling methods. A second issue for harsh environment electronics in general is the significant differences in boundary conditions that may be encountered by a system during its field life, or even during a single mission. LEDs used for aircraft external lighting, for example, may experience severe solar loading with poor (natural convection) cooling while the aircraft is on the ground, followed by extremely low temperatures and pressures while the aircraft flies at cruising altitudes, which may then be followed by high pressures and condensing moisture as the aircraft lands. Each of these specific environments must be addressed in the system design, and their often contradictory requirements will prevent the design from being optimal for any one condition. Finally, the potential for exposure to harsh environments such as chemicals, severe solar loading, temperature extremes, varying acceleration/gravity, and simply the long service life may limit the materials and thermal management technologies that may be used in a given application.

Customer acceptance can also lead to conditions that may be harsher for the LED system. For example, in many lighting applications current technology uses little cooling enhancements as heat energy is radiated to the environment. Using cooling methods for LEDs that add volume, noise or reliability concerns, often prejudice the customer away from using an LED design. As such, lighting designers may face difficult challenges in producing a lighting system that meets all technical requirements and other (often unwritten) customer desires.

Another example of customer expectations is found in the display industry. As described previously, as LEDs have become more efficacious and more powerful, fewer LEDs are required to provide the amount of light needed for a display than in previous generations of LEDs. This leads to wider spaces between the LEDs that require a longer light injection length to the backlight to avoid dark regions. For a side-lit display, this will cause a larger bezel to be required, and this is often undesirable from a customer's point of view. With styling today emphasizing thin edge bezels, this creates a difficult condition to use LEDs and effectively manage the dissipated heat.

In other cases, it is not the end customer but a governing agency (such as a regulatory one) that mandates requirements for newer systems such as lighting technologies. Frequently such requirements were not imposed or may have not even existed prior to the introduction of a solid state lighting (SSL) solution; however, regulations or requirements are often imposed today for various reasons. Such requirements (such as those shown earlier for SSL Energy Star mandates in the United States) are frequently appearing and must be accounted for by engineers.

15.4.2.1 Thermal Management Approaches

The following sections briefly describe a few developments and applications of technologies that have either been used in the thermal management of LEDs in harsh environments or may be viable candidates for future use in those applications. The order of these sections is roughly organized along the thermal path that begins at the LED active area and ends at the system thermal management interface at which heat is transferred to the surrounding environment.

Device and Package

As mentioned previously, one of the challenges of integrating any electronics into harsh environments is the reduced thermal budget (the difference between maximum component temperature and the ambient temperature) that results from high ambient temperatures. One method to increase this thermal budget is to reduce the effective ambient temperature experienced by the LED through active refrigeration, such as with thermoelectric devices. For example, one study showed that up to a relatively high power dissipation of 35 W, the junction temperature of an LED could be reduced if it were cooled with a thermoelectric module (TEM) and fan/heat sink rather than the fan/heat sink only [30]. The TEM used in that study was quite large so this application would presumably be for cooling an array of LED devices and its use did increase the overall power consumption of the lighting system.

An alternative approach for increasing the thermal budget of an LED used in harsh environments would be to increase the allowable junction temperature of the device. Recent work has shown that high temperature LEDs targeted at automotive and harsh environment optical communications equipment could be produced using linear cascade green LED arrays. In these devices, an InGaN layer was inserted between an n-type GaN cladding layer and InGaN–GaN multiple quantum wells to efficiently spread current. These devices were shown to operate at 200 °C with minimal impact on performance [31]. It should be noted however that in many cases the device junction temperature is not necessarily the weak point in a thermal design. In order to effectively use high-temperature wide-bandgap semiconductors, high temperature packaging approaches and materials are essential.

If the thermal budget for the LED cannot be increased by addressing temperature limits, then the thermal resistance must be reduced to meet the needs for a given power dissipation. At the device packaging level, this may include the use of a conductive electronics substrate such as Low Temperature Co-fired Ceramic (LTCC) with thermal vias in a solid state lighting array [32]. Another interesting package level thermal management technology that shows promise is the integration of dielectric fluids directly into the LED package to provide die-level thermal management through immersion cooling [33].

One other method for managing some thermal loads in white light systems is the use of remote phosphor conversion. A typical white LED uses a thin layer of appropriate phosphor placed directly over the LED die. The phosphor converts some blue light to a combination of longer wavelengths to form white light. The phosphors dissipate heat from this conversion process, which means the die is heated from this source as well. Some LED systems have moved to remote phosphor designs in which the phosphor is placed outside the LED package. The LED in this case emits blue light, and the conversion takes place millimeters or even centimeters away from the die. This reduces the die heating and lessens the burden on the immediate conductive path away from the LED; however it does lead to higher temperatures on the emitting surface.

Interfaces

High brightness LEDs may dissipate significant heat (in the range of 1 to as much as ~ 10 W) from packages with footprints in the order of 1 mm^2. This leads to extremely high heat flux values that will generate large temperature gradients if the thermal interface resistance is not well addressed. Metallurgical attachment of high power LEDs to carriers may produce sufficiently low interface resistances, but they do not provide stress relief for thermal expansion effects and materials must be compatible to prevent voiding that will significantly degrade thermal performance [6]. Direct bond copper (DBC) has been used to attach high power packaged LEDs to substrates, but the relatively high thermal resistance of the LED packaging minimized the benefits of this improved interface resistance [34]. The authors of that study concluded that DBC would be most appropriate for chip on board LED applications, particularly LEDs used in pulsed operation that could take advantage of the increased thermal inertia due to the mass of the copper during transient heating. New thermal interface materials (TIMs), which provide some strain relief between mating surfaces, that include high conductivity materials such as graphite are commercially available for use with LED cooling from multiple suppliers [35, 36]. A number of research institutions are actively developing TIMs that integrate carbon nanotubes (CNTs) into interface material to leverage both their high theoretical thermal conductivity and strength. At least one study reported specifically addressing the use of CNTs in the thermal interface between an LED and a heat sink, with the CNT-based interface showing a thermal resistance of one-third that of silver filled epoxy [37]. While CNTs do have significant theoretical potential, there are still challenges in ensuring consistent physical/thermal attachment of CNTs to surfaces other than those on which they have been grown. In addition to TIMs that are used to fill spaces due to surface roughness, thermal gap fillers may be used to fill spaces created by tolerance stack-up and physical implementations (such as the space between a circuit board and the outer cover of a module). Thermal gap fillers tend to be much thicker and generally need to be quite soft to avoid excessive force being applied to electronic components. The combination of thickness and softness, which both contribute to low thermal conductance, gives thermal gap fillers a relatively high thermal resistance. However, they can establish a thermal path that may be critical in conduction cooled systems such as those often used in aerospace applications. Many, but certainly not all, commercially available thermal gap fillers have been shown suitable for the long service life and environmental requirements of harsh environment military and aerospace applications. Stringent testing for factors such as environmental conditions, flammability, etc., to ensure that materials are suitable for a given application is essential, particularly for materials used in commercial airliners [38].

Another significant area of interface research is the use of graphene in TIMs. In one study, the use of single and bilayer graphene with thicker multilayers loaded at 2 % in a grease TIM matrix increased thermal conductivity by almost 2.5 times the base value [39]. Since the graphene was created by a well-known and simple exfoliation process, it is possible graphene could be a significant conductivity enhancement without the issues associated with CNTs.

As discussed previously, LEDs require a sound conductive path from the die to the ambient. Any introduction of high thermal resistances along the conductive path has a detrimental effect on the LED. Interfaces in lighting systems, especially those made from typical sheet metal joints, are highly suspect not only at initial manufacturing but over time as corrosion, fastener loosening and general loss of interface contact may happen. Careful attention to design of these interfaces along with a proper test plan for lifetime reliability are required to ensure interface failure does not occur.

Thermal Spreading

Thermal spreading, covered in-depth in another chapter of this volume, is the principle of reducing a conductive thermal path's resistance by increasing the area through which the heat passes. The elementary equation for thermal resistance, L/kA, shows that a low thermal conductivity path (k) may be offset by increasing the cross sectional area (A) through which the heat moves. To increase this cross sectional area, the heat energy must be spread.

This is accomplished by using a material of high thermal conductivity under the thermal load, usually of a thin but large area design such as sheet material. Suitable materials for this are usually copper, aluminum or graphite (natural or pyrolytic). In some design situations, materials with comparatively low thermal conductivity may work. It has been shown that an LED street light can be fabricated from thin sheet metal, which alone does not have sufficiently low thermal resistance to move heat throughout the structure, when the sheet metal is underlain with flexible graphite sheet material [35]. Such a design maintains a passive cooling approach (usually desirable) while significantly enhancing system cooling.

Over the past 2 decades, the use of heat pipes and vapor chambers has grown exponentially for cooling applications such as laptop computers, video games, etc. The use of heat pipes for cooling LEDs has been studied by many researchers and there are many commercially available heat pipe assemblies that specifically target LED cooling. Heat pipes have a long history of use in some harsh environments, particularly space and satellite applications, but many of today's commercial heat pipes are not necessarily produced with a goal of a long service life. While heat pipes are certainly a viable option for cooling LEDs in harsh environments, the design needs to address whether a particular heat pipe is suitable for the application. Some relevant issues for heat pipes in harsh environments include the need for long-term reliability, freezing and frozen-start operation, and the effects of acceleration and the direction of gravity, which both tend to vary in airborne applications

System Cooling

System level cooling is generally considered to be the portion of the thermal management path in which heat is rejected to the surroundings. For this review, this definition will be expanded somewhat to include transient thermal management as well as conventional system level cooling.

For the vast majority of electronics systems, system level heat removal relies on convection to air. As the power density increases, the complexity of air cooling likewise grows as larger finned heat sinks and more powerful fans are required. For any electronics, not just LEDs, fans and finned heat sinks can be unsuitable for certain harsh environment applications due to the noise and reliability of fans and the potential for long-term degradation of finned heat sinks due to fouling and corrosion. As discussed earlier in this chapter (and in this book), there is significant interest in nonrotating fans, such as synthetic jets [27], 'flapper' fans [40–42], that are actuated with electromagnetic or piezoelectric sources. These devices do not typically produce significant flow rates but instead primarily generate unsteady flow to increase the local heat transfer coefficient. These can be operated at frequencies that are outside of, or in low sensitivity regions of, the audible range and eliminate some potential failure points in conventional fans, such as bearings.

One suggested approach for system-level cooling of automotive headlights is the use of free-stream airflow (due to the motion of the automobile) that passes through a metal foam heat sink [43]. This analytical study estimated the required size and thermal performance of a heat sink that is cooled by ram air from the automobile at speed. While this approach is much simpler than more active approaches, such as liquid cooling, the authors of this study did not address critical issues such as the need to cool the headlights when the automobile is stopped and also the significant potential for a metal foam heat sink to become fouled by the sand, dust, and insects that are likely to be encountered in use.

A number of relatively exotic methods have been proposed for cooling automotive LED headlights. One study investigated a number of thermal architectures for a headlight that had 15 LEDs that dissipated 2.7 W each [44]. Analysis showed that air cooling, either with natural convection or with fan cooling, would lead to excessive LED junction temperatures. This was primarily due to the limited volume available for airflow to the headlight. Passive liquid cooling was determined to not be viable, due to gravity limitations associated with the orientation of the headlight surfaces and their orientation within the automobile. A loop heat pipe was considered, however the installation and maintenance requirements would have required the loop heat pipe to include flexible sections that would have made it cost prohibitive. The authors concluded that a self-contained, pumped water/glycol system represented the best solution. This system was designed with separate heat sinks (in parallel) for the high- and low-beam systems, with a flow-rate of approximately 3 L/min and an operating system pressure drop of $\sim$ 15 kPa.

Fluid cooling methods that have been investigated to cool high heat flux regions in harsh environment LED arrays include microjet cooling [45] and microchannel cooling [46].

One other system consideration for lighting designs is whether the overall system operates at steady state or in some transient manner. Some types of lighting operate in cyclic or irregular manners (for example, traffic signals, flashlights, and camera flash units). If regulations and other considerations allow, it may be possible to use smaller cooling systems and rely upon the thermal mass and appropriate thermal time constants of the system to control the temperature of the LEDs. An alternative to simply increasing system mass is the use of phase change materials (PCMs), which can increase a thermal time constant of a system to provide improved transient cooling. PCMs are typically organic materials (such as waxes) or low melting temperature solders. These materials absorb energy while melting and most materials tend to have a similar volumetric latent heat of fusion that typically is on the order of ($\pm$ 50 %) $\sim$0.06 W min/cm^3, i.e., 3.7 W dissipation will melt 1 cm^3 of material in approximately 1 min.

15.5 Summary

LEDs used in harsh environments face similar thermal constraints as those used in less severe conditions. The factors that introduce the most severe challenges for integrating LEDs into these environments are the higher ambient temperatures that reduce the overall thermal budget, external conditions such as dynamic loading and corrosion that may prevent the use of insufficiently robust materials and packaging/thermal technologies, and the long service life of products used in these harsh environments. With careful design and thorough qualification testing, these challenges can be met and LEDs will increasingly be used in harsh environment systems in the coming years.

References

1. Institute of Traffic Engineers (ITE) Vehicle traffic control signal heads: Light Emitting Diode (LED) Circular Signal Supplement from Jan 2005 (http://www.ite.org/standards/led.asp)
2. Acarlar MS, Plourde JK, Snodgrass ML (1990) A high-speed surface-mount optical data link for military applications, Digital Avionics Systems Conference. Proceedings, IEEE/AIAA/NASA 9th, pp 297–301
3. Chan EY, Beranek MW, Davido KW, Hager HE, Hong CS, St. Pierre RL (1996) Challenges for developing low-cost avionics/aerospace-grade optoelectronic modules, Electronic Components and Technology Conference. Proceedings, 46th, pp 1122–1129
4. Beranek MW, Chan EY, Hager HE, Le QN (1998) Status of optoelectronic module packaging for avionics/aerospace applications, Lasers and Electro-Optics Society Annual Meeting. LEOS '98. IEEE, Vol. 2, pp 329–330
5. Davis J, Tchon J (2011) Evolution of LED backlighting in avionics displays. Display Technologies and Applications for Defense, Security, and Avionics V; and Enhanced and Synthetic Vision 2011. Proc SPIE 8042
6. Wilcoxon R, Cornelius D (2006) Thermal management of an led light engine for airborne applications. SEMI-THERM, pp 178–185
7. www.iddaerospacecorp.com/faq_led.html
8. http://www.aarcorp.com/gov/Mobility/Shelters/Tactical_Mobile_Shetlers/PDFs/Opt05-17-ExpandableISO.pdf
9. www.lumecon.com/technology/military
10. Treichel T (2010) Environmental testing and analysis of ruggedized LED system designed for internal vehicular crew lighting. 40th International Conference on Environmental Systems, Barcelona, Spain, 11–15 July 2010, Paper No. AIAA-2010-6288
11. http://www.sti.nasa.gov/tto/Spinoff2010/cg_1.html
12. http://www.diodes.com/zetex/_pdfs/5.0/pdf/Automotive_lighting_issues_for_high_brightness_LEDs.pdf
13. Bielecki J, Jwania AS, El Khatib F, Poorman T (2007) Thermal considerations for LED components in an automotive lamp. 23rd IEEE SEMI-THERM Symposium, San Jose, USA, March 2007, pp 37–43
14. Hamada K (2010) Present status and future prospects for electronics in electric vehicles/hybrid electric vehicles and expectations for wide-bandgap semiconductor devices. In: Friedrichs P, Kimoto T, Ley L, Pensl G (eds) Silicon carbide, Vol. 2: Power devices and sensors. Wiley-VCH Verlag GmbH & Co. KGaA, Weinheim, pp 1–19
15. Hardy KR, Olsson MS, Lakin BP, Steeves KA, Sanderson JR, Simmons JE, Weber PA (2008) Advances in high brightness light emitting diodes in underwater applications, OCEANS 2008; Quebec City, Canada, 15–18 September 2008, pp 1–5
16. http://www.evcam.com/datasheet_US_Electric_Line.pdf
17. Kinzey B (2010) "LEDs in the real world", panel presentation at the transformations in lighting: 2010 DOE solid-state lighting R&D workshop, Raleigh, North Carolina,2–4 February 2010. *http://www1.eere.energy.gov/buildings/ssl/raleigh10_materials.html*
18. "Energy-saving traffic lights blamed in crashes", http://www.nbcnews.com/id/34436730/ns/us_news-life/t/energy-saving-traffic-lights-blamed-crashes/. Accessed 28 Jun 2012

19. Chang MH, Das D, Varde PV, Pecht M (2011) Light emitting diodes reliability review. Microelectron Reliab 52(5):762–782. doi:10.1016/j.microrel.2011.07.063
20. Riegler B, Bruner S, Thomaier R (2004) Optical silicones for use in harsh operating environments. Proc. SPIE 5590, 140 (2004), Optics East, Philadelphia, PA, 25–28 October 2001. doi:10.1117/12.568157
21. Petroski J, Norley J, Schober J, Reis B, Reynolds RA (2010) Conduction cooling of large LED array systems. 12th IEEE Thermal and Thermomechanical Phenomena in Electronic Systems (ITherm), Las Vegas, USA, 2–5 June 2010
22. IES LM-79 (Electrical and Photometric Measurements of Solid-State Lighting Products); http://www.ies.org/store/product/approved-method-electrical-and-photometric-measurements-of-solidstate-lighting-products-1095.cfm
23. IES LM-80 (Approved Method: Measuring Lumen Maintenance of LED Light Sources); http://www.ies.org/store/product/approved-method-measuring-lumen-maintenance-of-led-light-sources-1096.cfm
24. IES TM-21 (Projecting Long Term Lumen Maintenance of LED Light Sources); http://www.ies.org/store/product/projecting-long-term-lumen-maintenance-of-led-light-sources-1253.cfm
25. Energy Star® Program Requirements, Product Specification for Lamps (Light Bulbs); Eligibility Criteria—Version 1.0, DRAFT 1 (as of December, 2011)
26. Johnson R (2005) LED improvements summary report, Rockwell Collins Internal Report, Sept.
27. Schwickert M, Noska B (2011) LED cooling in harsh environments with synthetic jet technology. LED Professional Review, May/June 2011, ISSN 1993–890X, pp 44–49
28. www.aviationtoday.com/av/issue/feature/Product-Focus-Lighting_67139.html
29. Hillman D (2008) A reduced cleaning of printed wiring assemblies: the science of leaving flux residues alone. 2008 IPC/SMTA Cleaning Symposium, Chicago, USA, 28–29 October
30. Zhong D, Qin H, Wang C, Xiao Z (2010) Thermal performance of heatsink and thermoelectric cooler packaging designs in LED. 11th International Conference on Electronic Packaging Technology & High Density Packaging, Xi'an, China, 16–19 August 2010, pp 1377–1381
31. Jin-Wei Shi, Huang H-W, Kuo F-M, Sheu J-K, Lai W-C, Lee ML (2010) Very-high temperature (200 °C) and high-speed operation of cascade GaN-based green light- emitting diodes with an InGaN insertion layer. IEEE Photon Technol Lett 22(14):1033–1035
32. Shi PZ, Arulvanan P, Lu ACW (2007) Thermal performance evaluation of power SSL module in LTCC. Electronics Packaging Technology Conference, EPTC 2007; Singapore, Malaysia, 10–12 December, pp 285–288
33. Arik M, Utturkar Y, Weaver S (2010) Immersion cooling of light emitting diodes. Thermal and Thermomechanical Phenomena in Electronic Systems (ITherm) 2010, San Diego, USA, 2–5 June 2010, pp 1–8
34. Dehmel A, Schulz-Harder J, Roth A, Baumeister I (2007) Direct copper bonded ceramic substrates for use with power LEDS. Electronic packaging technology, ICEPT 2007, Shanghai, China, 14–17 August 2007, pp. 1–6
35. http://www.graftechled.com/HITHERM
36. www.industrial.panasonic.com/ww/i_e/00000/led_solution_e/led_solution_e/pgs_e.html
37. Kai Zhang, Yuen MMF (2009) Heat spreader with aligned CNTs designed for thermal management of HB-LED packaging and microelectronic packaging. Electronic packaging technology, ICEPT 2006, Shanghai, China, 26–29 August 2009, pp 1–4
38. http://www.ledsmagazine.com/features/6/1/3
39. Khan SMF, Balandin AA (2012) Graphene-Multilayer graphene nanocomposites as highly efficient thermal interface materials. Nano Lett 12(2):861–867
40. Petroski J, Arik M, Gursoy M (2010) Optimization of piezoelectric oscillating fan-cooled heat sinks for electronics cooling. IEEE Trans Compon Packag Technol 33(1):25–31
41. Kai Z, Xiao DGW, Zhang X, Fan H, Gao Z, Yuen MMF (2011) Thermal performance of LED packages for solid state lighting with novel cooling solutions. 2011 Thermal, mechanical and multi-physics simulation and experiments in microelectronics and microsystems (EuroSimE), Linz, Austria, 18–20 April 2011, pp 1–7

42. Ma HK, Chen BR, Lan HW, Lin KT, Chao CY (2009) Study of an LED device with vibrating piezoelectric fins. 25th IEEE SEMI-THERM Symposium, San Jose, USA, 15–19 March 2009, pp 267–272
43. Morkos B, Shankar P, Teegavarapu S, Michaelraj A, Summers J (2009) Conceptual development of automotive forward lighting system using white light emitting diodes. SAE World Congress & Exposition, April 2009, SAE Paper No. 2009-01-0593, doi:10.4271/2009-01-0593
44. Lai Y, Cordero N, Barthel F, Tebbe F, Kuhn J, Apfelbeck R, Würtenberge D (2009) Liquid cooling of bright LEDs for automotive applications. Appl Therm Eng 29(5–6):1239–1244
45. Liu S, Lin T, Luo X, Chen M, Jiang X (2006) A microjet array cooling system for thermal management of active radars and high-brightness LEDs. Proceedings of the 56th electronic components and technology conference, San Diego, USA, 30 May–2 June 2006, pp 1634–1638
46. Yuan L, Liu S, Chen M, Luo X (2006) Thermal analysis of high power LED array packaging with microchannel cooler. Electronic Packaging Technology, ICEPT '06.

Chapter 16
Future Directions in LED Applications

Robert F. Karlicek

Abstract This chapter describes possible future development directions of solid-state lighting from a high-level materials, devices, and system applications perspective. Since solid-state lighting is likely the highest volume future LED application and is a completely new form of lighting technology, advances in semiconductor and packaging materials, LED devices, and new lighting systems designs are convolved with the evolution of business models and new capabilities brought by new corporate entrants to the rapidly evolving global solid-state lighting business. Examining the development of future LED applications is, therefore, complicated, and it is helpful to briefly set the stage with a brief review of current and emerging LED technology trends and applications.

16.1 Introduction

Since the introduction of high-brightness GaN LEDs in the early 1990s [1], LED technology and applications have expanded rapidly, especially in the past 20 years. Today, LED technology continues to evolve to meet the demands of existing markets requiring efficient emission of optical energy from the deep UV to the near IR, with current technical and manufacturing efforts focused primarily on the development of efficient solid-state lighting systems based on blue LED/phosphor combinations [2]. In lighting markets, LED technology can be regarded as disruptive [3], as it will drive major changes in the design and operation of lighting systems, as well as create new applications uniquely enabled by compact, efficient solid-state devices. In some ways, LEDs are a disruptive technology in the same way as the development of solid-state electronics was disruptive in the electronics industry in the middle of the twentieth century [4]. In both cases, the solid-state electronics are (were) the logical successors to vacuum tubes (bulbs and tubes) with their hot filaments, plasmas, and sockets, but technologies and markets served by silicon are considerably different from those involved in lighting: The former created vast new markets (enabled by

R. F. Karlicek (✉)
Smart Lighting Engineering Research Center, Rensselaer Polytechnic Institute, CII7015,
110 8th Street, Troy, NY 12180, USA
e-mail: karlir@rpi.edu

C. J. M. Lasance, A. Poppe (eds.), *Thermal Management for LED Applications*,
Solid State Lighting Technology and Application Series 2,
DOI 10.1007/978-1-4614-5091-7_16, © Springer Science+Business Media New York 2014

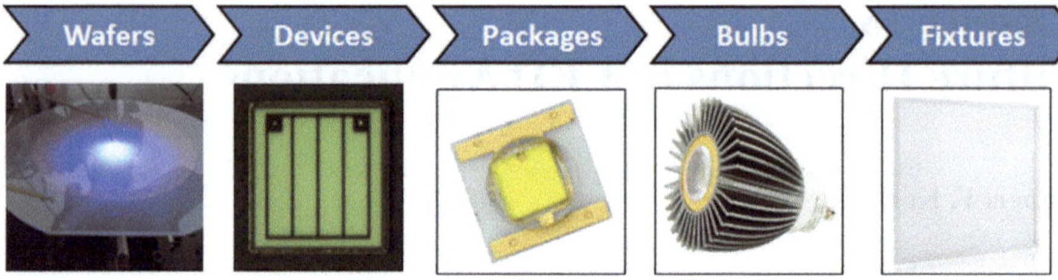

Fig. 16.1 The conventional supply chain used to describe the operation of the developing solid-state lighting industry. (Pictures courtesy of SemiLEDs Corp., used by permission)

large-scale and/or low-cost, high-speed computation) that previously did not exist while the latter is primarily focused on meeting the age-old human requirement for controllable, high-quality illumination. This distinction is important, because the evolution of solid-state lighting systems will be shaped by technical advancements in materials and LED device design driven by energy efficiency and system cost requirements, but will also be shaped by strong human factors considerations and will need to be integrated with architectural design, artistic expression, human physiological requirements, and cultural development. One need to only examine how fuel-based lighting design (candelabra and torchieres and table lamps) has impacted the design and performance expectations of modern electric lighting for the past century to see that human factors will remain an important part of solid-state lighting systems design. While initial aspects of solid-state lighting system adoption will be highly constrained by existing lighting design rules and consumer expectations based on current electric lighting properties, the migration from *electric* to *electronic* lighting enabled by LED technology will also enable new illumination form factors and integrated services, which, before LEDs, were not possible.

16.2 The Current State of LED Technology and Applications

The intent of this section is not to provide a detailed review of current LED technology, but rather to describe its current evolutionary status to demonstrate the unsettled technical landscape from which future LED applications will develop, and it follows a broad outline derived from the basic LED supply chain for solid-state lighting shown in Fig. 16.1.

Generally, even the most foundational LED die fabrication processes are technologically immature, undergoing significant development even as new, high-volume markets like solid-state lighting are developing. Continued development, even at the most upstream portion of the LED supply chain, will have an impact on the development of future LED-based applications, impacting efficiency as well as optical, thermal, and electronic designs that support those applications. Rapid technical evolution in all parts of this chain will ultimately rewrite how semiconductor technology for lighting and advanced illumination systems is designed and manufactured. Similarly, business models governing the manufacturing and delivery of illumination and lighting controls will change dramatically as system performance, new services, and reliability outstrip capabilities of conventional lighting systems.

16.2.1 Substrates

In the realm of compound semiconductors, device structures are fabricated from complex engineered semiconductor alloy thin films stacked epitaxially on substrates, which serve to insure a high degree of crystalline quality in the epitaxial layers. For the GaAlAs and InAlGaP alloy systems, the substrate of choice is GaAs, which allows the growth of high-quality epitaxial films. In the case of high-brightness visible AlInGaP LEDs, this substrate is typically removed during the LED fabrication process as it absorbs visible light. Though there continue to be incremental improvements in the growth and performance of AlInGaP visible LEDs, this can currently be regarded as a mature technology.

In the case of shorter wavelength, LEDs based on the AlInGaN alloy material system, substrate technology is still evolving. Historically, sapphire (and to a lesser extent, SiC) are used as substrates for the growth of AlInGaN LEDs emitting from the ultraviolet through the green range. Driven by quality challenges related to the use of less than ideal (where ideal would be GaN) substrates, as well as cost and production challenges of manufacturing large diameter sapphire and SiC substrates required for reducing the cost of LED chip fabrication, there are several different substrate technologies under development for GaN LEDs. Sapphire is the most widely used substrate for LED fabrication, and SiC is also used. Silicon, GaN and various types of composite substrates are also under development as substrates for LED fabrication. In most cases, the substrate technology chosen impacts the type of the LED semiconductor element that is produced for subsequent packaging, so the continuing evolution of substrate technology will impact several important aspects of future GaN-based LED applications. Beyond impacting the crystalline quality of the epitaxial layers grown to produce the LED die, the substrate chosen determines certain aspects of the optical, thermal, and electrical design of the LED package and, to a degree, the subsystem in which it is installed. This will be clearer in the next section where current LED die design platforms are considered.

For most substrate options, the substrate is removed during the LED chip (or in some cases, packaging) production process to facilitate the extraction of light from the epitaxial film. Key considerations in the selection of a substrate technology for LED manufacturing are driven by LED production costs, and those costs are largely determined by average wafer-level LED brightness, brightness uniformity, and wavelength uniformity. At the present state of LED substrate and epitaxial growth development, there is no clear substrate winner from a production yield or cost point of view. In general, the crystallographic quality of LEDs grown on any of the substrate options either in use or under development for LED epitaxial growth (with the exception of native GaN substrates) is poor when compared to LEDs grown in the AlInGaP material system on GaAs substrates. One possible exception (beyond native GaN substrates) may be the use of patterned sapphire substrates. In order to increase the light extraction from the LED die, growth is performed on processed sapphire wafers where light-extraction structures are etched into the surface before growth. Generally, epitaxial growth on nonplanar surfaces reduces the crystalline quality of the epitaxially grown surfaces, but since the crystallographic quality of GaN grown on sapphire is poor to begin with, improved light extraction and higher levels of light output from LEDs fabricated on patterned sapphire substrates has led to widespread use of this technique. Typically, the pattern feature size is on the order of a few micrometers, but recently, nanopatterning of sapphire surfaces before epitaxial growth suggests that both improved crystallographic quality and improved light extraction may result [5].

One substrate wildcard for the future evolution of LED chip technology is the development of native GaN substrates for LED growth. Today, GaN substrates are used for blue and green GaN-based laser diode production, but are typically considered too expensive for price-sensitive LED markets. This may change, as new technologies [6] enable the growth of lower cost, larger diameter substrates. An important consideration for LEDs stems from the use of GaN substrates for laser diode production: Much lower defect density (i.e., reduced crystallographic dislocation density) enables better light emission performance. Recent developments in LEDs grown on low dislocation density GaN substrates suggest that high-quality LED performance with much greater linearity of light output at high current density is possible [7], enabling high levels of light emission from smaller LED die.

This use of GaN substrates for very low droop,[1] near UV LEDs has been recently described by a startup company spun out of the University of California in Santa Barbara (Soraa) [8] and demonstrated in the development of a high-performance MR-16 style light bulb. The key to the Soraa development is the use of relatively large diameter (100 mm) GaN substrates with low dislocation density ($\sim 10^5$ cm^{-2}) to grow high-quality, low-droop short-wavelength LEDs. This achievement has resulted in several unique characteristics of the GaN die used in their lighting applications. First, it is quite small, operating at much higher current density than a typical high-power LED grown on other substrates. Second, in their current product, the LED die is triangular for improved light extraction. The size and shape of the die is important, and while the optical transparency of GaN substrates is probably not as good as in case of sapphire, the index match to the epitaxially grown LED structure aids in the extraction of light from the sides of the LED die. If this technology can prove cost-effective, then the use of smaller, higher power density LED die will significantly impact the thermal and optical design of light sources where a high-brightness point source is required (most spot lighting applications). Should the Soraa development be

---

[1] Droop refers to the drop in LED light production efficiency with increasing current density common in GaN based LEDs, particularly longer wavelength blue and green LEDs. The droop effect is different from efficiency reduction caused by LED heating due to higher operating current which can be minimized through adequate thermal design.

cost effectively extended to other wavelengths and integrated with improved thermal management technologies needed for high power densities enabled by reduced/no droop devices, it has the potential to significantly impact the future of solid-state lighting.

Another substrate wildcard is the successful development of GaN on Si devices. Currently, several companies are transitioning GaN on Si to prototype scale LED production even though top bin LED performance on silicon substrates is not quite as good as top bin LED grown on sapphire substrates. Less important than the simple LED on silicon, however, is the possibility that subsequent integration of electronic control and sensing functions in the silicon integrated with LED structures for light emission could lead to monolithically integrated driver and control functions as part of future LED designs. This functionality would most likely be implemented at the packaging step where processed Si wafers are used as a replacement for conventional lead frames or ceramic submounts. There are many technical challenges that need to be solved before such devices can be realized, including light extraction in the presence of silicon and compatibility of materials used for LED fabrication (Au contacts, Ag mirrors) with conventional silicon fabrication processes. If developed, it would start the transition from simple LEDs to LEICs, or *light-emitting integrated circuits*.

16.2.2 LED Die Performance and Design

Once a substrate technology is chosen for epitaxial growth, a complex set of alloy compositions are grown on that substrate. With current LED layer designs and epitaxial growth technologies, the number of distinct layers of different compositions grown can reach hundreds, each with carefully choreographed patterns of gas flow changes and temperature adjustments. The typical epitaxial growth time can range from 4 to 8 h, depending on the specific LED design. The resulting wafer is then processed into LED die through a series of additive (e.g., ohmic contact deposition) and subtractive (patterned etching) process steps. These processes are largely similar in concept to those used in silicon IC fabrication, but feature sizes are much larger and materials used to complete the LED die structure are different from those typically used in conventional IC fabrication. The fabricated dies are tested at wafer level before separation into individual parts using high-speed laser dicing processes. The cumbersome die by die testing of electro-optical performance is required because the nonuniformity of the distribution of the optical power and wavelength within the wafer is significant. Dies are sorted by brightness and wavelength (the binning process) for subsequent packaging.

The state-of-the-art (top bin) performance for AlInGaN-based LED devices versus emission wavelength is shown in Fig. 16.2, where estimated data for several consecutive years are shown [9]. It is clear that the highest efficiency is obtained for blue LEDs around 450 nm, and there are gradual improvements at discrete, commercially important UV wavelengths. At commercially important wavelengths in the green and yellow spectral regions, progress on improving the LED efficiency has

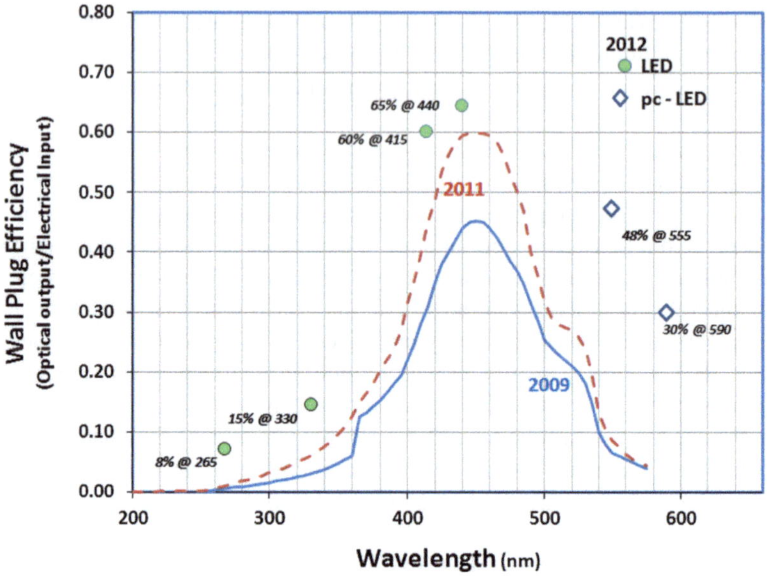

Fig. 16.2 Efficiency versus emission wavelength for GaN-based LED devices showing progress between 2009 and 2012

been slow, except for the structures using highly efficient blue LEDs to excite green and yellow-emitting phosphors (diamonds in Fig. 16.2).

More generally, phosphor conversion is used for essentially all-white LEDs in production today for solid-state lighting applications. A careful selection of blue LED die wavelength and blends of various phosphors are used to fabricate white LEDs with different spectral content for cool to warm white lighting applications, though some companies are evaluating the wafer-level deposition of phosphor blends during the chip fabrication process (before testing and separation into discrete LED dies) to reduce production costs and possibly tighten binning distributions.

The electro-optical design of the LED die is also continuing to evolve, not so much for increased light generation efficiency but for improved efficiencies at the packaging and system level, though the origin in the diversity of LED chip design steps from earlier efforts to improve LED light-extraction efficiency. Blue LEDs used for phosphor-converted white light applications are starting to approach practical limits of efficiency in lab demonstrations, so design is moving to platforms that permit easier package and subsystem-level integration.

It is possible to classify LED die into one of two different categories, depending on how the electrical current flows inside the LED device as shown in Fig. 16.3. The lateral design is produced in the highest volume for applications in small and large display devices, while the vertical design is used for most lighting applications, with one exception—the thin film flip chip design popularized by Philips Lumileds being in the lateral design category.

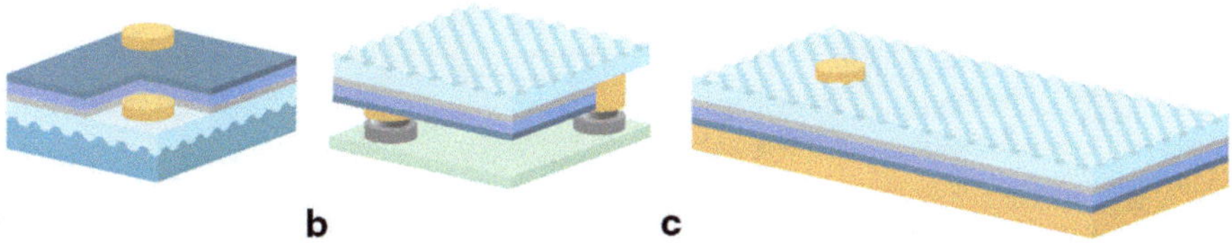

a b c

Fig. 16.3 **a** Conventional lateral LED grown on patterned sapphire substrate, with light emitted from five sides. **b** Lateral chip design flipped upside down, shown here with the sapphire removed. **c** Vertical style LED with a metal or semiconductor submount and the sapphire removed

Vertical LEDs are fabricated by growing the epitaxial LED structure on the substrate of choice (not including bulk GaN substrates). The general process is shown in Fig. 16.4 for a chip, though the process is performed at the wafer level. The figure also shows the use of a patterned sapphire substrate for the epitaxial process.

Typically, simple planar sapphire substrates are used, but the process can also be designed to work with the patterned sapphire substrate. The anode structure of the wafer surface (typically comprises a highly reflective metal stack) is fabricated and then bonded to a carrier wafer and the original epitaxial growth substrate is removed. Processing then continues to form cathode ohmic contacts and light-emitting structures prior to die testing and singulation into individual chips. This structure, first developed for GaN LEDs by Osram, has two key advantages over most lateral chip designs:

1. The carrier substrate is typically a good thermal conductor, so heat removal from the LED structure is more efficient than in case of most lateral designs.
2. The thin light-emitting layer facilitates better light extraction than is possible with most lateral LED designs.

Related to (2), the improved light extraction enables the fabrication of exceedingly large high-power LEDs with very high brightness per die [10].

One disadvantage of the vertical LED is that the bottom of the die is the anode and the primary heat extraction surface. For high-brightness lighting requirements, light outputs from multiple die frequently need to be combined, and the package design must allow for electrical isolation of the individual die. (Generally, the operation of LED die in parallel with common anodes and cathodes is prone to reliability issues associated with the nonuniform distribution of operating voltage between the different die in the package. This issue can be mitigated somewhat by careful forward voltage binning where the voltage is presumed to be stable over the life of the assembly.) Another, somewhat awkward issue for the vertical LED design is the need for a top surface wire bond in the optical path of the package, though this can be avoided with clever designs [11].

The lateral structure was the first commercially available GaN-based LED design and is currently the most widely used one, primarily for white LEDs used for mobile display manufacturing. Here, the die are packaged in an inexpensive conventional lead frame format and the sapphire base provides electrical isolation, enabling separate electrical and thermal conduction paths in the package and module design.

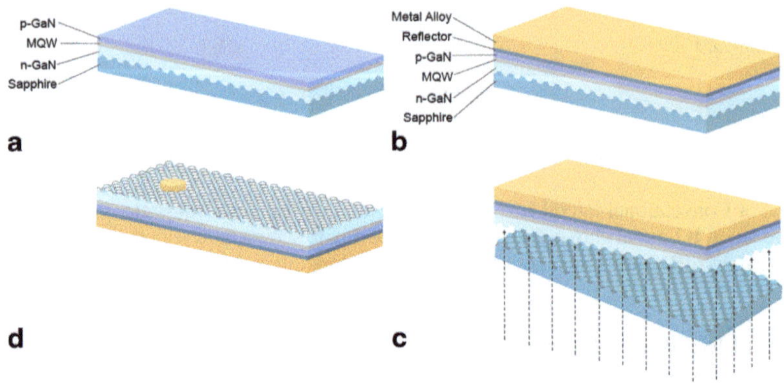

Fig. 16.4 Sequence used for vertical LED fabrication (typically performed at wafer level): **a** Normal LED structure with a patterned sapphire substrate. **b** A submount (metal or silicon) is bonded to the top of the LED wafer. **c** The structure in **b** is irradiated with a high-power laser, separating the sapphire from the bottom GaN. **d** The cathode side is completed with light-extraction electrical contacts

While vertical LEDs are currently widely used for solid-state lighting, the lateral chip design is starting to gain popularity because in a flipped chip format (Fig. 16.3c), multiple chips can be connected using circuitized ceramic carriers developed long ago for other electronic systems (e.g., radio frequency [RF] power amplifiers). In addition, arrays of die can be packaged without the need for wire bonds to the top surface of the LED die. Finally, these die designs are compatible with the latest wafer scale packaging technologies where advanced silicon fabrication technology (3D packaging, MEMS devices, etc.) can be used to create advanced LED packages that ideally integrate other functionality (from simple ESD protection diodes to more advanced driver and switching capability, as mentioned above).

There is one additional feature of lateral LEDs that is starting to draw interest and that is the use of LED fabrication designs, which permit several diodes (pn-junctions) to be fabricated on a single die. Depending on the ways in which these individual devices are interconnected, the integrated diode structure can be operated at higher DC voltages [12] or as an AC LED [13]. The advantages of AC LEDs are clear in that it, ideally, eliminates the need for AC/DC conversion and a driver, but testing methods for such solutions are not yet as advanced as for DC-driven LEDs. While conceptually clear, there are issues with harmonic distortion and flicker when AC-driven LEDs are used in lighting systems, but these issues can be managed at the subsystem level [14].

In some cases, the high-voltage DC LED helps to solve an issue with the design of drivers for higher current LEDs where the output stage of the driver typically requires a form of energy storage (inductors and capacitors) to achieve the current levels for power LEDs used for lighting. By designing power LEDs that operate at low current but high voltage, LED driver design can be simplified considerably [15].

The integration of many discrete diodes into a single die (HV or AC LEDs) comes at the expense of reduced light emission per unit area of LED die—especially in the case of AC LEDs where only 50 % of the LED area is emitting at any point in time. As substrate diameters go up, epitaxial growth yields improve, and economies of scale and competition in LED manufacturing continue to increase, what used to be considered precious GaN epitaxial material will be (and is currently being) rapidly commoditized. Other considerations, like thermal management, optical subsystems, and driver electronics, are currently the dominant cost factors in the design of most lighting applications that use LEDs. As LED die manufacturing continues to mature, LED die cost will continue to become less important in the cost structure of most advanced LED applications.

While there continues to be major research efforts on the design of efficient LEDs, research on improving the efficiency of blue LEDs used for solid-state lighting applications is approaching what is probably a point of diminishing returns. There are major improvements in LED design needed at both shorter and longer operating wavelengths, and ideally, the efficiency of AlInGaN LED structures at any UV or visible wavelength longer than about 250 nm will approach the current level of performance of blue LEDs at about 440 nm. If the efficiency of AlInGaN LEDs becomes nominally wavelength-independent, then there is a wide range of new applications that will become possible. Even in the field of solid-state lighting, the efficiency of light sources comprises discrete, efficient, red, green, blue, and yellow (RGBY) could then be higher than current phosphor-converted white LEDs. The theoretical luminous efficacy of radiation (LER) for narrow emission linewidth 100 % efficient RGBY sources delivering high-quality white light exceeds 400 lm/W [16].

Beyond wavelength considerations, evolution of LED die designs will be driven by cost and system integration considerations. LED die that lend themselves to higher speed packaging and designed to enable more efficient integration with thermal management systems and secondary fixture optics will bring down the cost of system fabrication, so in some sense, the roles for the mechanical, thermal, electrical contact, and optical designs of LED die will increase in importance.

One wildcard in the continued evolution in the development of LED die is the potential to integrate both electronics and optoelectronics on the same die. This was mentioned briefly in the consideration of LED die structures grown on silicon wafers and the creation of an LEIC. Another approach would be to explore monolithic integration of GaN electronics and LEDs, since GaN is now being developed independently for a host of power electronics and high-speed switching applications [17]. It may be possible to integrate simple electronics functionality and LED operation into the same wafer structure. In the limit that GaN epitaxial materials continue to get less expensive, monolithic integration of light emission and simple control electronics in the same structure would lead to system simplification, improved reliability, and reduced costs in much the same way as has been true for the integration of different functionality (e.g., combined analog and digital functions) in increasingly complex silicon integrated circuits. While the level of complexity of a monolithically integrated LEIC fabricated completely from AlInGaN materials would be many orders of magnitude less than current silicon ICs, the development of such integrated structures is quite feasible.

16.2.3 LED Die Packaging and Subsystem Integration

Continuing down the supply chain shown in Fig. 16.1, the next stop for the LED die is the packaging area. Technical details of LED packaging have been described elsewhere [18] and conventional packaging issues are well understood. In the previous sections on substrate and LED die considerations, status and future trends were highlighted and the future considerations in the selection of substrates and LED die designs would be increasingly defined by cost and ease of mechanical and electronic integration, as die efficiency was beginning to approach its practical limitations.

An LED package in a conventional sense is designed to protect the LED die, improve the light extraction and direct light in a manner required for the application, and facilitate integration with the subsystem (e.g., soldering onto a circuit board) with considerations for thermal management. Today, there is a very wide variety of packaging styles used for LED applications, depending on whether the die is integrated into a display system, an automotive tail light, or an LED replacement light bulb. Increasingly, LED end-users are developing new approaches to packaging that blur the lines of the conventional supply chain by both reducing the cost of the final product and developing new and innovative system designs using LEDs. Some of these new innovations for LED packaging are being adapted from advanced silicon and electronic packaging concepts like chip-on-board, wafer-level packaging, and system in package concepts. The former technology is designed to streamline packaging for reducing LED packing costs and developing integration of control electronics, with the latter being used for compact integration of light emission and electronic control and even sensing functions in a single subassembly [19]. In the case of chip-on-board technology, more LED end-users are bypassing conventional LED packaging in the supply chain to enable more creative integration of thermal and optical design. This is what might be called "die to fixture" integration and requires the applications end-user to become proficient in rudimentary aspects of LED die handling, encapsulation, and color conversion (phosphor deposition) technologies.

Current efforts to drive LED bulb replacement (also called retrofit) technology is driving most LED packaging development where the integration of optical, thermal, and AC/DC conversion into a compact structure has already led to the commercialization of a wide variety of LED replacement bulbs of varying designs (see Chap. 12). It is estimated that there are presently several thousand LED bulb companies chasing existing global incandescent bulb and fluorescent tube sockets [20]. While efforts are underway to standardize the electromechanical design of lighting modules that can be used for existing lighting sockets [21], both continued rapid innovation in LED die and packaging technology will remain a challenge for standardization. Another factor that will remain a challenge for standardization is what might be called socket saturation: As LED replacement units achieve the long lifetimes of which they are capable if properly designed, fewer and fewer new units will be required as available sockets are filled [22]. In addition to the fact that LED retrofit lighting systems cannot be optimally designed from a thermal management point of view, concerns regarding socket saturation (and perhaps, ultimately, socket extinction) will be one of the key drivers of future LED applications as lighting suppliers seek to add new features and capabilities to advanced lighting systems to maintain revenue growth.

Solid-state lighting will be the dominant driver of evolving LED packaging and module integration development. Future LED lighting systems will most likely move away from conventional LED packaging designs for reasons of cost, efficiency, improved integrated thermal management strategies and lighting system design evolution, and some of these trends are starting to emerge as lighting companies consider how to develop new lighting paradigms that enable delivery of illumination that balances the functional and artistic requirements of good lighting design. At the very highest level, it is possible to divide illumination systems into spot (bulbs) and diffuse (fluorescent) sources. Current LED packaging designs and the directional nature of LED die light emission are most suited for spot illumination applications, the market where solid-state lighting is making its most rapid penetration. It is in the area of diffuse lighting that significant LED packaging and systems integration innovation is needed.

While diffuse illumination applications have long been the promise of OLED technology, continuing efficacy, reliability, cost, and color stability challenges [23] of OLEDS are driving lighting companies to consider new approaches for diffuse lighting using LEDs. One platform currently under development is the use of LEDs to illuminate a waveguide with surface light- extraction features creating a diffuse light source. This is very similar, at least in concept, to how LEDs are used in edge- lit back light units used in LED–LCD TVs (see Chap. 14). Variations on this theme are being developed for fixture- level deployment of solid-state lighting that has the look and feel of fluorescent tube lighting fixtures. Called panel lights, these systems can be edge lit with linear arrays of white LEDs or back lit with 2D arrays of LEDs, and are more efficient than LED tube replacements designed as pin- compatible replacements for conventional fluorescent tubes. As LEDs and waveguide technology continues to improve, system- level efficacies should exceed those of high-end T5 fluorescent tube fixtures, but will never approach the magnitude of the efficiency advantages over fluorescent tube technology that LED bulbs demonstrate over incandescent light sources, so incentives to install LED replacement lighting will need to be driven by other considerations. For example, LED edge waveguide solutions offer interesting design flexibility (interesting shapes and patterns), which can mix both direct and indirect illumination with specially designed waveguide light- extraction features (Fig. 16.5). Approaches like these offer LED advantages (no warm up time, reduced maintenance, broader fixture design space) to lighting designers than are possible with fluorescent tube technology.

Beyond diffuse lighting with waveguide technology that couples conventionally packaged LEDs to the patterned waveguide, radically new packaging approaches are also being explored for LED die packaging to address the need for the low-cost fluorescent fixture replacement market. These nascent technologies are developing in response to the need to assemble and electrically connect thousands to millions of small LED die in a sheet (rigid or flexible) at high speed and low cost. Some approaches involve mesoscopic self-assembly concepts developed in the late 1990s [24]. Other approaches resembling high-speed, low-cost printing processes (screen

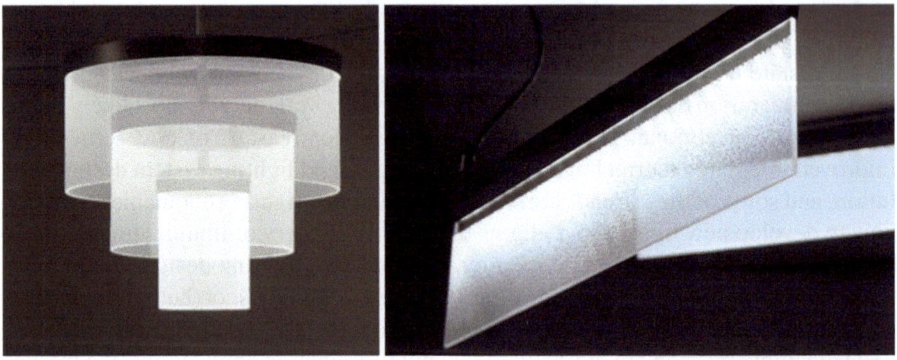

Fig. 16.5 High brightness LEDs can be coupled to specially designed waveguides to create new fixture designs. (Courtesy of Rambus, used by permission)

printing of LED die or high-speed pad stamping) are also under development [25]. These approaches will also drive the development of new materials (encapsulants, phosphors, conducting materials) and processes to create sheets of light at low cost. As suggested previously, these developments may also require differently designed LED die to facilitate high speed, low error rate placement of the light-emitting element into the diffuse illumination matrix, as well as integrated optical and thermal management solutions needed for operating parameters such as uniformity and luminance homogeneity, which might be considerably affected by thermal issues.

While the concept of millions of small LED die assembled at high speed to create the diffuse lighting systems of the future might seem farfetched, it is already being explored at several lighting companies today and may get assistance from the display sector, where concept displays employing millions of directly addressable, very small RGB LEDs have been demonstrated [26]. It is worth noting that because of the increasingly widespread adoption of LED technology in the display market, many display companies are entering the solid-state lighting market. It will be exciting to watch what new materials and processes are brought to bear on the development of future LED and lighting applications as commercial electronics technology enters illumination markets.

In this discussion of LED packaging and subsystem integration, there are several other important current considerations that may have a bearing on the development of future lighting applications. The first is phosphor development and the integration into LED packaging systems. Typically, phosphors are blended with encapsulants (typically silicones) or attached (using several different methods) directly onto the LED die in the package. Since energy is lost in the conversion of high-energy blue photons to lower energy yellow or red photons, thermal management of the phosphor is an important consideration, for both efficiency and long-term color stability of the LED. Typically, the phosphor is deposited on the chip or in the close proximity of the LED die or remotely, where phosphor thermal management occurs by conduction through the LED or by contact with the outside surface of the bulb/fixture. There

Fig. 16.6 The Philips L-Prize Bulb uses remote phosphors to achieve uniform illuminance and manage phosphor thermal issues. The bulb is *yellow* in the off state (shown), but completely white when powered. (Courtesy of Philips, used by permission of Philips Lighting)

has been some consideration to use LED arrays with remote phosphor coatings for various thermal and optical reasons. One disadvantage of remote phosphor coatings is that systems using them typically require larger quantities of phosphor, which is an expensive component if used in large quantities for LED fixture manufacturing. Another disadvantage is that of body color, or the color of the assembly when in the off state. An example (Fig. 16.6) is the first L-Prize bulb from Philips [27]. Depending on the fixture application, body color may not be an issue or may be hidden by suitable diffusive optics, with the concomitant loss of overall system efficacy. More research on phosphors and phosphor stability when integrated in LED packaging systems (e.g., encapsulants) is needed, especially, for the red phosphors used to create warm white (with a CCT of 2,700 K or so) lighting systems, where thermal quenching (loss of efficiency versus temperature) and narrow emission linewidth (ideally a few nanometers as opposed to many tens of nanometers), in addition to high-quantum conversion efficiency, are needed.

Related to the need for innovation on phosphors and their interactions with encapsulants and methods for incorporating them into the module or system design, is the development of lighting units employing both blue and red (AlInGaP) LED devices, with the red LEDs added to reduce the color temperature of the illumination source [28]. This is a logical approach to developing higher efficiency warm white solutions, but comes with the need to maintain color quality over the life of the product, as the aging rates of blue and red LEDs, as well as the color converting phosphor/encapsulant assemblies, may not be the same. While current systems may simply employ temperature monitoring to adjust red LED current (as the efficiency of red LEDs is more temperature-dependent than that of the phosphor/blue LED combination), some form of active sensing and correction might be required for longer operational life. Ultimately, this will be a price/performance tradeoff, but as the cost of solid-state lighting systems continues to drop, and new applications for solid-state lighting begin to develop, incorporation of active color control systems in multiwavelength LED lighting modules will likely become routine.

16.2.4 Setting the Stage for Future LED Applications

As LED light bulbs and LED panel lighting replacements for fluorescent fixtures begin to enter the market place, there continues to be rapid and disruptive technical innovation throughout the solid-state lighting supply chain. Almost all the LED supply chain technology components continue to evolve at a rapid pace and large, new industries previously foreign to the LED marketplace are entering the solid-state lighting arena with different sets of technical skills and business focus. These changes are convolved with other global and societal technology changes like the increasing demand for video information and bandwidth, the critical need for energy conservation and the deployment of sustainable technologies, and the growth of highly interconnected and diversified sources of alternate energy. It is against this complex technical and cultural backdrop that future LED applications will be developed, bringing a need to electronically control the spectral, spatial, and temporal properties of solid-state lighting systems to the marketplace.

Before describing potential future applications for LED technology, one must acknowledge that while energy savings have been the primary reason for developing LEDs for illumination, questions have been raised about whether or not LEDs will simply enable increased demand for lighting and new lighting applications with little overall impact on the global consumption of electrical energy. This question was raised in the late 1990s as the US Department of Energy launched its programs to accelerate the development of solid-state lighting for reasons of reducing energy consumption [29]. Sometimes referred to as the rebound effect, the fact that LEDs are used to bring color (and consume energy) in new spaces and applications never before possible, it is possible that LED technology will actually increase global energy consumption. Analysis of the historical consumption of light suggests that the human capacity for light consumption will continue to grow [30], and not ultimately saturate as improved solid-state lighting and a range of new functions and applications it uniquely enables drives even greater consumption of energy for new lighting applications.

16.3 Future LED Applications

The previous section paints a picture of rapid technological and business evolution caused by the development of LED technology. With unprecedented control of light properties enabled by LEDs, future LED applications, mostly convolved with illumination, will be developed to provide new markets and services. In some cases, these developments will be ushered in by the long lifetime of LED-based systems as suppliers seek to maintain revenue streams from their large investments in LED manufacturing technology. In other cases, the unique properties of LEDs will enable new agricultural, biological, and industrial applications resulting from the broad, tunable spectral capabilities of LEDs. Most future applications of LEDs will be offshoots of evolving markets for LEDs in illumination and display technology, but there will be other significant markets in industry and agriculture.

16.3.1 Advanced Lighting Systems

Solid-state lighting represents a move from electric to electronic lighting and will enable new applications and services that significantly broaden the role of future lighting systems. Future lighting systems will harness the ability to control spectral content, autonomously adjust the distribution of illuminance, transmit digital information, and interface with other energy management systems in power distribution grids and building controls. In some ways, these are not new concepts in lighting, but the nature of LEDs, as semiconductor components, makes the integration of advanced functions in lighting applications more practical than ever before. In the current popular jargon, these new systems are sometimes called "Smart" systems, though it is sometimes difficult to define exactly what smart systems are, and Smart Lighting (or also Intelligent Lighting) is similarly difficult to define.

It is taken for granted that LED lighting will ultimately replace most incandescent and fluorescent lighting technologies, and this transition will probably take longer than current market analysts predict, as rapid technical innovation, markets and new business models for lighting develop, and as human factors and lighting design considerations impact adoption [31]. The first wave of LED-based replacement bulb and tube technology will gradually give way to a much more sophisticated second wave of advanced lighting systems, with built-in intuitive controls, greater energy saving capability for simple (e.g., white) illumination, and added functionality with improved lighting quality. The development of high-quality, energy-efficient lighting designed with human factors considerations and artistic flexibility can be taken for granted, but beyond illumination, lighting systems will be increasingly convolved with communications and display technology. These trends are just starting to emerge in the market place and will enable a wide range of new LED-based applications for enhanced illumination.

16.3.1.1 Illumination with Communication

Communicating with light is commonplace, but lights that communicate with light are uniquely enabled by solid-state lighting as LEDs can be modulated at high speed. Referred to as Visible Light Communication (VLC) or more recently LiFi (like the RF-based WiFi), this additional functionality has been under development for some time [32]. Most VLC efforts focus on the wireless transmission of data at high bit rates as a means to avert limitations of RF-based wireless communications due to rapidly increasing demand for bandwidth and limited available RF space needed to meet the demand [33]. Dual-use (illumination and communication) lighting systems being explored with current phosphor-converted white LED systems use detection systems receiving modulated blue LED emission because the radiative lifetimes of efficient phosphors are too long to permit high-speed data transmission. When efficient color-tunable lighting systems using RGBY LEDs become available, VLC applications can be partitioned by color. Using experimental systems, high-speed data links using white LED systems have been demonstrated [34]. Advantages of VLC over RF-based wireless systems include security and unrestricted allocation of bandwidth, though many challenges need to be worked out on topics of mobility, integration with various fixture designs (combinations of spot and diffuse), and integration with RF uplink channels. Much of this work leverages IR free space optical communications work [35] and preliminary standards for VLC applications are under development [36]. Ultimately, integrated combinations of VLC and RF wireless communications will most likely be needed to meet the seemingly endless growth in demand for wireless data and mobile connectivity [37].

Beyond data communication, solid-state lighting systems that can illuminate an area and simultaneously transmit and receive data can enable a wide range of new and interesting functionality. Drawing on concepts of time-of-flight sensing and the use of structured light for coarse 3D mapping of surfaces [38], it will be possible to design illumination systems equipped with simple light detector networks to perform light-based occupancy sensing [39] and localization of occupants [40]. Even if each source in an illumination system simply transmits its identification code and setting information (dimming, color point, etc.), receivers in the lights will enable the lighting system to characterize the rough geometry of the illuminated space and determine the when, where, and type of illumination required. These types of applications will draw on concepts of light transport analysis developed in graphic arts research [41]. The ultimate goal of such systems is adaptive lighting control capability that will enable much more intuitive and even autonomous illumination control systems. In some sense, these types of applications are similar to earlier work where simple camera networks have been applied to issues of lighting control [42], but have seen limited application, possibly for privacy concerns.

More broadly, equipping illumination systems with sensors to enable low-resolution light field mapping will create lighting systems, which, for the first time, will "see" where the emitted light is going, coarsely image the surfaces that scatter or reflect it, and localize the illuminated objects in its field of view. Reflected light without a digital signature would suggest daylight and color-tunable systems could adjust the spectral output for circadian matching (or not, depending on the illumination requirements of the occupants detected by the lighting system), assuming the digital light detection system can adequately resolve color. Lighting systems could also autonomously interact with other light-emitting systems with their own digital light signature (e.g., video systems), adapting the illumination as required. As lighting continues its migration to the electronic age, even more novel applications of communicative illumination systems will emerge.

16.3.1.2 Illumination with Display Capability

LED technology has had a significant impact on display technology, enabling technologies like tablet computers and thin form factor LCD displays. Evolving development of gaming, immersive digital environments, and new lighting technology will lead to the fusion of illumination and display technology. Similar to the integration of illumination and communications, this trend has been under development for some time, and includes the use of LCD display technology for virtual windows and skylights [43], and the use of color-tunable LED lighting panel technology for illumination from an "open sky" source [44]. Significant technical development of both display and LED technology will be needed to fully develop this fusion, but it is likely that future lighting systems will more closely approximate natural lighting (and the scenes that accompany it). This trend will be analogous to trends observed in telephony, where what was once known as "plain old telephone service" (or POTS) to smart mobile communications systems with a fusion of communications, computation, and display technology.

The integration of display and illumination technology, if accomplished with inorganic LEDs in an emissive display format, will require significant improvements in the operating efficiency of green and yellow LED devices, and this has proven difficult for the LED die developers, though these wavelengths might start to receive more research attention as blue LED performance gains saturate. OLED technology may also be important in this vision if brightness, cost, and reliability gaps can be closed through continued technical development. Large research investments currently being made in both LED direct view display and color-tunable lighting as well as OLED technology for the same markets and applications will be interesting to watch.

16.3.1.3 Control of Illumination for Human Health and Well-Being

In the late 1990s, the discovery of nonvisual light receptors in the human retina [45] and their role in setting and altering human circadian rhythm [46] led to research suggesting that future lighting systems could play an important role in human sleep management. It is well known that lighting can mitigate some of the impact of seasonal affective disorder (SAD)— a well-studied condition linking depression with limited exposure to daylight [47], but since the spectral content of conventional incandescent and fluorescent lighting systems is fixed, detailed studies on the impact of lighting spectral content on human health have been somewhat limited. However, some interesting studies are beginning to emerge that suggest new approaches for control of light spectral content in the management of human performance. For example, recent studies on subtracting a narrow band of blue wavelengths from conventional lighting systems show that subtractive spectral management may have an impact on the circadian maintenance and alertness of shift workers [48]. Other studies show that the spectral qualities of light can impact cognition [49], educational performance [50], and the treatment of elderly patients suffering from Alzheimer's disease [51].

With the ability of solid-state light sources to tune the spectral content of the light source while maintaining high-quality illumination, there is growing interest in expanding research on the psychophysiological impacts of lighting on more fully understanding how solid-state lighting might impact human health and well-being, especially in healthcare environments. Some of this is based on research showing that exposure to sunlight can have a large impact on hospitalized patient health and well-being [52]. Tailoring the circadian color temperature of solid-state illumination systems is now being explored in a hospital setting to study the impact on patient sleep [53]. It will be interesting to see just how much impact spectral control of lighting in healthcare environments will impact patient recovery, and how these findings would migrate to improved worker health and on-the-job performance, where lighting is already a major factor in worker productivity and job satisfaction [54].

16.3.2 Other LED Applications

The efficient delivery of UV light has always been the province of various sorts of discharge lamps containing mercury, xenon, or deuterium, with a nominally fixed output spectrum characteristic of the active element in the discharge. The most common applications of UV sources are in the fields of metrology, dermatology, purification, and industrial curing of inks and coatings. Since UV radiation is difficult to detect with conventional light detection systems, it is also considered useful in covert communications. The most widely used traditional UV sources are low- and medium-pressure mercury arc lamps (Hg lamps).

In parallel with the development of high-performance blue LEDs, UV LEDs from AlGaN structures can cover the UVA (320–400 nm), UVB (300–320 nm), and most of the UVC (200–300 nm) spectral regions, though most commercially available UV LEDs today operate efficiently only at wavelengths longer than 365 nm. However, rapid progress is being made in the development of efficient and reliable UVC LEDs for primarily germicidal applications.

Industrial UV-based curing systems are used to "dry" inks, coatings and adhesives. Since this process occurs through photo-initiated polymerization, there are no solvents to evaporate and curing is almost instantaneous [55]. Almost all UV curing systems use Hg lamps, and the photochemistry is optimized for the UV. LED technology has the potential to revolutionize industrial UV curing because Hg lamps have relatively short lifetimes, cannot easily be turned on and off, produce a tremendous amount of heat, and, of course, contain mercury. Because of these limitations, the adoption UV curing technology as a more energy-efficient low-emissions technology has been limited, and conventional air or oven drying techniques are still widely used. The challenge for UV LEDs in industrial curing applications has been that only UVA LEDs can be operated efficiently at high optical output power, and many coatings and inks require some amount of shorter wavelength UV (easily obtained from Hg lamps) to achieve good curing at the surface of the film [56] where the chemical curing process can be inhibited by reaction with atmospheric oxygen. Continued progress on developing higher power UVB and UVC LEDs, as well as new ink and coating formulations that work well at UVA wavelengths will drive more rapid development of LED use for industrial curing applications.

The development of efficient, reliable deep UV LEDs is gaining increasing attention for use in germicidal applications [57], and continued progress on the efficiency, reliability of these sources could see broad applications in point of use water purification and food processing industries, replacing more commonly used mercury lamp and chemical disinfection means. The effectiveness of UVC radiation for killing spores, virus and bacterial contamination depends on wavelength and dose, and since UVC LED radiative efficiency drops as wavelength decreases, tradeoffs between higher doses at less effective wavelengths versus lower doses at more lethal wavelengths needs to be evaluated [58]. While the development of these markets is still in its infancy, the potential for improved human health and well-being is quite significant as exposure to contaminated food and water exacts an estimated cost of US$ 350 million in the United States alone annually [59]. It is reasonable to expect that the number of lives impacted and associated costs are much higher in developing nations.

Another application space that will gain in importance is the use of LED lighting in agricultural applications. LED lighting for horticulture [60] is continuing to be investigated as more continues to be learned about the specific effects of wavelength on plant growth. The presence or absence of certain wavelengths can impact the different parts of the plant's life cycle as well as the nutritional value of the produce the plant produces, and production yield under optimized illumination can be higher than under natural growth conditions, but the impact of spectral distribution can be complex and can depend on plant type [61]. Ordinarily, artificial light for plant growth requires high illuminance, and with the continued improvement in high-brightness LED systems for lighting applications, both the intensity and the price point of LED systems for horticulture are reaching the point where they can be considered for horticultural applications. While most produce will continue to be produced in open fields under sunlight, the use of LED lighting for extending the growing season in certain parts of the world, avoiding issues with bacterial crop contamination, bypassing increasing fuel costs for transporting produce long distances, and cultural trends favoring the use of locally produced or fresh crops suggest that this market will continue to grow.

A related emerging potential use of LED lighting for plant growth is in the use of genetically engineered plants for production of pharmaceutical products. Sometimes referred to as Pharming [62], research in the use of genetically modified plants and animals for low-cost production of specific pharmaceutical agents is continuing to increase. Since illumination wavelengths are already known to impact different phases of plant growth, genetically modified plants that utilize controlled spectrum conditions to optimize pharmacological yield will be an interesting application for future LED illumination systems since such transgenic systems would need to be isolated to prevent genetic contamination of their naturally occurring analogs.

16.4 Summary

The rapid development of LED technology, largely driven by the need for efficient illumination, is a truly disruptive technology that will have wide technological, commercial, and social applications beyond just illumination. While the photons originate in the semiconductor element, the efficiency and reliability of that element and the rest of the entire system will depend critically on thermal management, and in some sense, new and creative solutions to thermal management at the device as well as the system level must be developed to drive continued growth of LED markets. Since thermal management processes operate at the interface between the LED system and its environment, the development of future thermal management technologies will continue to be highly interdisciplinary, factoring thermal, electrical, mechanical, and optical performance requirements. In solid-state lighting, thermal management will also need to be factored into the artistic and esthetic considerations of LED systems design, trending to designs that can cool for the lifetime of the LED in any location and orientation, impervious to dust and dirt, and, ideally, without being seen or heard. The growth of new and existing LED applications will be convolved with the progress in new and innovative thermal management technologies, and continued research and development of thermal solutions is essential to developing the full potential of LED applications.

Acknowledgments The author would like to thank colleagues at Rensselaer Polytechnic Institute, Boston University and the University of New Mexico participating in the Smart Lighting Engineering Research Center (ERC) for extensive discussions relating to the future development of solid-state lighting systems and applications. This work has also been shaped by extensive discussions with many of the solid-state lighting industry members of the Smart Lighting ERC who are actively involved in developing new systems-level solutions based on advanced LED technology. This work was supported primarily by the ERC Program of the National Science Foundation under NSF Cooperative Agreement No. EEC-0812056 and in part by New York State under NYSTAR contract C090145.

References

1. Nakamura S, Senoh M, Mukai T (1993) High-power InGaN/GaN double-heterostructure violet light-emitting diodes. Appl Phys Lett 62(19):2390
2. Schubert EF (2006) Light emitting diodes, 2nd edn. University Press, Cambridge
3. Christensen C (1997) The innovators dilemma. Harvard Business School Press, Cambridge
4. Bassett RK (2007) To the digital age: research labs, start-up companies and the rise of MOS technology. The Johns Hopkins University Press, Baltimore
5. Li Y, You S, Zhu M, Zhao L, Hou W, Detchprohm T, Taniguchi Y, Tamura N, Tanaka S, Wetzel C (2011) Defect-reduced green GaInN/GaN light-emitting diode on nanopatterned sapphire. Appl Phys Lett 98:151102
6. Paskova T, Evans KR (2009) GaN substrates: progress, status and prospects. IEEE J Sel Top Quant Electron 15(4):1041
7. Zhao Y, Tanaka S, Pan C, Fujito K, Feezell D, Speck JS, DenBaars SP, Nakamura S (2011) High-power blue-violet semipolar InGaN/GaN light-emitting diodes with low efficiency droop at 200 A/cm^2. App Phys Express 4:082104

8. Whitaker T (2012) Soraa unveils GaN-on-GaN LEDs and MR16 lamps. http://www.leds-magazine.com/news/9/2/18. Accessed 28 Feb 2012
9. These data have been assembled from a collection of published and privately communicated information from LED manufacturers interviewed by the author
10. Karlicek R (2007) Photonic lattice LEDs are new class of light-emitting device. LEDs Magazine. Issue 15:20
11. Plößl AE, Bauer J, Eißler D, Engl K, Härle V, Hahn B, Heindl A, llek S, Klemp C, Rode P, Streubel K, Tangrin A (2010) Wafer bonding for the manufacture of high-brightness and high-efficiency light-emitting diodes. ECS Trans 33(4):613–624
12. See US Patent US 6,547,249 B2
13. See issued representative United States issued patents assigned to Seoul semiconductor: US 7,417,259, US 7,569,861, US 7,615,793, US 7,646,031
14. See US Patent Application Publication US 2011/0109244 A1
15. http://spectrum.ieee.org/semiconductors/optoelectronics/better-chips-for-better-led-bulbs
16. Tsao JY, Coltrin ME, Crawford MH, Simmons JA (2010) Solid-state lighting: an integrated human factors, technology and economic perspective. Proc IEEE 98(7):1162
17. Chow TP, Li Z (2012) Recent advances in high-voltage GaN MOS-Gated transistors for power electronics applications. In: Pearton SJ (ed) GaN and ZnO based materials and devices. Springer, Berlin
18. Liu S, Luo X(2011) LED packaging for lighting applications: design, manufacturing, and testing. Wiley, Singapore
19. See, for example, the use of LEDs for both illumination and light sensing for control: Knapp DJ, Ho HC, Lewis J. Lighting control system. US Patent Application 2012/0001567
20. Based on conversations between 2010 and 2012 with LED industry executive
21. See http://www.zhagastandard.org. Accessed 1 Mar 2012
22. Market analysts project that revenue growth from LED replacement bulb sales will begin to saturate by the end of the current decade (~2020). See Smallwood P, Global 60 W Replacement Market (2012), Solid state lighting design's 2012 SSL summit LA, Long Beach, CA (USA), April 3–4, 2012
23. Fu H, Cheng Y-M, Chou P-T, Chi Y (2011) Feeling blue? Blue phosphors for OLEDs. Mater Today 14:472
24. Whitesides MG, Boncheva M (2002) Supramolecular chemistry and self-assembly special feature: beyond molecules: self-assembly of mesoscopic and macroscopic components. Proc Natl Acad Sci U S A 99(8):4769–4774. doi:10.1073/pnas.08206589925. http://www.nthdegreetech.com/printed-lights.php. Accessed 2 Feb 2012
25. See http://www.nthdegreetech.com/printed-lights.php. Accessed 2 Feb 2012
26. See http://www.sony.net/SonyInfo/News/Press/201201/12-005E/index.html. Accessed 10 Jan 2012
27. See http://ledsmagazine.com/news/9/1/28. Accessed 10 Oct 2011
28. See, for example, US Patent 8,029,155 B2
29. Private communications with DOE officials in the course of setting the first DOE SSL technical roadmap for LED technology, 1999.
30. Tsao JY, Saunders HD, Creighton JR, Coltrin ME, Simmons JA (2010) Solid-state lighting: an energy-economics perspective. J Phys D: Appl Phys 43:354001
31. Ogburn WF (1922) Social change with respect to culture and original nature. B.W. Huebsch, New York
32. The Visible Light Communications Consortium was established in Japan in (2003). www.vlcc.net/modules/xpage3/
33. Kavehrad M (2010) Sustainable energy-efficient wireless communications using light. IEEE Commun Mag 48(12):66
34. Vucic J, Kottke C, Nerreter S, Langer KD, Walewski JW (2010) 513 Mb/s visible light communications link based on DMT-modulation of a white LED. J Lightwave Technol 28(24):3512
35. Williams S (2000) IrDA: past, present and future. IEEE Pers Commun 7(1):11

36. See the IEEE802.15.7 draft standard for visible light communications
37. Rahaim MB, Vegni AM, Little TDC (2011) Hybrid radio frequency and broadcast system, Proceedings of the 2nd IEEE globecom workshop on optical wireless communications
38. Geng J (2011) Structured light 3D surface imaging: a tutorial. Adv Opt Photon 3:128
39. Chiabrando F, Chiabrando R, Piatt D, Rinaudo F (2009) Sensors for 3D imaging: metric evaluation and calibration of a CCD/CMOS time-of-flight camera. Sensors 9(12):10080
40. Brida F, Duha J, Krasnovsky M (2007) Personal wireless communications. In: Simak B, Bestak R, Kozowska E (eds) IFIP International Federation for information processing, vol 245. Springer, Boston, pp. 423–432
41. Pattanaik SN, Mudur SP (1995) Adjoint equations and random walks for illumination computation. J ACM Trans Graph 14(1):77
42. Newsham GR, Arsenault CD (2009) A camera as a sensor for lighting and shading control, NRCC-50453 National Research Council Canada (a version of this report is published in Light Res Technol 41(2) 143)
43. Witherspoon B, Petrick M (2008) Scientific research and sky image ceilings, Sky Factory White Paper. http://www.skyfactory.com/files/SkyFactory_White_Paper_032408.pdf. Accessed 5 Jan 2012
44. Fraunhofer Press Release (2012) Sky light sky bright—in the office. http://www.fraunhofer.de/en/press/research-news/2012/january/sky-light-sky-bright.html. Accessed 5 Jan 2012
45. Berson DM, Dunn FA, Takao M (2002) Phototransduction by retinal ganglion cells that set the circadian clock. Science 295:1070
46. Stevens RG, Blask DE, Brainard GC, Hansen J, Lockley SW, Provencio I, Rea MS, Reinlib L (2007) Meeting report: the role of environmental lighting and circadian disruption in cancer and other diseases. Environ Health Perspect 115(9):1357
47. Rosenthal NE, Sack DA, Skwerer RG, Jacobsen FM, Wehr TA (1989) "Phototherapy for seasonal affective disorder". In: Rosenthal NE, Blehar MS (eds) Seasonal affective disorders and phototherapy. Guilford Press, New York, pp. 273–294
48. Rahman SA, Marcu S, Shapiro CM, Brown TJ, Casper RF (2011) Spectral modulation attenuates molecular, endocrine and neurobehavioral disruption induced by nocturnal light exposure. J Physiol Endocrinol Metab 300:E518
49. Knez I (2001) Effects of colour of light on nonvisual psychological processes. J Environ Psychol 21(2):201–208
50. Goven T, Laike T, Raynham P, Sansal E (2011) Influence of ambient light on the performance, mood, endocrine systems and other factors of school children. CIE 27th session, Sun City, South Africa, 112
51. Satlin A, Volicer L, Ross V, Herz LR (1992) Bright light treatment of behavioral and sleep disturbances in patients with Alzheimer's disease. Am J Psychiatry 149(8):1028–1032
52. Walch JM, Rabin BS, Day R, Williams JN, Choi K, Kang JD (2005). The effect of sunlight on postoperative analgesic medication use: a prospective study of patients undergoing spinal surgery. Psychosom Med 67:156–163
53. Philips Study at http://www.newscenter.philips.com/main/standard/news/press/2011/201111-22-healwell.wpd. Accessed 3 Apr 2012
54. Edwards L, Torcelli P (2002) A literature review of the effects of natural light on building occupants. National Renewable Energy Laboratory (NREL TP-550-30769)
55. Decker C (2003) Kinetic study and new applications of UV radiation curing. Macromol Rapid Commun 23(18):1067–1093
56. See US Patent 6,550,905
57. Crawford MH, Banas MA, Ross MP Ruby DS, Nelson JS, Boucher R, Allerman AA (2005) Final LDRD report: ultraviolet water purification systems for rural environments and mobile systems. Sandia report SAND2005-7245, Sandia National Laboratories
58. Wurtele MA, Kolbe T, Lipsz M, Kulberg A, Weyers M, Kneissl M, Jekel M (2011) Application of GaN-based ultraviolet-C light emitting diodes—UV LEDs—for water disinfection. Water Res 45:1481–1489

59. Ralston EP, Kite-Powell H, Beet A (2011) An estimate of the cost of acute health effects from food- and water-borne marine pathogens and toxins in the USA. J Water Health 9(4):680–694
60. Morrow RC (2008) LED lighting in horticulture. J Hortic Sci 43:1947
61. Hogewoning SW, Douwstra P, Touwborst G, Leperen WV, Harbinson J (2010) An artificial solar spectrum substantially alters plant development compared with usual climate room irradiance spectra. J Exp Bot 61:1267–1276. doi:10.1093/jxb/erq005
62. Abdullah MA, Rahmah AU, Sinskey AJ, Rha CK (2008) Cell engineering and molecular pharming for biopharmaceuticals. Open Med Chem J 2:49–61

Index

C. J. M. Lasance, A. Poppe (eds.), *Thermal Management for LED Applications*,
Solid State Lighting Technology and Application Series 2,
DOI 10.1007/978-1-4614-5091-7, © Springer Science+Business Media New York 2014

D

MIX
Papier aus verantwortungsvollen Quellen
Paper from responsible sources
FSC® C105338

If you have any concerns about our products,
you can contact us on
ProductSafety@springernature.com

In case Publisher is established outside the EU,
the EU authorized representative is:
Springer Nature Customer Service Center GmbH
Europaplatz 3, 69115 Heidelberg, Germany

Printed by Libri Plureos GmbH
in Hamburg, Germany